Avionics Training
Systems, Installation and Troubleshooting

Second Edition

Len Buckwalter

Avionics Communications Inc.
Leesburg, VA, USA

Copyright ©2005, 2007, 2009 by Len Buckwalter
ISBN 978-1-88-554421-6
All rights reserved. No portion of this book may be reproduced in any form or in any medium without the written consent of the publisher.

Library of Congress Cataloging-in-Publication Data

Buckwalter, Len.
　Avionics training : systems, installation, and troubleshooting / Len Buckwalter.
　　p. cm.
　Includes index.
　ISBN 978-1-88-554421-6
　1. Avionics-Textbooks. 2. Airplanes--Electronic equipment--Installation--Textbooks. 3. Airplanes--Electronic equipment--Maintenance and repair--Textbooks. I. Title.

TL693.B752 2005
629.135--dc22

2005045200

Avionics Communications Inc.
P.O. Box 2628
Leesburg, VA 20177 USA

Tel: 703/777-9535
Fax: 703/777-9568
E-mail: len@avionics.com

Visit: **www.avionics.com**

Preface

Avionics is changing rapidly, thanks to the computer revolution, satellites, digital electronics and flat panel displays, to name a few. These changes are also affecting the direction of avionics training.

Technicians a half-century ago were "radio mechanics," removing broken black boxes from airplanes, taking them to the shop and testing the circuit. They did "all-purpose" maintenance, equally at home on the flight line or work bench. But as avionics grew more complicated, the job was split in two. One person became the "installer" ---troubleshooting on the ramp, or mounting and wiring equipment in airplanes. The other person, trained in repairing circuits inside the box, became the "bench technician," skilled in troubleshooting down to the smallest component. For decades radio shops separated technical skills this way to service private aircraft in General Aviation.

In the airlines, the division of labor went further. Flightline maintenance was handled by radio mechanics scattered at major airports along their routes, supported by A&P mechanics. After a defective radio was pulled, it was sent back to the airline maintenance depot for repair by bench technicians. Among large airlines, it was usual to have different benches for specialists in each type of instrument or radio; autopilot, automatic direction finder, communications, etc.

By the 1990's, avionics took off in a new direction. Manufacturers began building radios with disappearing parts! Instead of resistors, capacitors and tubes, they populated them with integrated circuits encased in tough epoxy coatings that were difficult to remove. Other components no longer had wires, but were "surface mounted" directly to the board.

Other areas grew smaller. Radios had different sections to tune, amplify or produce some other function but much of that construction is now replaced by invisible software, which instructs the chips to become just about anything. It's the same idea as a personal computer and its applications software; it's a word processor, spread sheet or video game---at the press of a few buttons.

For the avionics shop, these developments reduce the need for bench technicians to repair down to the component level. Maintaining the new avionics requires expensive automatic test stations beyond the reach of most shops. Today's digital avionics are sent back to the factory or a major depot for repair. Some faults in this equipment, in fact, will not appear unless tests are repeated over many hours, often in a test chamber that runs hot and cold. These tasks must be done automatically, and not by a technician with a pair of test probes.

On the other hand, demand for installation technicians working on the ramp or flightline not only remains strong but will grow. Upgrades for old aircraft continue at a remarkable rate because new-generation equipment makes flying more economical, efficient and safe. Some avionics return their investment in as little as one or two years, then function another ten to twenty.

Airline and corporate aircraft must upgrade to fly in the coming air traffic system---to get more direct routes, altitudes with less headwind, fewer delays and better communication services, all of which repay the cost of avionics and keep passengers happy.

Beyond the flight deck. A whole new category called "cabin avionics" is spreading among airlines. Once called "in-flight entertainment," it adds Internet connectivity to every seat, e-mail, global telephone, video games and new forms of entertainment. An airliner typically has two or three radios per function in its instrument panel---but hundreds of passenger seats with equipment in the cabin that now fits under the heading "avionics".

Yet another growth area is the world-wide air traffic management system under construction. No longer will airplanes move point-to-point over land or on crowded tracks to cross the ocean. They will fly directly to their destinations in a concept called "Free Flight," a new mode which depends on satellite navigation and data communications.

The new technician. These developments call for the skills of a technician who understands avionics at the systems level---all the major functions and how they relate to each other. Finding trouble fast is critical in airline operations, where every minute of delay at the gate causes missed connections, lost revenue and angry passengers. In General Aviation, corporate aircraft provide vital transportation for industry. Even the private pilot needs competent servicing for the fleet of light aircraft fitted with the latest "glass" cockpits (electronic instruments). In the pages that follow, some 30 different systems describe a wide range of communications and navigation systems aboard aircraft of all sizes and types.

NFF. A systems understanding reduces one of the costliest errors in avionics maintenance. It's NFF, for "No Fault Found." The technician pulls a suspicious box and sends it back to the shop for repair. There, the diagnosis finds nothing wrong, and the radio is returned to service. Or it may be sent back to the factory or depot. After further testing, the radio is returned labelled "NFF." When the radio is re-installed on the airplane, the problem returns. Not only does it waste hours but often costs the airline over $8000 in diagnostics, labor and shipping. In the general aviation shop, the no-fault found not only incurs extra expense and wasted time, but an unhappy customer who loses confidence in the shop.

Simplified Diagrams. In describing these systems in this book, there are no schematics showing, resistors, capacitors or other small, internal components. Instead, simplified block diagrams illustrate the function and flow of signals with arrows. Where the shape of a signal is important, it is illustrated with graphic images.

Most of what is written on avionics is filled with abbreviations and acronyms---TAWS, EGPWS, MFD, TACAN, TCAS---and more. They make an unfair demand on the reader because even the most experienced avionics person must stop at each one and translate it to plain English. For this reason, abbreviations and acronyms are almost always spelled out in diagrams and in the text where they appear.

Gender. Throughout this book, a technician or pilot is referred to as "he." The avionics industry is populated by both genders and this should not be considered insensitivity. It avoids the awkward use of "his/her."

Maintenance Information. This book is not meant to be a "cook book"---with step-by-step instructions for maintenance. It is intended, rather, as a guide to understanding manufacturer's manuals. Also, it does not replace the FAA document on maintenance; Advisory Circular 43.13 1A-2B. This book is intended as a background to understanding manufacturer's manuals that cover specific equipment.

Appreciation. I want to thank the manufacturers who provided me with graphic material and documents. They are credited below their photo, drawing or text. If the reader wants further information, they are easily reached by inserting their name in a search engine along with the word "avionics."

Len Buckwalter
Leesburg, Virginia

Contents

Preface

Section 1 Systems

Chapter 1. The Meaning of "Avionics" 1
    First Instrument Panel 1
    "Blind Flying" 2
    All-Glass Cockpit 4

Chapter 2. A Brief History 6
    Sperry Gyroscope 7
    Turn and Bank 8
    Morse, Bell and Hertz 9
    First Aircraft Radio 10
    Lighted Airways 11
    Jimmy Doolittle; Beginning of Instrument Flight 12

Chapter 3. VHF Com (Very High Frequency Communications) ... 16
    Acceptable VHF Com Radios 17
    VDR (data radio) 17
    Navcom Connections 18
    VHF System 19
    Com Control Panel 20
    Com LRU 20
    Splitting VHF Channels 21

Chapter 4. HF Com (High Frequency Communications) 23
    HF Control-Display 23
    HF System 24
    SSB (Single Sideband) 24
    Line Replaceable Units 25
    HF Datalink 25
    Control Panel (Airline) 26
    HF Transceiver 26
    Antenna Coupler 27
    HF Antenna Mounting 27

Chapter 5. Satcom (Satellite Communications) 29
    Inmarsat 29
    Aero System 31
    Space Segment 32
    Cell Phones 33
    Ground Earth Station (GES) 34
    Aircraft Earth Station (AES) 35
    Satcom Antennas for Aircraft 36
    Steered Antennas 37
    High Speed Data 38
    Aero Services 39

**Chapter 6. ACARS (Aircraft Communication Addressing and 41
                   Reporting System)**
    In the Cockpit 41
    ACARS System 42

v

      Messages and Format 43
      ACARS Bands and Frequencies 45

Chapter 7. Selcal (Selective Calling) 46
      Controller, Decoder 46
      How Selcal Code is Generated 47
      Ground Network 47
      Selcal Airborne System 48

Chapter 8. ELT (Emergency Locator Transmitter) 50
      Search and Rescue 51
      ELT Components 52
      406 MHz 52
      406 ELT System 53
      Fleet Operation 54
      Cospas-Sarsat 55

Chapter 9. VOR (VHF Omnidirectional Range) 57
      Coverage 58
      VOR Phase 59
      VOR Signal Structure 60
      Subcarrier 61
      VOR Receiver 62
      Navigation 63
      Horizontal Situation Indicator (HSI) 64
      Radio Magnetic Indicator (RMI) 64
      Nav Control-Display 65

Chapter 10. ILS (Instrument Landing System) 67
      ILS System 68
      ILS Components and Categories 68
      Approach Lighting 69
      Flight Inspection and Monitoring 70
      ILS Signals 71
      Glideslope 72
      Glideslope Receiver 73
      Marker Beacon Receiver 74
      Marker Beacon Ground Station 74

Chapter 11. MLS (Microwave Landing System) 76
      Azimuth Beam 77
      Elevation Beam 78
      Time Reference 79
      Multimode Receiver 79

Chapter 12. ADF (Automatic Direction Finder) 81
      Radio Magnetic Indicator 82
      Sense 82
      ADF System 83
      NDB (Non-Directional Radio Beacon) 83
      Control-Display (Airline) 84
      Line-Replaceable Unit (Airline) 84
      Limitations 85
      Digital ADF 85
      EFIS Display 86

Chapter 13. DME (Distance Measuring Equipment) 88
      Obtaining Distance 89
      DME "Jitter" for Identification 89
      EFIS Display of DME 90
      Airborne and Ground Stations 91

Channels X and Y 92

Chapter 14. Transponder 94
Control-Display 94
Transponder Interrogator 95
Panel-Mount 96
ATCRBS and Mode S 96
Transponder System 97
Airline Control-Display 98
Line-Replaceable Unit 98
Mode S Interrogations and Replies 98

Chapter 15. Radar Altimeter 104
Antennas 105
Operation 106

Chapter 16. GPS/Satnav (Satellite Navigation) 108
GPS Constellation 109
Frequencies 110
Satnav Services 111
Panel-Mount Receiver 111
Time Difference Measurement 111
Finding Position 112
The Satellite Signal 113
GPS Segments 114
WAAS: Wide Area Augmentation System 114
Second Frequency for Civil Aviation 112
LAAS: Local Area Augmentation System 116
RAIM: Receiver Autonomous Integrity Monitoring 117
Galileo Constellation 118

Chapter 17. EFIS (Electronic Flight Instrument System) 120
Electromechanical to EFIS 122
Three-Screen EFIS 123
EFIS Architecture 124
Multifunction Display (MFD) 125
EFIS on the B-747 126
Airbus A-320 Flight Deck 127

Chapter 18. Cockpit Voice and Flight Data Recorders 129
CVR Basics 129
Underwater Locating Device (ULD) 131
CVR Interconnect 132
Flight Data Recorder: Solid State 134
Flight Data Recorder: Stored Information 135

Chapter 19. Weather Detection 136
Radar Color-Coding 137
Multifunction Display 137
Types of Detection 138
Radar Transmitter-Receiver 139
Weather Radar Control Panel 140
Lightning Detection 140
Radar Antenna 141
Datalink 141
Radomes 142
Radome Boot 142
Windshear 143
Lightning Detection 145
Windshear Computer 144

Satellite Datalink 145

Chapter 20. TCAS (Traffic Alert and Collision Avoidance System) 147
Basic Operation 148
Traffic and Resolution Advisories (TA, RA) 150
TCAS System 150
TCAS I and II 150
Coordinating Climb and Descend 150
TCAS Components 150
Whisper-Shout 151
Directional Interrogation 151
Non-TCAS Airplanes 152
TCAS III 152
Voice Warnings 152

Section 2 Installation

Chapter 21. Planning the Installation 154
Replacing "Steam Gauges" 155
Required Instruments and Radios 156
Flight Instrument Layout 157
Basic T 157
Large Aircraft EFIS 158
Flat Panel Integrated 159
Avionics Planning Worksheet 160
Manuals and Diagrams 161
Installation Drawing 162
Connectors and Pin Numbers 162
Pin Assignments 163
Schematic Symbols 164
Viewing Angle 165
Survey Airplane 165
Navcom Connections, Typical 166

Chapter 22. Electrical Systems 168
AC and DC Power 168
DC System 169
Airline Electrical System 171
Switches 172
Lighted Pushbutton 174
Circuit Breakers/Fuses 173

Chapter 23. Mounting Avionics 178
New or Old Installation? 179
Hostile Areas 179
Selecting Metal 179
Cutting Holes 180
Structures 181
Avionics Bay 182
Airlines (ARINC) MCU Case Sizes 183
ATR Case Sizes 184
Electrostatic Discharge 185
Cooling 186
Cooling for Airline Avionics 187
Locking Radios in Racks 188

viii

　　　　Panel-Mounted Radios 188
　　　　Remote-Mounted Radios (Corporate) 190
　　　　Airline Mounting 191
　　　　Locking Systems (Airline) 192
　　　　Indexing Pins 193
　　　　Integrated Modular Avionics 194
　　　　Instrument Mounting 195
　　　　Round Instruments: 2- and 3-inch 196
　　　　Airline Instrument Mounting 197

Chapter 24. Connectors 199
　　　　Typical Connectors 200
　　　　RF Connectors 200
　　　　How to Identify Connector Contacts 201
　　　　Contact Selection 201
　　　　Identify Mil-Spec Part Numbers 202
　　　　Coaxial Connectors, Typical 203
　　　　ARINC Connectors 203
　　　　Crimping Contacts 205
　　　　Releasing Connector Pins 207
　　　　Heat Gun for Shrink Tubing 207
　　　　Safety Wiring Connectors 208
　　　　Attaching Coaxial Connectors 204

Chapter 25. Wiring the Airplane 210
　　　　Swamp 210
　　　　High Risk Areas 211
　　　　Selecting Wire 213
　　　　High-Grade Aircraft Wire 213
　　　　Wire Sizes 214
　　　　Wire and Cable Types 215
　　　　Wire Stripping 216
　　　　Nicked or Broken Wires 217
　　　　Precut Cables` 217
　　　　Splicing Wires 217
　　　　Location of Splices 218
　　　　Ring Terminals 219
　　　　Terminal Strip (Block) 219
　　　　Marking Wires 220
　　　　Harnessing the Wire Bundle 222
　　　　Tie Wraps 223
　　　　Problems: Chafing and Abrasion 224
　　　　Clamping 224
　　　　Grounding to Airframe 227
　　　　Bending Coaxial Cable 228
　　　　Service Loops 229

Chapter 26. Aviation Bands and Frequencies 230
　　　　Radio Frequency Bands 231
　　　　Higher Bands (Microwave, Millimeter) 232
　　　　Low Frequencies 232
　　　　Skipping through Ionosphere 233
　　　　High Frequencies 233
　　　　Very High Frequencies 234
　　　　L-Band 234
　　　　From Hertz (Hz) to Gigahertz (GHz) 234
　　　　Line of Sight Communications 235
　　　　Control and Display of Bands and Frequencies 236

Chapter 27. Antenna Installation 239
　　　　Antennas for Airline, Corporate and Military Aircraft 240

  How to Read an Antenna Spec Sheet 241
  Antennas for Light Aircraft 242
  Airline Antenna Locations 243
  Antenna Types 244
  Location 245
  Bonding the Antenna to the Airframe 248
  Antenna Mounting 249
  Antenna Couplers 251
  Base Station and Mobile Antennas 252
  GPS Antennas 251

Chapter 28. Panel Labels and Abbreviations 254
  Silk Screen, Engraving, Tape 254
  Panel Abbreviations 255

Section 3

Chapter 29. Test and Troubleshooting 261
  ADF 262
  Antennas 263
  Antenna VSWR 263
  Autopilots 264
  Com Transceivers 264
  DME 265
  ELT-Emergency Locator Transmitter 266
  Glideslope Receiver 266
  Lightning Strikes 266
  Software Loading 267
  Transponder 267
  VOR 268
  Wiring and Connectors 270
  Fault Detection Device (Wiring) 271
  Precipitation (P) Static 273
  Avionics Checklist 274

About the Author 275

Index

Chapter 1

The Meaning of "Avionics"

The word "avionics" first appeared in the 1940's during World War II. Derived from "aviation electronics," it referred to fire control systems aboard U.S. Navy aircraft. During that time, the civilian world called it "aircraft radio" or "aviation electronics." Technicians who repaired them were known as "radio mechanics."

Avionics remained a military term for 30 more years. Civil aviation could not afford the systems aboard military aircraft. Not only was equipment built to military specifications, but each fighter and bomber had its own avionics suite that fit no other model.

But the world was rapidly changing as new components emerged from research labs; the transistor, integrated circuit, flat-panel display, solid-state memory and the "computer on a chip." Small in size and light weight, they consume little power, have few moving parts and, some believe, will operate a hundred years without wearing out. Millions of semiconductors within the size of a postage stamp created the microprocessor, which quickly became known as the "computer on a chip." It triggered the greatest technical achievement of the 20th Century; digital electronics. For the first time, an aircraft radio could not only receive, amplify, oscillate, filter and perform other simple functions; now it could perform logic, store large amounts of data, send thousands of pieces of information down one pair of wires, warn of problems, correct its own errors---and that's just the beginning.

First Instrument Panel

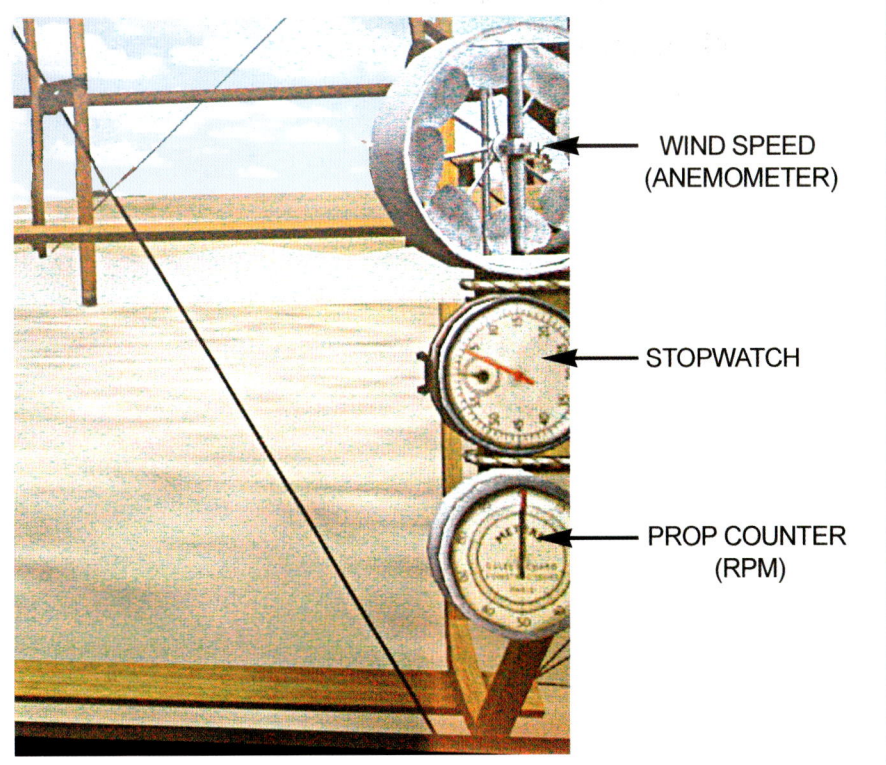

The three instruments shown here are ancestors of what will become "avionics" in 50 years. They were installed in the Wright Flyer that made the first successful powered flight in 1903. Although mechanically operated, these gauges will evolve into electronic instruments that comprise avionics on every type of 21st-Century aircraft.

Thus the Wright brothers not only deserve credit for inventing the first practical airplane, but the concept of an instrument panel in view of the pilot to provide valuable flight information.

1

Early Gauges

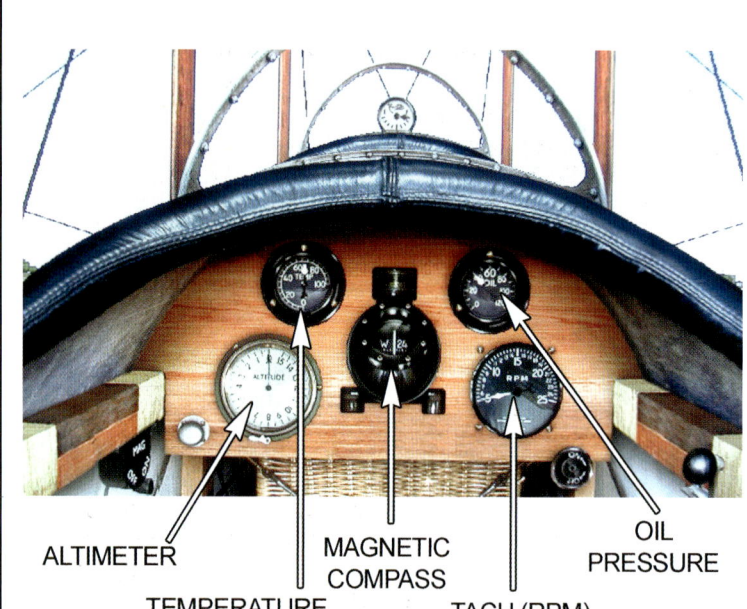

ALTIMETER
TEMPERATURE
MAGNETIC COMPASS
TACH (RPM)
OIL PRESSURE

The Curtiss JN ("Jenny") was the first airplane to fly the US mail in 1918. But a look at the instrument panel shows why pilots were killed while trying to live up to the Post Office motto; "Neither snow nor rain nor gloom of night...stays these couriers from their appointed rounds." No pilot can fly an airplane in very low visibility without "attitude" instruments to replace the sight of the horizon. Even flying at night was considered by an emergency by U.S. Army regulations.

First "Blind Flying" Instruments

TURN-AND-BANK INDICATOR
EARTH INDUCTOR COMPASS

1927 cockpit. Spirit of St. Louis

After the Wright brothers, Charles Lindbergh made the most famous flight in aviation history. In 1927 he flew solo from New York to Paris in a little over 33 hours. Although celebrated as a hero throughout the world, Lindbergh had more than skill and courage. His panel had a turn-and -bank, a gyroscopic instrument that indicated how rapidly the airplane turned left or right. Without such guidance, he could not have penetrated bad weather and low visibility (still the major cause of fatalities among low-time pilots). Lindbergh's airplane, the Spirit of St. Louis, had another important instrument; an earth inductor compass, shown in the panel. It was powered by an anemometer atop the fuselage (photo at right). This was an improvement over the simple magnetic compass, which is difficult to read in turbulence. Today, the earth-inductor compass is known as a "flux gate" and is standard on all but the smallest aircraft.

Wind-driven anemometer powered Lindbergh's earth-inductor compass

Higher Tech, Lower Cost

The new devices were snapped up, not only by the military but the telecommunications and consumer electronics industries. Semiconductors created hundreds of new products, from the personal computer and DVD, to data networks, cell phones and high-definition TV. "Chips" became building blocks of the Internet. Mass production reduced prices so far that a hobbyist could build digital projects with parts from the shelf of a local radio store.

These devices were embraced by aviation, which continuously seeks to reduce size, weight and power consumption. Old vacuum tubes were replaced by tiny integrated circuits that deliver many more functions. By the 1980's the term "aviation electronics," over a half-century old, no longer described a cutting-edge industry. Manufacturers, repair shops, parts distributors, airlines and general aviation sensed the need for a better term to replace "aviation electronics." And what better word than---"avionics?" During that period, "avionics" also appeared for the first time in volumes of FAA regulations on aircraft electronics.

The first generation of the new avionics was so successful, it began to outclass the military. Airlines, business jets and private aircraft were outfitted with flat-panel displays, anti-collision systems, flight management and GPS---long before they reached military cockpits. Recognizing the trend, the U.S. Department of Defense launched a cost-saving program known as "COTS," for Commercial Off the Shelf equipment. Today, many military aircraft are outfitted with civilian avionics of high capability.

Getting out the Mechanicals

An early example of how the new technology was applied is the King KX-170, a combined navigation and communications radio (or "navcom"). Despite rugged construction it contained large mechanical switches with dozens of contacts that inevitably failed.

When semiconductors became available, the manufacturer not only eliminated mechanical switching, but added functions to reduce pilot workload. A new model, the KX-155, could store frequencies and give the equivalent of two radios-in-one. An electronic display eliminated rotating mechanical drums and painted numbers, shrinking the size of the radio and freeing up valuable panel space.

Digital electronics also introduced systems that were impossible to build with old technology. The Stormscope appeared as the first practical thunderstorm detector for small aircraft. Other companies looked at the poor accuracy of fuel gauges, creating a digital fuel flow instrument that measures fuel consumption precisely, and also tells time and fuel to a destination.

Airline View of Avionics

- Line Maintenance
- Test Systems
- Communications
- Indicating Systems
- Navigation
- Autoflight
- Flight Controls
- Electrical Power
- Lighting

- Air Conditioning
- In-Flight Entertainment
- Engine Systems
- Fire Detection
- Landing Gear

These major topics are discussed each year at the Avionics Maintenance Conference, run by ARINC, the airline avionics organization. The left column shows traditional avionics. But as electronics creep into other systems, shown in the right column, they often become the responsibility of avionics maintenance.

In the airlines, the digital revolution just about eliminated the problem of "mid-airs." After two airliners collided over the Grand Canyon in 1956 the FAA investigated several anti-collision systems. Every design failed because of high cost, weight, size or an inability to detect small aircraft. One system required an on-board atomic clock, which cost more than most airplanes.

But as the price of computing power dropped, "TCAS" (Traffic Alert and Collision Avoidance System) became practical. It warns when two aircraft head toward each other with a closing speed over 1,000 miles per hour---and detects most small aircraft not equipped with TCAS.

Gauges: From Round to Rectangular

By the 1970's cockpits of aircraft began to lose their "steam gauge" appearance, where instrument panels resemble an 1830 railway locomotive. Instead of round dials and pointers, the new look became the "glass cockpit," where separate gauges are replaced by images on a CRT or flat panel LCD. The system is called "EFIS," for Electronic Flight Instrument System and it rapidly spread through every size aircraft.

Today, a blank screen may become any instrument--altimeter, airspeed, tachometer----or all simultaneously. It's done by modifying software, or simply changing the plug on the back during installation. This

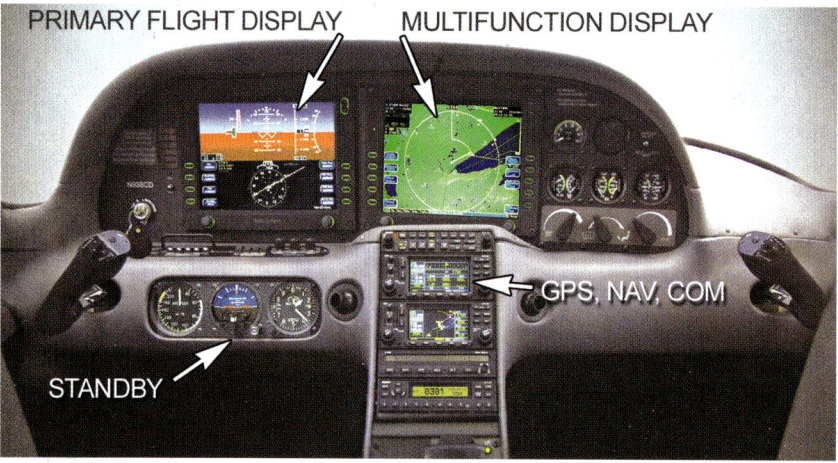

Toward the All-Glass Cockpit

The trend toward the "all-glass" cockpit is seen in this instrument panel for a Cirrus aircraft. Round gauges are replaced by large LCD screens which produce images of any instrument. What remains of the old-style panel is at the lower left, where standby instruments act as backup. In the center stack, GPS navigation and communications are integrated into one radio, with a backup below it. There are no large control yokes. They are replaced by sidestick controllers which give the pilot an uncluttered view of the instrument panel. The large Primary Flight Display shows all flight instruments, weather, moving map and traffic. The Multifunction Display can also show these functions. Having two such displays enables the pilot to put flight instruments on one screen and navigation and terrain on the other. This advanced cockpit is neither a military nor airline system, but in a kit-built airplane. The large panel displays are Avidyne's Flight Max.

tells the screen what it will be. This also reduces the number of spares needed on the shelf, a great cost benefit to airlines flying far-flung routes.

More than CNS

As the term "avionics" established itself in the civil world, it divided into three categories often called "CNS"---Communications, Navigation and Surveillance (the last referring to radar surveillance). CNS includes most avionics systems installed on the airplane. An autopilot, for example, falls under "Navigation," a transponder is a component of "Surveillance."

The list of avionics, however, grows longer. A look at the agenda of the Avionics Maintenance Conference (an airline organization) reveals more than CNS. One-third of the new items were never considered avionics or even aircraft electronics (see table "Airline View of Avionics). What happened is that engineers began using semiconductors to replace sections of mechanical and hydraulic systems. The nose wheel steering of a LearJet, for example, is by microprocessor. Engine control is no longer through levers and steel cable. It is done by FADEC, for Full Authority Digital Engine Control, which provides better fuel economy, precise engine settings and protection against excess temperature and pressure. Each year the aviation industry moves closer to what it calls the "all-electric airplane," a concept that will slash heavy oil-filled hydraulic lines, steel control cables and hundreds of miles of copper wire. In their place will be thin wires carrying multiple messages (the "databus") to electric actuators. These airplanes will fly farther on less fuel and in greater safety. Airliners are already equipped with the first of the "fly-by-wire" systems.

The growth of avionics is also reflected in the price tags of aircraft. In the transport aircraft of 1945-1950, about five percent of the cost was electrical, radio and lighting systems. Some 20 years later, that portion quadrupled to about 20 percent. More recently, airlines added the most costly and extensive electronics aboard aircraft. It is IFE, or In-Flight Entertainment, also called "cabin electronics." If an airplane has 300 seats, that means 300 IFE installations, each wired for audio, video, satellite phone, Internet and other services. In the military sector, the cost of a fighter aircraft rose to more than 40 percent for avionics.

These percentages can only increase. Airplanes divide into three main sections; airframe, propulsion and avionics. Airframes have grown larger but they still fly with the three-axis control system patented by the Wright brothers. In propulsion, the jet engine is a marvel of reliability and power, but it still works on a basic principle---action and reaction---defined by Isaac Newton 300 years ago.

Avionics, on the other hand, re-invents itself nearly every ten years, providing the industry with fresh solutions to rising fuel prices, fewer airports and crowded skies. To find answers for the 21st Century, two hundred countries of the world under the banner of ICAO (International Civil Aviation Organization) deliberated for 20 years. They agreed that technology is here and aviation is ready for its biggest change in moving more airplanes safely within limited airspace, and provide passenger services to make the flight enjoyable and productive. Nearly all the systems--- described throughout this book---are created from building blocks provided by avionics.

Review Questions
Chapter 1 The Meaning of Avionics

1.1 In the first solo across the Atlantic in 1927, how did Charles Lindbergh keep control of the airplane while flying in clouds and darkness?

1.2 Name three instruments used by the Wright Brothers in their first flight that marked the beginning of what would become "avionics".

1.3 What generated power for Lindbergh's earth-inductor compass?

1.4 Why do airlines consider the following systems part of "avionics": air conditioning, fire detection, landing gear?

1.5 What technology was widely adopted in avionics to reduce size and weight, as well as provide greatly increased function.

1.6 What system, made possible by digital electronics, greatly reduces the problem of mid-air collisions?

1.7 What replaces early "steam gauges" in aircraft instrument panels?

1.8 How can the function of an electronic instrument be easily changed?

1.9 What does "CNI," which describes basic functions of avionics, stand for?

1.10 What does the term "FADEC" mean?

1.11 Name the world body that deliberates future aviation technology?

1.12 "Avionics" is a contraction of _____ and _____.

Chapter 2

A Brief History

The invention of the airplane is tied to the beginning of radio. Both arrived at about the same time; the Wright brothers made the first powered flight in 1903, two years after Marconi sent the first radio messages 2100 miles over the Atlantic from England to Canada. Until then, people flew in hot air balloons or glided downhill in oversize kites. Radio was a laboratory curiosity and one of its early experimenters (Hertz) didn't think much would come of it.

Aviation and radio quickly grew together with the coming of World War I (1914), when airplanes proved to be deadly fighting machines. When the war ended, barnstorming pilots spread over the countryside, amazing people with stunts and joy rides in open bi-planes. But when the young industry attempted to get serious by transporting people and mail--- the results were disastrous. Many air mail pilots lost their lives in crashes where nothing went wrong with the airplane. Somehow, when fog or cloud obscured a pilot's view outside, even the most skilled pilot couldn't keep the airplane straight and level. For this reason, military pilots were warned that flying at night is an emergency. This inability to remain upright in less than visual conditions also held back early airliners. A passenger flying from New York to Los Angeles hardly gained time over riding the railroad. When darkness fell, he got off the airplane, boarded an overnight train---then reboarded the airplane in the morning.

First Radio Waves Over the Atlantic

The first practical use of aviation and electronics began at nearly the same time. The Wright Brothers' first powered flight was 1903. The first radio message was sent from England to Canada in 1901.

G. Marconi, after experimenting at his home in Italy, was first to communicate long distances by radio. In a 1901 demonstration, he sent signals over two thousand miles. The first message was three dots--- the letter "S" in Morse code. Early aircraft radio adopted the Marconi system, which consisted of a spark transmitter and magnetic detector for receiving. Although not known in 1901, the radio signals had travelled great distances by "skipping" from an electrical layer known as the ionosphere. Skipping is still used today by long-range aircraft with high frequency (HF) communications.

What went wrong? The aviation community discovered that, no matter how experienced the pilot, he cannot control an airplane when unable to see outside. Whether it's fog, cloud, blowing snow, dust or other obscuration, *any* pilot is about five minutes from losing control.

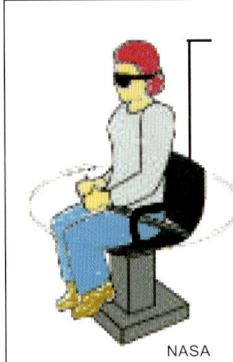

No pilot can outwit the Barany chair. Just a few slow turns and reversals while blindfolded remove the sense of which way is up. Without the eye, humans sense balance by an inner-ear mechanism, which is confused by motion of the chair.

Unless the pilot flies by instruments after entering a cloud, a "graveyard spiral" begins in about five minutes.

NASA

This is clearly demonstrated by FAA in its notorious "Barany chair," which is demonstrated at air shows and safety meetings. A pilot sits in the chair blindfolded. The instructor turns the chair (which is on a rotating base) at moderate speed. After several revolutions, the chair is stopped and the pilot asked, "Point to the direction that you're turning." As the pilots points, the audience breaks into laughter; he or she is pointing in the *opposite* direction. It is comical to watch, but is also the greatest killer of pilots. The accident report reads; "Continued VFR (visual) flight into IMC (Instrument Meteorological Conditions)."

The reason is that the eye is the primary organ for indicating "which way is up." When vision outside is blocked, however, the inner ear, which controls sense of balance, takes over. The problem is, the balance mechanism is easily fooled. When the Barany chair turns, the inner ear responds first to acceleration. When the chair is stopped the pilot senses deceleration. But the rotating motion of the chair confuses the inner ear and the pilot gives the wrong answer when asked which way he's turning.

Now transfer this scenario to an airplane entering a cloud. The untrained pilot looks out the windows and sees solid gray. Let's assume a gust of turbulence moves one wing down---then, a second or two later the wing slowly returns to level by itself. This causes the same phenomenon as in the Barany chair, causing the pilot to correct in the wrong direction. The airplane enters a tightening spiral from which there is rarely a recovery.

To worsen matters, there is another false clue. In straight and level flight, a pilot feels gravity pushing him into the seat. But in a turn, centrifugal force starts acting on his body---and it feels exactly like gravity. The pilot believes he is still sitting vertically and has no feeling the airplane is turning and descending. That's what confronted the budding aviation industry. Unless a pilot had artificial guidance inside the cockpit, airplanes would remain in the realm of barnstorming and air racing.

The breakthrough happened when Elmer Sperry invented the "turn and bank" indicator. Using a gyroscope as a stable platform, a needle on the instrument showed when the airplane entered a turn. If the pilot

Sperry Gyroscope (1914)

The greatest single device for aviation safety was the gyroscopic instrument, a spinning wheel that remains stable, even as the aircraft maneuvers. This provides the pilot with a reference within the cockpit when he cannot see outside. A gyro is shown here with Elmer Sperry, the inventive genius who applied it to the turn-and-bank indicator, the first life-saving device for instrument flight. Sperry went on to develop the artificial horizon, autopilot and other systems based on gyroscopes.

"Look...no hands!"

Sperry also used the gyroscope to design the first autopilot. A remarkable demonstration in 1914 is shown above. Sperry's son, Lawrence, is in the pilot's seat, holding his arms away from the flight controls. Standing on the wing to the left is a mechanic, whose weight should cause the wing to drop. The airplane, however, is stabilized by Sperry's gyro control system and remains level. The inventor wins France's Airplane Safety Competition (50,000 francs) and the distinguished 1914 Collier Trophy in the U.S.

It Started With Turn-and-Bank

This simple turn-and-bank indicator was a breakthrough that turned the flying machine into a practical airplane. Developed by Elmer Sperry and his gyroscope, the instrument began the quest for all-weather operations. A pilot could now fly with confidence inside clouds, approach airports during low visibility and fly safely on dark, moonless nights.

The instrument indicates if the airplane is turning. The turn needle remains centered so long as the wings are level. But if a gust lowers a wing, the airplane starts a turn, causing the needle to move to one side of center. Shown here is a turn to the right, as the needle moves under the right tick mark, usually called a "dog house." The pilot now knows he should apply left aileron to bring the wings back to level, which stops the turn.

The instrument does *not* show bank angle, or position, of the wings. It indicates only "rate of turn"---or how fast the airplane is turning. This is sufficient information to keep the wings level. If the pilot wants to turn, he lowers a wing with the aileron and puts the needle on the "dog house." The airplane now turns at the rate of 3 degrees per second.

The ball at the bottom is not a gyro instrument, but moves freely. It helps the pilot coordinate the turn with the rudder (or the airplane would slip or skid in the air). Keeping the ball centered with the rudder during a turn assures good control of the airplane when there is no view outside.

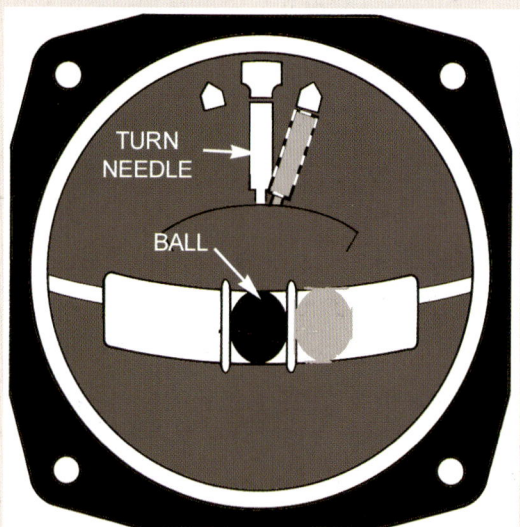

The effectiveness of the turn and bank instrument was shown during Charles Lindbergh's solo flight across the Atlantic to Paris in 1927. During the ocean crossing he spent several hours on the instrument to fly through bad weather.

The turn and bank is sometimes called the "turn-and-slip indicator," and is still required in many aircraft as a backup. A later version is called a "turn coordinator." Today, an improved gyro instrument, the artificial horizon, is the primary reference for instrument flying.

---

kept the needle centered, the airplane remained in level flight. Sperry's device removed a major obstacle to dependable flight operations.

Instruments were greatly improved by 1929 through the work of Jimmy Doolittle (who later became an Air Corps General in World War II). He came up with the idea of an artificial horizon that displayed the wings of an airplane against a horizon line. As the airplane maneuvers, the pilot sees miniature wings bank left or right, and rise and fall with the angle of the nose. By showing wing and nose position---roll and pitch--- on one instrument, the display is easy to fly because it recreates what the pilot sees through the windshield on a clear day. The artificial horizon was designed around Sperry's gyros.

Now that an airplane could be controlled in almost any visibility condition, aviation was ready for the next advance; the guidance required to fly to cross-country to a destination airport and make a safe landing. The first attempt placed a lighted beacon (like a lighthouse) every 10 miles along the route. It was an immediate success; airplanes could fly at night, speeding mail and passengers in less time. The day of navigating by compass, chart and timepiece seemed to be over.

But it soon became painfully obvious that light cannot penetrate fog, clouds and heavy snow. The answer was to abandon guidance by light and create airways formed by radio waves, which easily move through any form of precipitation.

The 1930's saw great advances in radionavigation and the growth of commercial aviation. Let's look at milestones that merged aviation and electronics into one of the fastest, safest forms of transportation.

Morse and the Code

Code keys were found aboard long-range aircraft of the 1930's. Radio operators were called "brass pounders."

Samuel FB Morse was first to transmit information by electrical signals. In an 1836 demonstration between Baltimore and Washington, DC, letters are encoded into dot and dashes. The first message: "What hath God wrought?"

Early aircraft used Morse code because voice was not possible until the invention of the vacuum tube.

Morse code survives today as the identifier for thousands of radionavigation stations. To avoid navigation error, pilots must listen to a station's identifier before using the signal (although many stations are also identified by voice).

First Voice Transmission

Alexander Graham Bell, inventor of the telephone, was first to transmit voice through wires (1876). The technique is later applied to wireless voice transmission and adopted by aviation for air-ground communications.

Bell announced his next project would be a "flying machine." He worked closely with Glenn Curtiss, who improved airplane design after the Wright brothers accomplished the first successful flights.

Hertz Demonstrates Radio Waves

Heinrich Hertz

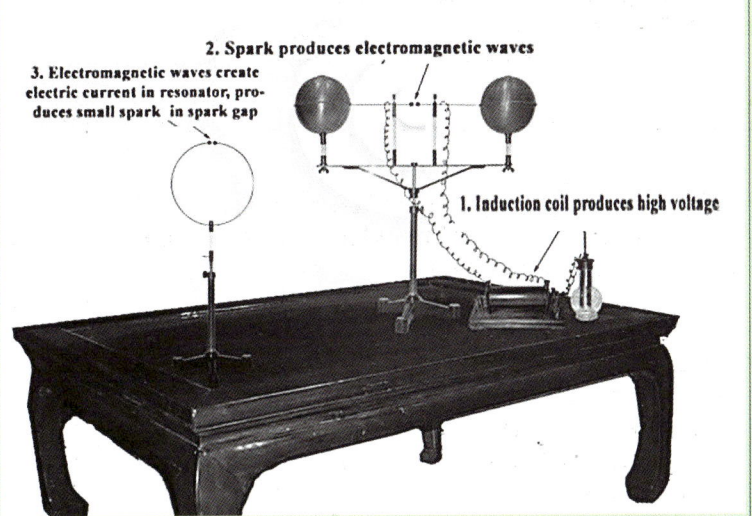

In a Berlin laboratory in 1887, Prof. Heinrich Hertz sends radio waves across a room. A transmitter (on the right) discharges sparks across a gap, creating radio waves. A receiver (left) responds by producing sparks (the received signal) across metal balls. The Professor is honored 160 years later when his name becomes the term to describe radio frequency as "hertz." Meaning the number of cycles per second, it's now written as kilohertz (kHz), megahertz (MHz), gigahertz (GHz), etc.

After the experiment, Prof. Hertz's students asked, "So what is next?" Hertz replied with the understatement of the century;
   "Nothing, I guess."

First Aircraft Radio

Carried aboard a Curtiss bi-plane in 1910, this rig made the first radio transmission from air to ground while flying over Brooklyn, NY. It weighed 40 lbs and mounted on a 2-ft-long board strapped to the airplane's landing skid. The pilot, James McCurdy, a Canadian aviation pioneer, transmitted with a Morse code key mounted on the control wheel.

The transmitter was a spark type. An induction coil created high voltage from a 6-volt battery (seen at far right). When the operator closes the code key, voltage jumps across a spark gap (much like a spark plug in an automobile). This sends current into the large coil at left. The coil is part of a tuning circuit which causes energy in the spark to circulate back and forth at a rapid rate. This is coupled to an antenna wire trailing outside the aircraft, which converts the oscillations into radio waves. In later experiments the aircraft carried a receiver to hear transmissions from the ground.

Spark transmitters were inefficient and emitted signals on many frequencies at the same time. Not until the invention of the vacuum tube, which could generate clean, powerful signals, did 2-way radio become practical in aircraft. The vacuum tube also made possible transmission of the human voice.

1910 Curtiss Biplane

Air-Ground Messages in England

Thorne-Baker in England holds a 1910 aircraft radio which used a Marconi electromagnetic detector for receiving. He communicated with a Farman biplane flying one-quarter mile away.

The radio aboard the airplane was a 14-lb transmitter fastened to the passenger seat. Pilot Robert Loraine transmitted with a Morse code key tied to his left hand. The antenna consisted of wires fastened along the length and width of the airplane.

Marconi Museum

Flying Machine Rescued by Radio

Lifting off from New Jersey in 1911, the airship *America* headed toward Europe. Encountering bad weather and engine problems 100 miles out, the crew abandoned the airship and took to a lifeboat. The wireless operator was able to communicate with the nearby Royal Mail Steamship *Trent*, which rescued the crew. The Marconi radio had a guaranteed range of 30 miles.

Note the cable dropping from the airship. It trailed in the seawater to provide a good electrical ground for the antenna.

Marconi Museum

Lighted Airways

Before 1926, air mail pilots could fly only during the day. That changed when lighted beacons were installed every 10 miles. A rotating light appeared to the pilot as a flash every ten seconds. Just below the beacon were course lights that pointed up and down the airway. Course lights also flashed a number code, the same number that appears on the roof of the building. "5" indicates it is the fifth beacon in a 100-mile airway. Although lighted beacons shortened the time for mail delivery, their effectiveness was poor in bad weather. Navigation by radio waves would provide the solution.

1927. Aviation receives own radio frequencies.

The International Radio Convention meets in Washington, DC to assign aircraft and airway control stations frequencies for their exclusive use.

1928. US Dept. of Commerce expands stations.

This year begins the rapid expansion of ground radio stations for transmitting weather and safety information to pilots. Old "spark gap" transmitters capable only of code are replaced by equipment that carries voice. Until this time, aircraft could only receive, but an increasing number install transmitters for two-way communication with the ground. By 1933 there were 68 ground stations. The next year, over 300 aircraft flying airways had two-way radio; over 400 could still only receive.

1928. First practical radionavigation.

U.S. Department of Commerce adopts the "four-course radio range," where pilots listen in headsets for audio tones that get them "on the beam." After beginning between Omaha, Nebraska and New York, the stations spread through the country. The last four-course range was taken out service in Alaska in the 1970's.

1928 was also the beginning of teletype machines to carry weather information throughout the US.

1929: First aircraft operational frequencies.

Until now, radio was only to send weather and safety information to pilots. This year, the Federal Radio Commission assigned frequencies to airline companies to allow them to speak directly with their aircraft in flight

This led to the creation of "ARINC" (Aeronautical Radio Inc), an organization of airlines which operates communications services today.

Another development during 1929: pilots flying airways were required to report their position, thus marking the beginning of Air Traffic Control.

1930: Airport traffic controlled by radio.

First installed at Cleveland, the system spreads to 20 more cities in the next five years (replacing the flagman on the roof).

1931: Weather maps begin.

Experiments transmit weather maps over the same teletype machines used in the Federal Airway System (which had been capable of operating only on paper tapes). By the next year, maps of the U.S. were transmitted six times per day to 78 cities. Briefers on the ground could now give pilots weather information.

Jimmy Doolittle and Beginning of "Blind" Flight

Jimmy Doolittle, an army lieutenant, was the first to take off, fly a course and land without seeing outside the cockpit. He controlled the aircraft solely by reference to instruments. Attitude information (pitch and roll) were indicated on an artificial horizon. A directional gyro, more stable than a magnetic compass, indicated direction, while a "sensitive" altimeter, which could be corrected for barometric pressure, replaced the conventional instrument.

Doolittle followed a radio course aligned with the runway created by a radio range station on the ground. Marker beacons indicated the airplane's distance from the runway.

The flight was the single most important demonstration of what would become "avionics." Because guiding aircraft to landing had been done only by light signals, which don't penetrate clouds, Doolittle's flight made commercial aviation a reality.

1931: Glideslope Appears

By tilting a radio beam vertically, experimenters at College park, MD created an electronic path that matched the glide angle of an airplane. Airplanes now had guidance for descending to a runway in low visibility. It later became the glideslope part of the ILS (Instrument Landing System).

1932: Instrument Rating Required

Air Commerce Department rules that air transport pilots must show an ability to use airway navigation aids and fly certain maneuvers guided entirely by instruments.

1933: Cross Country Instrument Flight

Bureau of Standards demonstrates a radio system for blind flight. Arriving at Newark, NJ, from College Park, MD, the airplane flew the first cross-country all-instrument flight.

1935: Radar

The Defense Research Council of Great Britain receives a report on a new system known as "radar" (for radio detection and ranging). It goes on to become a chain of stations along the British Isles for detecting hostile aircraft during World War II. To avoid shooting down friendly aircraft, a device known as IFF, Identification, Friend or Foe, is installed on British airplanes. Today IFF is known as the "transponder." Later in the war, the Massachusetts Institute of Technology scales down the size of radar for installation aboard aircraft, the first major instrument of electronic warfare.

1935 Air Traffic is Controlled

Airlines operate the first airway traffic control center at Newark, NJ, to provide safe separation for aircraft flying in instrument conditions. Chicago and Cleveland soon follow. It is the beginning of the en route Air Traffic Control system to separate traffic after it leaves the airport area.

1940: Radio for Oceanic Flight

Six powerful high-frequency radio stations are installed on Long Island, NY, to provide the first two-way radio communications for aircraft flying the Atlantic. The frequencies are in the HF band which, unlike lower aviation bands, "skip" great distances.

These stations also play an important role in ferrying military aircraft to England at the outbreak of World War II.

1944: ICAO is Born

Fifty-two countries meet in Chicago to launch ICAO, the International Civil Aviation Organization. The first global aviation authority, ICAO will publish standards to assure technical uniformity throughout the world. By 2004, ICAO had 188 member countries, which it calls "States."

1945: GCA Honored

The distinguished Collier Trophy is awarded to Dr. Luis W. Alvarez for his concept of Ground Controlled Approach. GCA uses a ground-based radar that emits two beams; one to indicate aircraft distance from the runway, the other to measure its height above ground. The radar operator watches the display and "talks" the pilot down to landing. Although successfully used by the military, GCA was never adopted for civil use. Airline pilots and government authorities preferred the Instrument Landing System (ILS), which became the standard for well into the 21st Century.

1946: Radar-equipped Control Tower

At Indianapolis, a demonstration of the first control tower equipped with radar for civil flying. Adapted from naval equipment, it could reduce ground clutter by ignoring targets not in motion.

Today, this type of radar is known as a Tracon, for Terminal Radar Control, and manages traffic in and out of airports at a range of 30-40 miles.

1947: VOR Commissioned

After experimenting on the New York-Chicago airway, the

During the 1920's, traffic in and out of airports was controlled by a flag man standing atop a hangar.

Civil Aviation Authority opens the first VOR (Very High Frequency Omnidirectional radio range) station. VOR grows rapidly to about 1,000 stations throughout the U.S. and replaces the obsolete four-course radio range.

1947: Navy Pursues TACAN

An effort to make VOR a common system for both military and civil navigation fails. The U.S. Navy selects TACAN (Tactical Air Navigation), a development which the Navy needed during the Korean War in 1950.

Since most military flying is done in civil airspace, military aircraft must also be equipped with VOR receivers. On the other hand, civil aircraft use a part of TACAN to operate their DME (Distance Measuring Equipment).

1948: Bell Labs Demonstrates the Transistor

In searching for a device to replace electromechanical switches in telephone systems, Bell scientists invent the transistor. It was tiny, had no moving parts, didn't wear out and generated little heat.

The transistor will become important in avionics for the same reasons.

1953: Transmissometer Installed

The first electronic device for measuring visibility on the ground is completed at Washington National Airport. Located

Before surveillance radar, air traffic controllers plotted aircraft from pilot position reports sent by radio. Today, traffic is viewed on radar screens. Note the four-course radio range station on the chart. It provided radionavigation during the 1930's, but was later replaced by VOR and ILS.

alongside the runway, a transmitter sends a light beam to a receiver several hundred feet away. The receiver measures the loss of light due to fog or other obscuration and converts it to RVR, or Runway Visual Range. This value, in feet, is more accurate than a human looking at a distant point and estimating visibility.

1956: Airliners Collide

A TWA Constellation and United DC-7 collide over the Grand Canyon (Arizona) killing 128 people. Both airplanes were flying VFR (visually) on a sunny day in wide open airspace.

The response by authorities is to require all aircraft flying over 18,000 feet to fly IFR (instrument flight rules), keeping them under positive control, in radar surveillance and safely separated.

The accident starts development of an on-board anti-collision system. The search continues for 40 years---until the introduction of TCAS (Traffic Alert and Collision Warning System).

1957: Narrow Band Receivers

The Civil Aeronautics Administration (CAA) begins installation of new radios designed to double the number of aviation channels. Until that time, radio channels were 200 kHz apart. The new radios "split" channels for a spacing of 100 kHz.

1956: Flight Recorders

The CAA rules that air carrier and commercial aircraft over 12,500 lb must have a flight data recorder by 1958. The FDRs record airspeed, time, altitude, vertical acceleration and heading.

1957: Boeing 707 first flight

After building 857 airplanes, 707 production ends in 1991.

1958: FAA and NASA are Born

During April and May, military aircraft collided with civil airliners in two separate accidents. The collisions raised a storm of protest to eliminate the Civil Aeronautics Agency, which controlled only the civil sector. Legislators call for a unified agency to control both military and civil aircraft when flying in civil airspace.

Later that year, Congress passes the bill that creates the FAA (Federal Aviation Administration). A major responsibility is to control airspace in the US and develop a common system of air traffic control for civil and military aircraft.

In October of the same year, Congress creates NASA (National Aeronautics and Space Administration). Although NASA is identified with space exploration, it also justifies the "aeronautics" part of its name. NASA will contribute to airline avionics in the form of databuses, displays, synthetic vision and human factors. It will solve problems in small aircraft that limit their usefulness in bad weather. There are NASA programs on safer cockpits, airframe icing and low-cost anti-collision devices.

1959: DME Approved

The International Civil Aviation Organization chooses DME (distance measuring equipment) as the world standard to complement VOR navigation.

1959: Transponders Begin (ATCRBS)

Known as "secondary radar," the transponder not only provides more powerful returns than conventional radar, but encodes aircraft identification as well. It is based on World War II "IFF" (Identification, Friend or Foe).

The system is "ATCRBS," for Air Traffic Control Radar Beacon System, and triggers replies from an aircraft transponder. The first ground interrogator is installed in New York and expands to 19 more air route traffic control centers. A pilot can dial up to 64 codes (for ID), which is expanded to 4096 codes in later equipment.

1960: Airborne Weather Radar

FAA requires US airliners to carry airborne weather radar. It is phased in over several years and, in 1966, expanded to cover large cargo aircraft.

1960: More Com Channels

The first increase in VHF communications channels in the aircraft band since 1946. It adds 5 megahertz to the band, with 100 more channels for air traffic control. The new channels, in MHz: 126.825 to 128.825, 132.025 to 135.0

1964: Cockpit Voice Recorders

FAA requires CVRs in large turbine and 4-engine aircraft. In the event of an accident, the recorder provides cockpit conversation during the 30 minutes preceding the crash.

1964: Inertial Navigation Systems (INS)

Pan Am installs INS on most of its jet aircraft to provide accurate navigation over oceans and remote areas where ground stations are not available.

1964: Category II Landings

FAA announces requirement for Cat II (ILS) instrument landings, another step toward all-weather operations Decision height is lowered to 100 feet and runway visibility range (RVR) of 1200 feet. United Airlines is first to qualify, with its DC-8s (in 1965).

1964: Helicopter Certified for IFR
Sikorsky S-61 becomes first civil helicopter to be certified for IFR (Instrument Flight Rules).

1964: Single Sideband Radio (SSB)
FAA begins operating first single sideband (SSB) ground station in Alaska for air traffic control over the North Pole. SSB, which operates in the high frequency (HF) band, will eventually replace older, less-efficient HF radio for oceanic and remote communications.

1964: First Automatic Touchdown
A British European Trident at London makes the first automatic touchdown of a scheduled commercial airliner carrying passengers.

1965: DME
FAA requires large foreign transports flying in U.S. to carry Distance Measuring Equipment (DME)

1966: Satellite Communications
FAA reports "voice messages of excellent clarity" during first test of a satellite for long-range communications. The vehicle is NASA's Applications Technology I. "Satcom" will eventually replace High Frequency equipment.

1967: Satcom Datalink
FAA and NASA demonstrate a datalink system using a satellite for transmitting navigation data from aircraft to ground stations. A Pan AM cargo jet sends data to the ATS-1 satellite, which relays it to a ground station in California. It is also the first test of an aircraft antenna designed to transmit satellite messages.

1968: Altitude Alerting
FAA requires an altitude alerting system on turbojet aircraft because of their rapid climb and descent rates. Pilot presets an altitude and receives aural and visual warning in time to level off. Also alerted is straying from an assigned altitude.

1970: Microwave Landing System (MLS)
Secretary of Transportation forms group to study development of MLS, the eventual replacement for ILS

1970: Satcom for Air Traffic Control
FAA establishes communications via satellite between San Francisco and Hawaii. The first full-time satcom for air traffic control uses Intelsat. Because of superior communications, FAA closes down High Frequency station in California.

1970: Advanced Flight Data Recorder (FDR)
FAA requires advanced FDR's for large transport aircraft. The new type records over three times more information.

1972: Category IIIa Landing
TWA receives first authorization to operate Cat IIIa (ILS) weather minimums. It allows Lockheed L-1011 to operate down to visibility of 1000 feet (Runway Visual Range), then as low as 700 ft RVR.

1973: Mode C Transponders
FAA requires aircraft flying in Terminal Control Areas (which surround major airports) to carry transponders capable of Mode C (altitude reporting).

1973: Public Address System
FAA issues a rule requiring aircraft carrying more than 19 passengers to have public address and interphone systems to keep crew and passengers informed during an emergency.

1974: Radio Range Shut Down
The last four-course radio range, located in Alaska, is decommissioned by FAA. Replaced by VOR and ILS, it was the first radionavigation system for "blind" (instrument) flight in bad weather. The four-course radio range was important for the growth of aviation during the 1930's.

1974: Ground Prox Installation
A rule requiring Ground Proximity Warning Systems on airliners is published by FAA. GPWS warns when the aircraft is below 2500 feet and in danger of closing too rapidly with the ground.

1980: Avionics and Two-Person Crews
Boeing plans on two-person crew for its new B-757-767 airliners. Digital systems in these aircraft reduce the need for third person (flight engineer). It's made possible by new EFIS (Electronic Flight Instrument System), which centralizes instruments and displays, as well as automatic monitoring of engine parameters.

1981: FAA Selects TCAS
FAA adopts the Traffic Alert and Collision Avoidance System (TCAS). Compatible with existing and future transponders, there are two versions: TCAS I, which delivers only a traffic alert, and is practical for small aircraft; and TCAS II, which adds vertical escape maneuvers and is required for airliners.

Radar surveillance room of an Air Traffic Control facility. Pilots call in position reports by radio; the airplane appears on the "PPI," or Plan Position Indicator, seen at lower left. "Flight Strips" are print-outs which inform the controller of flights arriving in his sector. These paper strips are replaced in future ATC with electronic displays.

FAA

TCAS III, which adds horizontal maneuvers, proved difficult to develop and was dropped.

Future anti-collision systems will be based on satellite surveillance.

1981: Search and Rescue Satellite

U.S. launches weather satellite carrying Search and Rescue Satellite-Aided Tracking (SARSAT). It is capable of receiving signals from an aircraft ELT (Emergency Locator Transmitter). A similar satellite called COSPAS is launched in 1982 by the USSR (now Russia).

1983: GPS Nav Across the Atlantic

A Rockwell International Saberliner is first to cross the Atlantic guided only by GPS.

1984: Loran Approved

FAA approves Loran as an area navigation system for IFR (Instrument Flight Rules) flight.

1988: Wind Shear

Turbine-powered airliners with 30 or more passenger seats must carry equipment that warns of low-altitude wind shear. Guidance for recovery from wind shear is also required.

1991: Loran Gap Closed

Complete coverage of the US by Loran signals results from new station constructed to fill the "mid-continent gap."

1991: Mode S Interrogators

The first two Mode S systems are delivered to FAA. It's the beginning of the new radar beacon ground interrogator system that will eventually number 137 in U.S. airspace. Aboard aircraft, Mode S transponders will replace the ATCRBS system.

1993: GPS Approach Approved

Continental Express flies approved non-precision approach using GPS into two Colorado airports.

1994: MLS Halted

FAA will no longer develop the Microwave Landing System (MLS). Future effort at all-weather landing systems will be done with GPS.

Following this announcement, FAA cancels plans to purchase 235 new ILS's (Instrument Landing Systems). Rapid development of GPS is the reason.

1994: Free Flight

In one of the most sweeping changes in air navigation, FAA begins study of "Free Flight." Aircraft will fly with greater freedom, enabling pilots to choose the most favorable routing. Air traffic controllers would intervene only to assure safety or avoid crowding in the airspace. Because of this, the term "air traffic controller" will become "air traffic *manager*."

A two-year trial of Free Flight begins in 1999 in Alaska and Hawaii.

1995: FANS Trial

FAA and Australia's Qantas Airlines complete first trials of new satellite-based communication, navigation and surveillance system recommended by the International Civil Aviation Organization (ICAO). Called "FANS" (Future Air Navigation System), it improves communications with aircraft flying in oceanic and remote areas. This is the beginning of a global changeover to the next-generation of air traffic control.

1996 : Flight Recorders Expand

FAA proposes increase in the amount of information collected on Flight Data Recorders (FDR). The number of parameters would increase--from as few as 29 to as many as 88, depending on when the airplane was manufactured. Airlines would retrofit over about four years from the effective date of the rule.

1996: Enhanced Ground Prox

American Airlines is the first carrier to receive FAA approval for the Enhanced Ground Proximity Warning System (EGPWS). Installations are on all American's B-757's.

Review Questions
Chapter 2 A Brief History

2.1 Radio frequencies are measured in Hertz (Hz), after Heinrich Hertz. What was his contribution to communications?

2.2 What was the first system for marking cross-country airways? How was it limited?

2.3 What was the first instrument to enable pilots to maintain control of an airplane without seeing outside the cockpit?

2.4 What component led to the artificial horizon and autopilot? Name the developer of these early systems.

2 5 What type of transmitter sent the first radio message from an airplane to the ground?

2.6 What was the first radionavigation system for guiding airplanes?

2.7 Who was the pioneer who flew the first instrument flight, sometimes known as "blind flying," in 1929?

2.8 What system in Air Traffic Control replaced position reports by voice?

2.9 In 1980, manufacturers began designing airliners without a third crew member. What avionics development made it possible?

Chapter 3

VHF Com
Very High Frequency Communications

Communications move information in and out of an airplane for air traffic control, airline company operations and passenger services. The earliest "com" radios sent and received Morse code, then advanced to voice as technology became available. Today, voice messages are also headed for extinction, as digital information travels more efficiently on "datalink," a technology spreading through aviation.

VHF-Com. Radios for communication may be labelled "Com, Comm, VHF-Com" or simply "VHF." They receive and transmit in the VHF com band from 118.00 to 136.975 MHz. When a radio is a *navcom* both communications and navigation are combined in a single case or housing. Because the com half transmits and receives, it is a "transceiver."

The VHF band is under great pressure because of the growing number of aircraft. Frequencies are assigned by international agreement and difficult to obtain because many non-aviation services compete for limited space in the radio spectrum. These include public-safety (police, fire, emergency medical and other government activity). VHF is also in demand by "landmobile" services such as taxi, and delivery vehicles. As a result, avionics engineers have developed new techniques for expanding communications inside the existing VHF aviation band.

Splitting. One method for squeezing in more channels is "splitting." As the accompanying chart shows, the VHF band has been split four times, resulting in an increase from 90 channels to over 2,280. This became possible with advances in digital signal processing, especially to make the com receiver respond very selectively to the new, narrow channels.

A large number of old-technology avionics could not operate with such tight spacing and, in 1997, radios with 360 channels or fewer were banned (see chart).

VHF Data Radio. In the coming years, there will be a dramatic drop in the number of voice transmissions on the VHF aircraft band. It is due to the rise of

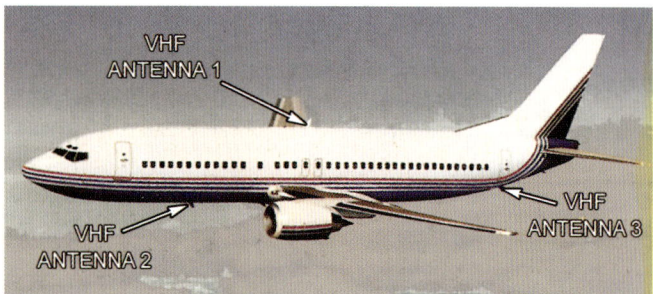

Three VHF com radios are often carried aboard airlines; the antenna locations are shown above. VHF com 1 and 2 are for communicating with air traffic control. VHF com 3 operates in the ACARS system for what is commonly called "company communications."

Acceptable VHF Com Radios

NO. OF CHANNELS	SPACING	ACCEPTABLE?	NOTES
90	100 kHz	No	January 1, 1997 banned radios with 360 or fewer channels
180	50 kHz	No	The increase from 180 to 360 channels resulted when the band was expanded from 126.90 to 135.9 MHz
360	50 kHz	No	Doubling of channels resulted from "channel splitting" (moving frequencies closer together).
760	25 kHz	Yes	Again channels were split, but new technology produced selective receivers which could separate close-spaced channels. Also, another megahertz was added to the band, providing 40 more channels.
2280	8.33 kHz	Yes	Further channel splitting tripled the channels to accommodate increasing air traffic. This spacing, 8.33 kHz, was first used in Europe, where frequency congestion became critical.

"datalink," where messages are sent and received in digital coding. The human voice delivers information at a slow rate---about 300 words per minute. Compare this to an e-mail message building on a computer screen. Three hundred words appear in about one-tenth of a *second!* Not only will datalink take one channel and split it more than four ways, it operates faster, and eliminates misunderstood words.

VDR. Airliners and other large aircraft are equipping with a new generation known as VDR, for VHF Data Radio. Because many years are required to transition to a new system, the VDR must operate on both existing and future systems.

Voice. This is the traditional air/ground communications where the pilot talks over a microphone. It is known as AM, or amplitude modulation.

ACARS. An automatic system that reports via VHF radio to an airline company when its aircraft take off and arrive, and carries messages about company operations (described in the next chapter).

VDL. Yet another mode is VDL, for VHF datalink. Many airliners have equipped with VDL because their wide-ranging flights must be prepared to communicate with systems everywhere.

VDR Radio

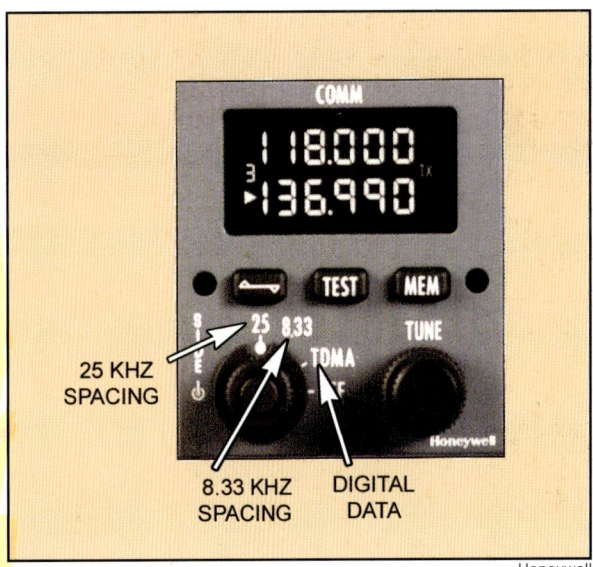

Control-display head for the VDR (VHF data radio). The knob at the lower left enables the pilot to select either channel spacing; 25 kHz or the newer 8.33 kHz. The next position is "TDMA" which can send digital voice and data.

Basic VHF-Navcom Connections: General Aviation

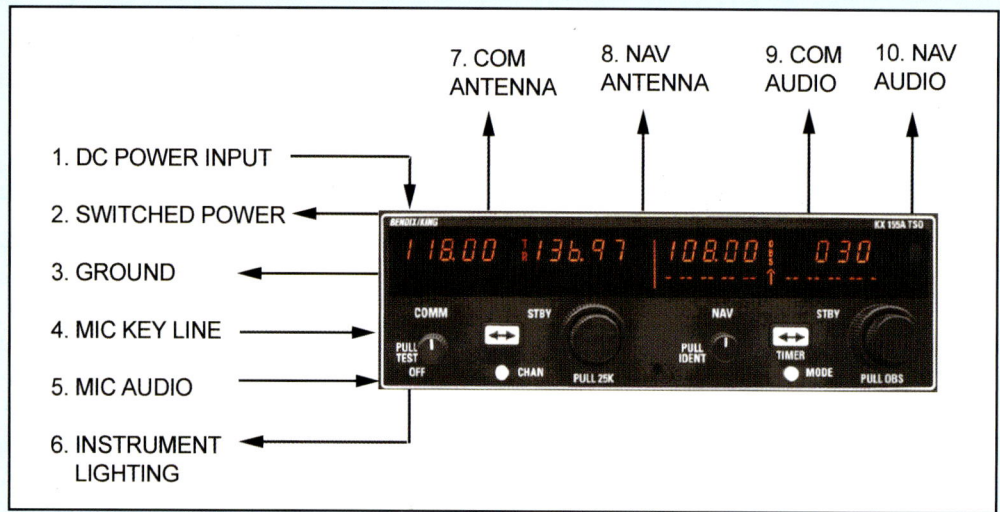

1. DC Power Input
Depending on the airplane's electrical system, this is primary power to the radio; 14- or 28-volt DC. It from a circuit breaker or fuse designated for a navcom. In some diagrams, DC power input is also called the "A" lead.

2. Switched Power
When the radio is turned on, this connection sends power from the radio to certain accessories, such as an external indicator that displays navigation information. An example is the VOR instrument that displays left-right, up-down steering commands.

3. Ground
The negative side of the electrical system, it can be any part of the airplane's metal structure that goes back to the negative side of the battery.

In composite (non-metal) airplanes, the ground is a "bus bar," or heavy copper wire or braid that extends the negative battery lead through the airplane.

4. Mic Key Line
Turns on the transmitter when a microphone button is pressed. The button may be on the mike, or mounted on the control yoke. Releasing the button switches the radio back to receive.

5. Mic Audio
This is the voice signal from the pilot microphone brought into the radio through a microphone jack or audio panel. Mic audio is applied to the transmitter, and drives the pilot intercom or passenger address system.

6. Instrument Lighting
At night, the pilot may dim lights on the panel with one control. When this connection is wired to the dimmer, radio lighting is controlled along with all other illumination.

7. Com Antenna
Coaxial cable that runs to the VHF com antenna.

8. Nav Antenna
Coaxial cable to the VOR nav antenna.

9. Com Audio
Audio received from an incoming signal. In simple installations, this line connects to the pilot's headphone jack. Audio at this point is "low level," meaning it can only drive a headphone, and not a cabin speaker. Although some radios have built-in amplifiers, many aircraft add an audio panel. It not only provides amplification for the cabin speaker, but boosts and mixes low level audio from other sources.

10. Nav Audio
This enables the pilot to listen to and identify navigational signals, which broadcast an ID in Morse code and voice.

VHF-Com System

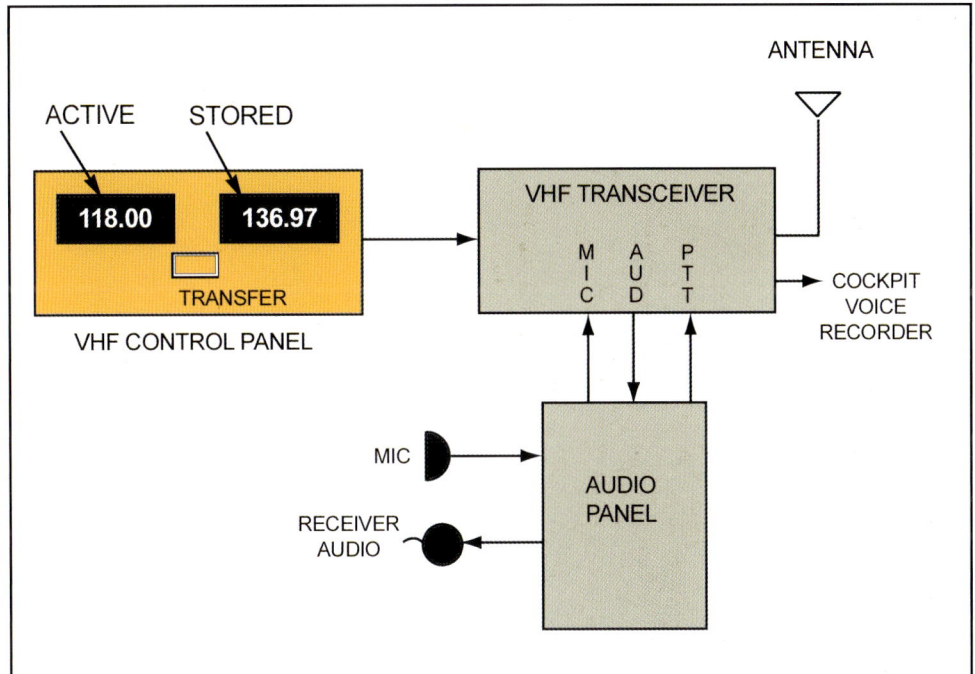

A com radio typically found in airliners and large aircraft. The pilot operates the VHF Control Panel, while the main unit of the VHF transceiver is in a remote electronics bay.

Two frequencies may be selected at one time; the *active* channel sends and receives, the *stored* channel remains inactive. When the transfer button is pressed, the two channels exchange places.

The Audio Panel connects pilot microphone and headset or loudspeaker. "PTT" is the push-to-talk button that switches the radio between transmit and receive. The button is on the microphone or the control yoke.

The transceiver also provides an audio output to the Cockpit Voice Recorder to retain radio messages in the event of a safety investigation.

There are usually three VHF com radios aboard an airliner. One radio, however, is operated in the ACARS system, as described below.

Third Com Radio

This is the same as the other two com radios, except for modifications to operate on ACARS (Aircraft Communication and Reporting System) described in a later chapter. There is no pilot control panel because the frequency is pre-set to an assigned ACARS channel. ACARS automatically receives and transmits messages about company operations.

A pilot may also use voice on this radio through the mic and receiver audio connection.

VHF com radios in large aircraft typically operate from a 28 VDC power source. The transmitter is often rated at 25 watts of radio frequency output power.

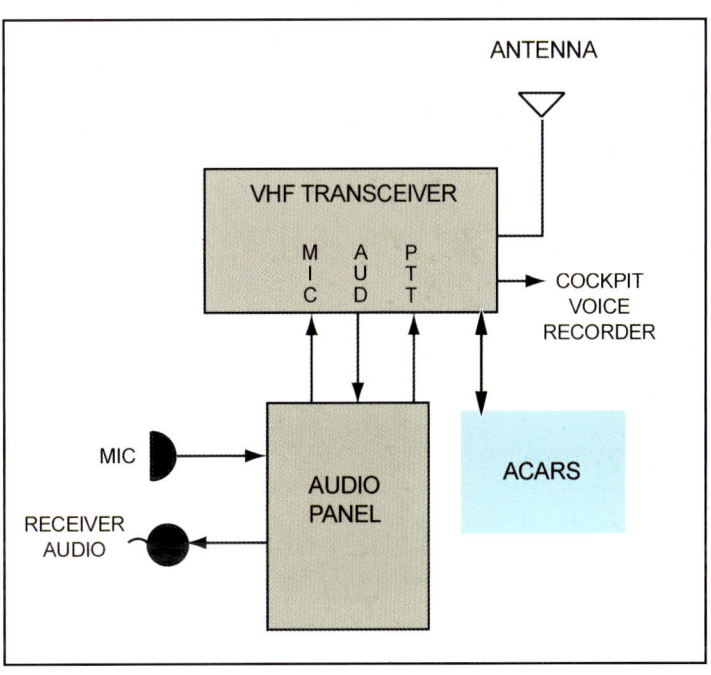

VHF-Com Control Panel

Typical airline control head for one VHF transmitter-receiver (transceiver). The pilot is communicating on the left display; the frequency transfer switch is pointing left. He has stored the next frequency on the right side. A flip of the transfer switch activates the next frequency. This panel-mounted unit controls a remote transceiver in the electronics bay.

The "Com Test" button at the bottom disables the automatic squelch. This allows atmospheric noise to be heard, which is an approximate test of whether the radio is operating.

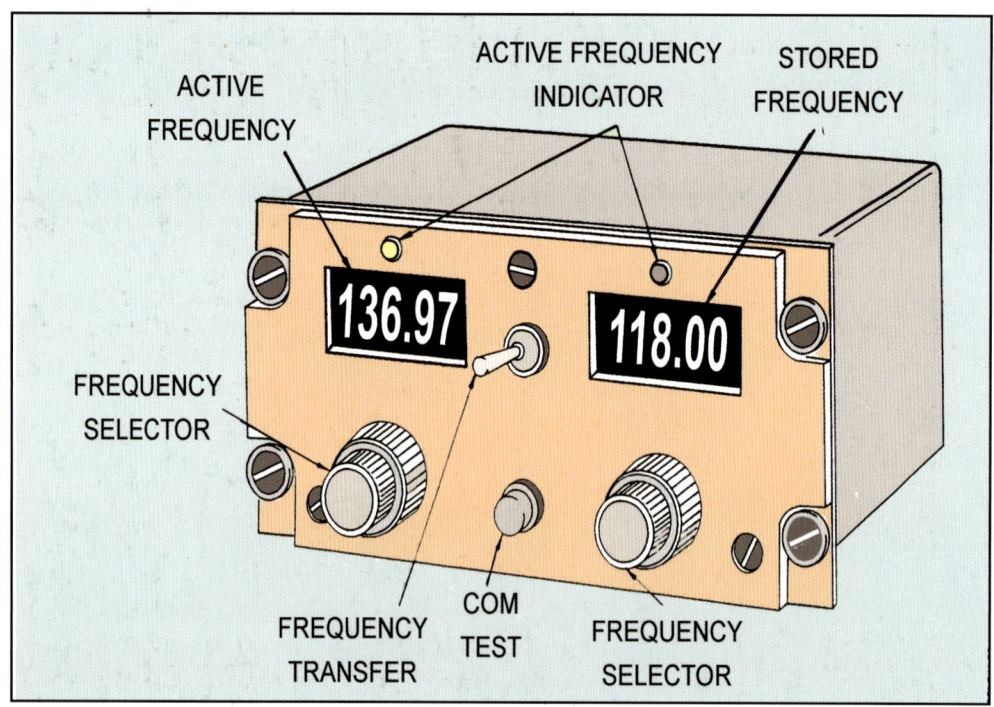

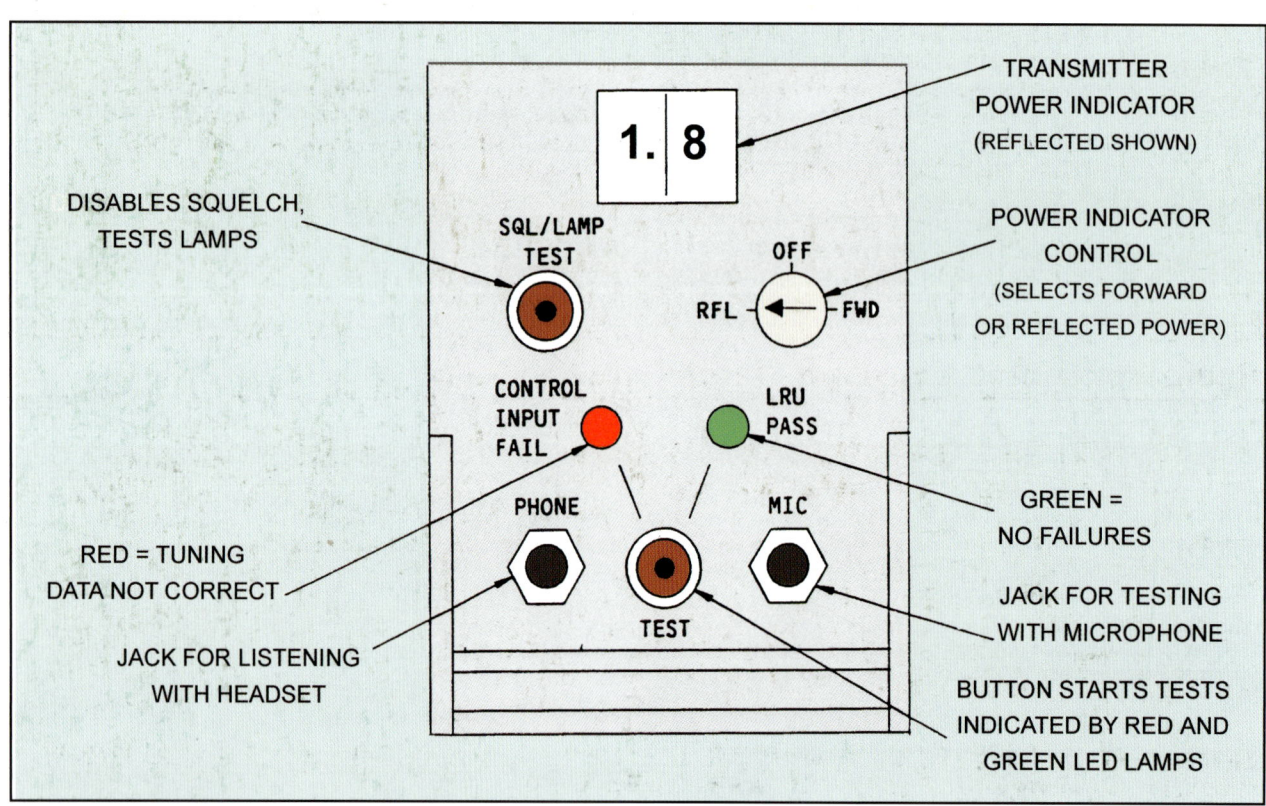

Located in the electronics bay, the VHF transceiver is remotely tuned by the control head in the instrument panel. This LRU (line replaceable unit) has several test features built in. The indicator at the top shows transmitter power in the forward direction (toward the antenna) or power reflected back to the transmitter. If reflected power is high, there's a problem in the antenna or cable. This is covered in the chapter on test and troubleshooting. The jacks at the bottom enable the technician to talk and listen while testing in the electronics bay.

"Splitting" VHF Channels

Almost every decade for the last 50 years, VHF com channels have been "split," dividing the space occupied by one frequency. Another way of viewing it is that channels are moved closer together within the same band.

The first aircraft radios had "200 kHz" spacing. (Note that "200 kHz" can be written as ".2 MHz" by moving the decimal three places to the left.) Thus, the dials of early com radios appeared as:

120..........120.2..........120.4..........120.6..........120.8..........etc.

This spacing divided the VHF com band into 70 channels. As aviation grew, the band was increased in size and 20 more channels were added (still with the 200 khz spacing.

But aviation was growing and demanding more frequencies. Fortunately, the avionics industry was also advancing with techniques that made receivers more "selective," enabling them to separate two channels that are closely spaced. The progress of splitting went like this:

Spacing	No. of Channels
200 kHz	90
100 kHz	180
50 kHz	360
25 kHz	760
8.333 kHz	2280

For channel-splitting to work, transmitting frequencies are held to tight tolerances to avoid drifting and causing interference to other channels. Because early radios could not comply, they were outlawed on the aviation bands. Most radios now operate on 25 kHz and 8.333 kHz spacing.

The last split, to 8.333 khz, marks the beginning of a new-type com radio that handles both voice and digital data. It is VDL (Very High Frequency Data Link). Using digital signals, each 8.333 frequency can operate simultaneously with up to four channels of information; four voice and two digital messages.

Radio Management System

Chelton Avionics

CONTROL-DISPLAY UNIT | RECEIVER/TRANSMITTER | NAV (VOR) RECEIVER | ADF RECEIVER | DME | TRANSPONDER

A Radio Management System eliminates numerous knobs, buttons and separate control heads for operating com and nav radios. It's less of a workload to operate and saves space on the instrument panel. The pilot sees only the control-display unit (at the left) and selects or stores frequencies, transponder codes, etc. All the other units are mounted in remote racks and are controlled through a databus (ARINC 429).

This system, the Chelton RMS 555, is used by corporate, regional airline and military aircraft.

Review Questions
Chapter 3 VHF Com

3.1 What frequencies define the Very High Frequency (VHF) band?

3.2 What is the frequency coverage of the VHF com band?

3.3 What is "splitting" channels?

3.4 What development greatly reduces the number of voice reports on VHF com?

3.5 What is the narrowest spacing for channels in the VHF com band?

3.6 What is the purpose of a mic key line?

3.7 A typical com radio has two frequency displays; one for the _____ frequency, the other for the _____ frequency.

3.8 What is the purpose of a squelch?

3.9 What function does the "com test" control provide?

3.10 Where is the LRU (line replaceable unit) for a com transceiver of a large aircraft located?

3.11 What is the benefit of a radio management system?

3.12 What is the third VHF com radio of an airliner often used for?

Chapter 4

HF Com
High Frequency Communications

SUPERSEDE by SATCOM

When an airplane leaves the coastline for a transoceanic flight, it moves into a polar region or ventures over a remote area, it loses VHF communications. VHF signals are line of sight and cannot curve over the horizon. For long-range flight, the airplane switches to HF---high frequency---communications.

In a band from 2-30 MHz, HF radio travels 2000-6000 miles by "skipping" through the ionosphere, an electrical mirror that reflects radio waves back to earth.

HF has never been a pilot's favorite radio. Early models didn't have the reliability of VHF because the ionosphere is always changing---between day and night and season to season. It is struck by magnetic storms from the sun which repeat over an 11-year sunspot cycle, interrupting communications for hours, even days.

The cure is the eventual elimination of HF radio by satellite communications. Nevertheless, thousands of aircraft will continue to fly with HF for decades before the transition is complete. Fortunately, HF has enjoyed several improvements.

Early HF radios were difficult to operate. Most antennas for aircraft measure from inches to several feet long. The length tunes the antenna to one-quarter wavelength, which is standard for aircraft. A VHF antenna at 120 MHz, for example, has a quarter-wavelength of only two feet, easy enough to mount as a small whip or blade on the airframe. But as operating frequency goes lower, wavelength grows longer. A quarter-wave HF antenna on 2 MHz would have to run

HF Control-Display

Bendix/King

Control-Display for an HF radio for General Aviation. With output power of 150 watts (Single Sideband, or SSB) it tunes 280,000 channels. An antenna coupler (remotely located with the receiver and transmitter)) automatically tunes the antenna when the microphone button is pressed. Some models, like this one, have a "Clarifier" control (lower left) to fine tune the incoming voice.

The radio also solves the problem of hunting for a workable frequency under changing ionospheric conditions. With circuit called "automatic link establishment" it searches for the best available channel. This model is the Bendix/King HF-950. Weighing about 20 lbs, it is also used by helicopters, which often fly in remote areas beyond VHF communicating range.

23

HF System

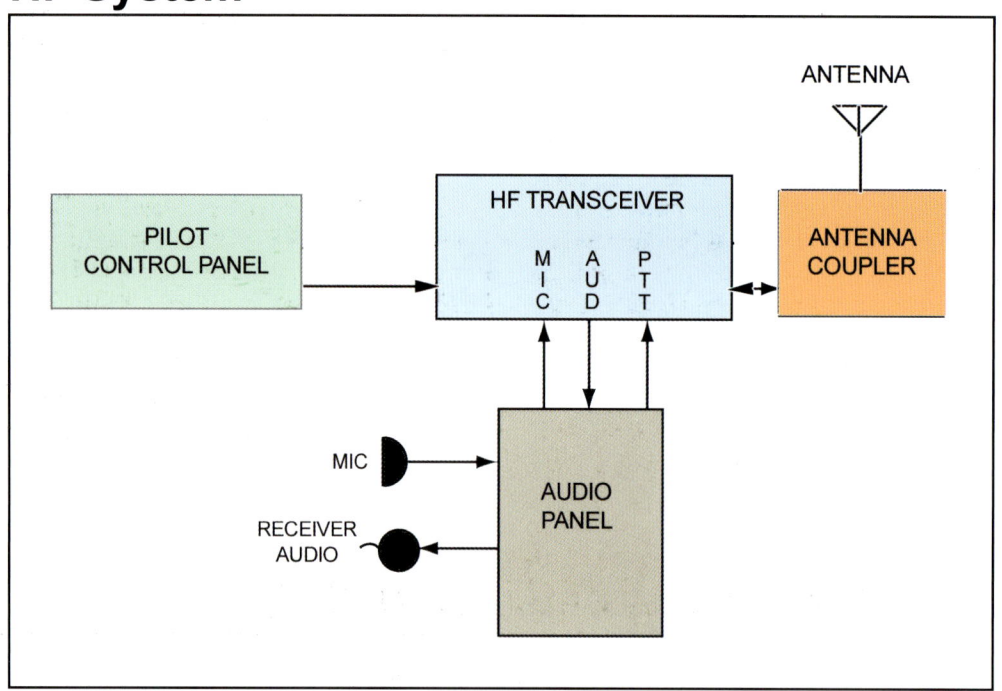

When the pilot keys the transmitter on a new channel, a 1000 Hz (audio) tone is modulated onto the radio wave and sent to the antenna coupler. This enables the coupler to match the antenna to any HF frequency. A tone is used because, unlike voice, it produces a steady radio-frequency output for the coupler to measure. Antenna tuning usually takes less than four seconds.

over 100 feet long!

Radio pioneers solved this with a "trailing wire" antenna----reeling it out to float behind the airplane. If radio conditions changed, they hunted for a new frequency and changed antenna length. They also had to manually adjust an antenna tuner.

A breakthrough happened when Arthur Collins (founder of a company that produces air transport avionics) came up with an improvement called "Autotune." It is a tuning unit that matches a short, fixed antenna on the airplane to the wavelength of any HF frequency. It's done automatically when the pilot selects a channel.

The concept of automatic antenna tuning is based on "reflected power." If an antenna and antenna tuner are adjusted for, say, an operating frequency of 12 MHz and the pilot changes to 5 MHz, there will be a large electrical mismatch between the antenna and feedline from the transmitter. This causes radio frequency power to reflect back from the antenna and be lost. The Autotune system measures the reflected power and operates tuning elements in the antenna coupler to reduce the reflection to the lowest possible value. (The concept of reflected power reappears in the chapter on test and troubleshooting, where it's called VSWR, for voltage standing wave ratio.)

In today's HF radios, changing frequencies and retuning the antenna can occur in less than a second from the time the pilot turns the channel selector.

Automatic HF antenna tuning, which greatly reduced pilot workload, was followed by a development that improved HF radio's ability to avoid fading signals and poor radio conditions caused by variations in the ionosphere.

SSB HF radio originally transmitted in the AM (amplitude modulation) mode, the same as AM broadcast radio today. An AM transmitter generates three components; a radio-frequency (RF) carrier, an upper sideband and lower sideband. The audio (or voice) is found only in the sidebands. This was discovered in the 1920's, along with the observation that the RF carrier served only to create the sidebands inside the transmitter. The carrier doesn't "carry" the sidebands. Sidebands travel just as well with or without a carrier. Because the sidebands lie just above and below the carrier frequency, they are termed USB and LSB (for upper and lower).

In regular amplitude modulation, more than two-thirds of the transmitter power is lost in the carrier. What's more, the upper and lower sidebands carry the identical information. So all that's required for transmitting the voice is a "single sideband."

It took several decades for the electronics industry to develop stable transmitters and receivers and sharp filtering to make "SSB" practical. As a result, today's

HF-SSB transceiver places nearly all transmitter power into a single sideband, producing a powerful signal that punches through worsening ionospheric conditions.

HF Datalink. Despite the improvement of SSB, pilots were not yet completely satisfied; HF still didn't provide the solid reliability of VHF communications. In seeking further improvement, the avionics industry considered digital communications to handle routine messages. The first experiments failed as researchers discovered that digital signals barely survived the turbulent ride through the ionosphere. Too many digital bits were lost in transmission.

At about this time, the first communications satellites were rising in orbit, offering solid long range communications to the aeronautical industry. This threatened to kill further development in HF, but the airline industry wasn't ready for "satcom." Satellite installations at the time proved too expensive for many carriers, which motivated researchers to design a workable HF datalink.

Today, HF datalink is a reality. The new radios perform "soundings"---listening for short bursts of data from ground stations around the world. A link is established to the best one for communications. If conditions deteriorate, the radio automatically searches for, and switches to, a better channel. If there are errors in transmission, the ground station senses them and automatically calls for repeats until the data is correct.

Remote Line Replaceable Units (HF)

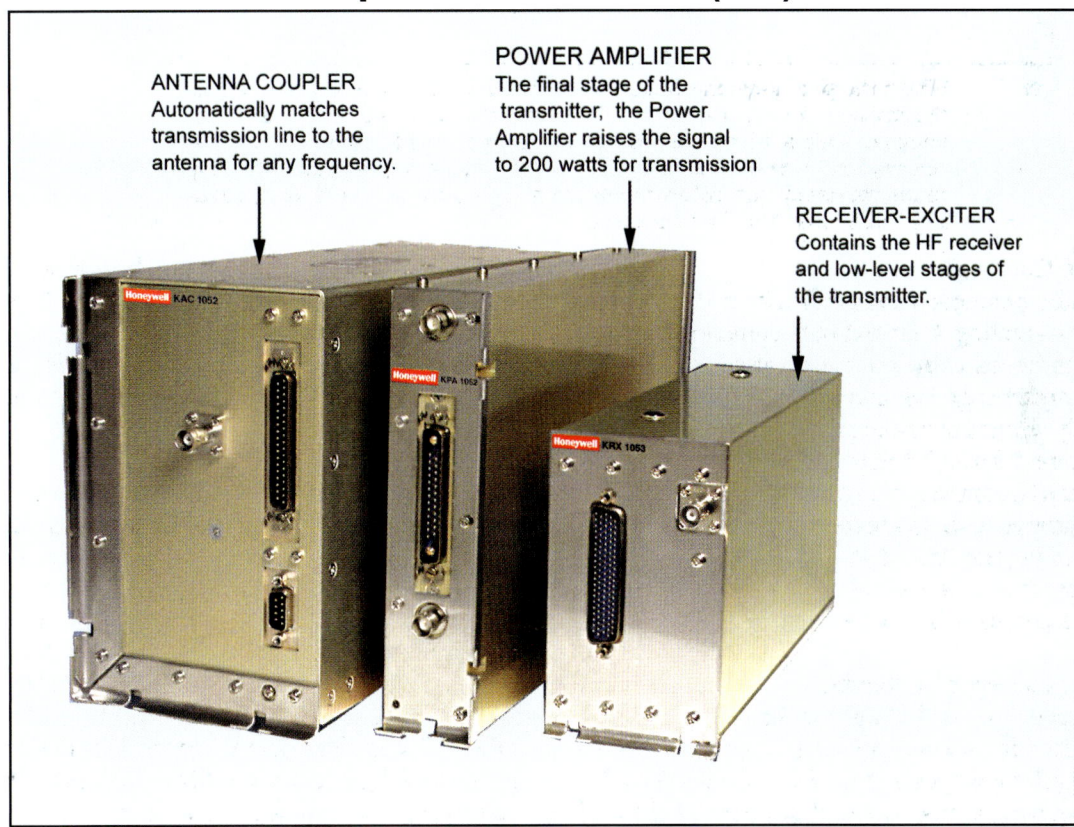

Three remote-mount boxes for an HF radio installed on business aircraft. They are controlled by the pilot on the flight deck.
    The radio tunes 280,000 channels and stores 99 user-programmable frequencies (for quick retrieval). For sending distress calls, the international maritime distress frequency on 2.182 MHz is pre-programmed. The model shown here, the Primus HF-1050, is upgradeable for HF datalink.

25

HF Control Panel: Airline

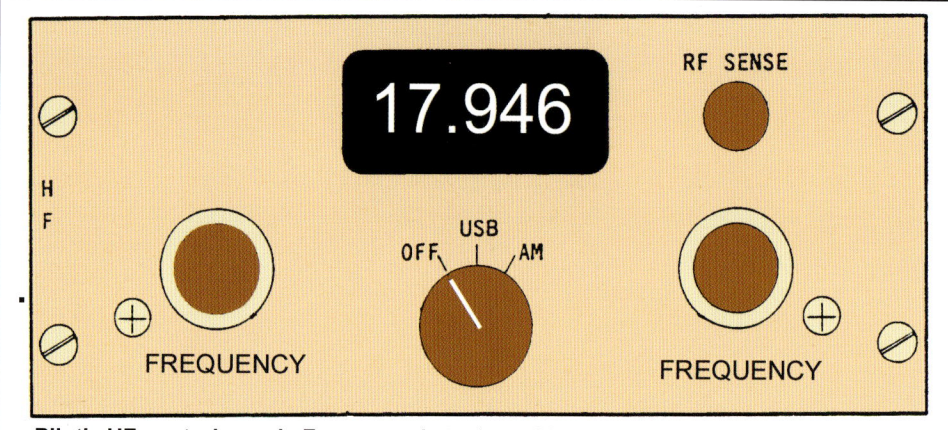

Pilot's HF control panel. Frequency is selected by two outer knobs. RF SENSE adjusts receiver sensitivity. The knob at bottom---OFF - USB - AM---selects mode of operation. Most HF communications for aircraft are on USB (Upper Sideband). Lower sideband is not permitted in aeronautical service. The AM knob selects old-type Amplitude Modulation, which is much less effective than SSB, but enables pilot to talk to ground stations not equipped for SSB,

HF Transceiver

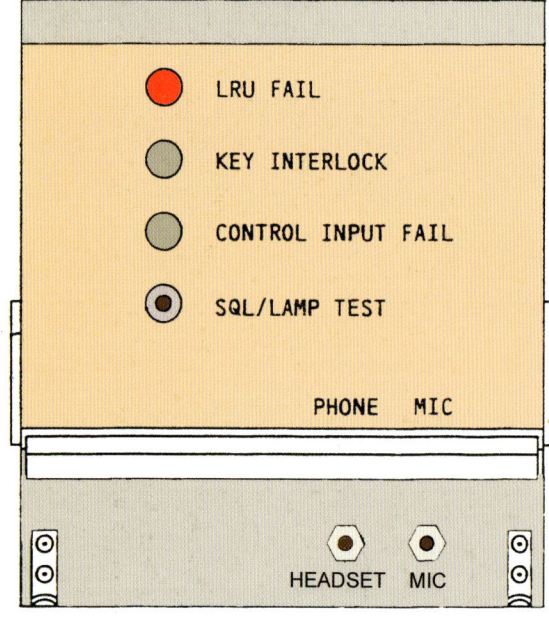

Mounted in an electronics bay, the HF transceiver is operated from the pilot's HF control panel. It has several provisions for testing. Three lights show system status (the red lamp is indicating "LRU FAIL," meaning this transceiver, a line-replaceable unit). The button "SQL/LAMP" is pressed for two tests; all lamps should light, and the squelch is disabled. During a disabled squelch, the technician should hear atmospheric noise, which is an approximate test that the receiver is working. He can also plug a microphone into the "MIC" jack and talk on the radio during troubleshooting.

The transmitter in airline service is usually rated at 400 watts of radio frequency power during single-sideband (SSB) transmission; 125 watts in the AM mode.

Advantages of HF Datalink

- Lower Pilot Workload
- Shorter Message Transmission Time (less than 3 seconds vs more than 1 minute)
- Channel Access Time (less than 60 seconds vs up to 10 minutes)
- Less Operational Training for Flight Crew
- Data Relieves Congestion on Voice Frequencies
- Automatic Selection of Frequency and Data Rates
- Voice Is Prone To Human Error and Interpretation
- Data Detects Errors and Automatically Retransmits
- Data Extracts Signals in Noisier Environments (3dB/10dB)
- Increased HF Traffic Capability
- Assured Communication Link Automatic Air/Ground HF linkage With Less Acquisition and Message Cost (Compared to Satcom)
- Improved Voice/Data Quality
- Data Link Messages Are Not Written or Sensitive to Verbal Language

(Based on a Honeywell report)

HF Antenna Coupler

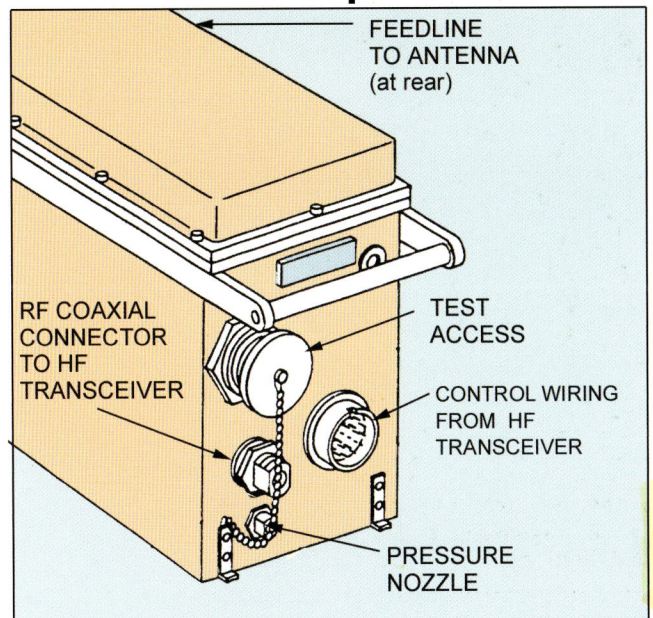

Mounted below antenna in rudder fin, HF antenna coupler tunes antenna to the frequency in less than 4 seconds after pilot selects a channel.

Pressure Nozzle at the bottom pressurizes the coupler enclosure. Otherwise, low air pressure at altitude would cause high voltage in tuning coils to arc over and short.

HF Antenna Mounting

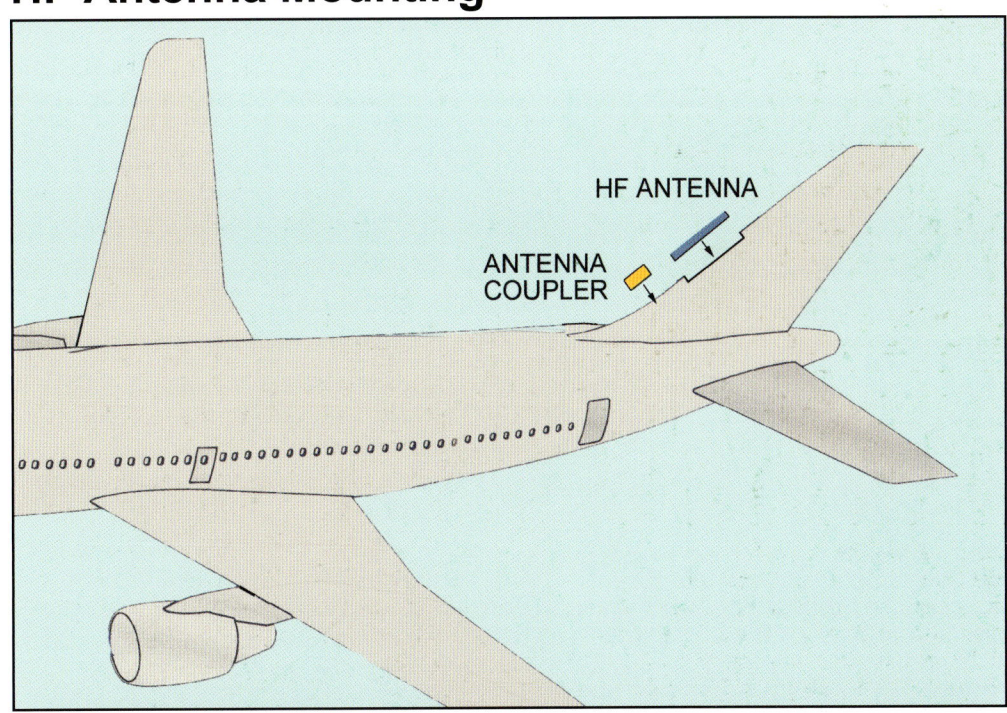

The HF antenna on a typical airliner is located in the vertical tail fin. The radiating antenna is inside a U-shaped fiberglass leading edge. The antenna coupler is just below, inside the rudder fin. The coupler matches any HF frequency (2 - 30 MHz) and sends it through a feedline to the antenna.

Review Questions
Chapter 4 HF Com

4.1 Why are High Frequency communications not as reliable as those of VHF?

4.2 What is the advantage of "Autotune."

4.3 Why is SSB (single sideband) more efficient than conventional AM radio?

4.4 What made HF datalink successful?

4.5 What are three major components of an HF line-replaceable unit (LRU)?

4.6 Name two advantages of HF datalink?

4.7 What is the purpose of an HF antenna coupler?

4.8 Where is the HF antenna mounted on many airliners?

Chapter 5

Satcom
Satellite Communications

Satcom provides communications between aircraft and ground through most of the world. Free from atmospheric interference and limited bandwidth, it is the replacement for High Frequency (HF) as the band for long-range communications. Satcom signals penetrate the ionosphere without bending or reflecting and are unaffected by electrical noise or weather. As satcom avionics build through aircraft fleets, they will eventually replace VHF com, as well. The signals of satcom are digital, not only for data communications, but voice, as well. This means voice messages can be encrypted for security.

Satcom is also the foundation for the next generation of air traffic control. After a half-century of aircraft confined to narrow routes and tracks, a changeover is beginning to a new architecture known as FANS, for Future Air Navigation Systems. More airplanes will fly safely within the same airspace under a concept known as "Free Flight." Satcom makes it possible, as well as providing information, entertainment and other services for passengers in the cabin.

Inmarsat

The London-based organization providing satellites and ground support is Inmarsat (International Maritime Satellite). Consisting of more than 60 member

(Inmarsat)

A Ground Earth Station communicates with orbiting Inmarsat satellites, which relay messages to and from aircraft. The Ground Earth Station receives and sends those messages through telephone companies and other telecommunications services throughout the world.

Generations of Inmarsat Spacecraft

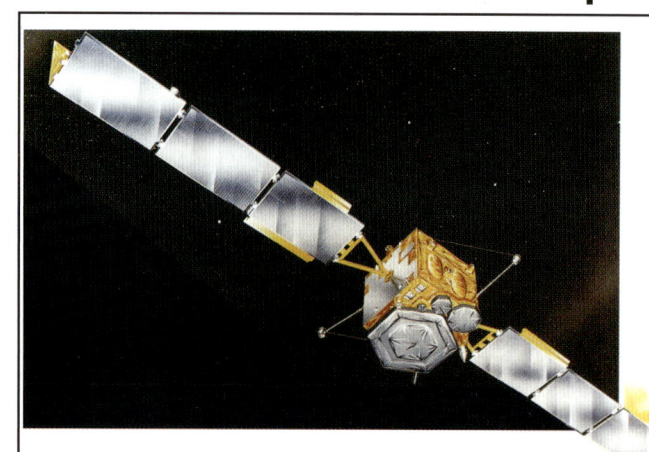

Inmarsat-2 (called I-2) was launched in 1991 after the first generation raised the demand in aviation for more satellite services. I-2 provides four times more capacity than I-1. The constellation consists of four active satellites, with four spares in orbit to assure continuous service.

All satellites are monitored by control centers on the ground. As gravity causes a satellite to drift from orbit, the vehicle's attitude and orbit are adjusted by a control station. When the satellite moves into the dark side of the earth at night, its solar cells are eclipsed. Batteries provide power in the dark. Controllers monitor the battery backup to be sure satellite power is sufficient.

Inmarsat-3 has ten times the capacity of I-2. Each of the four spacecraft has one global beam which covers a wide area of the earth. Each also has 7 spot beams which concentrate power over a narrower area (usually where demand is high, along heavily traveled routes). Backup for I-3 is the previous generation of I-2 spacecraft.

The next constellation is I-4, designed to be 100 times more powerful and have ten times the communications capacity.

---

countries, it provides the space segment known as "Inmarsat Aero". Using four satellites, it provides two-way voice and data (fax, Internet, e-mail, ATC) over most of the world. Because the satellites hover over the equator, their beams cannot extend into the North and South Pole regions. Future systems will add polar orbits to fill in these limited areas. The four Inmarsat satellites:

Pacific Ocean Region (POR)
Indian Ocean Region (IOR)
Atlantic Ocean Region West (AOR-W)
Atlantic Ocean Region East (AOR-E)

Each satellite is backed up with a spare orbiting in same vicinity. The other two major components of the satcom system are:

Ground Earth Station (GES). These radio stations around the world operate large dishes for communicating with satellites. They receive messages sent to a satellite by an aircraft, then pass them to a telecommunications company for relaying them to any telephone or data terminal in the world.

If the message is intended for an aircraft in flight, the ground earth station receives it through telecommunications networks and beams it up to a satellite for relay to the airplane.

Aircraft Earth Station. This is the avionics system aboard the aircraft for communicating with satellites. It must conform to Inmarsat standards and the specification for ARINC 741.

Satcom antennas. A component of the airborne system is the antenna, which must always aim directly at the satellite (to receive all of its services). Although Inmarsat satellites appear never to move (they are in geostationary orbit), the airplane often cruises over 500 mph, rapidly changing position. This is solved by a beam steering unit on the airplane that operates a motor-driven antenna or an electronic system known as a "phased array." Consider the satcom antenna categories:

Low Gain. Various communications via satellite require a different amount of power. A message consisting only letters and numbers moving at a slow rate (300-1200 bits per second) uses relatively little power and can operate on a "low gain" antenna on the airplane. The antenna is simple, little more than a blade, and picks up signals from any direction.

Inmarsat-Aero System

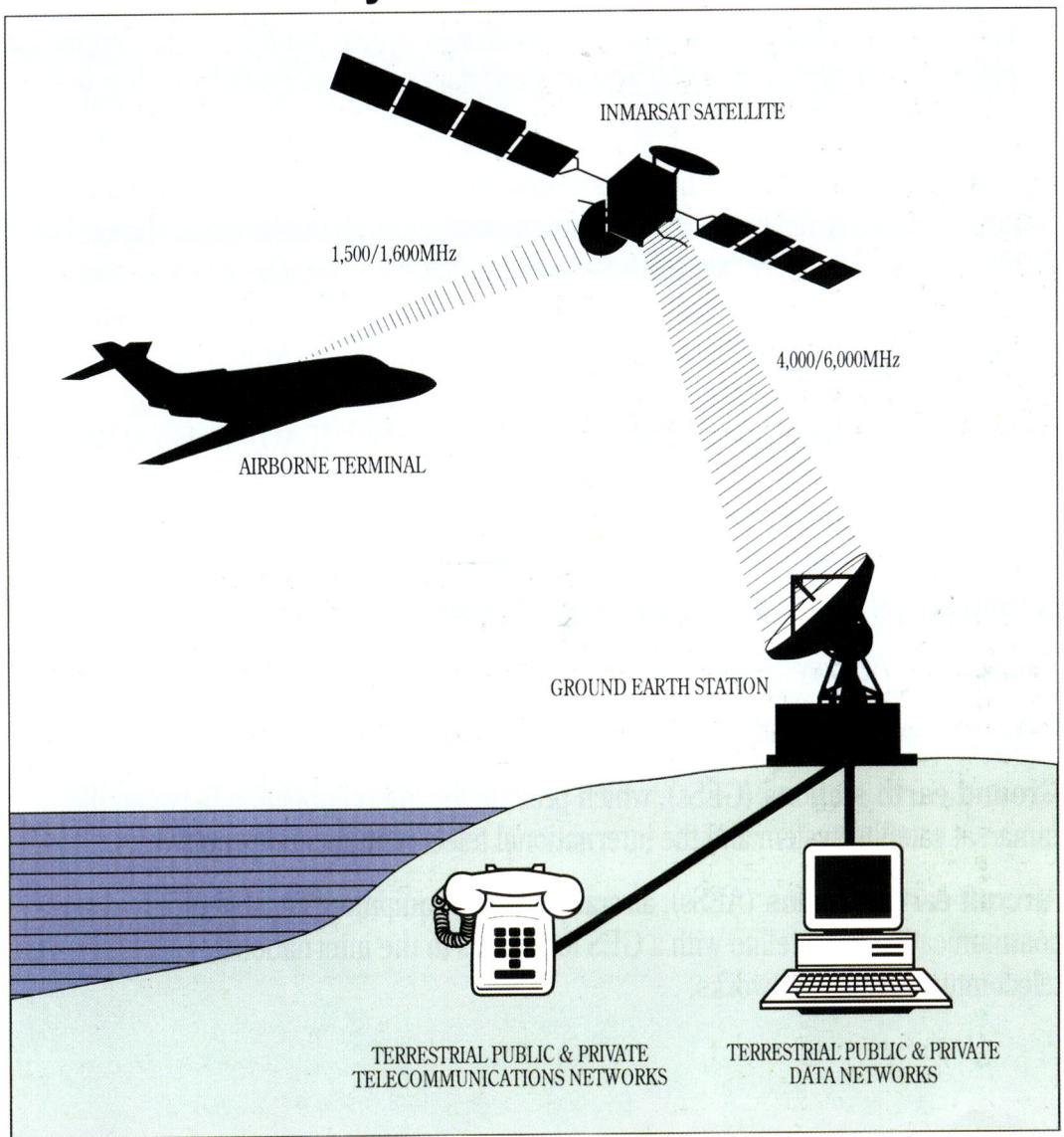

The system for satcom consists of three basic elements; satellites (space segment), airborne avionics (aircraft earth station) and the connection into telecommunications networks (telephone companies, for example.) This last element is known as the ground earth station.

Four Inmarsat satellites cover nearly all the world (coverage falls off at the poles). Four additional satellites are in orbit as spares, ready to take over during a malfunction.

An airplane communicates with satellites, not ground stations. There are ten ground stations throughout the world for relaying aircraft communications into telecommunications networks.

High Gain. This antenna supports the full range of satcom services, which require more power than is possible with a simple blade. The "high gain" antenna is more complex and expensive. The improvement in power ("gain") is achieved in two ways. First, the antenna is made highly directional with additional elements to focus signals into a beam. The narrow beam, however, must always aim directly at the satellite.

Steering the beam is accomplished in two types of high gain antenna, shown on the following pages. One is the electromechanical; the antenna is rotated in azimuth and elevation by electric motors (much the same as done in an astronomical telescope).

An airplane in cruise is always moving with relation to the satellite. To keep the high gain antenna pointing toward the satellite, the airborne satcom steers the beam using information from the airplane's navigation system. As shown in the illustration, the electromechanical high gain antenna fits in the tail cap of the airplane.

Conformal. The second type of high gain antenna is the "conformal," which fits the curve of the fuselage and protrudes less than a half-inch. The radio signal is shaped into a narrow beam and aimed electronically. Inside the conformal radome are many small micros-

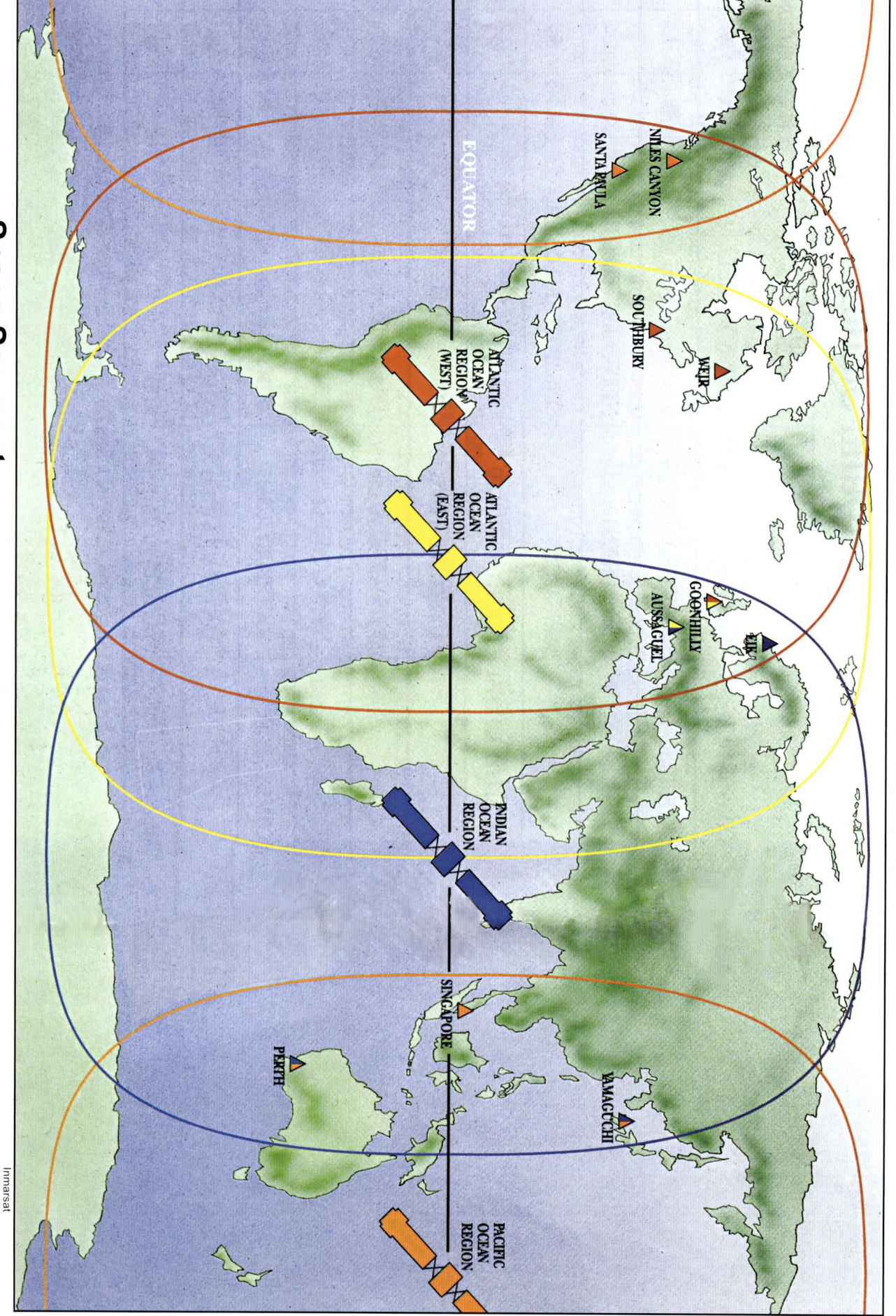

Space Segment

Four Inmarsat satellites provide global coverage. All are directly overhead the equator (at 22,300 miles, or 36,000 kilometers) and in geostationary orbit. Because they complete one orbit in 24 hours (also one rotation of the earth), the satellites appear to remain fixed in position. Each satellite is backed by a spare orbiting nearby.

Inmarsat

trip antennas. The beam steering unit adjusts the signal in each microstrip antenna so its energy adds or subtracts according to a pattern that forms a beam. The energy is focussed and steered in a technology known as "phased array."

Either conformal or electromechanical, high gain antennas support all satcom services. Data rates begin at 10.9 kilobits per second, which handles fax, voice and high speed data, but this rate is increasing.

Swift64. A recent service is Swift64, which communicates at 64 kilobits. With a high gain antenna on the airplane, this rate accommodates such wideband services as Internet, e-mail and video conferencing.

Intermediate Gain. A more recent development affects the gain of signals from the space vehicles. The first two satellite generations broadcast "global" beams to cover as much of the earth as possible---and the latest generation still does. Recent satellites, however, add "spot beams," which concentrate power over a smaller area (but total earth coverage is still about 75%). Because of this added power, a third type of aircraft antenna emerged; the "intermediate gain" type, which falls between the low and high gain models. It is less costly and simpler, yet provides a wide range of satcom services.

Cell Phones in the Cabin

Inmarsat

Cell phones were banned in aircraft because they contact too many ground stations simultaneously while at altitude. But intense passenger interest is producing new systems that will almost certainly be adopted. They work with the passenger's own cell phone and billing is done on his regular cell phone account.

The technology places a base station aboard the aircraft that commands cell phones to operate at low power and avoid raising regular ground stations. The base station relays the calls through satellites, then into the regular landline telephone system.

Ground Earth Station (GES)

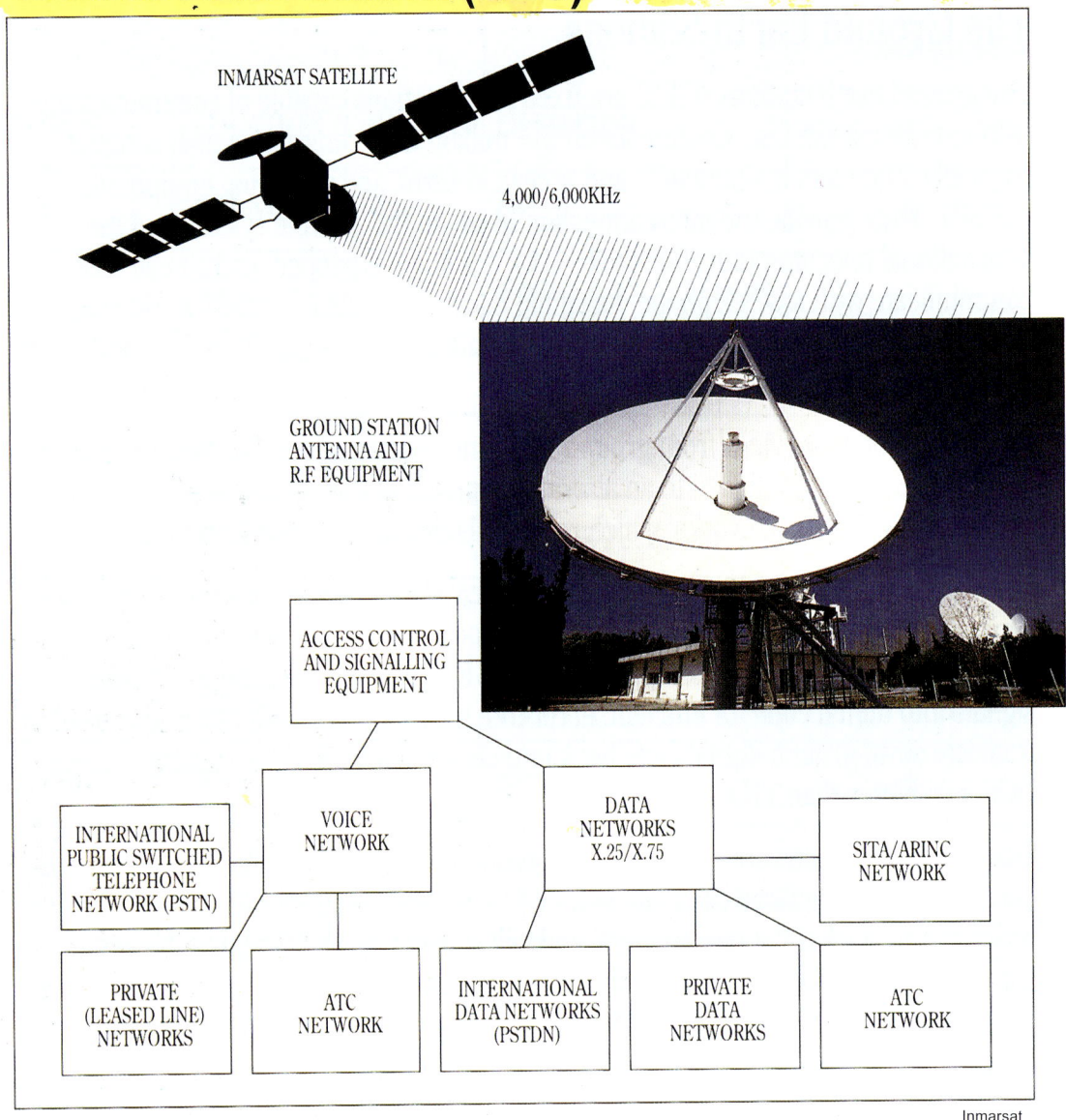

INMARSAT SATELLITE

4,000/6,000KHz

GROUND STATION ANTENNA AND R.F. EQUIPMENT

ACCESS CONTROL AND SIGNALLING EQUIPMENT

INTERNATIONAL PUBLIC SWITCHED TELEPHONE NETWORK (PSTN)

VOICE NETWORK

DATA NETWORKS X.25/X.75

SITA/ARINC NETWORK

PRIVATE (LEASED LINE) NETWORKS

ATC NETWORK

INTERNATIONAL DATA NETWORKS (PSTDN)

PRIVATE DATA NETWORKS

ATC NETWORK

Inmarsat

Ten ground stations like this one are located around the world for communicating with aircraft via satellite. The ground station connects to international telecommunications networks to route calls and messages to any telephone, fax machine or data terminal in the world.

The station's dish antenna is typically 10 meters in diameter and operates in the satcom band between 4 and 6 GHz.

Voice sent via satellite uses "codec," for digital voice coding and decoding. Digitizing the voice reduces error in transmission and speech is high in quality.

The various blocks seen at the bottom of the illustration reveal a wide range of satellite services for aircraft, including air traffic control, passenger telephone, airline operations and data.

Aircraft Earth Station (AES)

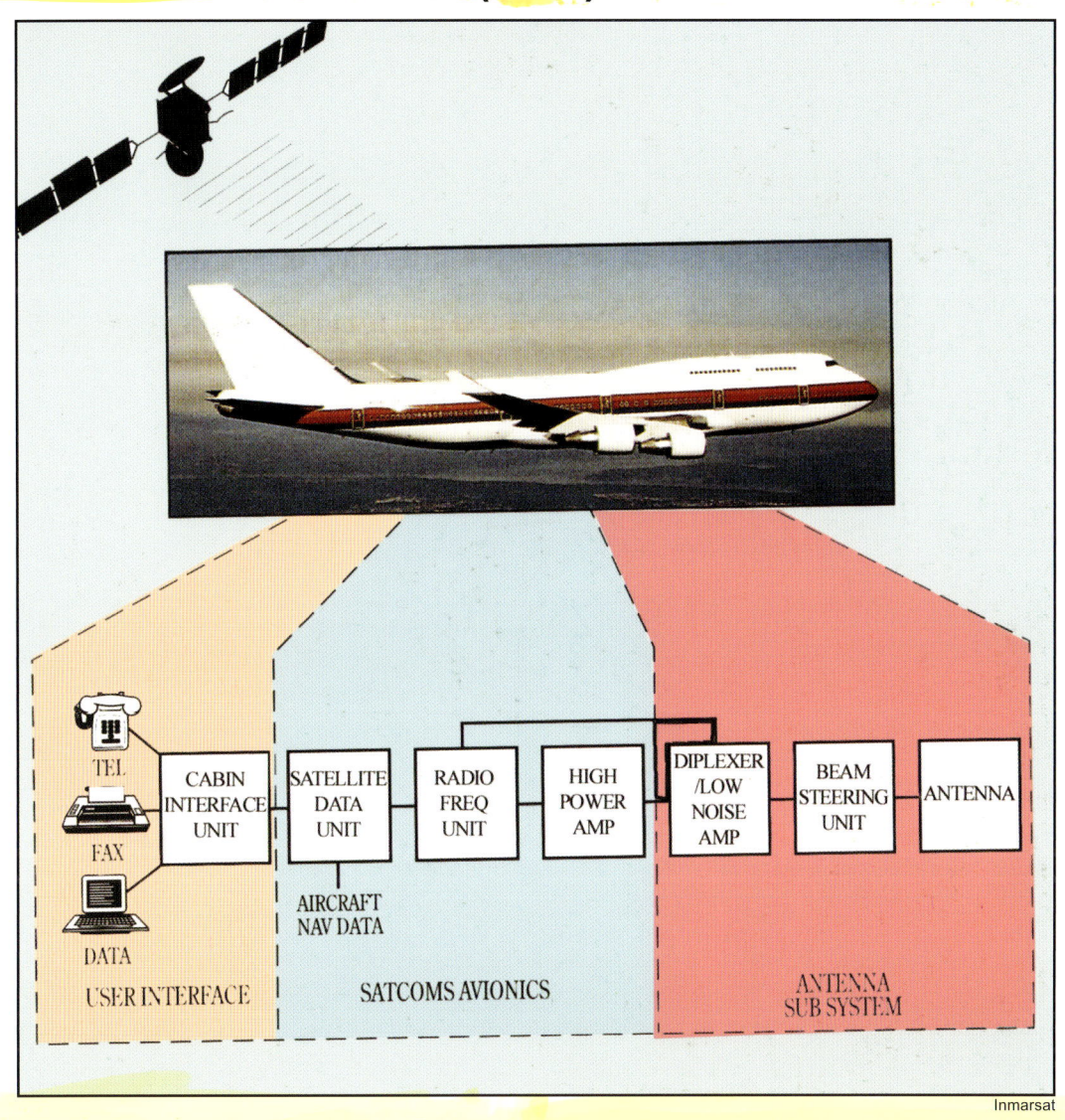

Avionics aboard the aircraft for communicating via satellite are known as the Aircraft Earth Station. It sends and receives radio frequency signals to and from the satellite in the L-band (1.5-1.6 GHz). It provides interfaces to various systems aboard the aircraft for voice, data, fax, video, etc.

This equipment conforms to ARINC 741, as well as standards from Inmarsat, the satellite service provider.

The basic components of an Aircraft Earth Station:

Satellite Data Unit (SDU)

As the heart of the airborne equipment, the SDU interfaces with the airplane navigation system. Airplane location is required to steer antennas, select satellites and report position to air traffic services.

The satellite data unit also processes all message data, protocols and digital coding and decoding.

Radio Frequency Unit (RFU

For data ready for transmission, the RFU converts signals from an intermediate frequency up into the L-band. They are sent to the High Power Amplifier (HPA) for transmission to the satellite.

During receive, the signal from the antenna first passes through a Low Noise Amplifier, then is applied to the Radio Frequency Unit. Signals are converted down to a lower, or intermediate, frequency and sent to the Satellite Data Unit.

The Low Noise Amplifier boosts the radio frequency signal received from the antenna.

Beam Steering Unit (BSU)

For high gain antennas, the beam steering unit keeps the elements pointed at the satellite as the airplane position changes. Position information is received from the Satellite Data Unit.

The antenna may be steered in two ways. For electromechanical types, the antenna array is positioned by a motor. For electronic antennas, different combinations of fixed elements are selected to focus the beam.

Low and High Gain Satcom Antennas for Aircraft

Low Gain Antenna

Low-gain model has conformal antenna mounted on fuselage. Because of its simplicity it operates only on services with slow data rates (600 bits per second) such as air traffic control and airline operational messages. Such data may be a stream of characters (letters and numbers) displayed on a screen. Low in cost, the low gain system operates in the "Aero L" service.

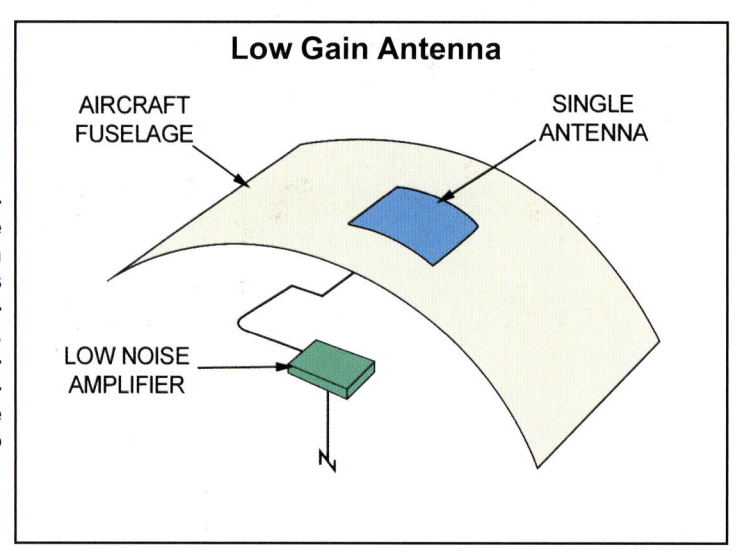

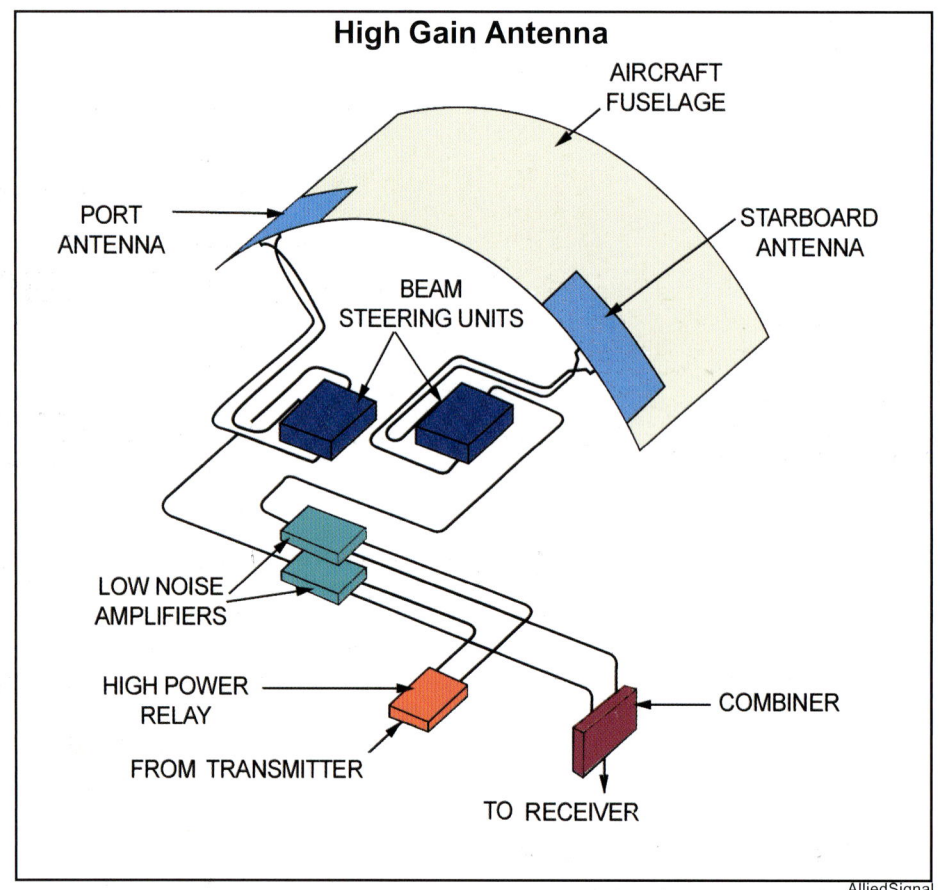

High Gain Antenna

The high gain satcom antenna supports services with higher rates, such as "Swift 64," which handles multichannel voice, data, fax and Internet connectivity. The transmission rate is 64K bits per second.

High gain is achieved by an array of antenna elements formed into a beam that focuses on the satellite.

Although the satellite is stationary, the airplane is moving. The antenna, therefore. needs the "beam steering unit," which keeps the beam aimed at the satellite. Steering information is obtained from the airplane's navigation system.

Electronically Steered Conformal Antenna

High gain satcom antenna, the "Airlink" by Ball, measures 32-in x 16-in, with a depth of only .29-in. It is a conformal antenna, sufficiently flexible to curve to the aircraft body. It is attached by fasteners around the edges of the antenna. Frequency range is 1530-1559 MHz and 1626.5-1660.5 MHz, for communicating with Inmarsat satellites.

Antenna circuits inside the housing use microstrip technology, with no active electronic components. The outer assembly is fiberlass laminate.

Conformal antenna location for a B-747. Antenna is positioned so mounting holes along edges and two holes for the RF cables do not interfere with structure of the airplane.

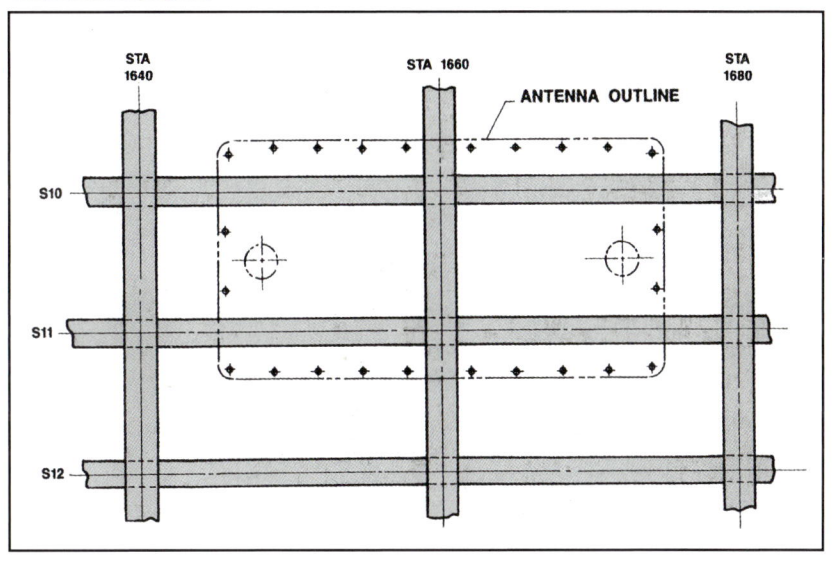

Electromechanically Steered Antenna

Two steerable satcom antennas mounted in a rudder cap. They operate electromechanically, under control of a beam steering unit. The antennas are aimed toward the satellite, regardless of aircraft position on earth, and provide high gain performance.

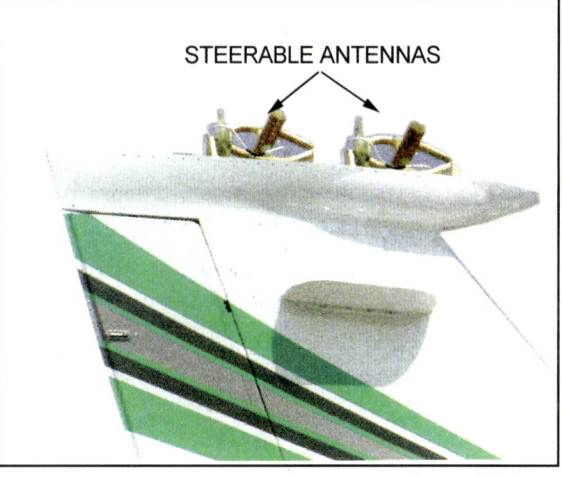

Honeywell/Thales

In-flight High Speed Data System

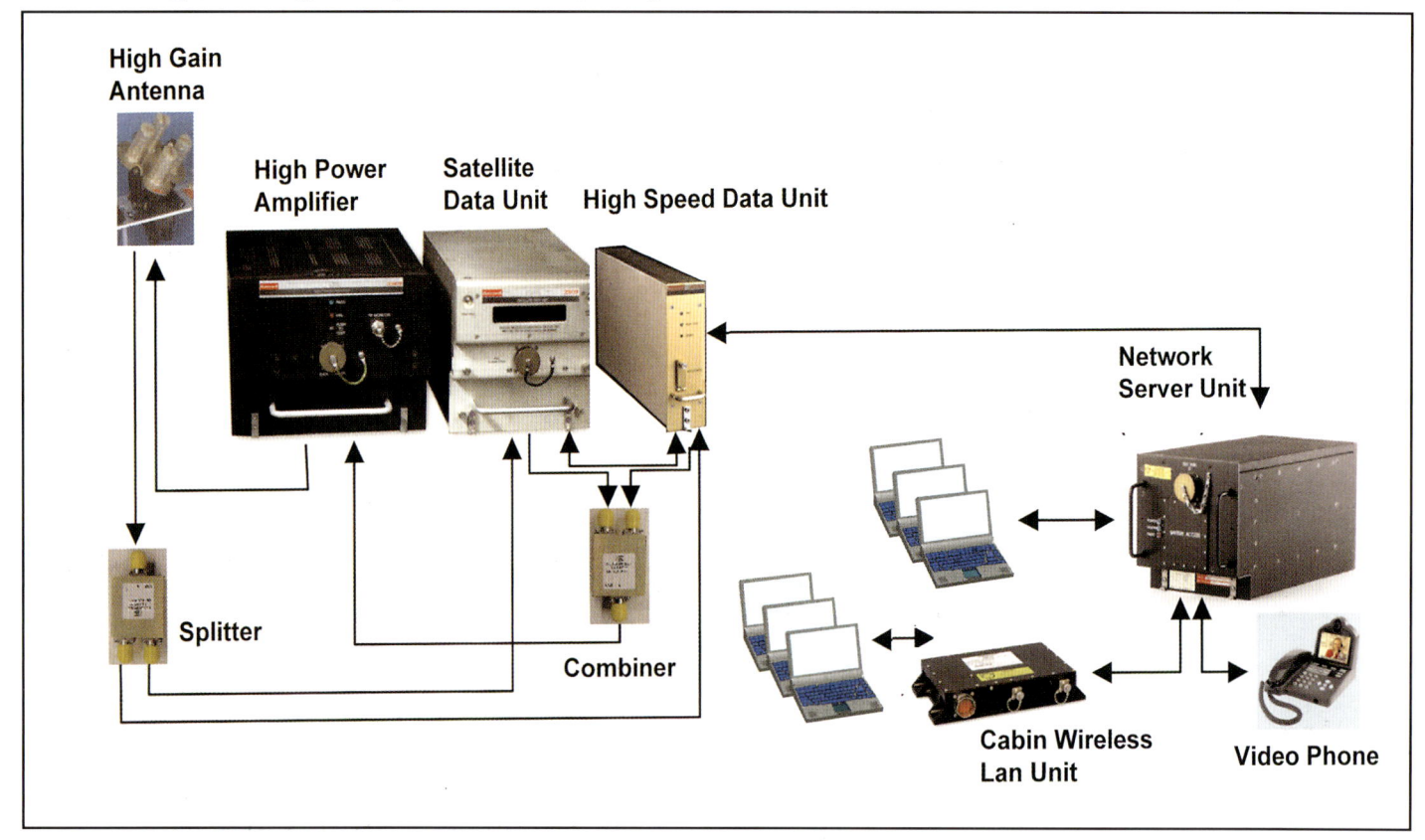

This system, developed for business jets by Honeywell and Thales, accesses private or corporate computer networks anywhere in the world from the passenger cabin. Using the Inmarsat Swift64 service, the passenger uses his own laptop to access the cabin network pictured above. The interface between cabin network and satcom avionics (at the left) is a High Speed Data Unit (top center). The Network Server Unit (top right) is a server and provides file storage and other resources such as an ISDN/Ethernet router and hub switch. It complies with ARINC 763.

In addition to data services, the system provides voice, fax and live video teleconferencing.

Intermediate Gain Satcom Antenna

The "intermediate gain" antenna is a more recent satcom type. Operating in the Aero I service, it can operate with voice, fax and data. It is simpler and lower in cost than a high gain model because of stronger "spot beam" transmission from Inmarsat-3 satellites.

Voice is digitized and encrypted to keep communications secure. Quality exceeds that of passenger telephone systems based on network of ground stations.

Inmarsat Aero Services

Swift64
Based on Inmarsat's Global Area Network (GAN), Swift64 offers Mobile ISDN and IP-based Mobile Packet Data Service (MPDS) connectivity at a basic rate of 64 kbit/s to support high-quality voice, fax and data communications for air transport, corporate and VIP and government users.

Aero H
The original Inmarsat voice and data service, Aero H supports multichannel voice, fax and data communications at up to 9.6 kbit/s anywhere in the satellites' global (hemispherical) beams for air transport, corporate and VIP and government users.

Aero H+
An evolution of Aero H. When an Aero H+ equipped aircraft is operating within a high-power spotbeam from an Inmarsat I-3 satellite it can receive Aero H levels of service at lower cost. Outside the spotbeams the terminal works with the global beam as if it were a standard Aero H system.

Aero I
Exploiting the spotbeam power of the Inmarsat I-3 satellites. Aero I brings multi-channel voice, fax and data at up to 4.8 kbit/s to corporate aircraft, military transports and regional airliners through smaller, cheaper terminals.

Aero L
Low-speed (600 bit/s) real-time data, mainly for airline ATC, operational and administrative communications.

mini-M Aero
Single-channel voice, fax and 2.4kbit/s data for small corporate aircraft and general aviation.

Aero C
The aeronautical version of the Inmarsat C low-rate data system, Aero C allows non-safety-related text or data messages to be sent and received by general aviation and military aircraft operating almost anywhere in the world. Aero C operates on a store-and-forward basis; messages are transmitted packet-by-packet, reassembled and delivered in non-real-time.

Compact, lightweight Aero C equipment, with an antenna similar in size to a VHF blade, can be installed in corporate and general aviation aircraft and helicopters.

Aero C supports:
- Globally available two-way 600 bits/s data communications, messaging, polling and position-reporting for non-safety-related purposes
- Interfaces with international X.25 networks
- Integrated Global Positioning System (GPS) capability through a common antenna.

Aero C aircraft equipment comprises an antenna, a duplexer and a transceiver. The transceiver is connected to a flight deck data terminal or a laptop and, optionally, to a printer.

Capable of handling messages up to 32,000 characters long, Aero C is typically used for weather and flight plan updates, maintenance and fuel requests, in-flight position reporting, and business communications.

Aero C is based on store-and-forward technology. Messages entered into the aircraft terminal are subdivided into data packets and transmitted to the ground earth station, where they are reassembled into the complete message and sent to the ultimate addressee via the national and international telecommunications networks. The process is reversed for messages to the aircraft.

The packets are error-protected: if errors are detected, retransmission of the affected packages is requested. The complete messages are transmitted to destination only after all error-free packets have been recompiled.

Four Inmarsat satellites are in geostationary orbits, 22,500 miles above the earth. Spread around the globe, they all follow the line of the equator. Because one orbit equals one rotation of the earth, they appear fixed in position. Each satellite has one backup spare in orbit.

Early satellites produced "global" coverage, spreading their power over the greatest area. The present generation, Inmarsat-3, also broadcasts "spot beams," which concentrate power over a narrower area. In the illustration, the spot beam of each satellite is shown in blue.

Spot beams illuminate the busiest air traffic regions and simplify equipment on the aircraft.

Ground Station Location	
United Kingdom	Norway
Singapore	Australia
France	Canada
USA (3)	Japan

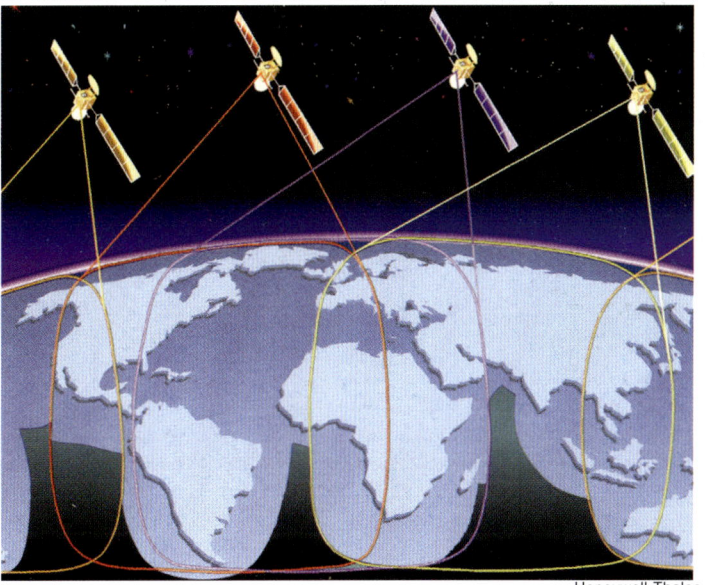

Honeywell-Thales

Review Questions
Chapter 5 Satcom

5.1 What are three advantages of satellite communications?

5.2 What is the name of the next generation air traffic control system based on satellites?

5.3 How many Inmarsat satellites provide global coverage, and where are they located?

5.4 Name two types of stations used in satellite communications.

5.5 Name one advantage and one disadvantage of a low gain satcom antenna.

5.6 What is an advantage and disadvantage of a high gain satcom antenna?

5.7 What is a "conformal" antenna?

5.8 What is the term "Space Segment"?

5.9 After satellite messages are received at a Ground Earth Station, how do they get to their final destination?

5.10 On what band does the aircraft send and receive satellite communications?

5.11 What is the purpose of a Beam Steering Unit?

5.12 What is the typical location on the airplane for an electromechanically steered antenna?

5.13 Why do communications satellites appear to remain fixed in one position?

Chapter 6

ACARS
Aircraft Communication Addressing and Reporting System

Most aircraft communications fall into two categories; ATC and AOC. The first, "Air Traffic Control," is about safely separating aircraft by providing route and altitude clearances, radar tracking, weather advisories and other subjects dealing with airplanes in the same airspace. ATC ground stations are nearly all government-owned and operated.

Airlines are also a business. Each company needs tactical information about when its flights take off, when they arrive, whether maintenance will be needed, fuel remaining, diversions, crew hours and dozens of other items. Communications in this category are AOC, "Airline Operational Control." Pilots call it "company communications."

As airplanes grew more numerous, airspace became more congested and cruise speeds approached

ACARS in the Cockpit

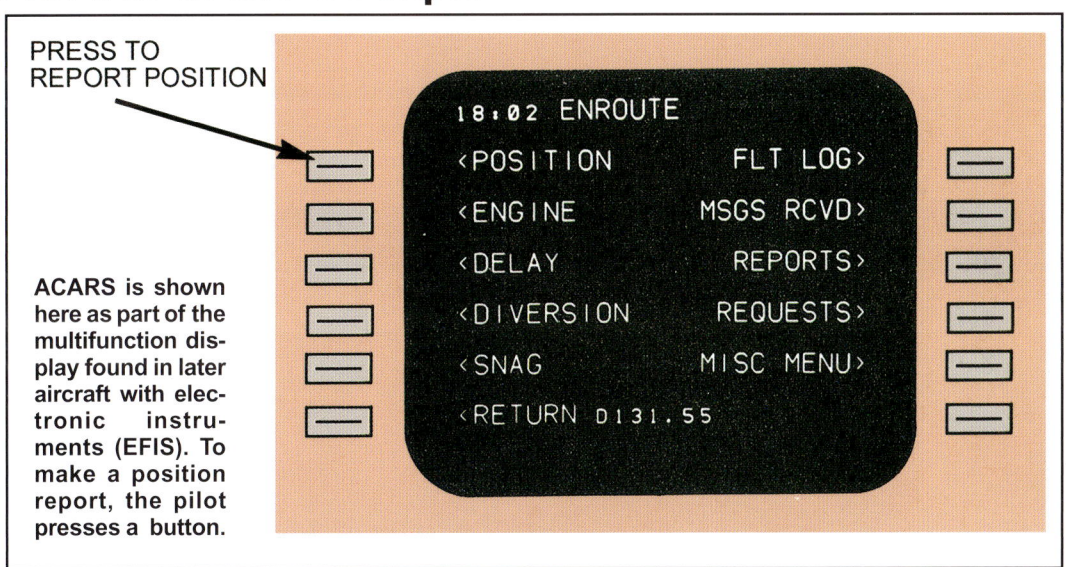

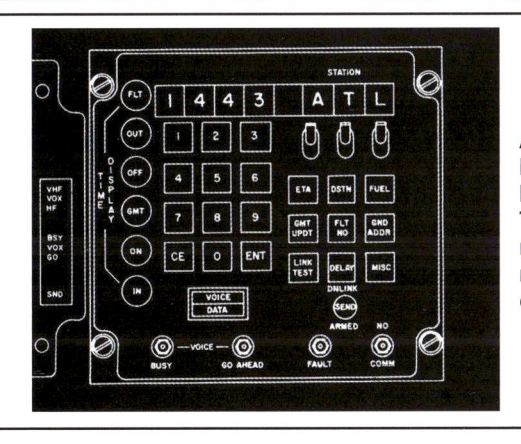

PRESS TO REPORT POSITION

ACARS is shown here as part of the multifunction display found in later aircraft with electronic instruments (EFIS). To make a position report, the pilot presses a button.

```
18:02  ENROUTE
<POSITION        FLT LOG>
<ENGINE          MSGS RCVD>
<DELAY           REPORTS>
<DIVERSION       REQUESTS>
<SNAG            MISC MENU>
<RETURN D131.55
```

Another type of ACARS control unit. Most data is automatically collected but pilot may also key in messages. This controller is an early type; more recent models are built into a flight management system or multifunction display shown above.

41

Mach 1, it was clear that a pilot communicating with both ATC and his company raised the workload to intolerable levels. In 1978, a system called ACARS was introduced to automate most company messages. Ground and satellite networks that support ACARS are operated by organizations such as ARINC in North America and SITA in Europe.

Meaning "Aircraft Communication Addressing and Reporting System," ACARS is used by airlines of all sizes, corporate aircraft and government agencies. Because it operates on digital messages, it is one of the earliest forms of "datalink" in commercial aviation. Not only does it eliminate voice for routine messages, but sends data automatically from sensors aboard the aircraft without assistance from the pilot.

OOOI. ACARS' first job was automatically communicating to the airline company the time each flight pushes back from the gate, takes off, lands and when it arrives at the destination gate. Put those functions together---Out, Off, On, In---and they form the abbre-

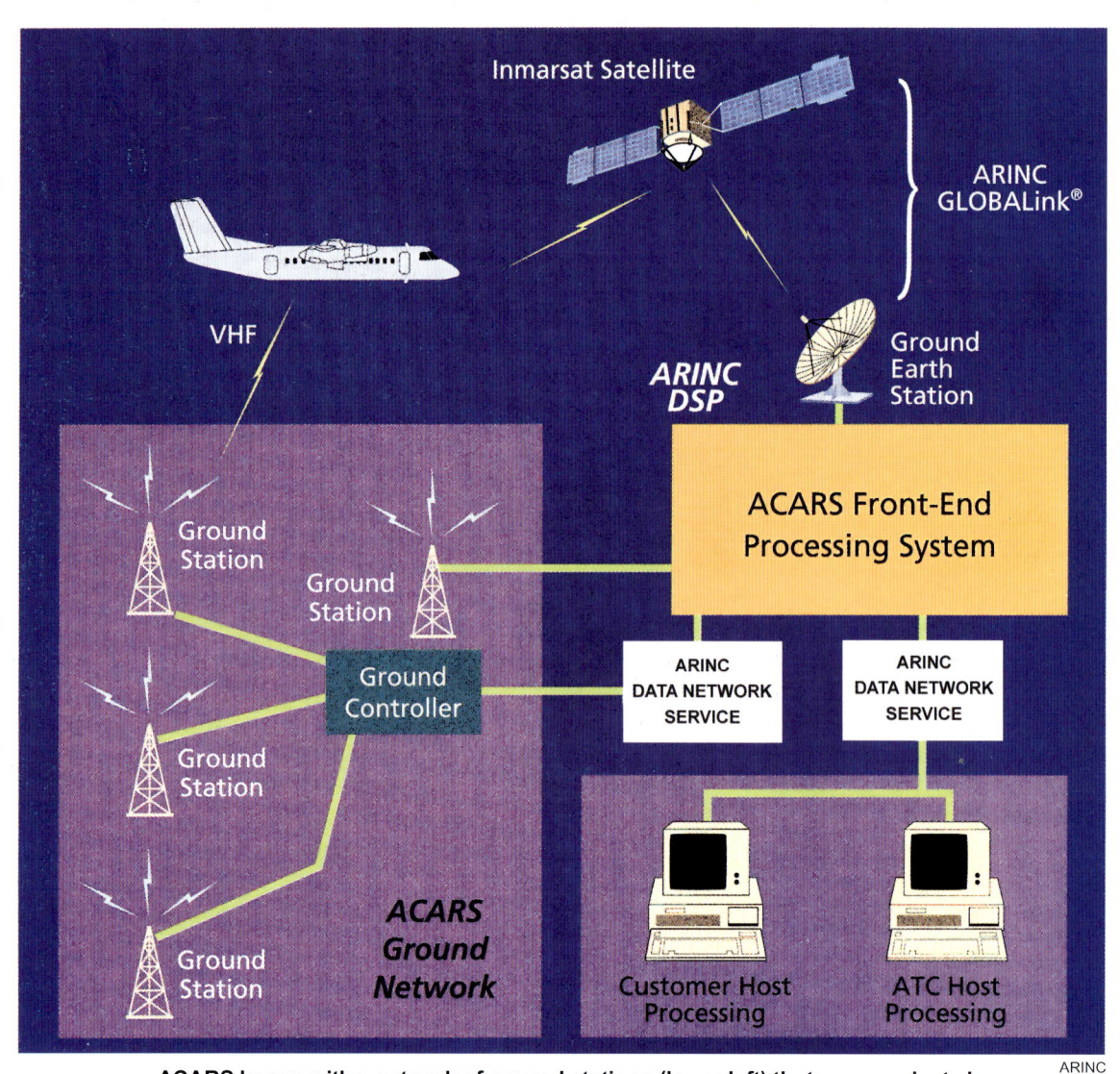

ACARS began with a network of ground stations (lower left) that communicate by VHF radio to aircraft mainly in North America. Full coverage is at altitudes above 20,000 feet, with additional stations at about 300 airports for on-the-ground communications.

More recently, ARINC extended ACARS worldwide by a satellite-based system known as "GLOBALink" (upper right). This service requires satcom equipment aboard the aircraft.

ACARS Messages

"OUT"	"OFF"	"ON"	"IN"
Preflight and Taxi	**Takeoff and Departure**	**En Route**	**Approach and Landing**	**Post-Landing and Taxi**

From Aircraft
Crew Information
Fuel Verification
Delay Reports
OOOI Out

To Aircraft
PDC   Aero C
ATIS
Weight and Balance
Runway Analysis
Flight Plan
Dispatch Release
Remote Maintenance Release

From Aircraft
OOOI Off
Destination ETA
Fuel Remaining
Special Requests

From Aircraft
Position Reports
ETA Updates
Voice Request
Engine Parameters
Maintenance Reports
Provisioning

To Aircraft
ATIS
ATC Oceanic Clearance
Weather
Ground Voice Request (SELCAL)
Gate Assignment

From Aircraft
ETA Changes
OOOI On

To Aircraft
Hazard Reports
Weather Advisories

From Aircraft
OOOI In
Gate Coordination
Final Maintenance Status
Fuel Verification

ARINC

Many different messages are transmitted by ACARS datalink. The service is used by airline and corporate aircraft in much of the world. Although most traffic is for airline company operations, ACARS also handles air traffic clearances when government radio services are not available, such as oceanic regions.

ACARS Message Format

PREAMBLE	TEXT	BLOCK CHECK

The preample contains the address of the aircraft (flight or tail number). If address is intended for another airplane, the message is rejected.
The preample also synchronizes the characters transmitted.
There is an "ackknowledgment character to indicate the message is being received.
A label identifies the message and how it will be routed. There are labels departure, fuel, ETA, diversion and about two dozen others.

Up to 220 characters can be transmitted in this block. They contain report information (departure time, arrival, etc.) which need only a few characters. However, more characters are included for "free talk," sending and receiving longer messages.

This sequence detects errors. If the system is operating properly, it generates characters for "ACK" (acknowledge) or "NAK" (negative acknowlledge

Three building blocks of an ACARS message. Characters that make up the message are comprised of digital bits (ones and zeroes). They are, however, not transmitted digitally, but in analog form as two audio tones; 1200 and 2400 Hz. Transmission is through the aircraft VHF transceiver; downlinked or uplinked from an ACARS ground station.

At the receiving end, tones are decoded back into a digital signal.
This system will remain in operation until it is eventually replaced by all-digital ACARS signals and transmission through satellites.

ACARS Message on Aircraft Take-Off

N1234	QB	1	2804	RAL 5322
ADDRESS (Tail number of airplane)	**MESSAGE LABEL** "QB" means "Off Time"	**DOWNLINK BLOCK IDENTIFIER**	**MESSAGE SEQUENCE** (Minutes and seconds past the hour)	**AIRLINE & FLIGHT NO.**

viation OOOI (pronounced "*Ooe*"). About eight million such messages are sent every month via ACARS.

A pilot does not have to receive the large volume of ACARS messages transmitted to other aircraft. If a message is not intended for the airplane it is not selected. Each ACARS system aboard the aircraft accepts only its unique address.

A message that requires several minutes to send by human voice moves through ACARS in milliseconds. A position report, for example, is done with the push of a button; the data is picked up from the airplane's navigation sensor. Other messages may be keyed in by the pilot.

Not only does ACARS reduce congestion in crowded com bands, but avoids the garble and error when two airplanes transmit on the same frequency at the same time. ACARS avoids collisions with other transmissions and checks each message for accuracy.

Another benefit is that pilots can flight-plan in a dispatch office but don't have to wait for clearances to come back from air traffic control. The information is sent to the cockpit via ACARS.

ACARS is expanding to other services. It reports engine performance to the ground while in flight, so problems are recognized early, often before they've caused major damage. By using the data to show normal performance, airlines obtain extended warrantees from engine manufacturers. Weather information uplinked to the cockpit via ACARS can be evaluated while pilots are not in a high workload phase of flight. Over 60 applications, shown in the chart, are supported in the ACARS system.

SITA

There are two major organizations providing air-ground company communications for the air transport industry. One is ARINC, which mainly serves aircraft flying over North America. Similar services for Europe are provided by SITA (Société Internationale de Télécommunications Aéronautiques). On the VHF bands, the SITA service is called AIRCOM, which operates through ground stations. Increasingly, ARINC and SITA provide a full range of services via satellite, rather than a network of ground stations on VHF and HF bands.

Text of an Actual ACARS Message

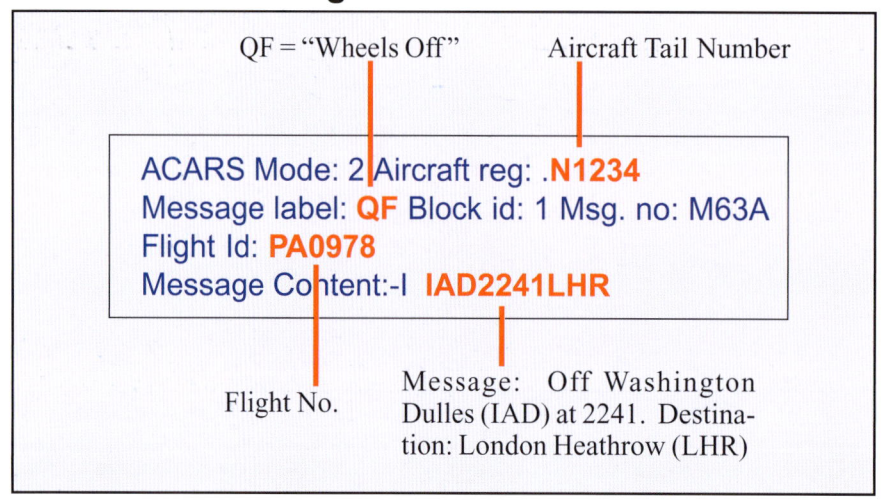

QF = "Wheels Off"   Aircraft Tail Number

ACARS Mode: 2 Aircraft reg: .N1234
Message label: QF Block id: 1 Msg. no: M63A
Flight Id: PA0978
Message Content:-I IAD2241LHR

Flight No.

Message: Off Washington Dulles (IAD) at 2241. Destination: London Heathrow (LHR)

ACARS Bands and Frequencies

VHF (Very High Frequency)

REGION	VHF CHANNELS
USA, Canada	129.125, 130.025, 130.450 MHz
USA, Canada, Australia	131.550 MHz (Primary)
USA	131.125 MHz
Japan.	131.450 MHz (Primary)
Air Canada	131.475 MHz
Europe	131.525, 136.900 MHz
Europe	131.725 (Primary)

These channels, at the upper end of the VHF band, carry ACARS messages to and from ground stations. Channels shown in red are original ACARS frequencies, which have expanded with increasing air traffic.

New forms of transmission are multiplying the number of messages that can be carried on a single channel. Known as VDL---VHF datalink---it enables one channel to carry up to 30 times more data than the conventional ACARS.

HF (High Frequency)

GROUND STATION	HF CHANNELS
Shannon, Ireland	8843, 11384
Hot Yai, Thailand	5655, 13309
Islip, New York	2887, 5500, 8846, 17946
Kahalelani, Hawaii	2878, 4654, 6538, 21928
Johannesburg, S.Africa	8834, 13321, 21949

A sampling of frequencies and stations in the High Frequency band used during long-range flights over oceans and remote areas. Each ground station has channels throughout the band in order to select one according to changing radio propagation conditions.

Review Questions
Chapter 6 ACARS

6.1 What is the meaning of the abbreviation "ACARS"?

6.2 What type of communications occur on ACARS?

6.3 Who operates ground and satellite services for ACARS?

6.4 What is the meaning of the ACARS message, "OOOI"?

6.5 How is an ACARS message received only by the aircraft it's intended for?

6.6 What two bands carry ACARS services?

6.7 What satellite-based system carries ACARS services worldwide?

Chapter 7

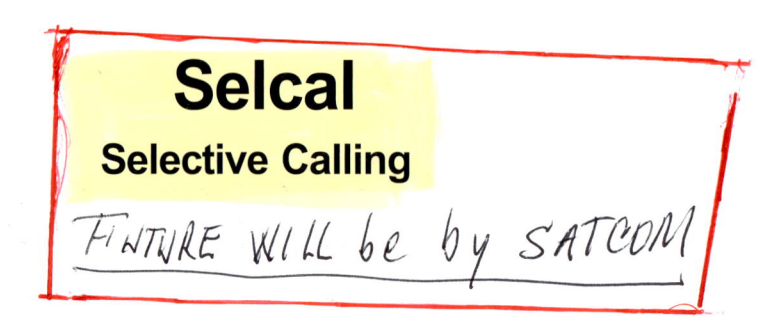

Selcal
Selective Calling

FUTURE WILL be by SATCOM

During oceanic flights, aircraft monitor a HF (high frequency) radio for clearances from a ground controller. Because HF reception is often noisy, and many messages are intended for other airplanes, a pilot prefers to turn down the audio He will not miss calls intended for him, however, because of Selcal---selective calling. The ground controller sends a special code that sounds a chime or illuminates a light to warn the pilot of an incoming message and to turn the volume up. Because it's selective, Selcal "awakens" only the HF receiver with the appropriate code.

This Selcal controller, located on the instrument panel, monitors two radios simultaneously (VHF or HF). An incoming tone code lights a green lamp and sounds an aural warning (chime). The pilot turns up the audio volume on the radio. Pressing the RESET button arms the system to receive the next call.

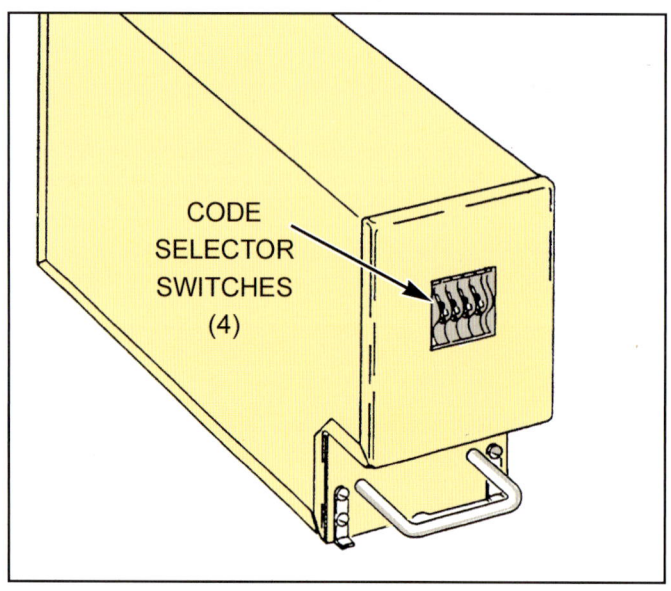

Selcal decoder is an LRU (line replaceable unit) located remotely in the airplane's electronic bay. The four-letter code assigned to that airplane is programmed manually by four thumb wheels (code selector switches).

The four-letter code (EG-KL, for example) is drawn from the letters A through S (I, N and O are excluded).

Some aircraft have two decoders, one to receive Selcal tones for up to four radios (2 VHF and 2 HF). The same assigned letters, however, are entered into the decoders.

How Selcal Code is Generated

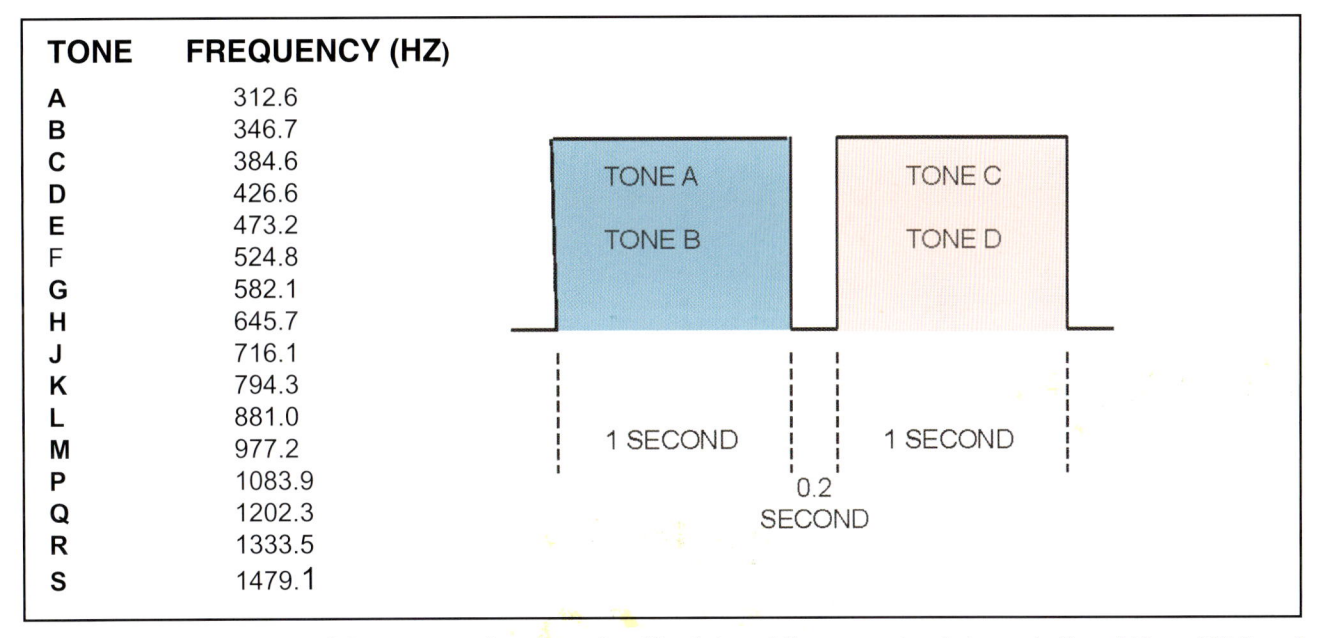

TONE	FREQUENCY (HZ)
A	312.6
B	346.7
C	384.6
D	426.6
E	473.2
F	524.8
G	582.1
H	645.7
J	716.1
K	794.3
L	881.0
M	977.2
P	1083.9
Q	1202.3
R	1333.5
S	1479.1

A Selcal code consists of four tones taken from the 16 audio frequencies shown at the left. In this example, the code is AB-CD. As seen in the diagram, they are sent in two pairs. A and B are mixed together (312.6 and 346.7 Hz) and transmitted for one second. After a .2-second interval the second pair is sent; C and D, or 384.6 and 426.6 Hz. (The technique is similar to touch-tone dialing for telephones.) Because the tone signals are audio in the voice range, they can be detected by a conventional VHF or HF communications transceiver.

Selcal Ground Network

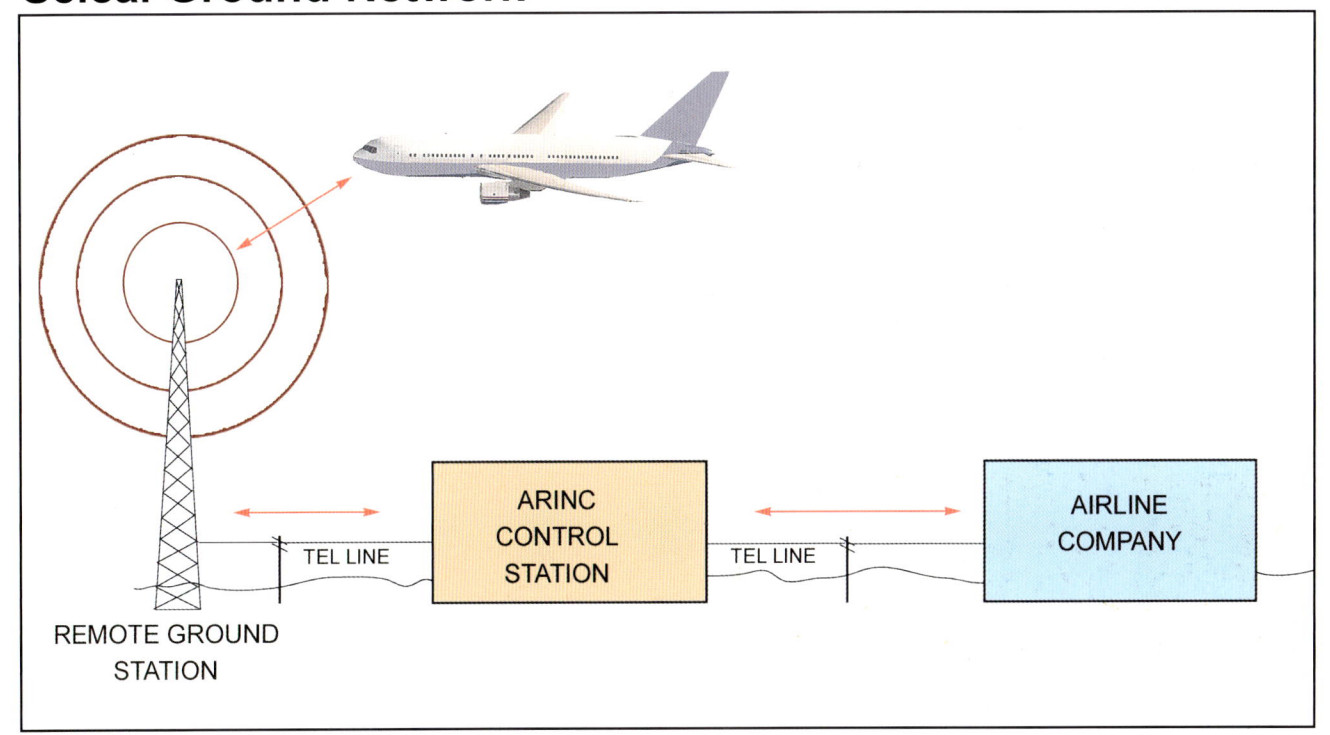

When Selcal must operate on VHF, where maximum range is about 200 miles, it is done through a network of remote ground stations. The airplane, always within range of some ground station, transmits and receives Selcal messages through an ARINC control station (in the U.S.). ARINC relays the message to the airline company. The link between stations is usually through telephone lines.

Selcal over oceanic routes is done on HF, where range from airplane to ground may be several thousand miles. The future of Selcal will be satcom; the airplane will communicate with satellites for relay to the ground.

47

VHF. Selcal also operates with VHF radios, used by aircraft flying within a country or continent. Not only does Selcal reduce pilot workload, but extends the communication distance of VHF. If an airline company in Denver, for example, wants to talk to one of its airplanes in flight over Chicago, this is far beyond the range of VHF. Instead, the message is sent through a telephone line to a network of ground stations. A VHF ground station near the aircraft transmits to the airplane, and the pilot is signalled. He replies on VHF to the ground station and the message reaches the airline company through the network.

Coding. Selcal is based audio tones, as shown in the illustration. Each airplane has a code of four letters set into the Selcal decoding unit aboard the airplane. The code is entered into the flight plan so controllers can address it.

Although there are nearly 10,000 possible four-letter codes, they are in short supply. The demand is so high that more than one aircraft may be assigned the *same* Selcal code. To avoid answering a call intended for another airplane, identical codes are assigned in widely separated parts of the world. There is also an attempt to assign the same code to airplanes with different HF channel assignments.

It is important to warn pilots that it's possible to receive a Selcal alert not intended for them. This can be corrected by the pilot by clearly identifying his flight to the ground station.

Selcal Airborne System

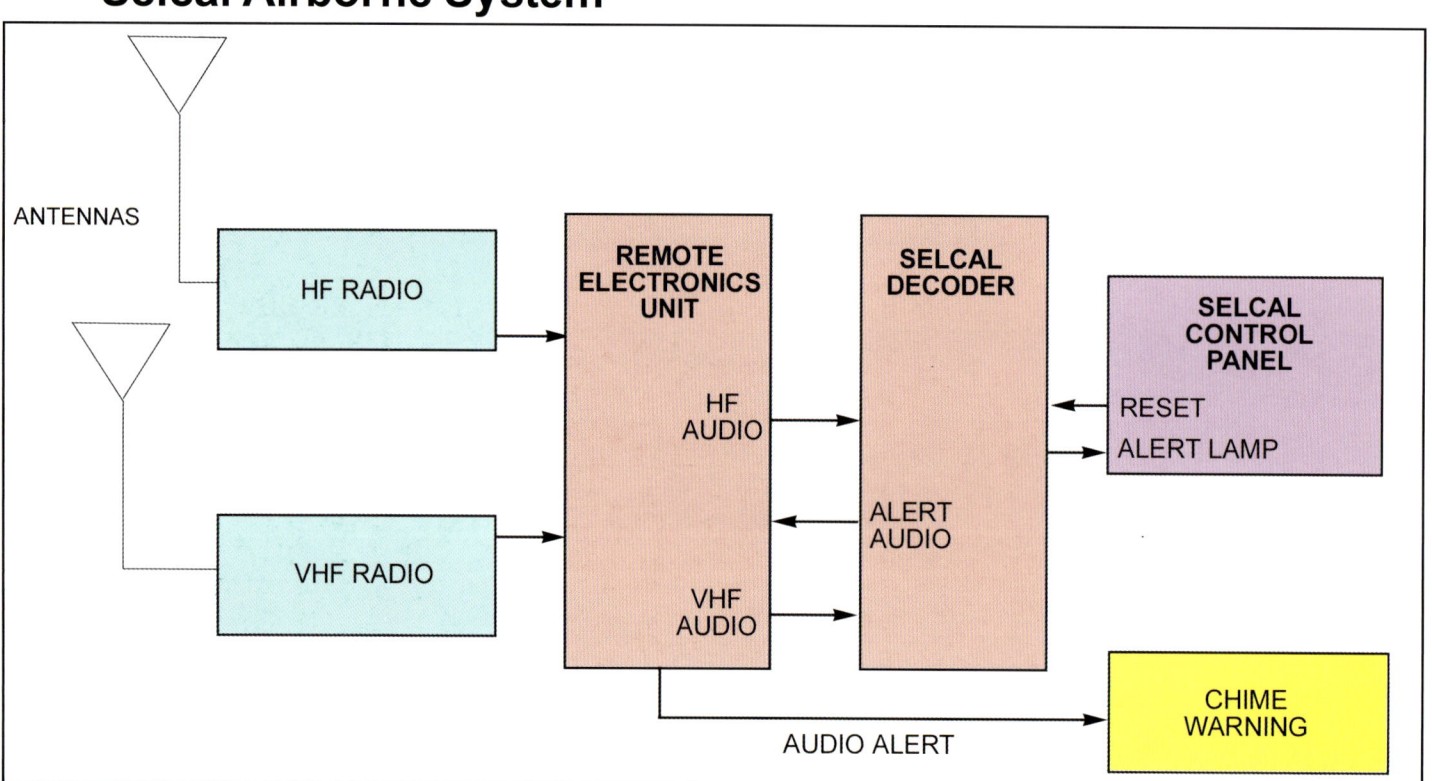

Block diagram of Selcal system. Signals are received from ground stations through the aircraft HF and VHF transceiver. They are processed by the Remote Electronics Unit and sent to the Selcal decoder for delivery to the pilot (on a screen or printer).

An incoming signal with the correct code illuminates a green panel light in the Selcal Control Panel and sounds a chime (aural alert).

A single system is shown here, but many aircraft have dual Selcal installations.

Review Questions
Chapter 7 Selcal

7.1 What does the contraction "Selcal" mean?

7.2 Give two reasons why Selcal is used.

7.3 How many tones are in a Selcal code?

7.4 How many Selcal tone pairs are transmitted simultaneously?

7.5 Can two aircraft have the same Selcal code?

7.6 What precaution is necessary if a pilot receives a Selcal intended for a different airplane?

7.7 How is the problem reduced where two aircraft have the same Selcal code?

7.8 How is the pilot warned of an incoming Selcal message?

Chapter 8

ELT
Emergency Locator Transmitter

Two U.S. Congressmen were missing in an Alaskan snowstorm in 1972 and never heard from again. Search and rescue forces flew over 3000 hours looking for the downed airplane but found nothing. Even if the congressmen survived the crash and called for help there was no assurance that anyone was listening or within radio range.

Congress responded with a law requiring aircraft to carry a "beacon" to automatically sense a crash and send out emergency signals on 121.5 MHz, the distress frequency. The theory was that other airplanes flying in the vicinity would monitor 121.5 (found on all VHF com radios) and report a beacon signal to a ground station. The new law required General Aviation airplanes (Part 91) to be equipped with an ELT. For the airlines (Part 121) ELT's were required for extended flight over water and uninhabited areas.

Flaws in the system soon appeared. First, there was no guarantee a distress call would be heard by a passing airplane or ground facility. What is more, the number of false alarms rose so high that only a few percent resulted from actual crashes. Despite an enor-

A beacon, like this Artex C-406-N, sends three separate ELT signals to the antenna through one coaxial cable; a warbling tone on 121.5 and 243 MHz, and an encoded digital message on 406 MHz. Output power on 406 is 5 watts, with a lithium battery rated for 5 years.
   Note the precaution about mounting the ELT with respect to the direction of flight, which assures proper operation of the crash sensor (a G-switch).
   An ELT for a helicopter has a different G-switch, which responds in six different directions.

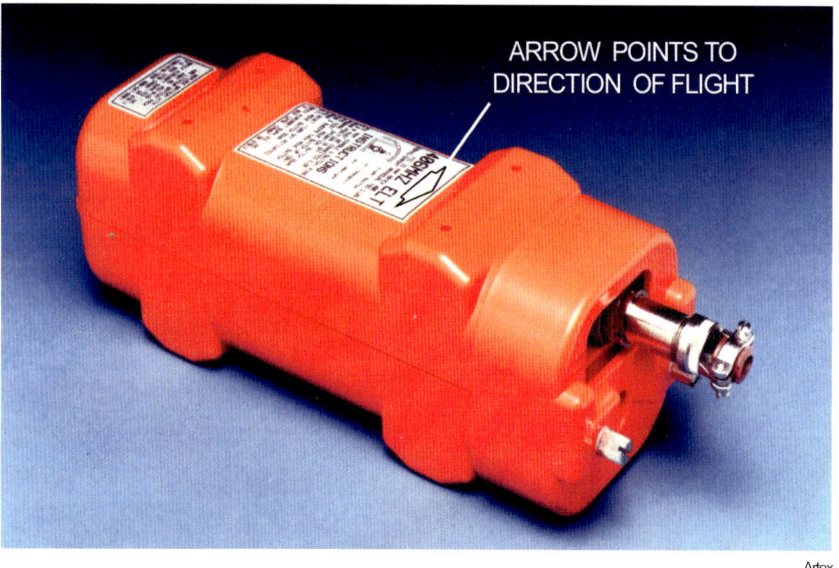

ARROW POINTS TO DIRECTION OF FLIGHT

Artex

50

mous waste of search and rescue resources, there was agreement that the system should not be abandoned, but improved.

Changes came in the form of tighter standards and better design. The ELT industry also gained experience and learned that failure to activate during a crash was often due to poor ELT installation, corroded internal parts, defective G-switches, faulty antennas and cables and dead batteries.

In 1995 all ELT's under the original certification (TSO C91) would be replaced by the next-generation ELT (TSO 91a). The regulations also tightened maintenance requirements; once a year, an ELT must be inspected for proper installation, battery corrosion, operation of controls and crash sensor, and sufficient signal radiated from the antenna.

Cospas-Sarsat

While the new rules improved ELT hardware, there was still the question; "Who's listening for distress signals?" The answer arrived with earth-circling satellites. By listening from orbit, satellites increase the chance of intercepting an ELT distress signal.

The satellite system, known as Cospas-Sarsat, consists of satellites provided by the United States and Russia. "Cospas" is a Russian term meaning "Space System for Search of Vessels in Distress." These satellites are primarily for the Russian navigation system, but with added instruments for search and rescue. They operate on 121.5 MHz, the civil aviation distress frequency, and 243 MHz, the military equivalent.

U.S. satellites are "Sarsat," for "Search and Rescue Satellite Aided Tracking." The primary role is weather survey, with search and rescue instruments added on. As shown in the illustration, the satellites are supported by a network of ground stations, mission control and rescue coordination centers.

Location. In the era before satellites, rescuers found downed aircraft by radio-direction finding. Using an attachment to a VHF radio and a directional antenna, searchers "home in" on the ELT signal.

Satellites use a different technology, the "Doppler shift." As a satellite rises over the horizon toward the crash site, its forward speed "squeezes" the ELT radio waves. Instead of receiving 121.5 MHz, the satellite hears a slightly higher frequency. When the satellite moves away from the crash site, 121.5 appears to stretch out---producing a lower frequency. These changes (Doppler shift) reveal the position of the crash after several satellite passes from different directions. Although Cospas-Sarsat solved the monitoring problem, it actually *increased* the number of false alarms by its global coverage.

Search and Rescue Satellites

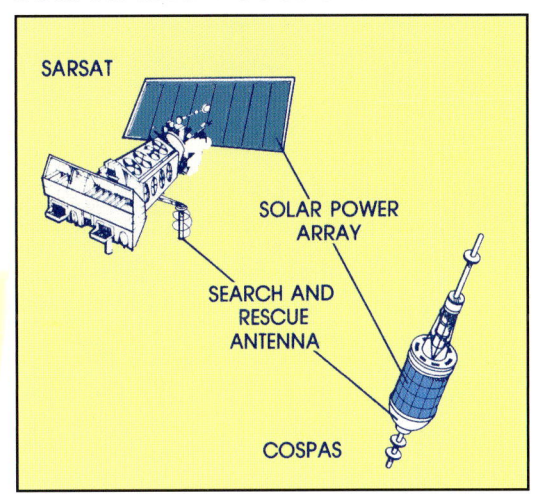

The U.S. satellite, SARSAT, is operated by NOAA (National Oceanic and Atmospheric Administration). It is in polar orbit at an altitude of 528 miles, circling the earth once every 102 minutes.

The Russian satellite, COSPAS, circles the earth every 105 minutes at an altitude of 621 miles.

The US satellites' primary mission is observing weather and the environment, and is also equipped for receiving search and rescue signals.

COSPAS is part of the Russian spacecraft navigation system, with the search and rescue function added.

Payloads on both satellites (for search and rescue) are provided by France and Canada.

Ground Stations

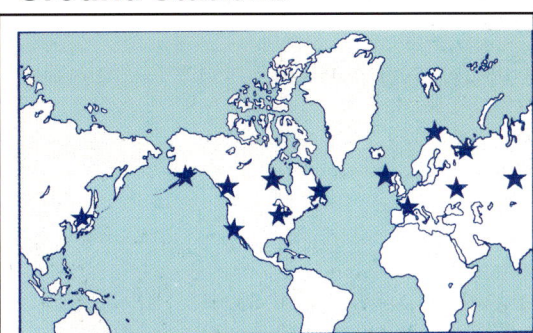

There are ground stations over the world for the search and rescue system. Known as "Local User Terminals," they receive emergency transmissions picked up by satellites from downed aircraft. Almost half the world is covered for ELT's operating on 121.5 MHz; the entire globe is covered on the 406 MHz frequency.

406 MHz ELT

By the year 2000, more than 180 countries voted to end the 121.5/243 MHz generation of emergency beacons. The cut-off date would be 2009. The replace-

ELT Components

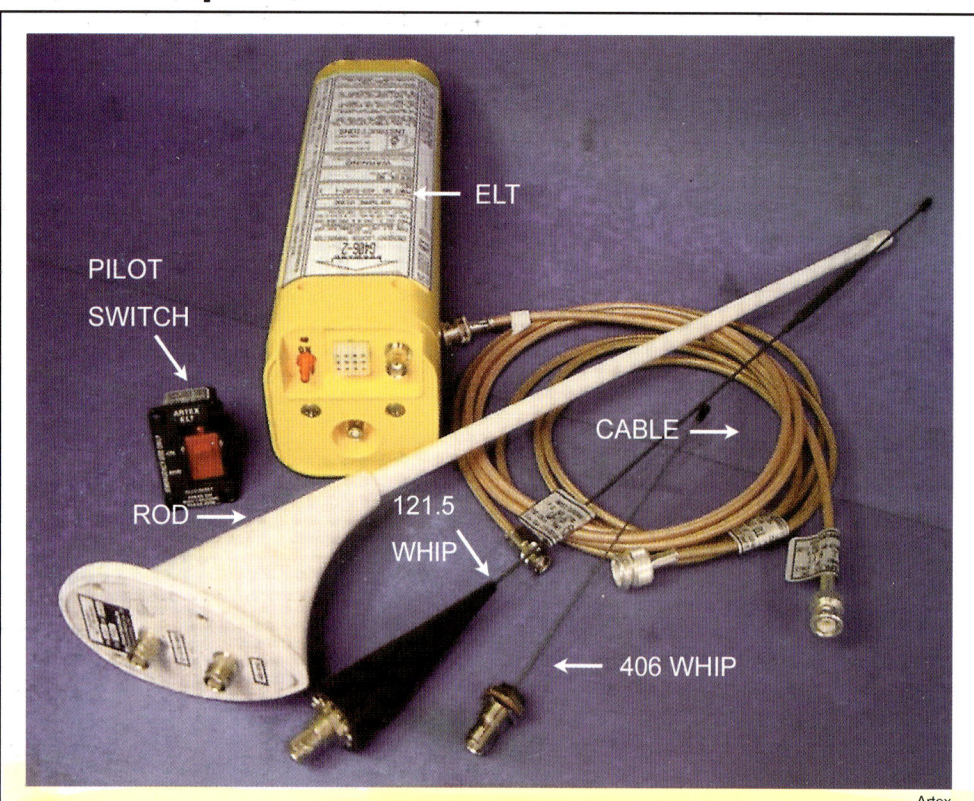

Major components of an ELT. The system broadcasts on three emergency frequencies; 121.5 MHz, the original distress channel; 243 MHz, the military distress frequency and the newer 406 MHz. When a crash activates a G-switch inside the ELT a varying audio tone is broadcast (up to 50 hours) on 121.5 and 243.

The antennas are chosen according to speed of the aircraft; the rod is for greater than 350 kt, the whip for slower aircraft.

Although the ELT activates automatically, it can also be turned on manually by the pilot switch. If the ELT is activated accidentally on the ground, it sounds a buzzer to alert the ground crew.

For the ultimate in accuracy, the ELT can broadcast latitude and longitude (on 406) if this data is provided from the airplane's navigation system.

(Shown is the Artex G406-2.)

ment is 406 MHz, with numerous improvements to reduce false alarms and raise location accuracy. (406 is operating now and can handle 121.5 and 243 MHz).

The 406 system is a mixture of Leosar and Geosar satellites. Leosar ("low earth orbit search and rescue") completely covers the globe and can "store and forward" messages. The satellite does not have to see both crash site and ground station at the same time, but stores the distress message, then replays it when a ground station comes into view.

Leosars, however, do not provide continuous coverage; an airplane in distress must wait for the satellite to come into view. This gap filled by additional satellites known as Geosars ("geosynchronous orbit search and rescue"). Parked 22,500 miles above the equator in geosynchronous orbit, they appear stationary and provide full earth coverage, except over North and South poles.

Because Geosars are stationary, they cannot find beacons by the Doppler shift. They must receive a distress message that contains the airplane's position. This information is provided by a GPS receiver that is part of the ELT or from an external GPS receiver on the airplane.

The 406 system is far more capable than the first-generation ELT, which guided rescuers only within about 15 miles of the crash site. They had to narrow the search with a homing receiver. The 406 brings rescuers within 1 to 3 miles of the target using improved Doppler shift detection. The most precise guidance is when the 406 MHz ELT is coupled with a GPS, where accuracy becomes 300 feet or less.

The transmitting power of a 406 ELT is 5 watts,

versus one-tenth watt for 121.5.

When an airplane crashes, the occupants' chance of survival rapidly drops with the passage of time. Nevertheless, search and rescue forces do not respond to the first alert from a 121.5 ELT. Because so few signals are from actual crashes, rescuers face unnecessary hazards. They don't start the search until the alarm is verified. With the 406 system, however, they will respond to the first alert, which saves an average of six hours in reaching a crash site.

Registration. Much of the benefit from 406 is from an ELT registration system. No longer will an ELT broadcast anonymously, but transmits its identification as a digital message on the radio signal. Each user of a 406 ELT must register (at no charge) with Sarsat authorities, giving telephone numbers and other contact information. Each 406 ELT is issued a serial number that is broadcast with the signal.

Now when a distress call is received by search and rescue, they make a telephone search. The pilot may be at home or work (unaware the ELT had a false activation). Searchers speak with an airport manager who checks the ramp for the airplane, or make additional phone calls to verify whether the airplane actually made the trip and is in distress. The registration program should reduce false alarms by 70 percent.

406 MHz ELT System

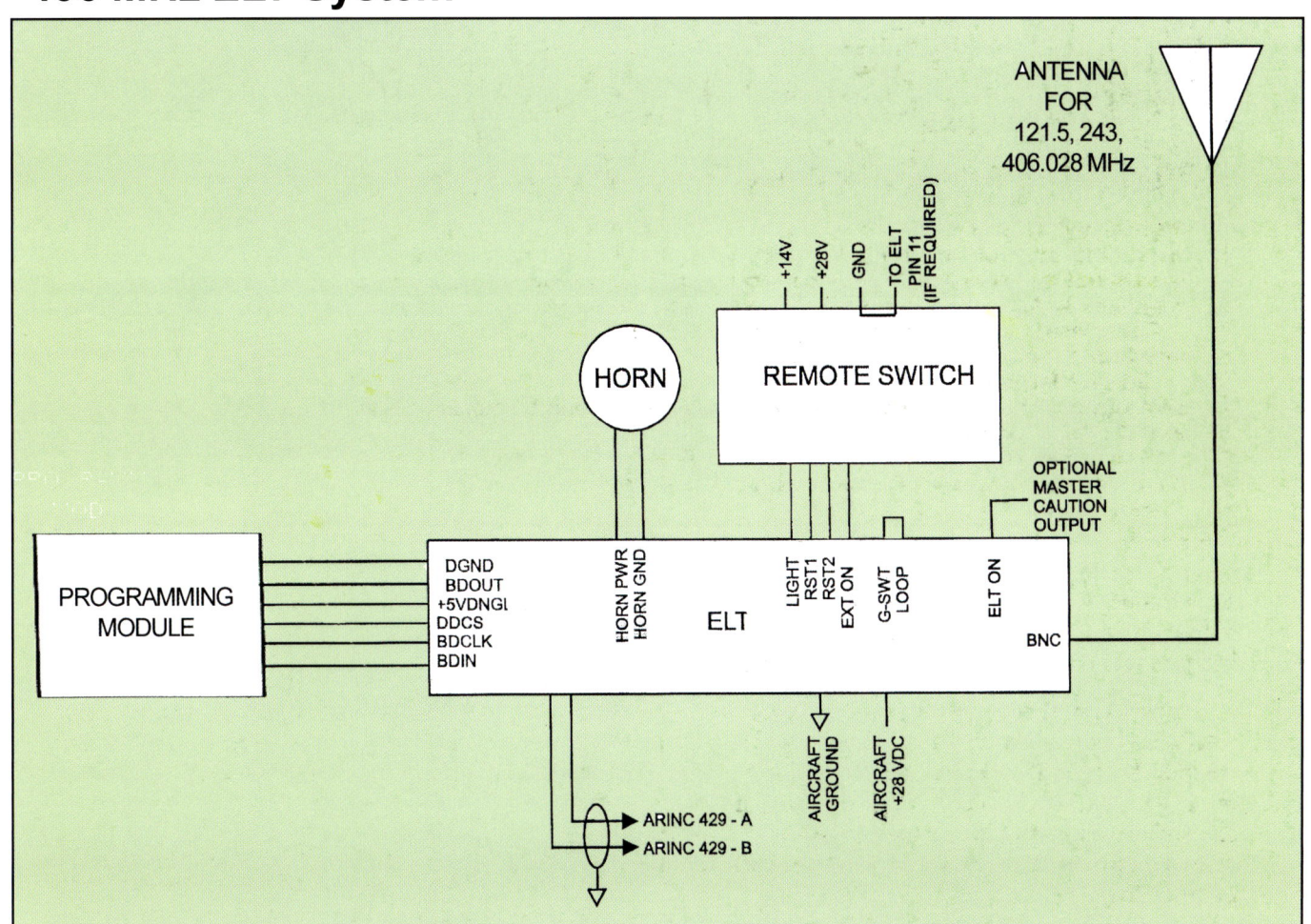

The Programming Module (lower left) sets up the 406 ELT for its unique code; a 24-bit address or aircraft tail number. This is required of all 406 ELT's. At top center, the Horn sounds to warn the pilot of a false activation. The Remote switch controls the ELT from the cockpit. At the bottom center, the ARINC 429 connections bring a signal from the airplane navigation system into the ELT. This transmits an accurate location of the downed aircraft.

At upper right, the single antenna radiates three ELT frequencies through one cable; 121.5, 243 and 406 MHz.

53

ELT for Fleet Operation

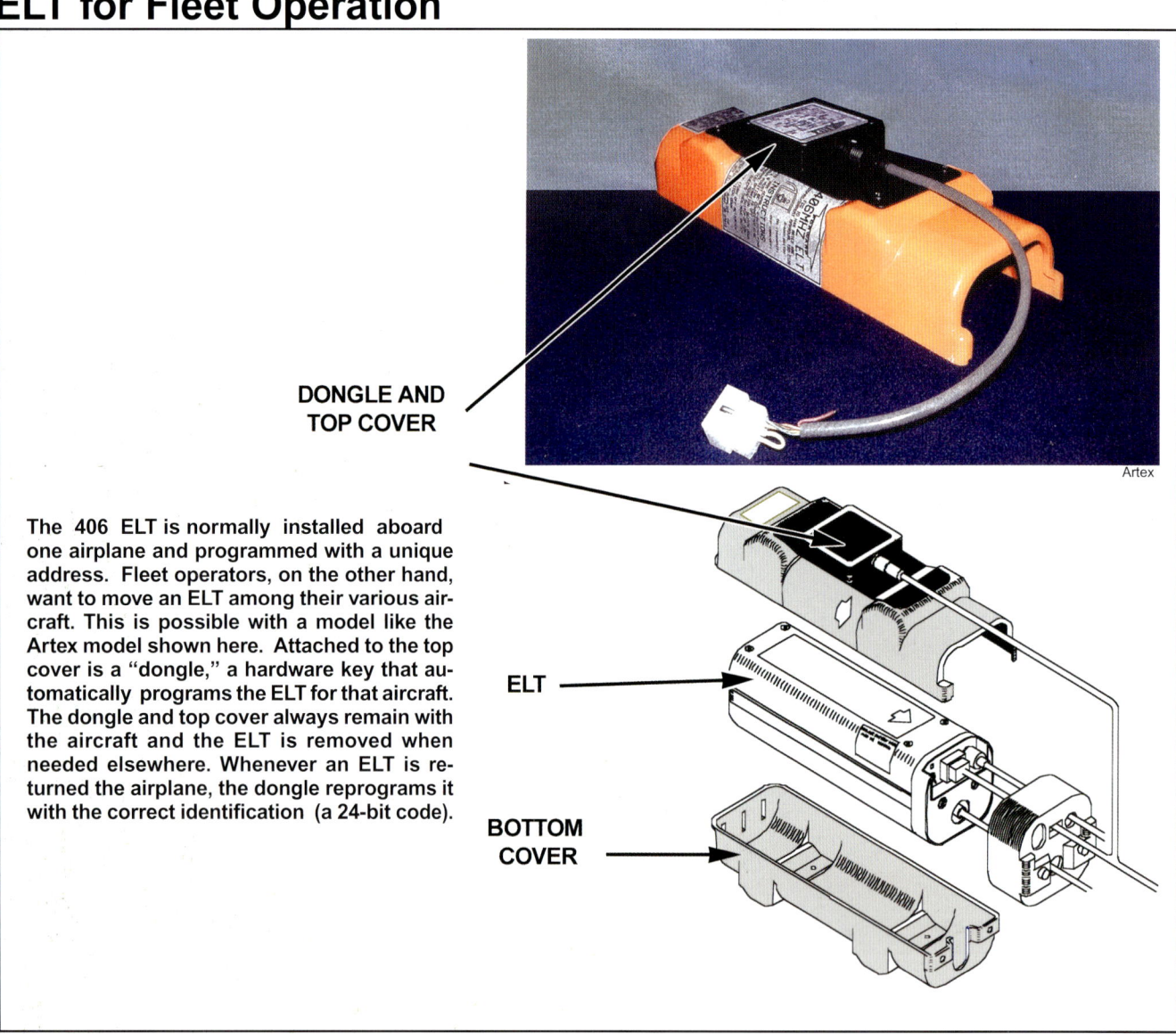

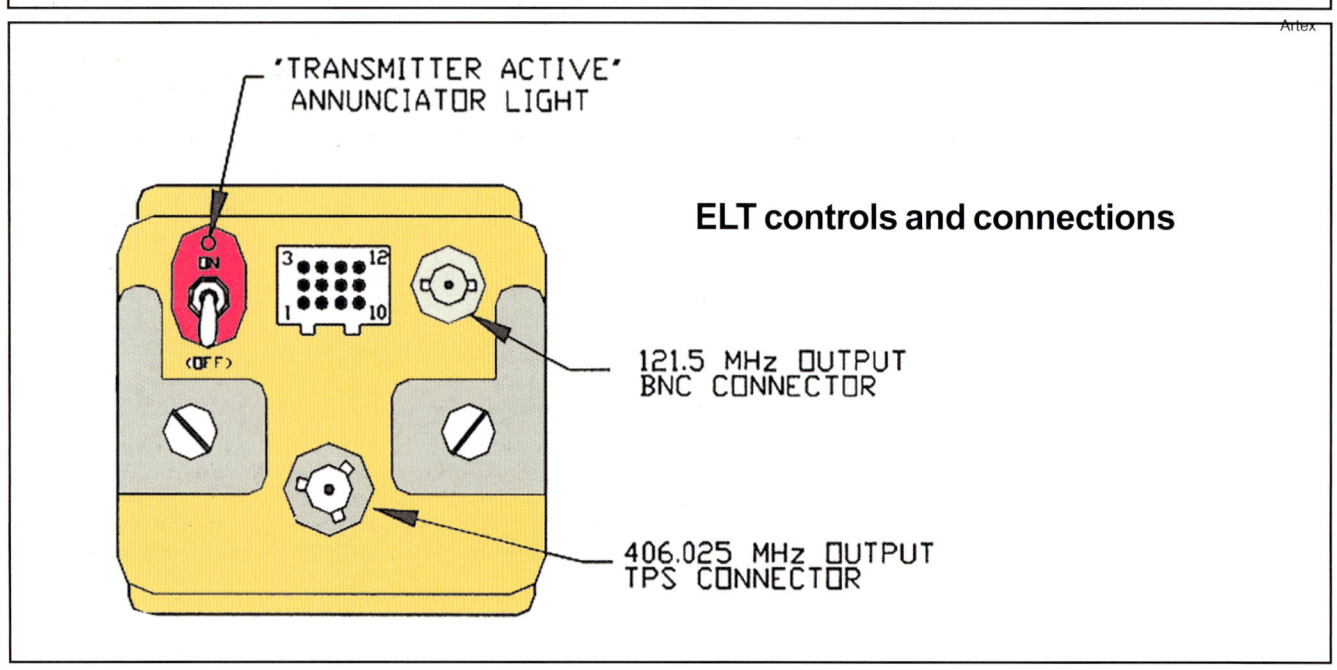

DONGLE AND TOP COVER

ELT

BOTTOM COVER

The 406 ELT is normally installed aboard one airplane and programmed with a unique address. Fleet operators, on the other hand, want to move an ELT among their various aircraft. This is possible with a model like the Artex model shown here. Attached to the top cover is a "dongle," a hardware key that automatically programs the ELT for that aircraft. The dongle and top cover always remain with the aircraft and the ELT is removed when needed elsewhere. Whenever an ELT is returned the airplane, the dongle reprograms it with the correct identification (a 24-bit code).

ELT controls and connections

'TRANSMITTER ACTIVE' ANNUNCIATOR LIGHT

121.5 MHz OUTPUT BNC CONNECTOR

406.025 MHz OUTPUT TPS CONNECTOR

Cospas-Sarsat System

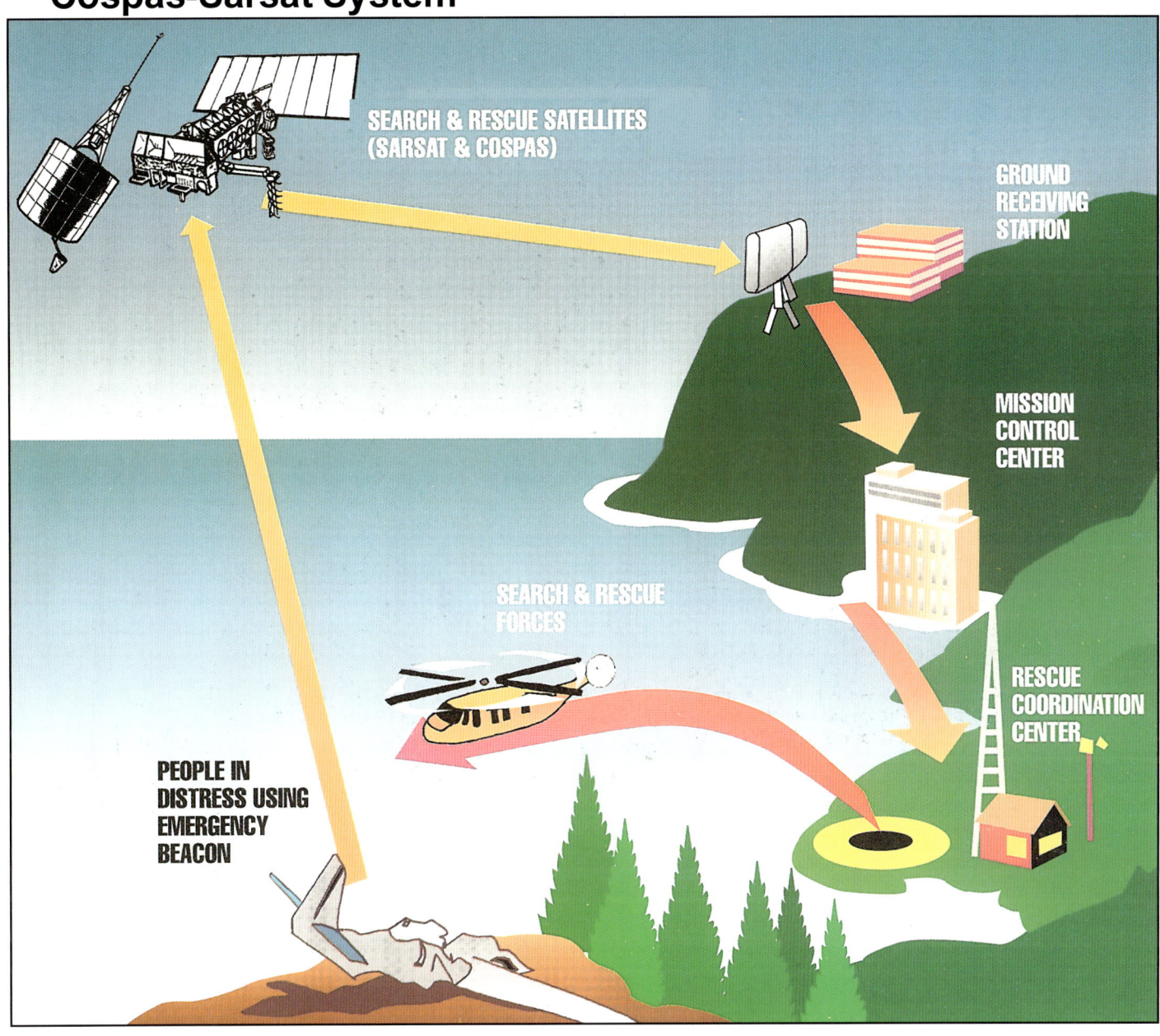

There are two types of satellites in the COSPAS-SARSAT system (see upper left). One is LEO, for low earth orbit. The other is GEO, for geostationary earth orbit. Because LEO's circle over North and South poles, they provide coverage in these regions. LEO's also are better able to pick up signals when the distress aircraft is surrounded by trees and other obstructions. This is because the LEO moves rapidly across the sky and views the distress aircraft from many different angles.

GEO's, on the other hand, are stationary over the equator to cover large areas of the earth. The advantage is that a GEO picks up a distress call almost immediately. Thus, the two types---LEO and GEO work well together.

406 ELT Registration

Unlike first-generation ELT's, it is important to register a new 406 MHz ELT. This data is used by authorities to identify aircraft type, ownership, telephone number, home base and other information. It enables searchers to discover most false alarms before taking off on a dangerous and costly rescue mission.

Review Questions
Chapter 8 ELT (Emergency Locator Transmitter)

8.1 What three radio frequencies are sent out by an ELT during a distress call?

8.2 Why must an ELT be mounted in line with the direction of flight?

8.3 Name the satellites that pick up and relay ELT signals?

8.4 Name one method satellites use to locate a downed aircraft transmitting an ELT signal.

8.5 Where do satellites relay the location of downed aircraft?

8.6 What is the most accurate method for identifying the location of an ELT signal, as used in the 406 MHz system?

8.7 What is the main benefit of registering ELT's, giving aircraft ID, and ownership?

8.8 How accurately can searchers locate a 406 MHz ELT coupled to a GPS source?

Chapter 9

VOR
VHF Omnidirectional Range

Navigating by radio signals is one of the most successful and reliable aviation systems ever developed. Every day many of the 2,000 flights in the U.S. alone fly through clouds, darkness, rain and fog---reaching their destination in greater safety than driving an automobile. When an accident happens, investigators almost never find that a faulty navigational aid misled a pilot into a hazardous situation.

Not long ago, radionavigation began its greatest change in 75 years. "Navaids," as they're called, consist of thousands of ground stations emitting guidance signals for at least eight avionics navigation systems. In a transition now in progress, ground stations will be replaced by signals from space, broadcast by orbiting satellites. The changeover will occur over the next 20-30 years, with both ground-based and satellite navigation existing side by side. The world made a decision through the International Civil Aviation Organization that future air navigation will be "GNSS"---for Global Navigation Satellite System. To keep air travel safe during the changeover, early ground stations must remain operational over the long transitional period. Those years will also enable aircraft operators to get full value out of their large investment in today's avionics systems.

VOR and DME Radiate from Same Station

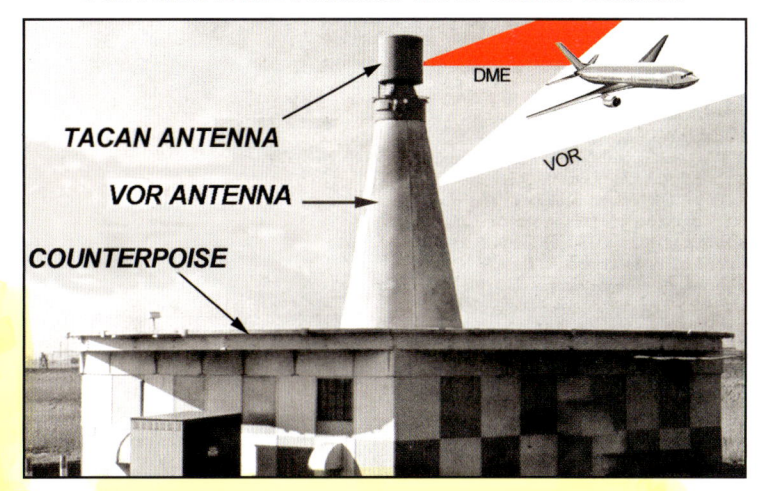

Vortac (VOR + Tacan) ground station. The VOR antenna atop the shelter transmits the VOR signal. The Tacan antenna is a military navigation system which transmits the DME signal (distance measuring equipment). Tacan means "Tactical Air Navigation." Only military aircraft use the azimuth (directional information) from the Tacan station.

The "counterpoise" at the base is a grid of metal that acts as an electrical ground to improve efficiency of the signals.

VOR Coverage

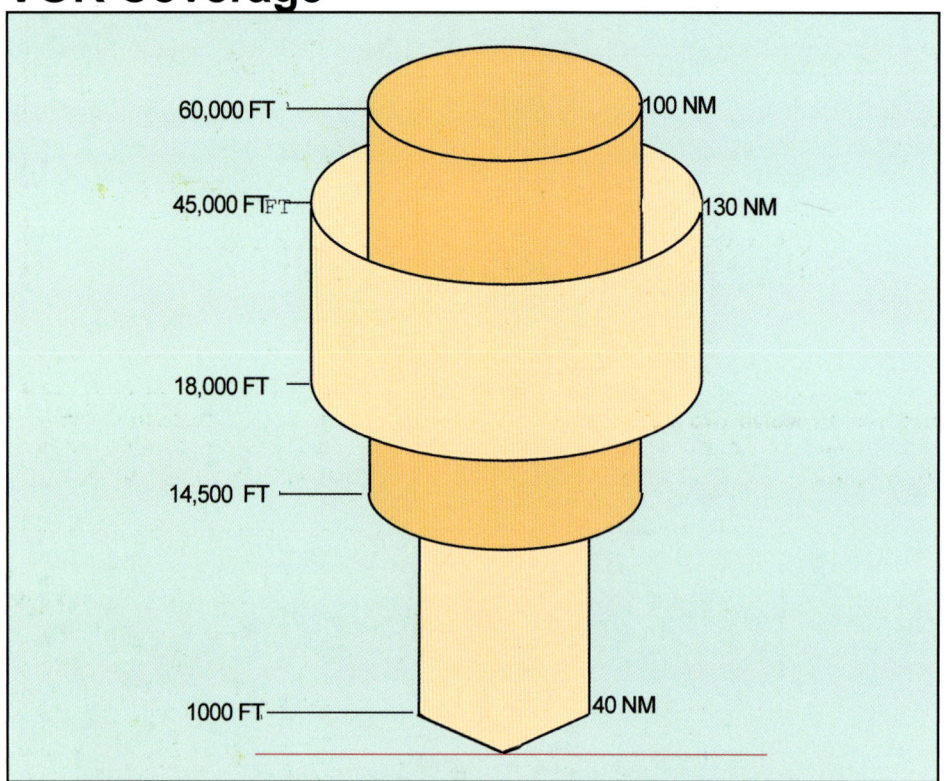

The service volume for a High Altitude (H) Vortac. Greatest range, where good signal reception is assured, is 130 naut miles between 18,000 and 45,000 ft. These are flight levels flown by turbine-powered aircraft during the en route phase. The three types of VOR are shown in the chart below.

VOR

VOR is the short-range radionavigation system for much of the world. When introduced in 1946, it eliminated interference problems of earlier systems. Radionavigation from 1920 to 1940 operated on low frequencies, where energy from lightning strokes are received over 100 miles away. Low frequencies are also susceptible to other natural and man-made sources.

Not until World War II could designers produce an airborne radio that operated at VHF (very high frequency) where there is little electrical interference. VOR frequencies start just above the FM broadcast band and run from 108 to 117.950 MHz.

VHF is immune to other problems of lower frequencies. The waves travel in straight lines like light, which is important for creating accurate courses. A well-designed VOR receiver can be accurate to within one compass degree.

VOR / DME / TACAN Service Volumes

T (Terminal VOR)
From 1,000 feet above ground level (AGL) up to and including 12,000 feet AGL, at distances out to 25 NM.

L (Low Altitude VOR)
From 1,000 feet AGL up to and including 18,000 feet AGL at distances out to 40 NM.

H (High Altitude VOR)
From 1,000 feet AGL up to and including 14,500 feet AGL at distances out to 40 NM. From 14,500 AGL up to and including 60,000 feet at distances out to 100 NM. From 18,000 feet AGL up to and including 45,000 feet AGL at distances to 130 NM.

VOR Signal Has Two Navigation Components: A "Reference" and "Variable" Phase

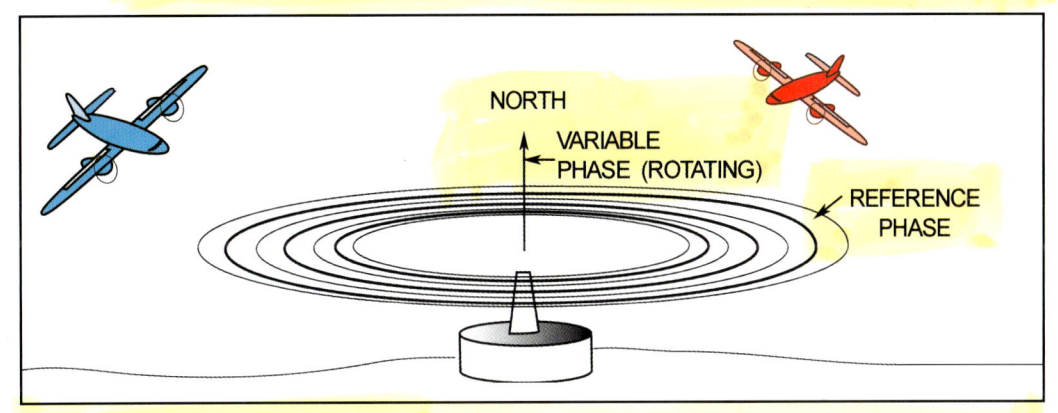

Each aircraft must receive two VOR signal components. One is the "reference phase," which is broadcast in all directions. This is picked up by all aircraft lying out in any direction from the station. The reference signal is broadcast 30 times per second.

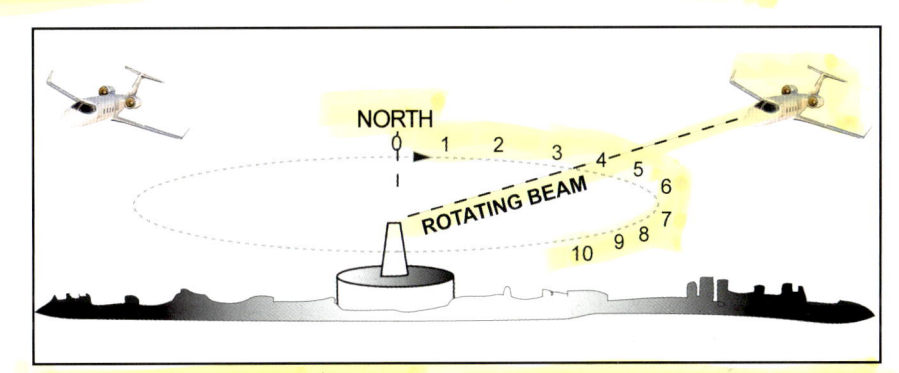

The VOR station also transmits a rotating beam that turns full circle. Because the beam is narrow It is intercepted only when the aircraft is aligned with the beam, as shown at the right. This "variable phase" signal is compared with the "reference phase" in the receiver.

As shown on the next page, the airborne receiver compares fixed and variable phases to determine the number of compass degrees, or bearing, from the station.

Short Range and Doglegs

VOR signals cover up to about 130 miles from the station. To travel from Los Angeles to New York, therefore, a pilot flies to and from about a dozen VOR stations. In continental U.S. there about 1000 VOR stations on the ground. Because stations may not lie in a straight line along the route, the trip might have a "dogleg."

RNAV. The delay of flying a dogleg was of not much concern when jet fuel was 17 cents a gallon, but as world prices rose in the 1970's a new type of VOR navigation emerged. Called "RNAV," for "area navigation," it could receive a VOR off the straight line course and electronically move it on a desired course.

VOR Principles

A VOR station sends out two separate signals. One rotates like the narrow beam of a lighthouse. Imagine sitting on a beach at night, watching the beam go around; you see a bright flash only when the light points directly at you. At that moment, begin counting to see how much time it takes for the beam to flash again. Let's say the beam takes 10 seconds for one rotation, or 360 degrees, and assume you're sitting north of the lighthouse. Now you can convert the number of seconds into where the beam is aimed at any time. By counting five seconds from the flash, for example, you know the beam moved half-way around---180 degrees---and is aimed south.

59

VOR Signal Structure

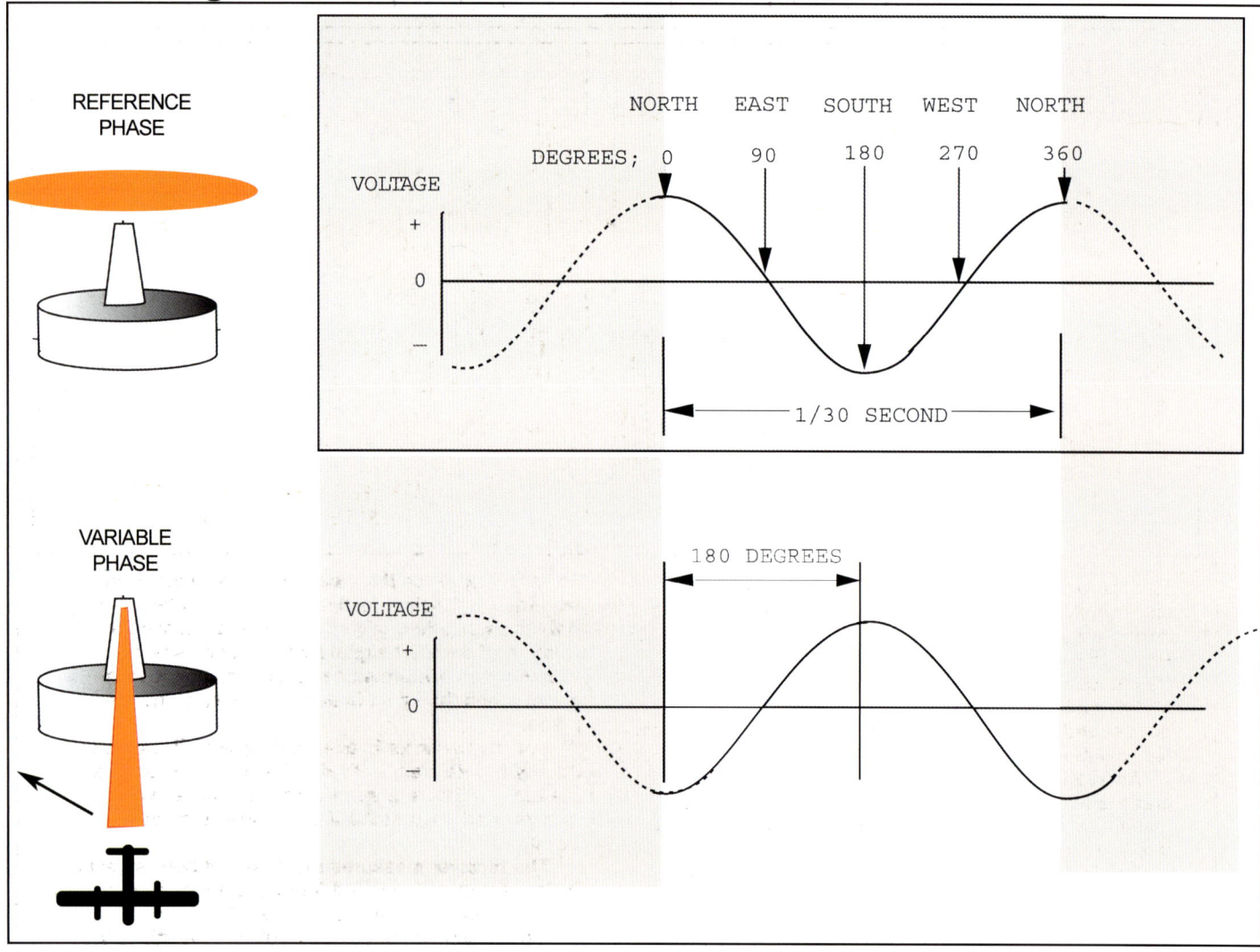

By comparing reference and variable phases, the VOR receiver determines a difference in degrees. This also becomes the magnetic bearing from the station.

In this example, the reference phase is at 0 (or 360) degrees. The airplane, south of the station, receives a variable phase signal of 180 degrees. The difference (360 - 180) is 180 degrees, or south.

The reference signal always goes through its 0 degree phase at the instant the variable signal rotates through magnetic north. This provides the correct reference for comparing the two signals.

A VOR also transmits two additional signals for station identification. One is an audio tone keyed in Morse code, enabling the pilot to identify the station. The tone is 1020 Hz.

The fourth signal is voice. Many VORs also broadcast voice to announce the ID, and enable the pilot to listen to the voice of a flight service station (for weather and flight plans.) The pilot, however, never transmits his voice on a VOR frequency because this would interfere with navigational signals. He transmits on another channel, and receives on the VOR frequency.

VOR Broadcasts Two Navigational Signals

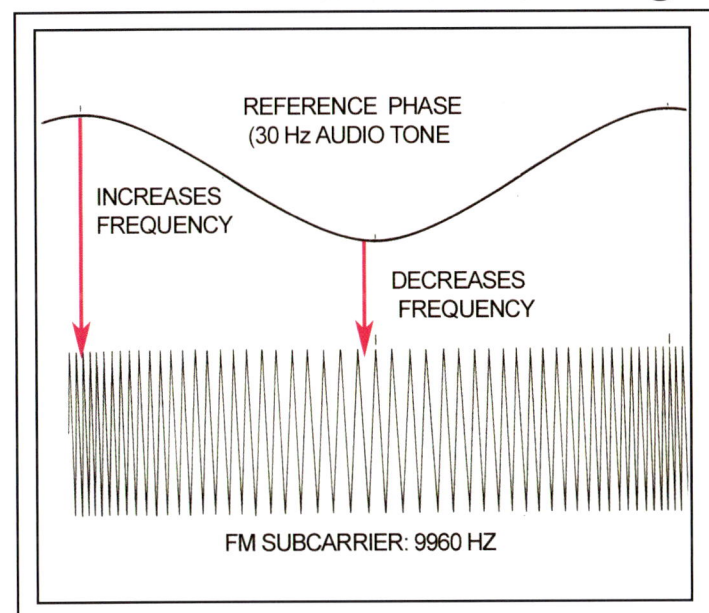

REFERENCE PHASE (30 Hz AUDIO TONE)

INCREASES FREQUENCY

DECREASES FREQUENCY

FM SUBCARRIER: 9960 HZ

30 Hz AMPLITUDE MODULATION (VARIABLE PHASE)

30 Hz FREQUENCY MODULATION (REFERENCE PHASE)

The two navigational signals from a VOR—Reference and Variable Phase—cannot be allowed to mix during transmission. To keep them apart, the Reference Phase is placed on a "subcarrier." At a resting frequency of 9960 Hz, the subcarrier is shifted up and down in frequency by the Reference Phase 30 times per second. The Reference Phase, therefore is transmitted by FM—Frequency Modulation.

As seen in the illustration, the subcarrier increases in frequency as the Reference Phase goes maximum positive (upward) and decreases the subcarrier frequency when it goes full negative.

The information about North occurs at the positive peak of the Reference Phase, shown by the red arrow at the left. The subcarrier rises in frequency to 10,440 Hz. South is shown by the second red arrow, where the subcarrier lowers in frequency to 9480 Hz.

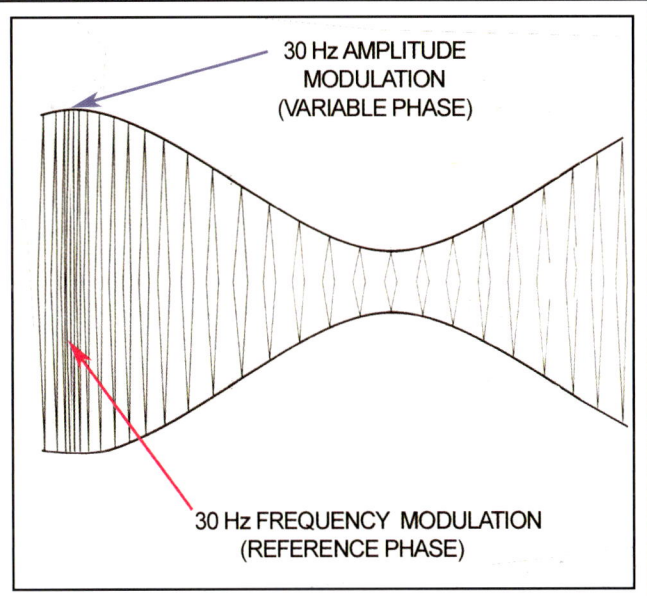

The VOR signal as it appears in the receiver. Note how the carrier rises and falls in strength (which is AM, or amplitude modulation). This is caused by rotation of the VOR signal by the ground station, producing maximum signal (blue arrow) in the receiver when the beam is aimed directly at the airplane.

The same carrier is also undergoing FM modulation by the Reference Phase. The red arrow shows the highest FM frequency of the subcarrier, which always occurs when the Variable Phase moves through north.

The receiver measures the phase of both signals, makes a comparison, and indicates the difference as a magnetic course from the VOR.

In the example above, the airplane is north of the VOR. The Variable Phase is at a positive peak, while the Reference Phase is at its highest frequency (meaning north). Because the phase difference between signals is 0, the airplane is directly north of the VOR.

When describing VOR, it is convenient to visualize 360 "radials" spreading out from the station like spokes of a wheel. In air traffic procedures, a "radial" always moves *outward* from the VOR.

The VOR receiver needs one more piece of information; when to start counting. This is the purpose of the second VOR signal ("reference phase"). When the first beam ("variable phase") moves around and points to magnetic north, the second beam flashes in *every* direction at once. All aircraft within receiving range of the VOR, no matter where they're located, will "see" that North-identifying beam. Now when they receive the rotating beam some time later, they can calculate a magnetic direction to the station.

VOR Receiver

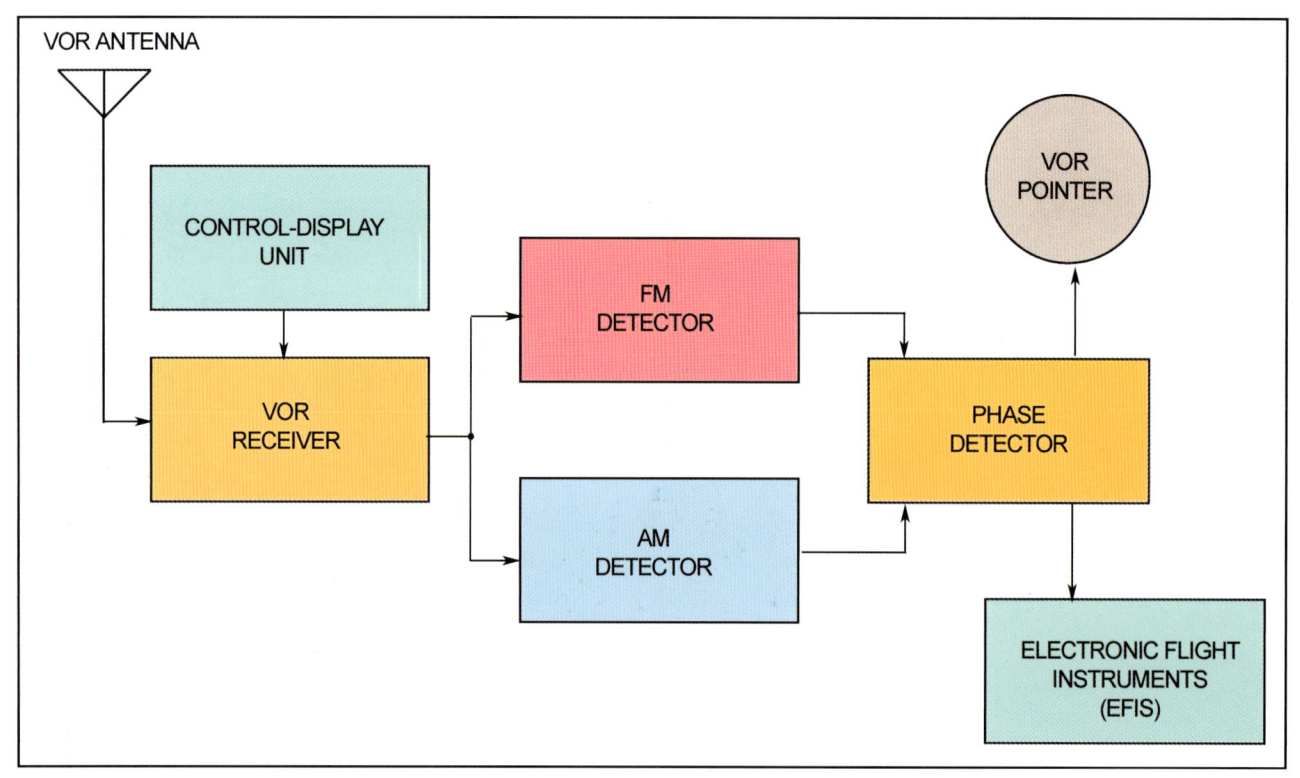

An airborne VOR receiver. Signals (from 108 - 117.95 MHz) enter the VOR antenna and are applied to the receiver. The receiver is tuned to a desired channel by the control-display unit. The FM and AM detectors process the two major signals transmitted by the VOR station; one on FM, the other on AM. The AM signals carry the "variable phase," the narrow beam which sweeps in a circle. The FM signal carries the "reference phase," which is broadcast in all directions. Each time the variable phase passes through north, the reference phase is at 0 degrees. The phase detector compares their phase and the difference is the number of degrees, or bearing from the VOR station. This information is displayed to the pilot on a VOR pointer or deviation bar on other instruments.

VOR Navigation

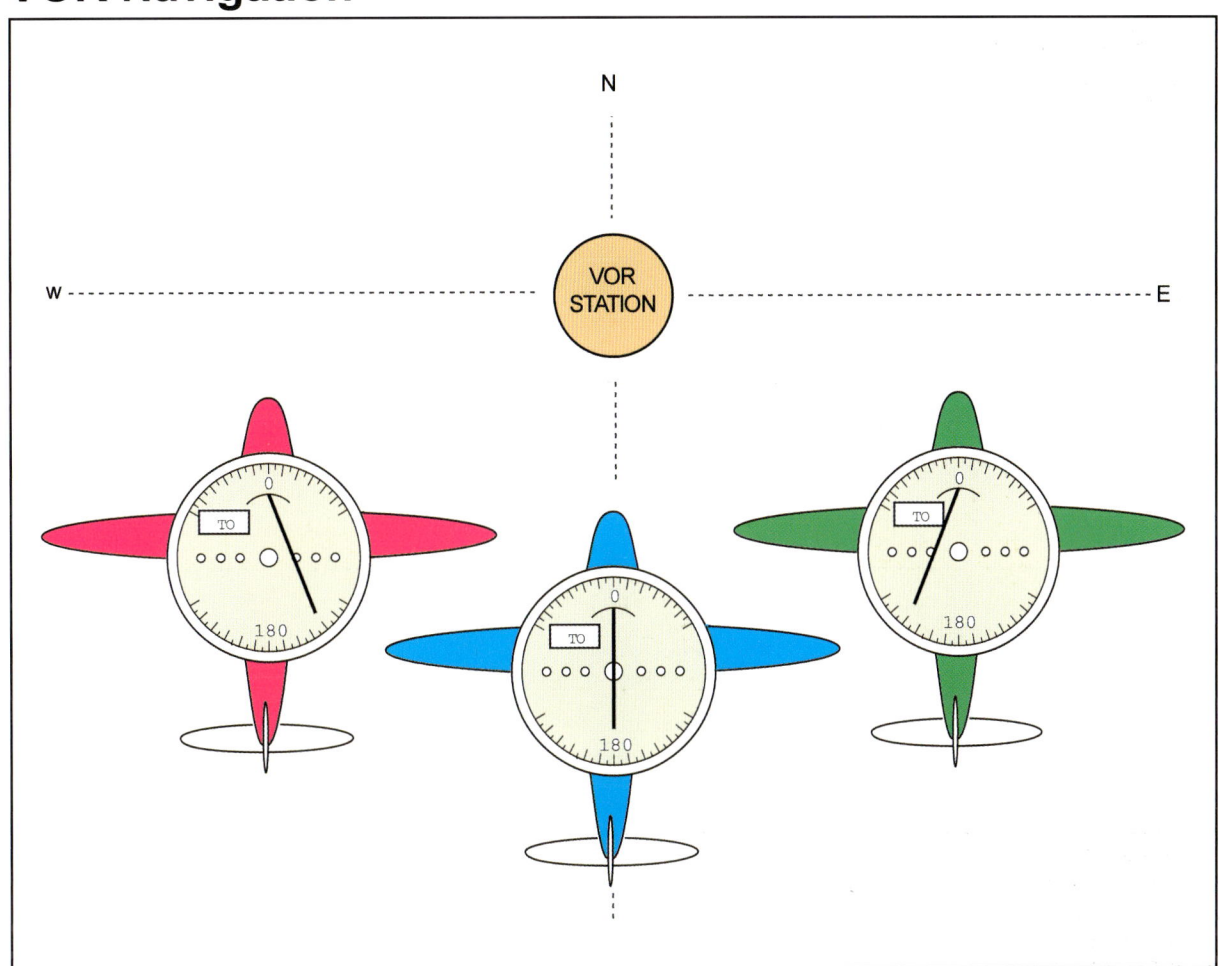

In VOR navigation, the pilot selects a desired course, in this example North (0 or 360 degrees). The airplane is south of the VOR station so the To-From flag (at upper left of display) indicates "To".

The airplane in the center is on course, so the needle is centered. The needles in the other airplanes show the direction to fly when the airplane is left or right of course.

These indications are not related to the heading of the airplane, as in Automatic Direction Finding (ADF).

There was early confusion over how to view the needle. Some pilots saw it as the airplane and steered toward the center circle----which is incorrect. The industry determined that, regardless of the instrument, the pilot should always "fly toward the needle" to get back on course.

When each of the airplanes crosses the East-West line, their To-From flags flip to "From."

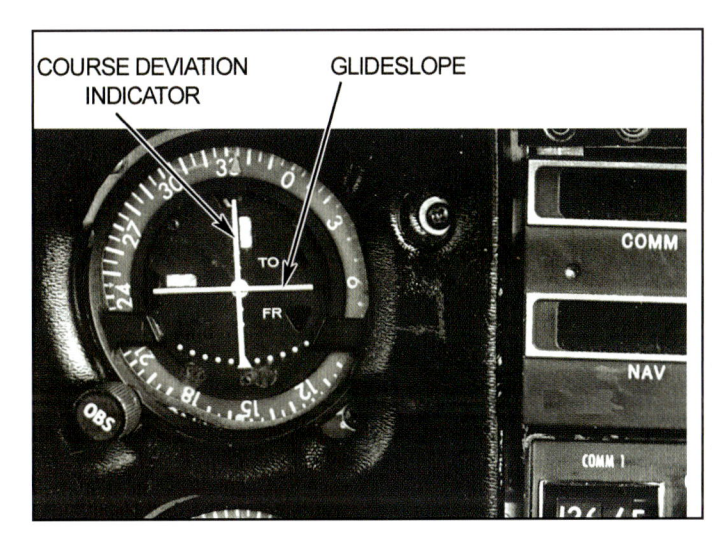

VOR indicator in a light aircraft. The course deviation indicator (CDI) gives left-right steering commands. Note "To" and "FR," which indicate whether the aircraft is flying to or from the station. For this to be correct, the course selected (shown here as 334 degrees) must generally agree with the course shown on a magnetic compass.

The VOR course is selected by the OBS knob at lower left (Omni Bearing Selector). The two white rectangles are flags which indicate if there is loss of reception.

The horizontal indicator, separate from the VOR system, is a glideslope needle used for ILS (Instrument Landing System).

63

VOR on Horizontal Situation Indicator (HSI)

AlliedSignal

VOR information is displayed to the pilot on an HSI (Horizontal Situation Indicator) found on large and high-performance aircraft. It gives a pictorial view of the airplane in relation to the VOR ground station. Note that VOR information is shown in green, for example; the pilot selected the No. 1 VOR receiver, shown at the left. He also adjusted the green course pointer to 20 degrees on the compass card, which is the desired course to the VOR station.

The airplane, however, is not yet on course to the station. This is shown by the green deviation bar split off from the course pointer. By turning the airplane to the right, the bar should move to the center and show the airplane on course to the station.

Radio Magnetic Indicator (RMI)

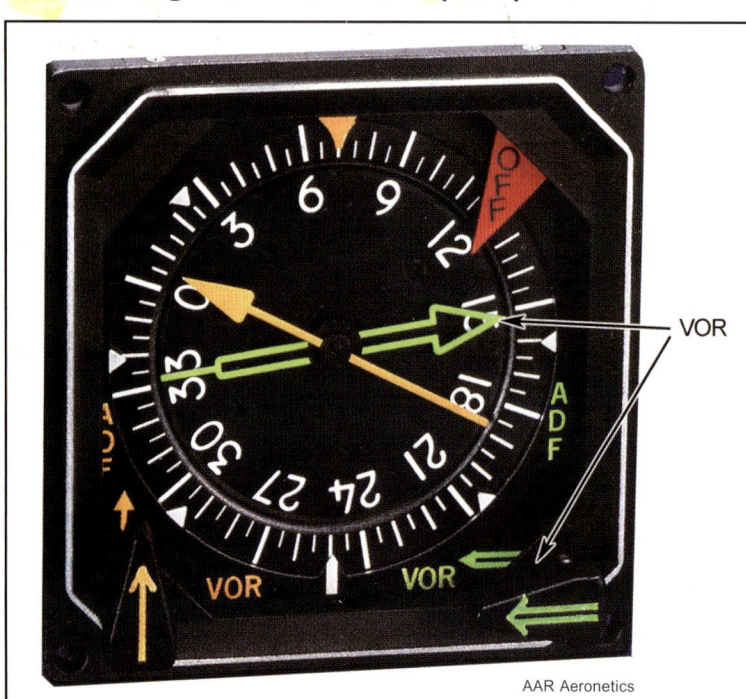

AAR Aeronetics

The RMI displays VOR and ADF information, or any combination of the two (VOR 1 and VOR 2 or ADF 1 and ADF 2). Displayed against a compass card, the needles simplify navigation by always pointing in the direction of the station (VOR or ADF).

In this illustration, the pilot turned the lower right knob (green arrow) to line up with "VOR" on the display. Now the green arrow will point to the station. (The orange needle is selected for ADF).

The more advanced Horizontal Situation Indicator at the top of the page has replaced many RMI's in aircraft but RMI's are often found as a backup to the HSI.

Nav Control-Display

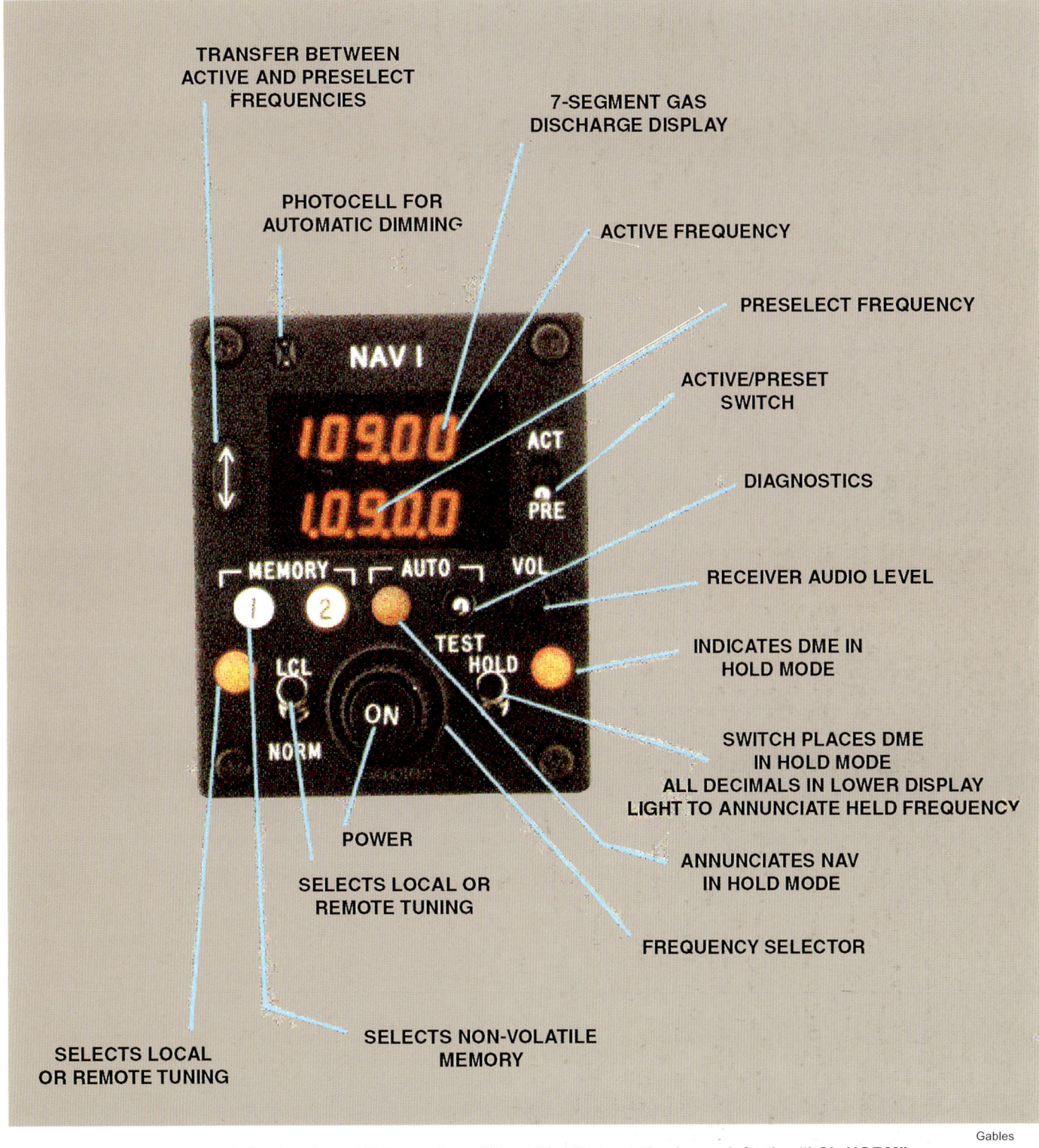

TRANSFER BETWEEN ACTIVE AND PRESELECT FREQUENCIES

7-SEGMENT GAS DISCHARGE DISPLAY

PHOTOCELL FOR AUTOMATIC DIMMING

ACTIVE FREQUENCY

PRESELECT FREQUENCY

ACTIVE/PRESET SWITCH

DIAGNOSTICS

RECEIVER AUDIO LEVEL

INDICATES DME IN HOLD MODE

SWITCH PLACES DME IN HOLD MODE ALL DECIMALS IN LOWER DISPLAY LIGHT TO ANNUNCIATE HELD FREQUENCY

ANNUNCIATES NAV IN HOLD MODE

FREQUENCY SELECTOR

POWER

SELECTS LOCAL OR REMOTE TUNING

SELECTS NON-VOLATILE MEMORY

SELECTS LOCAL OR REMOTE TUNING

Gables

A control-display for a VOR receiver ("Nav 1"). Note at the lower left, the "LCL-NORM" switch, for local and remote tuning. The Remote position allows the radio to be tuned automatically by a Flight Management System.

65

Review Questions
Chapter 9 VOR

9.1 What is the name of a combined VOR and Tacan navigational station?

9.2 What problem of early radionavigation did VOR overcome?

9.3 VOR waves travel _____

9.4 Name the two major components of a VOR signal

9.5 The reference phase broadcasts in what direction?

9.6 The variable phase rotates _____ times per second.

9.7 What happens when the variable phase moves through magnetic north (0 degrees)?

9.8 How does the VOR receiver know its bearing from the VOR station?

9.9 Besides fixed and variable phase signals, what other information is broadcast by a VOR station?

9.10 Why is it necessary to place the reference phase signal on an FM subcarrier?

9.11 What is the purpose of the course deviation indicator (CDI) on a VOR receiver?

Chapter 10

ILS
Instrument Landing System

The ILS is responsible for the ability of airliners and other aircraft to reach their destination more than 95 percent of time in bad weather. The system improves safety to such a degree that most airlines will not operate into airports without an ILS. In the business world, many corporations will not base their airplanes at airports without an ILS.

The ILS isn't only for bad weather. While descending into an airport at night at a brightly lighted city, pilots see a "black hole" where the runway surface should be. But a descent along an ILS glideslope clears all obstacles and brings the airplane safely within feet of the touchdown zone. That guidance is also needed on bright summer days when an airport is hidden in haze.

Another benefit of ILS is that it provides a "straight in" approach. As airplanes become heavier and faster, there is more danger in maneuvering close to the ground at low airspeed. A 70-ton airliner cannot nimbly bank and turn through the right angles of an airport traffic pattern. But flying the ILS, the airplane "stabilizes" on the approach 30 or 40 miles from the airport and flies straight "down the slot."

An array of antennas launches the localizer signal near the end of an ILS runway. The beams are pointed toward an airport runway at the right and reach out to about 50 miles.

Airport Systems Int'l

ILS System

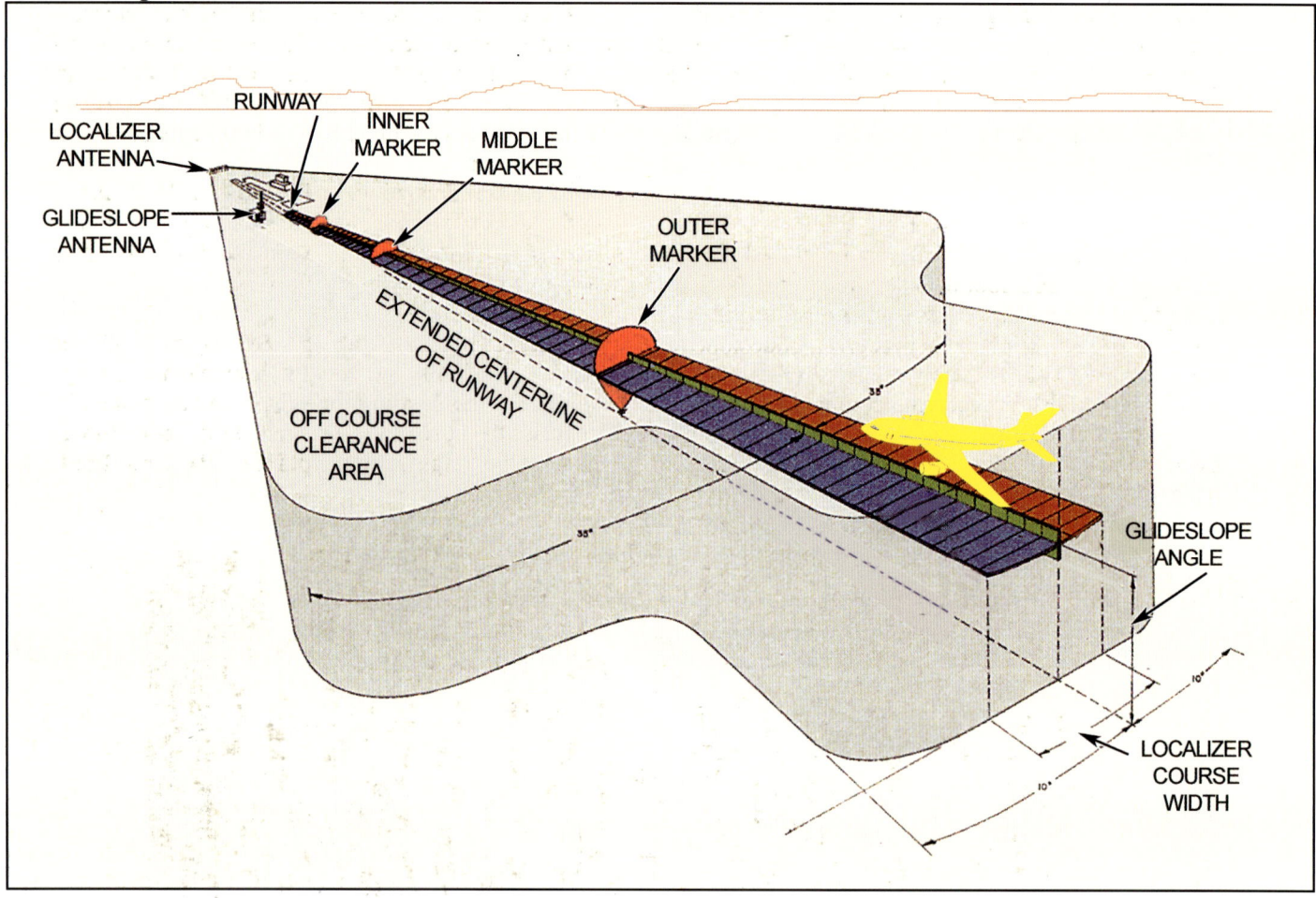

A three-dimensional path leads an airplane to the runway threshold at upper left. After intercepting the localizer (right), the airplane receives left-right guidance. At the outer marker the airplane begins a descent on the glideslope. At the middle marker the pilot decides whether there is sufficient visibility to land, or perform a missed approach.

ILS Components

The ILS consists of more than a half-dozen systems, both aboard the airplane and on the ground. Each ILS fits in a category, depending how low the airplane may fly---known as "minimums"--- before seeing the runway and deciding to land. Even a few dozen feet have great impact on airline operations. If the ceiling, for example, is 150 feet and ILS minimums require a descent no lower than 200 feet, the airplane may have to fly to an alternate airport, deal with hundreds of unhappy passengers, miss connecting flights and disrupt schedules over the country. Similar problems face the overnight express industry (Fed Ex, Airborne,etc.). But with sufficient investment in avionics, training, maintenance and ground facilities, airplanes are unable to land at their destination only four or five days a year! (In the US, this usually happens when a low pressure area with clouds, fog and rain cover the East Coast.

ILS Categories

Because avionics in the airplane and ground stations must be equal to the ILS to be flown, consider the major divisions. The categories are based on ceiling and visibility at the airport when the airplane arrives. For ILS operations they are known as "Decision Height" (DH) and RVR (Runway Visual Range).

Decision Height. When the airplane descends to decision height (shown on an instrument approach chart) the pilot must decide whether to continue and land, declare a missed approach or go to an alternate airport. To continue the approach, he must be in a position to land (without excessive maneuvering) and see the approach lights or other visual component on the surface.

RVR (Runway Visual Range). Visibility is usually estimated by a weather observer and stated in miles. But the person may be more than mile from where the aircraft touches down, and visibility changes drastically over short distances. To give the pilot an accurate report, visibility is measured electronically where the airplane touches down. It's done by a transmissometer, which sends a light beam between a transmitter and receiver. By measuring the loss of light (caused by haze or fog), it provides an RVR in feet. This informs the pilot whether the runway is below landing minimums for visibility. Here are the ILS categories:

Category I. By far the most common, there are approximately 400 Cat. I ILS airports in the U.S. The minimums are 200-foot decision height and RVR of 2400 feet. Typical equipment aboard the airplane to do this approach is a localizer receiver, glideslope receiver, marker beacon receiver and automatic direction finder (ADF) for receiving compass locators.

How the runway is lighted affects minimums. Lower visibility is allowed---an RVR of 1800 feet---if the runway has touchdown zone and centerline lights.

Distance Measuring Equipment (DME) is required for ILS procedures at some airports.

Category II. This ILS is installed at most international and large metropolitan airports. It brings decision height down to 100 feet and RVR to 1200 feet. In addition to avionics required for Cat. I, the airplane requires a second localizer and glideslope receiver and radar altimeter.

The last ILS category is divided into IIIa, b and c.

Category IIIa. Decision height drops to 100 feet and RVR to 700 feet. Additional avionics include an autopilot.

Category IIIb. Decision height drops to 50 feet and an RVR of 150 feet. A highly capable autopilot is required for this landing, one that can automatically flare (raise the nose of the airplane before touchdown) and decrab (straighten the airplane on the centerline).

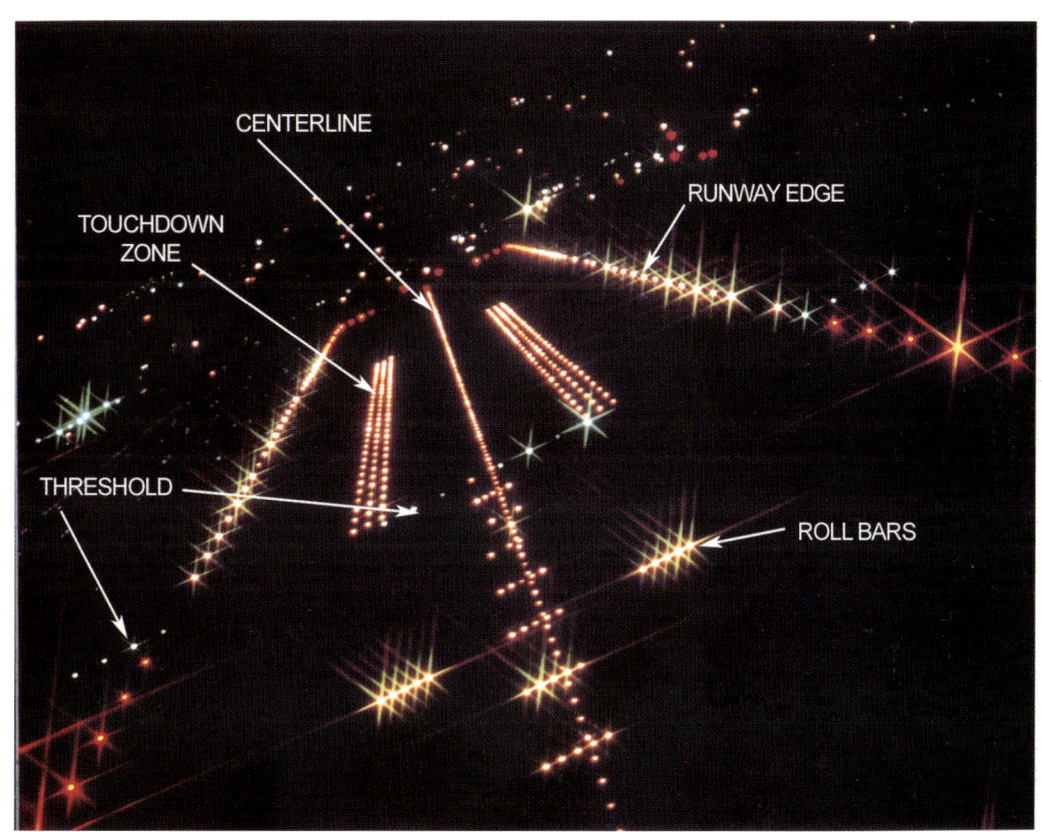

Besides avionics, an ILS requires an approach lighting system. The pilot must be able to see the runway environment and make a visual touchdown and roll-out. Visual references are supplied by approach lights and runway edge lights, a requirement for a Category I landing.

For the runway to support Category II ILS (with lower minimums), two systems are added; touchdown zone lights and the lighted centerline. Note the "roll bars," which serve as an artificial horizon during the last few seconds before landing.

Category IIIc, the fully "blind" landing, is where the pilot sees no lighting on the runway.

69

Flight Inspection and Monitoring

A flight inspection aircraft is over the runway checking accuracy of localizer beams formed by the antennas below. The beams are also monitored by nearby receivers, which sound an alarm if accuracy is lost. In advanced ILS systems, a defective localizer transmitter is switched off the air and a spare turned on.

The system is known as "autoland."

Although a Cat. IIIb landing is done in dense fog, there is just enough remaining visibility (150 feet) to roll out on the runway, then taxi to the terminal.

Category IIIc. This is the fully "blind" landing. Conditions are now "zero-zero" (for ceiling and visibility, meaning no decision height and no RVR). Even the weather report looks unusual; it reads "WOXOF," symbols that mean; "ceiling indefinite, sky obscured, visibility zero in fog." On this day, it is said, "Even the ducks are walking."

But let's assume the autopilot lands the airplane safely. Now the pilot has another problem; he cannot see to taxi to the terminal.

As this is written, there are no Cat. IIIc airports in the U.S. The fully "blind" landing, however, is not far in the future. Enhanced and synthetic vision guide a pilot without him seeing beyond the windscreen.

ILS Components

An ILS includes airborne and ground systems:

Localizer. From an antenna array on the ground, a localizer transmitter projects radio beams aligned with the centerline of the runway. The beams extend at least 18 miles out and to an altitude of 4500 feet. (Signals can be received much farther away, but are not guaranteed for navigation.) In the cockpit, the pilot is guided to the runway centerline with an indicator that shows "fly right" or "fly left," until the needle is centered.

Forty channels in the VHF band are set aside for localizers. Because they lie from 108.1 - 111.95 MHz, they fall within the tuning range of the VOR receiver. ILS, however, is only on *odd-tenths* of the frequency, for example; 108.1, 108.3, etc. Localizer signals, therefore, are processed through much of the VOR receiver, then split off to their own detectors. When the pilot selects a localizer frequency, the receiver automatically configures for localizer processing.

The localizer transmits an audio ID for the pilot to verify the correct station. There is a Morse code identifier (which always begins with the letter "I").

Glideslope. The glideslope provides vertical guidance by sending beams at a typical angle of 3 degrees (to match the glidepath of the approaching aircraft).

There are no pilot controls for the glideslope receiver or audio ID. When a localizer is selected, the correct glideslope frequency is automatically channeled. There are 40 glideslope frequencies, each paired with a localizer. Operating in the UHF band, glideslope frequencies extend from 329.15 MHz to 335 MHz. The pilot knows the glideslope is operating by movements of the horizontal needle on the display or a no-signal warning from an indicator flag.

Marker beacon. Lying along an extended centerline of the runway, marker beacons give the pilot visual and

ILS: Two Audio Tones on a Carrier

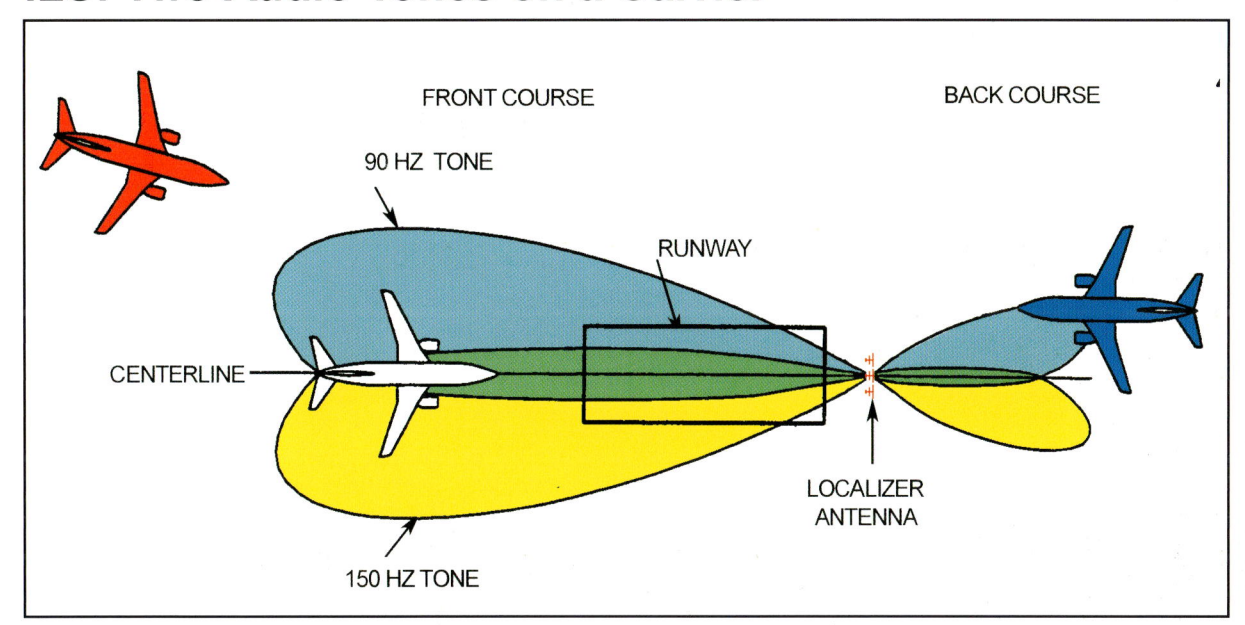

The localizer antenna focuses the radio carrier into two narrow lobes, shown here as blue and yellow. The yellow lobe is modulated with a 150 Hz tone; the blue lobe by a 90 Hz tone. If the airplane flies along the center of the overlapping (green) area, the pilot sees a centered ILS pointer.

An ILS receiver does not measure the difference in *strength* of the two radio lobes. Rather, the receiver compares the *difference* in strength between the two audio tones. This is known as "DDM," for difference in depth of modulation.

Many localizer antennas also launch signals off their back end (to the right in this illustration) and form a "back course." This can be used for limited guidance, but in simple localizer receivers, the needle indications are reversed; the pilots flies *away* from the needle to get back on course (known as "reverse sensing"). Because this is confusing to the pilot, most localizer receivers have a back course switch ("BC") to keep the same sensing as on the front course.

Back courses are present at all localizers, but should never be flown unless there is a published procedure. Also, there is no glideslope with a back course approach.

Many localizer displays use the blue and yellow colors shown in the above illustration. The trend, however, is not to use these colors on an instrument because they don't provide useful information. The needle provides all the guidance.

Localizer Indications

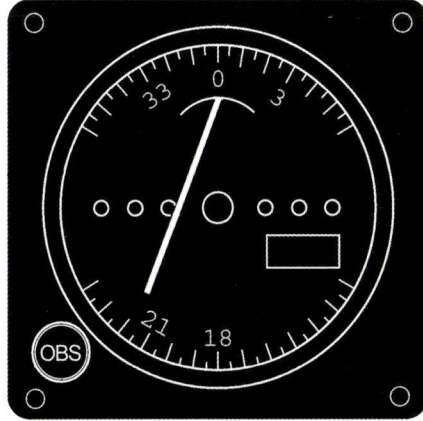

The localizer needle indicates "fly left" to intercept the centerline of the localizer course. When tuned to a localizer frequency, the OBS (omnibearing selector) is disabled. Most pilots, however, set it to the localizer course as a reminder. This example is Runway 36 (the same as 0 or 360 degrees).

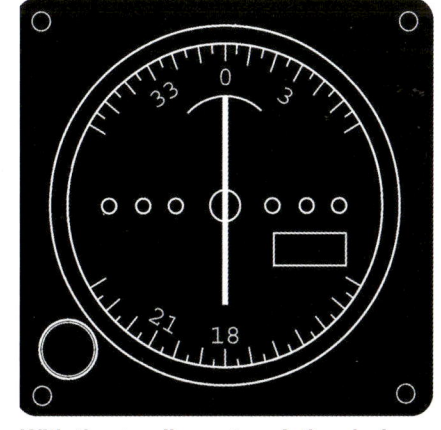

With the needle centered, the airplane is on an extended centerline of the runway. The same indicator is used for VOR navigation, but when a localizer frequency is selected, the needle become four times more sensitive. This achieves the higher accuracy required for an ILS approach.

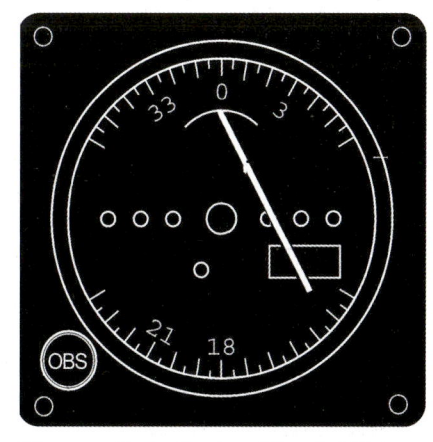

The needle indicates "fly right" to get on the localizer course.

The overall width of a localizer course is usually 5 degrees. Thus, a needle deflecting full right or full left indicates the airplane is 2.5 degrees off the centerline.

Glideslope Guidance

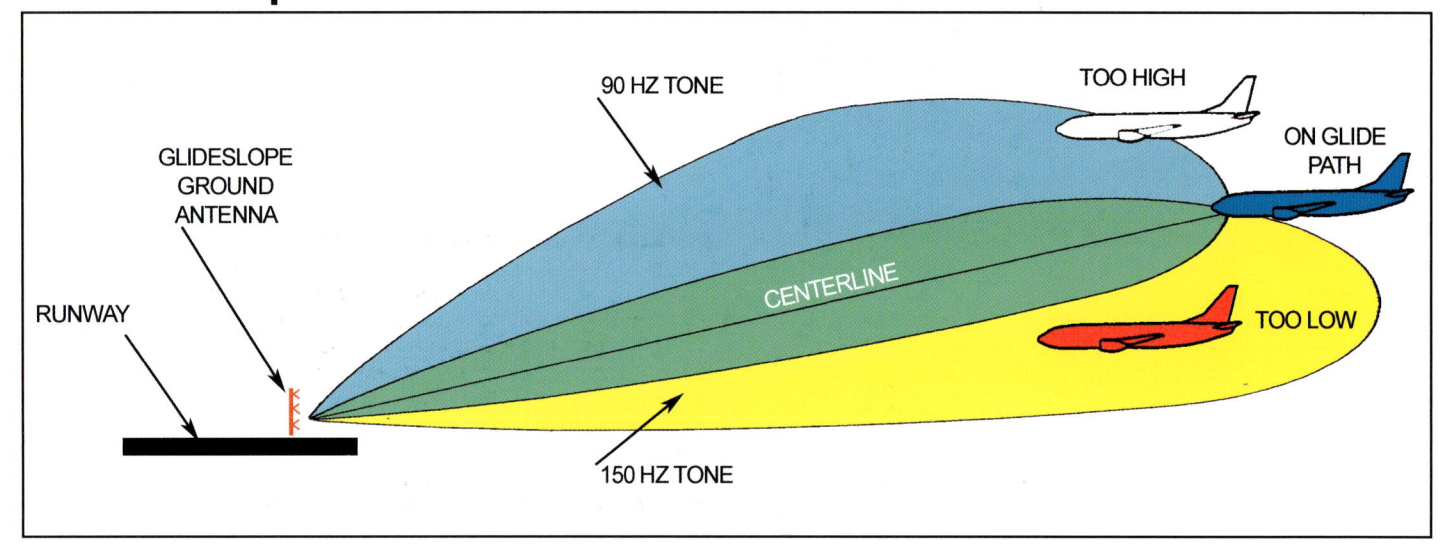

The glideslope is formed by a radio carrier aimed upward into the glidepath. The lobe seen in blue is modulated by a 90 Hz tone; the yellow lobe modulated by a 150 Hz tone. When the glideslope receiver in the airplane receives equal signal (green area) it is on the glideslope. Most glideslopes are designed for glide angle of 3 degrees.

Glideslope Indications

Horizontal glideslope needle is below center of instrument, commanding pilot to "fly down" to intercept the glideslope.

Centered glideslope needle shows the airplane is descending on the correct vertical approach path to the runway.

The high position of the horizontal glideslope indicator is telling the pilot to "fly up" to intercept the glidepath. This is a dangerous situation because the airplane is too close to the ground.

audible cues on his distance to the airport over the last 4 to 7 miles. They are located to mark important phases of the approach, such as glideslope intercept, decision height and when to begin a missed approach.

Compass Locator. Some ILS approaches have a compass locator at the outer marker. This is a low power station picked up by an ADF (automatic direction finder receiver) to guide the airplane, arriving from any direction, to the outer marker. Its operation is described in the chapter on ADF.

Glideslope Station

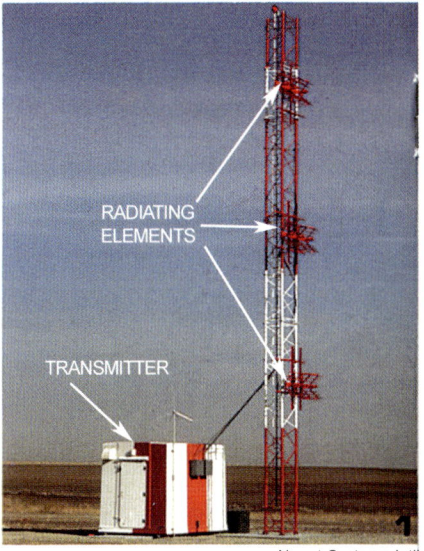

A glideslope antenna array, found alongside runways with an ILS, radiates signals that angle upward. When the airplane is about 5 miles away and 1000 feet above ground, the pilot intercepts the glideslope beam and flies it down toward the runway. Glideslope frequencies, which lie between 328 and 336 MHz are not selected by the pilot; they are automatically tuned when he selects the localizer frequency for that ILS approach. The glideslope receiver is "channeled" by the localizer receiver.

Airport Systems Int'l

Glideslope Receiver

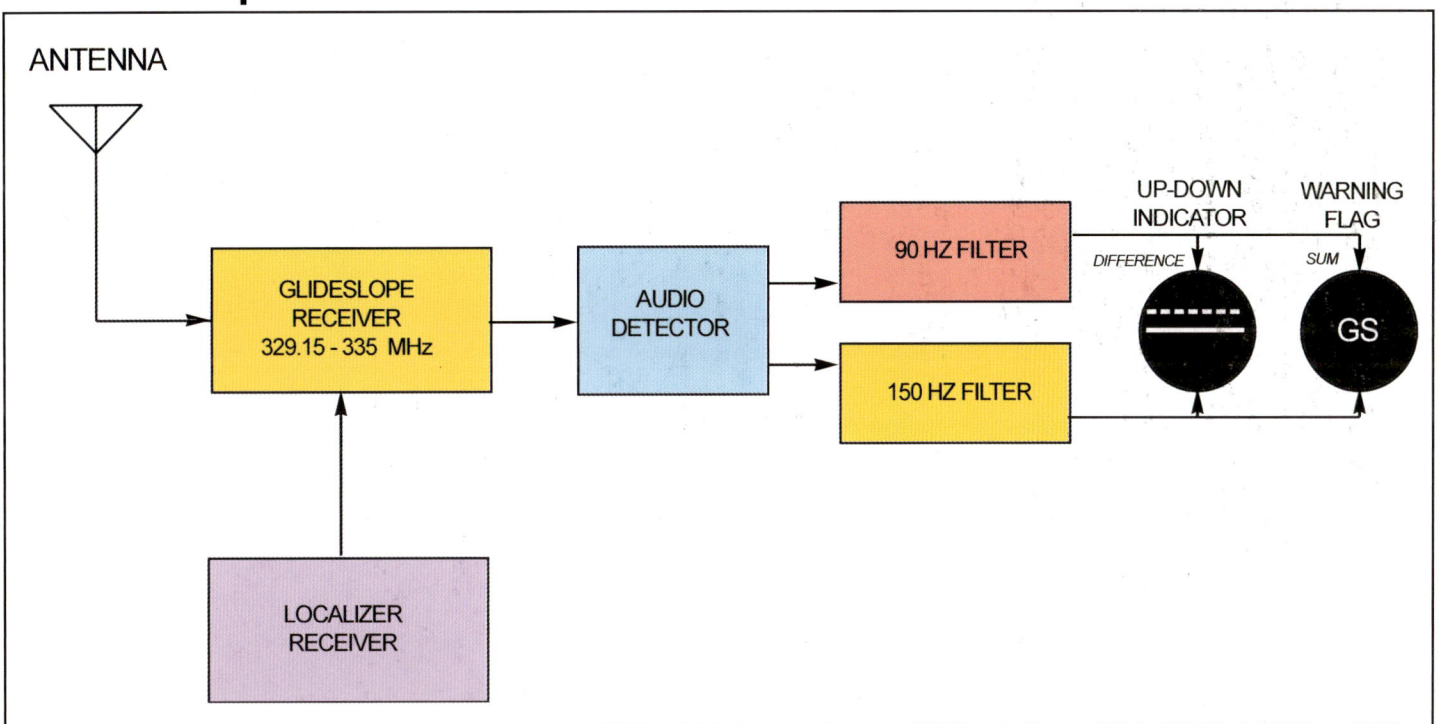

Up-down information is sent as two audio tones—90 and 150 Hz—which are recovered from the radio carrier by the audio detector. Two filters separate the signals.

The difference between the signals is presented to the up-down indicator. If the airplane is on the glideslope, equal amounts of 90 and 150 Hz are applied to the indicator, but out of phase (opposite polarity). The two tones cancel each other and the needle remains centered. This is correct for an airplane on the glideslope. If the airplane rises higher, however, the 90 Hz signal grows stronger and moves the needle above the center position—telling the pilot to "fly down."

An important part of the system is the warning flag. If a malfunction causes loss of signal, the needle returns to the center position. This could be dangerous because the pilot might believe he is precisely on the glideslope. This is avoided by the warning flag. It receives the *sum* of the two signals; if they are not sufficiently strong for navigation, the flag appears. The warning usually shows the letters "GS" or a barber pole symbol (red stripes).

73

Marker Beacon Receiver

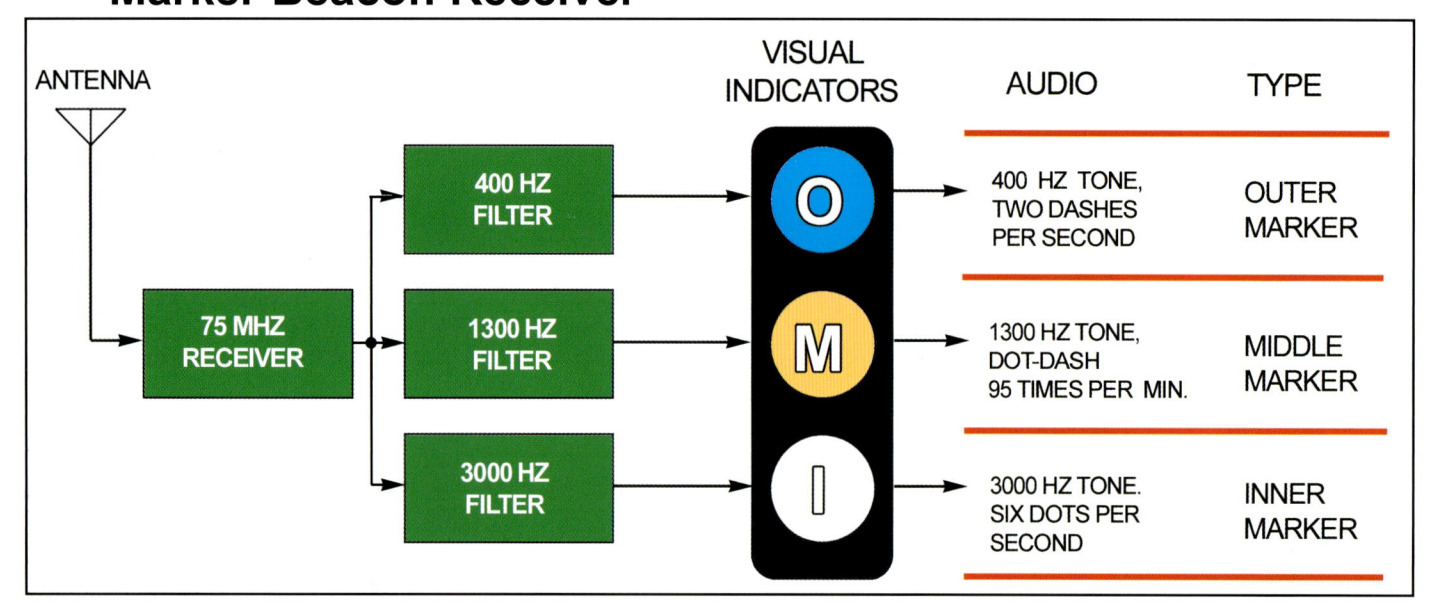

Visual Indicator	Audio	Type
O (blue)	400 HZ TONE, TWO DASHES PER SECOND	OUTER MARKER
M (amber)	1300 HZ TONE, DOT-DASH 95 TIMES PER MIN.	MIDDLE MARKER
I (white)	3000 HZ TONE, SIX DOTS PER SECOND	INNER MARKER

Signal path: Antenna → 75 MHZ RECEIVER → 400 HZ FILTER / 1300 HZ FILTER / 3000 HZ FILTER → Visual Indicators and Audio.

The marker beacon receiver is fixed-tuned to 75 MHz, the carrier frequency for all marker ground stations. The pilot identifies the marker by viewing the 3-light indicator (blue, amber and white) and listening for an identifying tone.

The audio rises in pitch and sounds faster as the airplane passes each marker and is closer to the runway.

Most ILS's have only two markers; Outer (blue light) and middle (amber light). The Inner marker is for Category II systems, which are at few airports.

On older aircraft, the white inner marker indicator may show the letter "A" instead of "I." "A" is for "airways," once used for cross-country navigation, but no longer needed because of numerous VOR stations. Its position is now occupied by the inner marker.

Marker Beacon Ground Stations

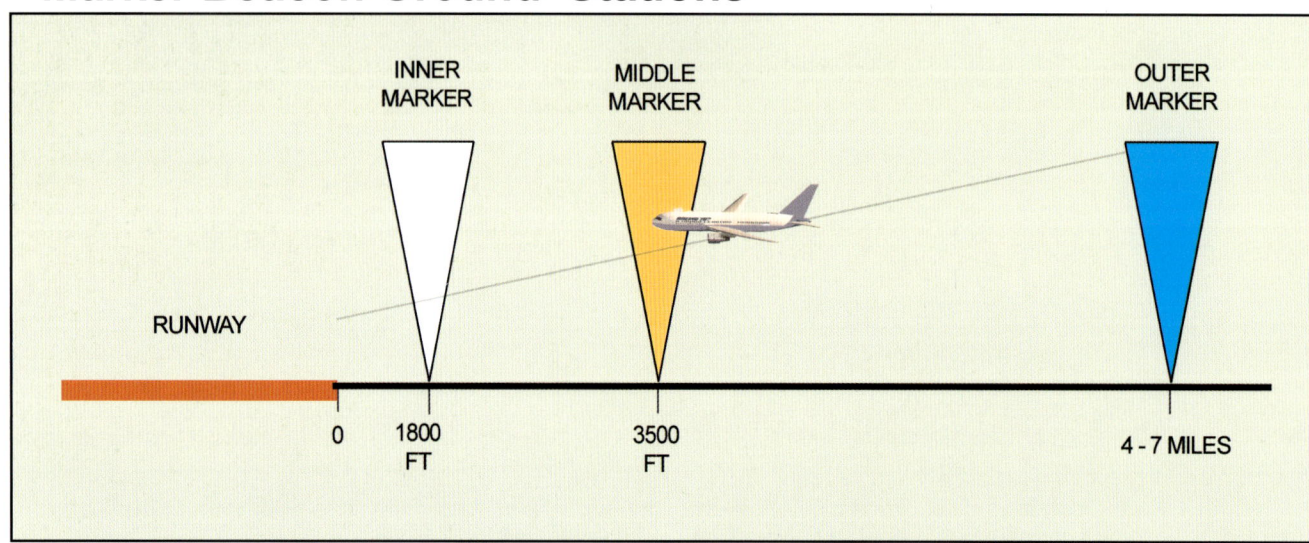

Ground station layout from runway: INNER MARKER at 1800 FT, MIDDLE MARKER at 3500 FT, OUTER MARKER at 4-7 MILES.

The three ground stations--- inner, middle and outer markers---broadcast on 75 MHz, but with different tone codes. Distances of the stations from the runway vary according to the airport location, but typical distances are shown in the diagram.

Markers broadcast very low power (about two watts) to keep their radiation close in and to limit their radiation to a small area. During an ILS approach, the airplane intercepts the glideslope signal at the outer marker. When arriving at the middle marker, this is usually the decision height for a Category I landing.

Another function of the outer marker is to prevent the pilot from flying down a false glideslope. All glideslopes produce false signals above and below the correct one. The pilot knows he is on the correct one by checking on the approach chart for correct altitude when marker tone and light are received.

Review Questions
Chapter 10 ILS

10.1 Name three markers along an ILS.

10.2 How is RVR (Runway Visual Range) measured on an ILS runway?

10.3 Name the categories of ILS.

10.4 What component of an ILS provides an extended centerline to the runway?

10.5 Name the ILS component that provides vertical guidance to a runway.

10.6 How many channels are allocated to localizers?

10.7 The localizer frequency is selected on the _____ receiver.

10.8 The frequencies of the two audio tones that provide left-right guidance oa localizer are _____ and ____.

10.9 When a localizer frequency is selected on the VOR receiver, the indicator needle becomes _____ times more sensitive than for VOR navigation.

10.10. The frequencies of the two audio tones that provide up-down guidance on a glideslope are _____ and _____.

10.11 The compass locator of an ILS is received on the _____.

10.12 How are glideslope receiver frequencies selected?

10.13 What is the frequency for all marker receivers?

Supersede by MMR = Multimode Receiver = (ILS – MLS – GLS or GPS)

Chapter 11

MLS
Microwave Landing System

Space Shuttle: Early MLS User

One of the first users of MLS was the Space Shuttle orbiter. Accuracy is important because approach and landing are "deadstick"---there is no engine power for the 300,000-pound "glider" to try a second time.

If necessary, the MLS system can perform an "autoland," bringing the orbiter to touchdown and roll out on the runway. The crew may also fly manually with reference to instruments, guided by MLS signals. The approach begins at 18,000 feet and 10 miles from touchdown.

The orbiter carries three independent MLS receivers whose output is continuously compared and averaged. Distance to the runway is provided by precision-DME near the touchdown point.

MLS allows the orbiter to land in either direction on the runway.

For over a half-century, ILS proved the most dependable system for landing airplanes in bad weather. But by 1960 there were signs that ILS could not keep up with a growing aviation industry.

Few channels. ILS has only 40 frequencies in the VHF band, with almost no chance for expansion. There are simply not enough frequencies to satisfy growth at large metropolitan and international airports.

Limited Capacity. An ILS serves one runway with a single course. When weather is bad, en route traffic headed for the ILS is strung out hundreds of miles, forcing each airplane to wait its turn for the approach.

Interference. The rise of powerful FM broadcast stations during the 1980's further threatened ILS. The FM band ends at 108 MHz, just under the beginning of the ILS band. One early complaint came from an Air Force pilot flying an ILS and monitoring the audio ID; "I hear music" he reported to the controller. An FM radio program was breaking into his ILS receiver. As more FM stations went on the air, it forced major design changes in ILS receivers to harden them against interference.

Terrain Problems. Installing an ILS at an airport is not simple. For ground antennas to function, they need a wide area clear of obstructions. ILS signals reflect and cause "multipath" error. (It's been said the cost of moving earth for ILS construction can cost more than the ILS equipment itself.) Some airports in moun-

MLS Azimuth Beam Sweeps Wide Area

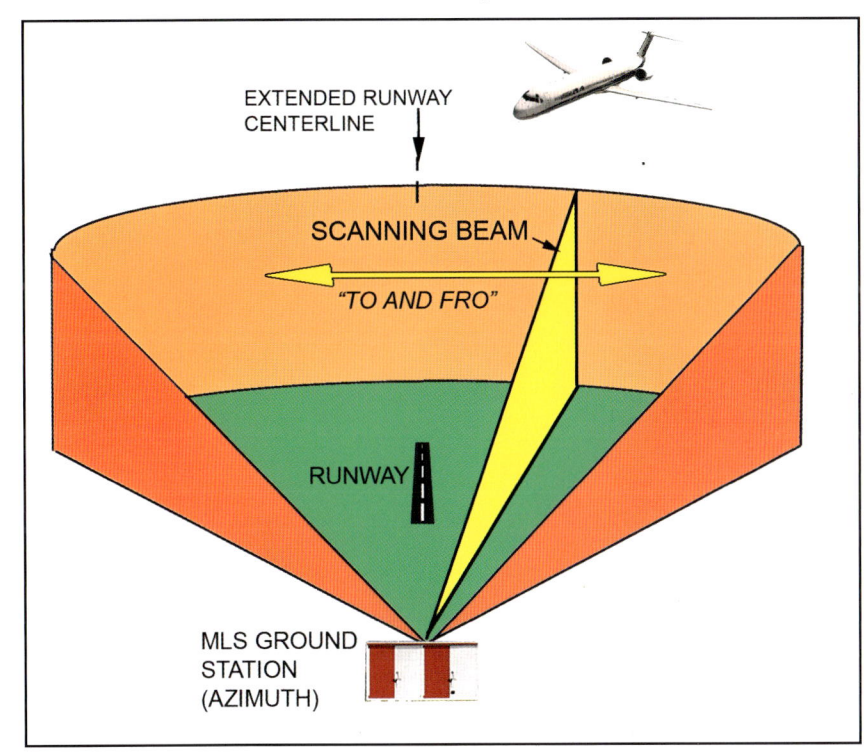

A narrow scanning beam from the MLS station sweeps back and forth beyond either side of the runway. An arriving aircraft picks up the sweeps ---"To" and a "Fro"--- and determines the time difference between them. From this information, the centerline of the runway is computed.

The approach to the runway does not have to be straight in. If the MLS receiver has flight management capability, a curved path may be computed and flown according to MLS guidance.

tainous regions can never have an ILS, which limits emergency medical aircraft, flights into ski resorts and other services. ILS cannot create the steep descent angles needed by a helicopter landing inside a city.

Military Operations. During the 1950's military services sought a new instrument approach system to fill their special needs. For tactical reasons, they relocate to new areas, clear the ground for a runway and quickly begin air operations. Air traffic must operate in and out of these remote fields under all weather and lighting conditions. It called for a new landing system that would fit in few portable cases, be flown in, set up, ready to land airplanes in about 15 minutes---and do it in all weather.

ILS fell far short of the goal. Its half-century-old technology required acres of open land and much signal tweaking to get the courses correct. ILS was also unsuitable for landing airplanes on aircraft carriers. Responding to military requirements, the avionics industry came up with MLS, the Microwave Landing System.

Shorter Waves. The first benefit of MLS arises from its wavelength. ILS frequencies have full wavelengths of about 8 feet. Much higher in the microwave region, signals are only 2.5 *inches* long (full wavelength). Not only are microwave antennas smaller, they are easier to form into a narrow beam and steered electronically.

As shown in the illustrations, the operating prin-

Transmitting MLS Azimuth Signal

One of two major components of MLS is a ground station that transmits the "azimuth" signal. This is similar to the localizer of an ILS, but MLS sweeps a wide area beyond left and right sides of the runway. This provides many inbound courses to the runway.

The station is about 400 feet beyond the far (or stop) end of the runway, as seen by an arriving aircraft. The airplane on the runway is flight inspecting to check accuracy of the MLS signal.

The white tower at the left is a field monitor. From its position between the station and runway, the monitor continuously samples the MLS signal and warns of a malfunction.

77

MLS Elevation Beam Scans Vertically

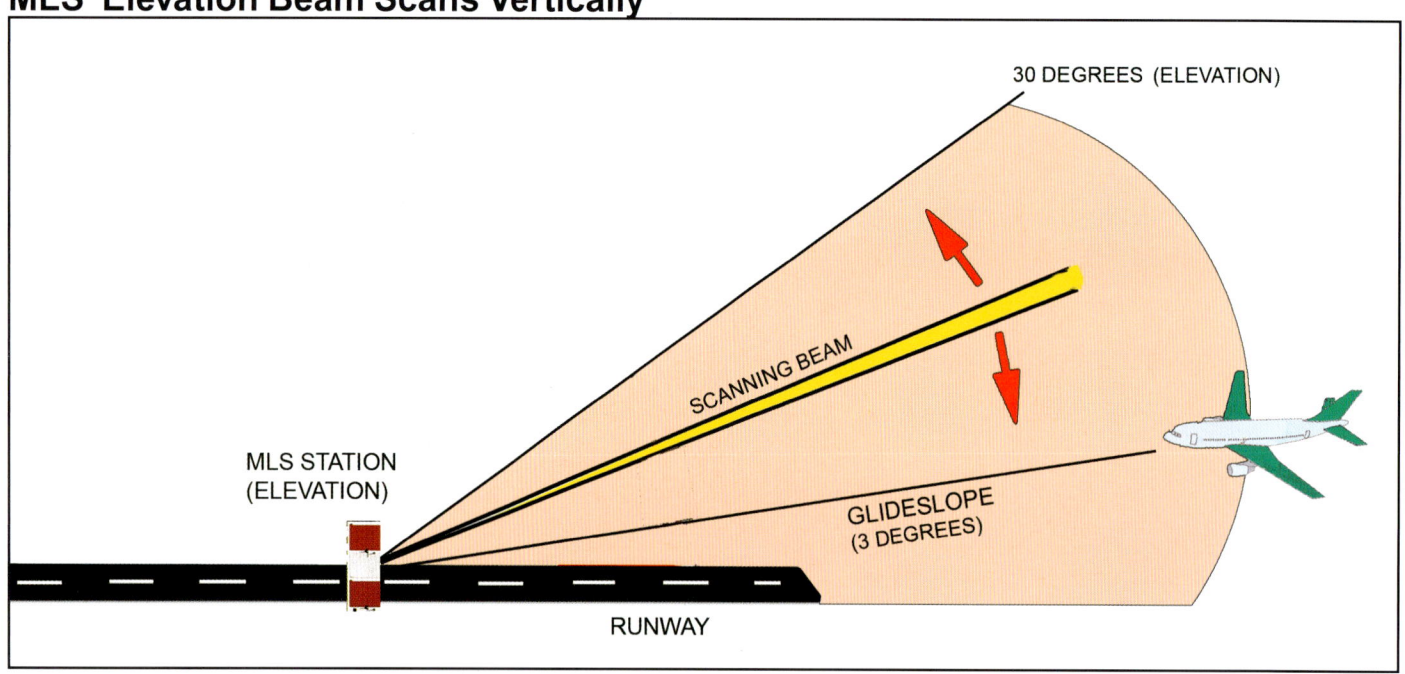

Scanning up and down, the MLS elevation beam is similar to an ILS glideslope. MLS, however, sweeps a greater area, creating many selectable glidepath angles. The arriving aircraft seen here chose a 3-degree glideslope, a typical approach angle for fixed-wing aircraft. A helicopter, on the other hand, may choose a steeper glidepath in order to land on a rooftop in a city.

The MLS elevation station is located alongside the runway, near where aircraft touch down (about 400 feet from the approach end).

ciple is simple. The MLS station beams a signal that swings back and forth. An airplane receives the beam as the signal travels in the "To" direction, then again when the beam returns in the "Fro" direction. By measuring the time in between, the receiver calculates where the airplane is located with respect to the centerline of the runway. This is the "azimuth" function of MLS and is equivalent to the localizer of an ILS.

The same technique obtains the glideslope. Here, the MLS beams a signal up and down the approach path, and the receiver calculates a vertical glideslope.

Curved Approaches. Unlike ILS, the MLS signal provides three-dimensional navigation and new types of approaches. Those dimensions are elevation (glideslope), azimuth (localizer) and range, or distance, from the runway with precision DME. These dimensions are available over a wide volume, enabling the aircraft to arrive from many directions. It also provides all the data required for an onboard computer to create a curved or segmented (stepped) approach, which increases traffic capacity at an airport.

More Channels. In the microwave region, 200 channels were set aside by international agreement for MLS---which is five times as many as the 40 assigned to ILS. Relief from growing frequency congestion seemed at hand. Microwaves, too, are removed from interference of megawatt FM broadcast stations.

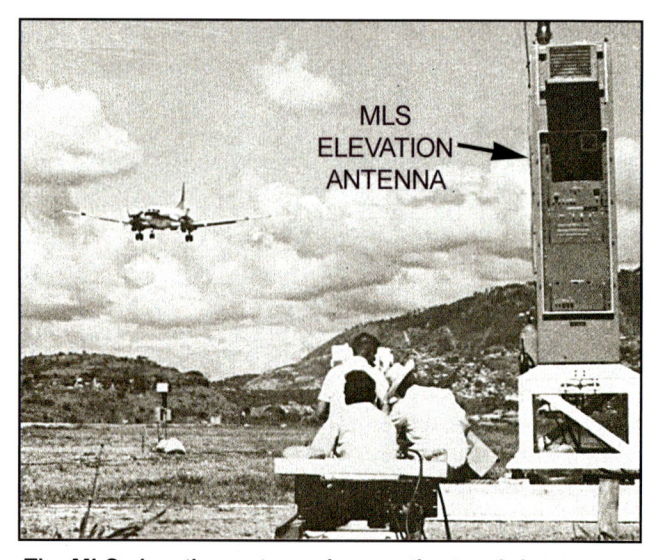

The MLS elevation antenna is near the touchdown zone of the runway. Also located here is P-DME, precision distance measuring equipment that accompanies an MLS installation. P-DME provides range to touch down and is ten times more accurate than conventional DME.

Another advantage of MLS is a pilot-selectable glideslope. A pilot may choose the glidepath to match his airplane performance.

The pilot is protected against selecting a glidepath that could lead to a collision with a mountain. Part of the transmitted signal contains a digital message which prevents the receiver from following a dangerous path.

78

MLS Time Reference

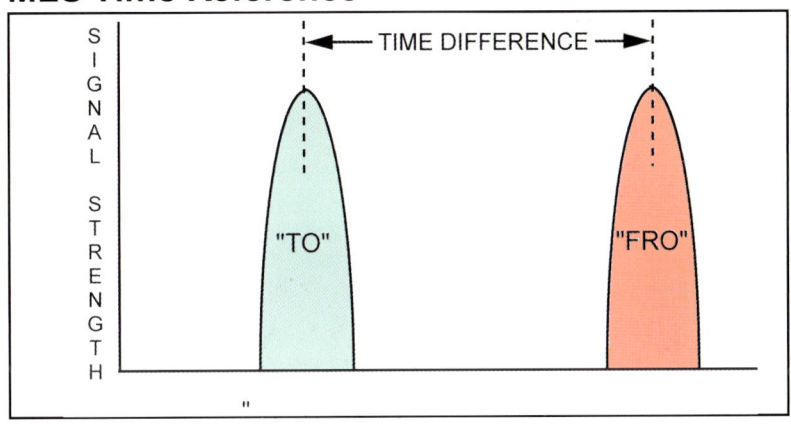

MLS signals arriving at the airplane produce two peaks as the beam sweeps back and forth over the receiver antenna. The centerline of the runway is determined by computing the elapsed time between the pulses. This method is used for both azimuth (AZ, or runway direction) and EL (elevation or glidepath).

The receiver knows whether it's receiving AZ or EL because each pulse is preceded by a short identification known as a "preamble."

The azimuth signal sweeps at the rate of 13.5 scans per second. The elevation signal scans 40.5 times per second.

Why the difference? A loaded Boeing-747 weighs 400 tons. If it lands six feet left or right of the runway centerline, it's probably not serious. But if the airplane flares 10 feet too high, the airplane may stop flying and drops to the runway for a hard landing. To avoid that, MLS updates the elevation (glidepath) signal three times faster than the azimuth for greater accuracy.

Preparing the Site. Microwave signals are less affected by terrain and nearby buildings. MLS ground stations do not need a large localizer array or tower for glideslope antennas. In one demonstration, a manufacturer loaded an MLS system aboard an airplane and flew it to an airport several thousand miles away. On arrival, the station was erected and instrument landing demonstrations followed almost immediately.

World Standard. That demonstration happened when the aviation world was deciding which of several MLS systems to approve as the international standard. The winner was TRSB, for "time referenced scanning beam," using the To and Fro method described earlier.

MLS was hailed as the replacement for ILS. Optimism ran high and, in 1978, the International Civil Aviation Organization adopted MLS as the new world standard. ILS would be phased out after a transition period.

Enthusiasm for MLS was so great that states like Michigan and Alaska did not wait for the government to install MLS at local airports and bought their own. A town in Colorado purchased an MLS as the only practical system to fly in skiers during bad weather. MLS would be the answer to precision approaches almost anywhere. At the same time, a new system appeared that changed the future of avionics.

Satnav. While the world awaited final approval of MLS, the US Department of Defense was examining a navigation system that had nothing to do with airplanes. The Department was seeking a method to *rendez-vous* troops in the field---that is, troops finding each other in unfamiliar territory. The researchers envisioned a device to pick up satellite signals and provide guidance to down to 30 meters or better.

An important requirement was the receiver. It had to be lightweight, inexpensive and accurate. In a system called Navstar (Navigation with Timing and Ranging), they achieved those goals only today it's called GPS, the Global Positioning System.

MLS Survives. As GPS proved its value, aviation authorities dropped plans to replace ILS with MLS. Airline operators flying into congested airports, however, couldn't wait for a changeover to GPS, which could take ten years. As a result, MLS has enjoyed a limited revival, with installations in Europe, where high-volume traffic operates into international airports. There are MLS systems in the United Kingdom, Holland, Germany and France. Other countries in or near Europe are also planning to purchase MLS. On the military side, MLS continues to prove its worth as a tactical landing aid.

Multimode Receiver. Because airlines outside Europe must be equipped to land at any international airport, a new type of avionics appeared. It's the MMR, for "Multimode Receiver." It's a single radio that operates on ILS, MLS and GLS (for GPS Landing System). Once the radio is tuned to one of these services, the pilot sees the same guidance on the instrument panel.

Review Questions
Chapter 11 MLS

11.1 What were reasons for approving the Microwave Landing System?

11.2 MLS creates inbound courses to runways by a scanning beam which moves _____. This is called the _____ signal.

11.3 Glidepaths are creating by a scanning beam which moves _____.
This is called the _____ signal.

11.4 How does an MLS scanning beam determine the centerline of a runway?

11.5 Why are there so few MLS installations at airports?

11.6 What type of receiver can use ILS, MLS and GPS signals?

Chapter 12

ADF
Automatic Direction Finder

The Automatic Direction Finder is one of the few air navigation systems that still operates in the low end of the radio frequency spectrum. The ground station, known as an NDB (non-directional radiobeacon) transmits from 190 to 1750 kHz, which spans the Low and Medium Frequency Bands. Despite many shortcomings, ADF is still an important component in instrument operations and will not soon be taken off the air.

ADF was a great step forward when Bill Lear (known for the LearJet) manufactured the first automatic direction finders during World War II. Before that time, a pilot or navigator turned an antenna loop by hand to find a bearing between the airplane and ground station. It's the same effect you hear on a portable AM radio; if the radio is rotated, there is a point where the station fades. You also notice that during one full rotation, there are *two* directions where the signal fades---which are opposite each other. The reason is, an antenna loop can only indicate a *line of position;* it cannot tell on which side of the station the airplane is located. This is known as an "ambiguity" and must be eliminated in order to fly toward the station.

Early navigators solved ambiguity by drawing lines of position from two different stations and fixing their position where the lines intersect. These systems, known

ADF on Fixed Compass Card

The simplest display is a fixed compass card and ADF pointer. To fly to the station, the pilot turns right and brings the pointer straight up. The fixed compass card is of limited value, only telling the pilot the relative (not actual) bearing to the station.

81

Radio Magnetic Indicator

A major improvement is the RMI, for Radio Magnetic Indicator. The compass card is coupled to a heading reference (such as a horizontal gyro), so the top of the card is always the actual heading of the airplane.

The RMI has two pointers, enabling the pilot to select any combination of VOR and ADF stations. In this illustration, selected are: yellow pointer for ADF, green pointer for VOR.

as MDF (manual direction finder) or RDF (radio direction finder) were crude and time-consuming for obtaining position fixes.

Lear's design made direction-finding automatic by recognizing that a radio wave consists of *two* components; an electric and magnetic wave. Radio waves, in fact, are known as "electromagnetic" energy. Before Lear, direction finders worked only on the magnetic portion of the wave, which is picked up by a loop-shaped antenna. If the loop is held with its wide, or open, side toward the station, the waves hit left and right sides equally. This generates little signal in the antenna because *equal* voltages occur on both the sides of the loop. There is no voltage difference, which is required to drive signal current down to the receiver. When the loop faces this way, its low-signal condition is known as a "null."

Next, turn the loop so one edge faces the station. Now the signal first strikes the forward part of the loop, then hits the back part. This causes a voltage difference in the loop and current flows to the receiver. This condition (high signal strength) is called the "peak."

The reason for the voltage difference is that a radio wave is rapidly changing, or alternating. In the distance between front and back parts of the loop, the wave moves through a different phase of its cycle, resulting in unequal voltages in the loop.

Sense

The ability of a direction-finding receiver to know which side of the station it's on uses the electric portion of the wave. To pick it up, a second antenna, called the "sense" antenna is added to the receiver. Unlike a loop, it picks the electric portion of the signal from all

Analog ADF

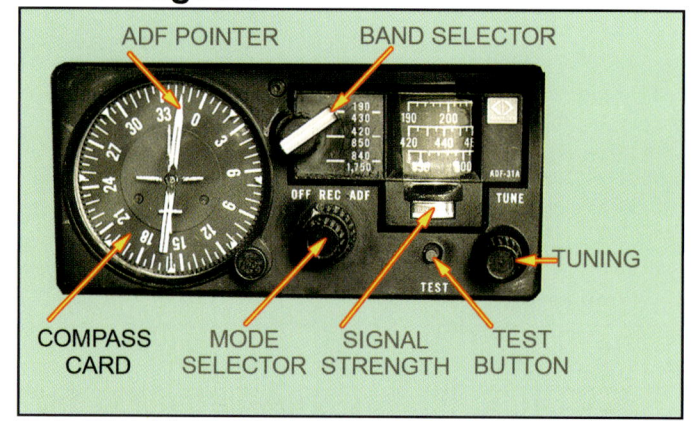

Early analog ADF receivers are still aboard many aircraft. The major items on the panel

ADF Pointer: When a station is tuned, the indicator points directly to the station, regardless of aircraft heading or flight path.

Compass Card: In this simple radio, the compass card does not move unless turned by the pilot.

Band Selector: Most aviation beacons are in the first band (190 - 430 kHz), and some fall in the second band (420 - 850 kHz). Most of the second band is for standard AM broadcast, which also occupies most of the third band (840 - 1750 kHz).

The ADF receiver navigates on both aviation stations (Non-DIrectional Radio Beacons) and AM broadcast stations.

Mode Selector. When the pointer is on REC (receive), the radio uses only the ADF sense antenna, and the radio acts as a conventional AM receiver. When placed on ADF, the receiver uses both sense and ADF loop antennas and operates as an automatic direction finder.

Signal Strength. As the pilot selects a station with the Tune knob, the meter helps find the strongest signal.

Test Button. To determine if the ADF pointer is not jammed or inoperative, the button is pressed. The test causes the pointer to swing at least 90 degrees and return when the button is released.

ADF System

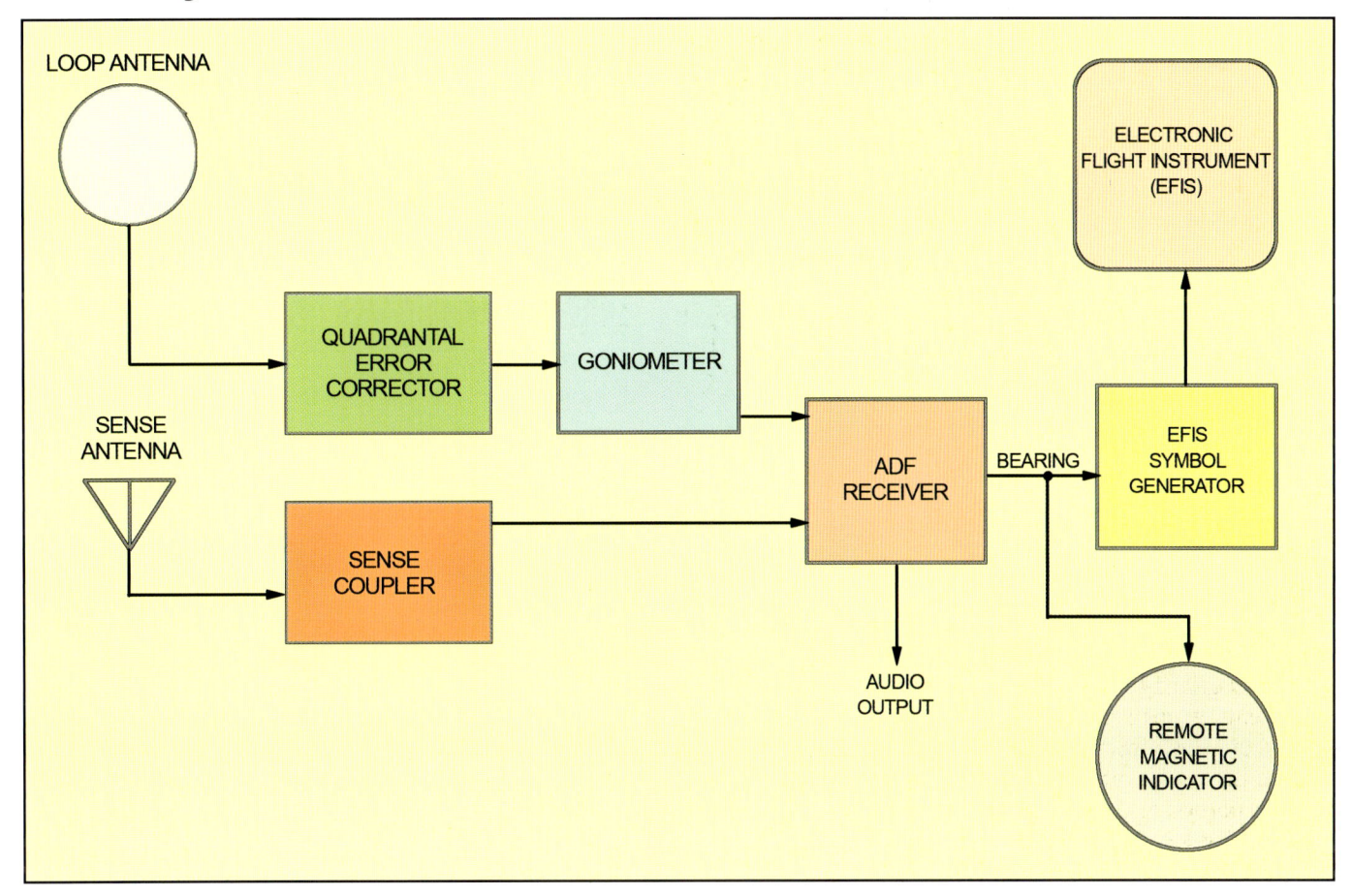

Diagram: Loop Antenna → Quadrantal Error Corrector → Goniometer → ADF Receiver. Sense Antenna → Sense Coupler → ADF Receiver. ADF Receiver outputs Audio Output and Bearing → EFIS Symbol Generator → Electronic Flight Instrument (EFIS), and also to Remote Magnetic Indicator.

The Loop Antenna is highly directional, but metal areas on the airplane distort its pattern. This is corrected by adjusting the Quadrantal Error Corrector.

The Goniometer captures the incoming signal and produces angle information which is fed to the ADF Receiver.

Also feeding the receiver is a Sense antenna. It is not directional, but mixes with the Loop Signal to remove the "ambiguity." Otherwise, the loop would indicate only a line of position that could run to or from the station.

The sense system eliminates the incorrect direction.

Bearing-to-station information developed by the receiver is fed to an EFIS symbol generator. This creates the symbol of an ADF pointer on the electronic display.

Bearing information may also be applied to a Remote Magnetic Indicator (RMI), an older, electromechanical instrument.

Audio from the beacon station (voice, Morse Identifier) is sent to the aircraft's audio panel.

directions. By combining loop and sense signals in the receiver, the correct direction of the loop signal is selected and the indicator points toward the station.

The receiver uses the null portion of the loop signal, not the peak, because it produces a sharper, more accurate directional indication.

Lear's first ADF's automatically turned the loop to home in on a station, but required motors and mechanical components. Today, the antenna is made of "crossed loops," two coils on ferrite cores placed at right angles to each other. They feed their signals to a *goniometer,* a device which compares them and produces bearing information. The loops now remain stationary and have no moving parts. Instead of a large circle of wire inside a dome, today's small loop antennas barely protrude from the surface of the airplane.

The sense antenna began as a long wire strung above the fuselage, a system that would hardly work at the speed of jet aircraft. The sense antenna on an airliner is often part of a plastic fairing near the wing root. The fairing is covered with a thin coat of metal to form the sense antenna.

NDB Station

There are several classes of ground station for non-directional radiobeacons (NDB), mainly depending on broadcast power The lowest wattage NDB is part of the Instrument Landing System (ILS) where it is known as a "compass locator." It guides the pilot to the outer marker or final approach fix at about 30 miles from the airport. For cross-country flight, NDB stations generate 100-400 watts and reach out 50-100 miles. When an NDB is located near a coast, power may rise over 1000 watts to provide long-range guidance over water for several hundred miles. One NDB on Bimini Island near Florida, for example, covers most of the Caribbean.

ADF Control-Display: *Airline*

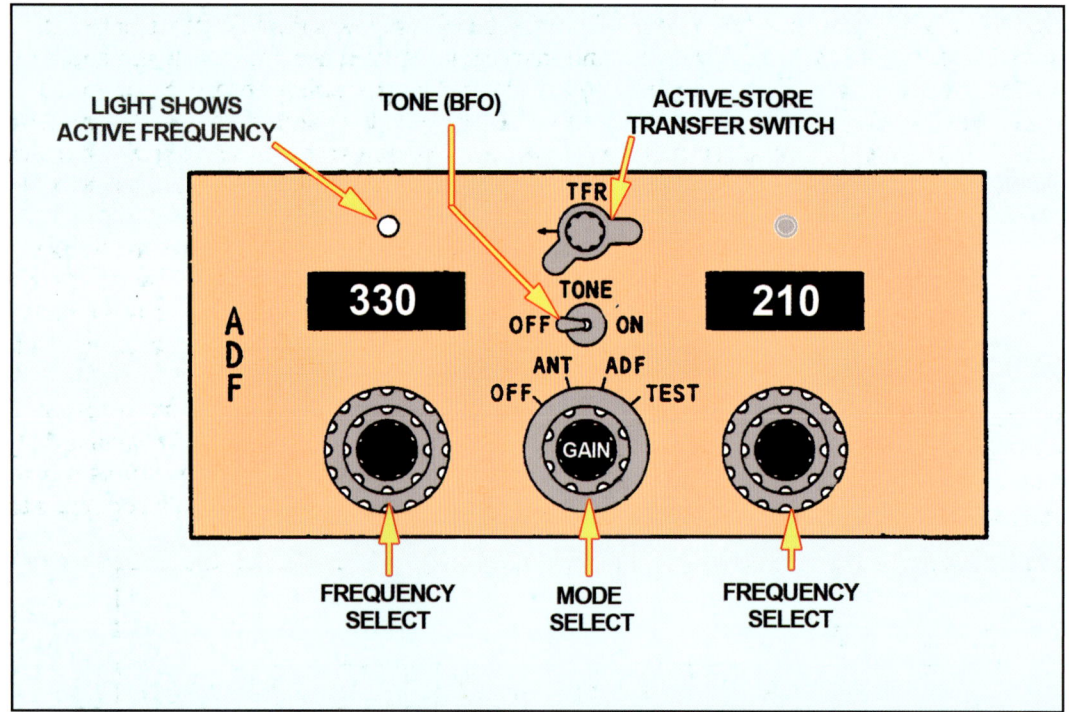

- LIGHT SHOWS ACTIVE FREQUENCY
- TONE (BFO)
- ACTIVE-STORE TRANSFER SWITCH
- FREQUENCY SELECT
- MODE SELECT
- FREQUENCY SELECT

The ADF display has two sides; for active and stored frequencies. The indicator light (above left) illuminates to show which side is active. A transfer switch provides an instant changeover.

The "Tone" switch near the center is the same as the "BFO" (beat frequency oscillator) on other ADF displays and serves the same purpose; to make the ID audible on stations that don't transmit audio tones.

ADF Receiver: *Airline LRU*

The remotely located ADF receiver has a test function that checks for correct movement of an ADF indicator.

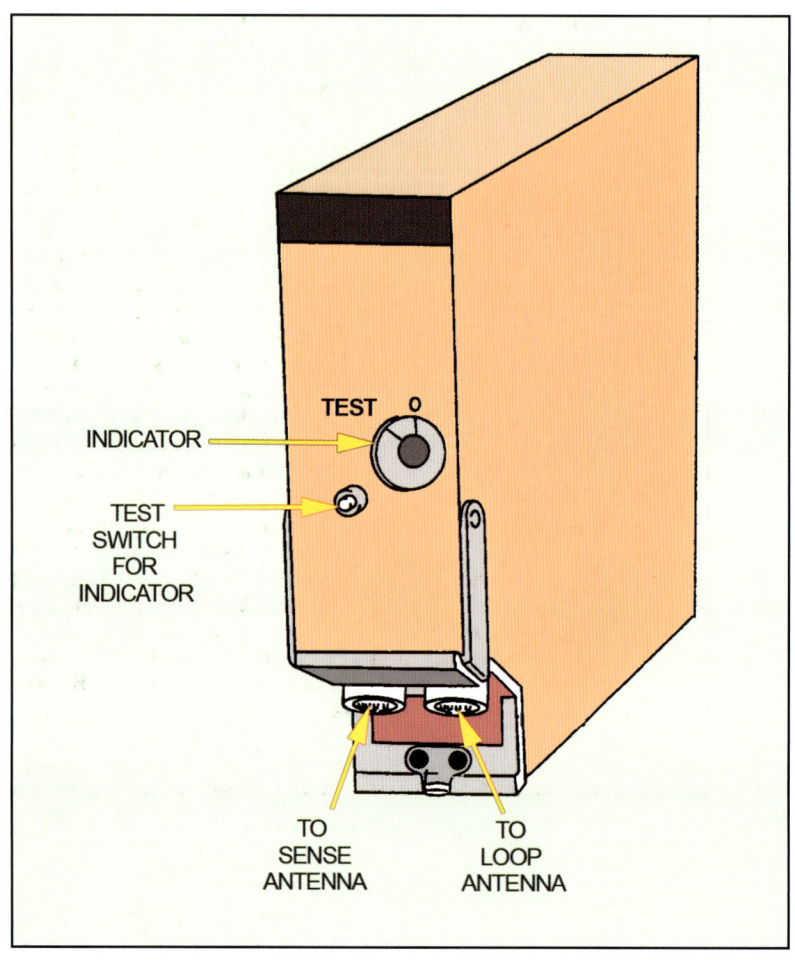

- INDICATOR
- TEST SWITCH FOR INDICATOR
- TO SENSE ANTENNA
- TO LOOP ANTENNA

84

NDB stations transmit their identification in Morse code (two or three letters) by modulating the radio carrier with an audio tone. Some high-power NDBs carry continuous aviation weather reports by voice. There may be NDBs in remote areas with no audible tone for identification. Instead, they key the carrier on and off to form Morse code. ADF receivers have a circuit (BFO, for beat frequency oscillator) to generate audio for these stations.

Broadcast Stations

Because an ADF receiver can tune the standard AM broadcast band (530 - 1700 kHz) it can also home in on those stations. The signals may not serve as airways in instrument operations and not be depended on for navigation. AM stations are difficult to identify because they announce their call letters at wide intervals. Also, many AM stations shut down at sunset.

ADF Limitations

Electrical Interference. Operating on low and medium frequencies subjects the ADF receiver to natural and man-made interference. It is also susceptible to noise from rotating machinery aboard the aircraft, such as alternators, generators and magnetos. Several techniques are needed to suppress it, such as shielded cables, filters, bypass capacitors and grounding, as described in the chapter on troubleshooting.

Natural sources of interference include lightning, which can occur hundreds of miles away. If the aircraft moves through precipitation, there is a build-up and discharge of energy that produces buzzing in the receiver.

Some pilots say an ADF needle points in the direction of lightning, and this is useful for avoiding thunderstorms. The Stormscope, in fact, is an instrument that operates on that principle. Using a conventional

Digital ADF

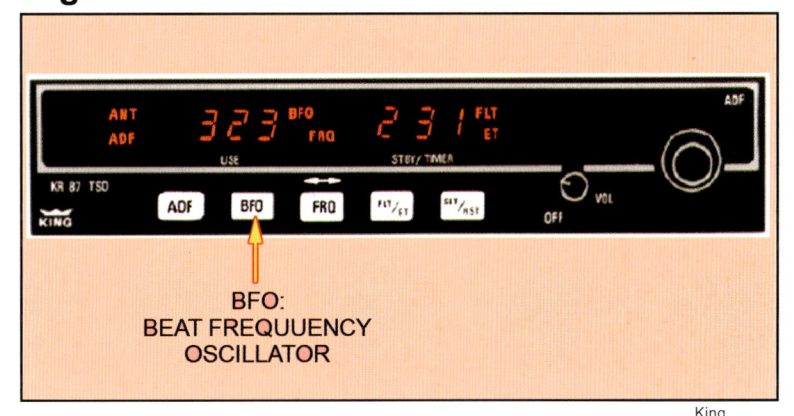

A panel-mounted digital ADF receiver. With the digital display, tuning is fast and precise. Note the BFO (Beat Frequency Oscillator). This is used for two reasons. Most NDB (non-directional beacon) ground stations broadcast an audio identifier for the pilot to verify he has tuned the correct station. This is known as "MCW," for modulated continuous wave. Some stations in remote areas, have no audio, and apply the identifier by keying the radio carrier (the operating frequency). To convert this to audio, the pilot switches on the BFO, which "beats" against the incoming carrier. This produces an audible tone, caused when the BFO and incoming frequency "beat" against each other, producing a difference frequency which is the audio tone. The pilot hears this as Morse Code.

Another use for the BFO is to locate very weak NDB stations. The BFO produces an audible tone that makes the carrier easier to find. This is most useful for older analog ADF receivers which are more difficult to tune than digital receivers.

ADF, however, is dangerous because the needle and its drive cannot swing rapidly and change direction, making it a poor indication of storms.

Night Effect. Lower bands "skip" great distances at night through the ionosphere and bring in distant stations. This causes the ADF needle to wander, and accuracy is poor if signals are weak.

Skipping is mostly a problem at night. Radio waves are bent at a shallow angle during the day, when the ionosphere reaches down to a low altitude. The reflected wave in daytime never returns to earth and is lost to space. With the setting sun, the ionosphere thins out (and appears to rise), causing radio waves to reflect back to earth---and cause interference. Thus the term "night effect."

Coastal Effect. Much of the signal from an NDB station travels by ground wave, hugging the surface of the earth. When the signal crosses between water and land, however, it is slightly bent, which decreases ADF accuracy. "Coastal effect" is most pronounced when the airplane is tracking at a small angle with respect to the coast (as opposed to moving directly from water to shore).

Attitude Error. The bearing to an NDB station is measured with respect to the nose of the airplane. To keep the bearing accurate, the ADF antenna is installed along a fore and aft line of the airplane. This provides good accuracy when the airplane flies straight and level. But during a turn, when wings are banked, the loop antenna is no longer aligned with the direction flight and accuracy suffers. This is not problem so long as the pilot is aware, and doesn't calculate his bearing to the station while in a turn.

Quadrantal Error. A loop antenna in free space works equally well on signals arriving from any direction. But on the airplane, loops are affected by engines, wings, fuselage and other masses of metal that lie unequally around the loop. They distort the receiving pattern by "quadrantal error." To reduce that effect, ADF receivers have a quadrantal error corrector, a device which usually mounts atop the loop antenna. The loop signal first passes through the quadrantal error corrector before proceeding down the transmission line to the receiver.

Loop Swing. A maintenance procedure to reduce quadrantal error is the "loop swing." The ADF is tuned to a station and bearings recorded as the airplane is rotated through a number of compass degrees, using a magnetic compass as a reference. A chart is made showing each magnetic heading, the bearing shown on the ADF and the error between them. Depending on the type and size of aircraft, this is done on the ground or while in flight. When the data is completed, the inaccuracies are compensated by the quadrantal error corrector, following the manufacturer's instructions.

ADF Display: EFIS

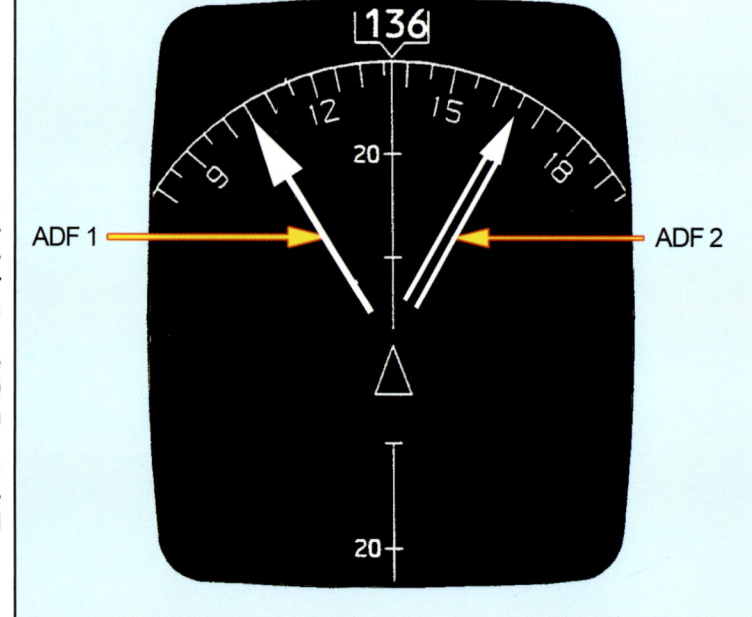

ADF is presented on various instruments. Shown here is the navigation display of an airline-type electronic flight instrument (EFIS) system. The pilot selected two pointers, one for each of the ADF receivers. (VOR information may also be selected for these pointers.)

The two "20" marks along the vertical line show the scale of the display; they indicate 20 miles above and below the airplane symbol (a triangle at the center).

In some aircraft, the ADF is shown on an electromechanical RMI (Radio Magnetic Indicator). The RMI has a 360-degree compass card and two pointers.

Review Questions
Chapter 12 ADF

12.1 On what bands does ADF operate?

12.2 An ADF with a fixed compass card can only indicate _____ bearing to an NDB (non-directional beacon) station.

12.3 When the edge of an ADF loop points toward the station, strongest signal is received. This is known as a _____.

12.4. When the flat side of the loop faces the station, the received signal is weakest. This is known as a _____.

12.5 Which is used by the ADF receiver for determining direction, the peak or null? Why?

12.6 What is the purpose of a sense antenna?

12.7 What is "quadrantal error."

12.8 How can the sense antenna be selected by the pilot?

12.9 How is direction-finding selected?

12.10 What is the function of the switch marked "BFO" or "Tone"?.

12.11 What methods reduce interference to ADF reception?

12.12 What type of interference may occur from distant stations?

12.13 What device in an ADF receiver reduces the effect of metal masses on the airplane?

Chapter 13

DME
Distance Measuring Equipment

Navaids such as VOR, NDB and localizer guide an aircraft along a *course*. But they do not help a pilot fix his position because they don't show *distance* to a station. That information is provided by DME, Distance Measuring Equipment. DME lowers pilot workload, and air regulations require DME for aircraft flying at or above 24,000 feet (Flight Level 240).

The principle of DME is that an airplane sends out an interrogating pulse to a ground station and the station replies. The DME aboard the airplane measures elapsed time to compute distance to the station. Time is multiplied by speed of the signal, which is close to the speed of light. A DME signal takes just over 6 microseconds to travel one nautical mile.

DME is usually part of a military system known as TACAN (tactical air navigation). TACAN provides course guidance (like a VOR) and DME distance for military aircraft. By agreement between military and civil authorities, TACAN stations are located on the same site (and in the same structure) as a VOR station. This benefits civil aircraft; they follow courses from the VOR station, while using the TACAN's DME for distance.

When a VOR houses a TACAN, the facility is known as a Vortac. Another combination---VOR-DME---provides only course and distance functions and not the complete TACAN facility. A DME may also be teamed with a localizer for an instrument approach.

DME Channeling

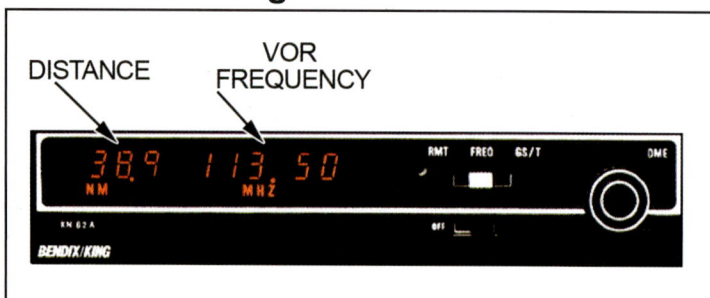

The pilot never sees the DME frequency. When the VOR (113.50 MHz in this example) located with the DME is selected, the DME receiver is channeled to its correct frequency. Note the distance indication shown to within one-tenth mile.

Ground Speed and TTS. Information developed by an airborne DME isn't only distance-to-station. By calculating how rapidly distance is changing, it also displays aircraft ground speed (GS). By knowing ground speed and distance, the DME also reads out TTS, or time to station.

DME ground speed and time-to-station are accurate only when flying directly to or from the station. The airplane, however, may fly in any direction and see the correct distance-to-station. In one instrument approach, the "DME arc," the pilot flies a circle and maintains a fixed DME distance from the station. This

Obtaining DME Distance

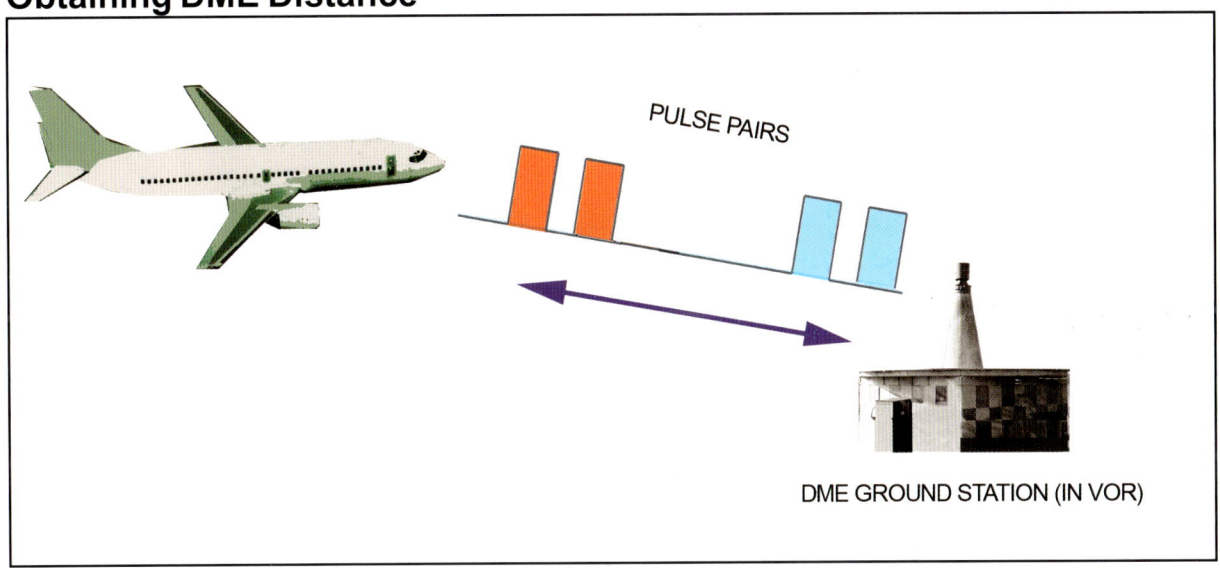

The airplane DME sends pairs of interrogating pulses to the DME ground station. After a delay, the ground station replies by retransmitting the pulses back to the airplane. The round trip time is divided in half and computed as one-way distance to the station.

Most DME ground facilities are housed in VOR stations, and are part of the military TACAN system.

guides the airplane to a safe position from which to begin the inbound course to the airport.

Slant Range. DME is very accurate, but has an error known as "slant range." Because signals follow a slanting path from the airplane to the ground, altitude is included in the distance measurement. It is not a factor when the airplane is many miles from the station; at 35 miles at an altitude of 4000 feet, the error is only several dozen feet. When overflying the station, the DME reads the altitude of the airplane.

There are ILS approaches which require DME to provide distance fixes to the airport. In this application, the pilot need not be concerned about slant range error. He uses DME distances shown on the approach chart, which has been verified by flight inspection aircraft.

Scanning and Agile DME. In recent years, it's become possible to process more than one DME station at a time. Known as "scanning" DME, the airborne system looks for up to five DME stations within a 300-mile range. When it locks on to three good signals, it continuously fixes the position of the airplane by triangulation. Each station is automatically identified by a 3-letter Morse code ID. The pilot does not have to listen to the code; the dots and dashes are electronically detected and identified.

Random Spacing ("Jitter") Identifies Each DME Signal

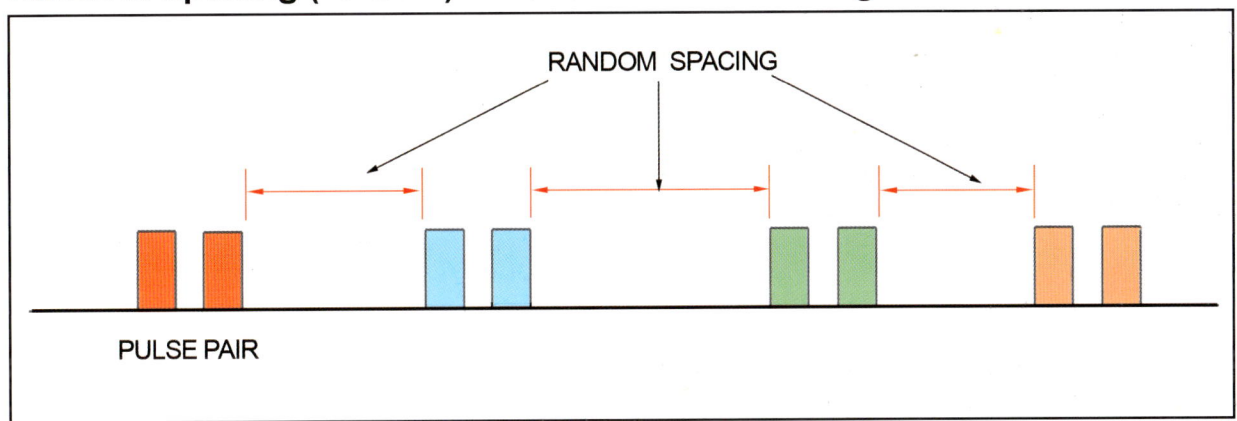

The DME interrogator aboard the aircraft sends out pulse pairs with random spacing ("jitter"). The ground station replies with the identical spacing. This enables the airborne DME to select its replies from those of other aircraft using the same DME station.

89

DME Readout on EFIS Display

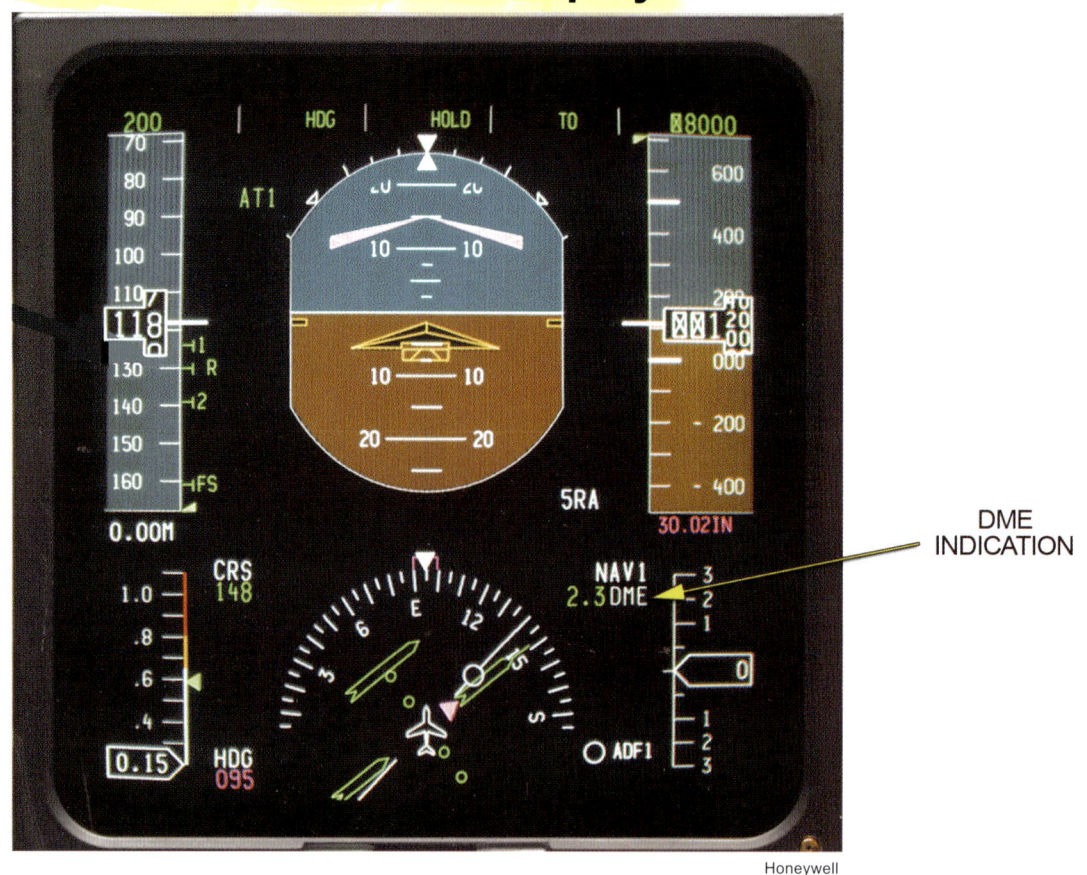

DME distance appears on the EFIS screen, near the compass rose (horizontal situation indicator). "NAV 1" shows it's reading data from the No. 1 VOR receiver.

Scanning DME is an important navaid in airline operations, especially in Europe where there is a shortage of navigational aids, too few frequencies and congested air traffic. European authorities made a special effort to distribute DME ground stations over a pattern that favors scanning DME. So long as the aircraft processes three DME's simultaneously, it can navigate with high accuracy and require no other navaids.

DME Channeling. A pilot does not directly tune a DME frequency; this is controlled by the VOR receiver. When the pilot chooses a VOR or localizer frequency, the radio automatically channels the correct DME frequency. (This is similar to the pairing of localizer and glideslope frequencies.)

Some DME control-displays do not tune VOR/LOC stations---only DME. However, the pilot still selects a VOR/LOC channel to obtain the DME station paired with the VOR frequency.

DME Jitter and Overload. The DME system has several enhancements to make it work in high traffic environments. In the area of a major airport, dozens of aircraft may be interrogating a single DME ground station. Because all aircraft receive all replies, each needs to sort out and identify its own reply. As described in the illustration, each aircraft varies the spacing of its interrogations in a random pattern known as "jitter." When replies arrive, each airplane looks for its unique jitter pattern and locks on to it.

Another problem is overloading the system. This happens when more than about 100 airplanes interrogate one ground station. To protect itself, the station reduces its receiver sensitivity and will not reply to airplanes at the outer edge of its range.

To prevent electrical interference from sending false pulses to a DME, all signals are sent in pulse *pairs,* measured precisely in microseconds. It would be very unusual for lightning strokes or other disturbances to emulate the pulse pair of a DME signal.

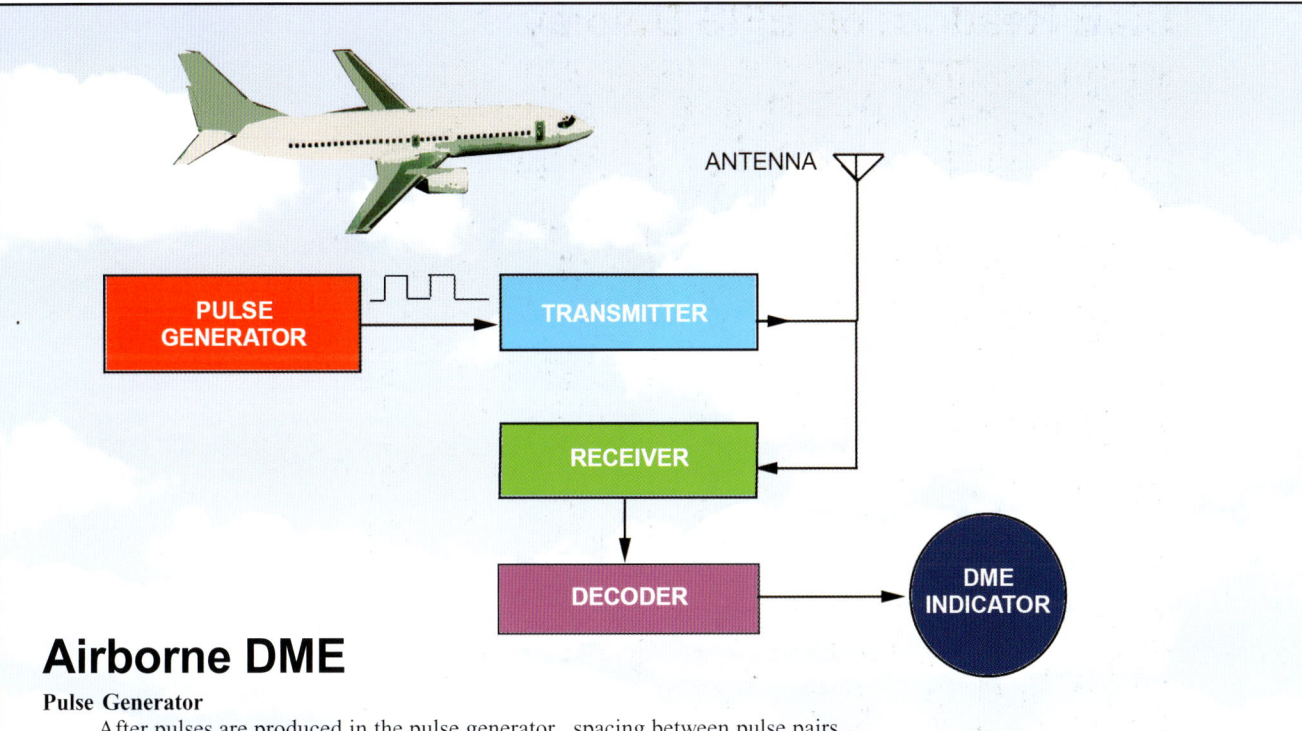

Airborne DME

Pulse Generator
After pulses are produced in the pulse generator, spacing between pulse pairs is varied in random fashion. This imprints the signal with its own identity, in a process known as "jitter." Each aircraft will have its own jitter pattern.

Transmitter
Pulses modulate the transmitter, then are emitted by the antenna as radio signals. They are DME interrogations on a channel between 978-1213 MHz.

Receiver
After interrogations are received from the ground station, they return to the antenna as replies. Note that both transmitter and receiver are connected to the same antenna. Outgoing and incoming pulses don't conflict because they are sent and received on different frequencies (63 MHz apart).

Decoder
Replies from the ground station for every airplane in the area are received at the antenna. The decoder in the airplane recognizes its own signal after searching for, and locking on to, its unique jitter pattern.

DME Indicator
After measuring the transit time for a reply from the ground station, the DME computes distance and time to station. Ground speed is determined by the rate of change of the distance signal.

DME Ground Station

The ground station receives, decodes and replies to interrogations from the airplane.

50 Microsecond Time Delay. When the airplane is close to the DME station, outgoing pulses may not allow enough time for replies to arrive from the ground station. To avoid interference, the ground station delays transmitting the reply by 50 microseconds.

Squitter If the ground station receives no interrogations from any aircraft, it "squitters"---that is, freely broadcasts pulses. This "awakens" any aircraft within range; and their DME's go from "automatic standby" to an interrogating mode.

Audio ID. Every 30 seconds, the ground station sends a Morse code identifier on 1020 Hz. The pilot can identify the DME, or tones are decoded electronically by scanning DME's.

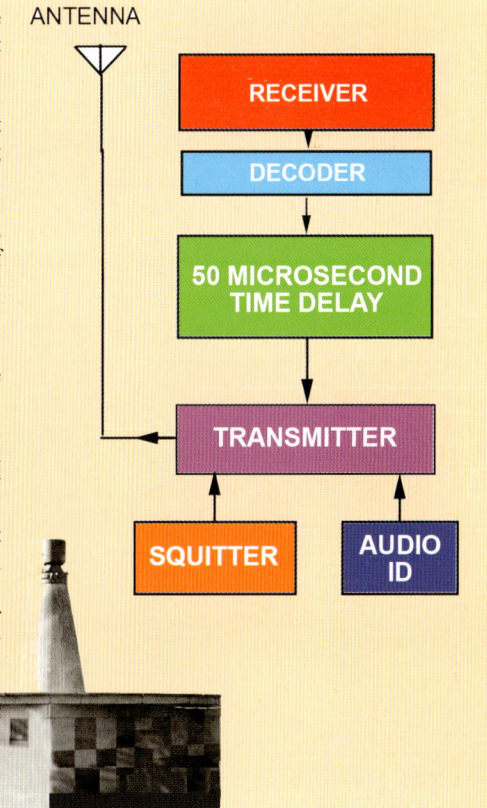

DME Channels: X and Y

VOR Freq	Airborne DME Freq	Type	Pulse Spacing	Ground Reply Freq	Ground Reply Pulse Spacing
117.20 MHz	1143 MHz	X channel	12 microsecond	1206 MHz	12 microseconds
117.25 MHz	1143 MHz	Y channel	36 microseconds	1080 MHz	30 microseconds

When DME began, there were only 100 VOR frequencies spaced .1 MHz apart; for example, 117.20, 117.30, etc. The original DME stations paired with these VOR's are named "X" channels. As shown above, the VOR on 117.2 is paired with a DME frequency of 1143 MHz.

As air traffic increased, the number of VOR's was doubled by "splitting" the channels; 117.20, *117.25*, 117.30, etc. This doubled the VOR channels from 100 to 200. The added frequencies created the new DME "Y" channels. There are two differences between X and Y channels:

Pulse Spacing
Note in the table above, 117.20 (an X channel) sends out interrogations on 1143 MHz, with a pulse pair spacing of 12 microseconds.

The interrogations for the Y channel, paired with 117.25, has a pulse spacing of 36 microseconds, three time longer.

Next, notice the different pulse spacing for the reply from the ground station; for the X channel it is 12 microseconds, and 36 microsecond for the Y channels.

Reply Frequencies
The DME system requires two separate radio frequency carriers so interrogations and replies do not interfere with each other. For this to work, their frequencies must be widely separated (so the transmitter does not overload the receiver). This is done by separating the radio carriers by 63 MHz.

Consider the X channel example above:

    Reply frequency (ground station): 1206 MHz
    Interrogation (from airplane)   1143
                    Difference   63 MHz

Next, the Y channel:
    Interrogation   1143 MHz
    Reply   1080
            Difference   63 MHz

Regardless of whether the airplane is on an X or Y channel, the interrogation always goes out on 1143, MHz. However, for an X channel, the reply comes back 63 MHz *higher* than the interrogation. For the Y channel, the reply is 63 MHz *below* the interrogation.

By these techniques of changing the space between pulse pairs and a different position for the reply frequency, the system doubles the amount of DME stations and allows tight spacing of the channels.

One DME for Two VOR Receivers

"N1" SELECTS VOR 1   "N2" SELECTS VOR 2

A single DME display can connect to two VOR receivers. The pilot selects either "N (nav)1" or "N2." Note that "1" appears near the top, above "NM," to indicate VOR 1 is the source of the DME.

Review Questions
Chapter 13 DME

13.1 An airborne DME sends out a pulse known as an _____.

13.2 DME is a component of a military system known as _____.

13.3 A DME station is located as part of a _____ ground station. Together they are known as a _____.

13.4 In addition to distance-to-station, an airborne DME computes _____ and _____.

13.5 A distance error in DME is called _____.

13.7 All aircraft interrogating the same DME ground station are on the same frequency. How does an aircraft identify its replies from all others?

13.8 How is a DME station tuned in?

13.9. What happens when more than about 100 aircraft interrogate the same DME ground station?

13.10 Why does the DME ground station delay its reply by 50 microseconds?

13.11 Does the DME station transmit an ID?

Chapter 14

Transponder

A transponder ("transmitter-responder") receives a signal from a ground station (an "interrogation") and automatically transmits a reply. Transponders were developed for the military at a time when radar could locate airplanes but couldn't tell the friendlies from the enemy. The reply of a transponder provides that information; the airplane's ID, altitude and other data.

When first introduced, transponders were called "IFF," for Identification, Friend Or Foe. The term is still used, but mostly by the military. In civil aviation, it is in a system called SSR, for "Secondary Surveillance Radar." It is *secondary* because primary radar simply sends a signal from the ground that reflects from the metal surface of the airplane and receives an echo called a "skin return."

In the airline world the transponder is labeled "ATC," referring to Air Traffic Control.

Squawk. When a pilot is instructed by ATC to set his transponder to a code (say, 1234), the controller says:

"Squawk 1234."

The pilot selects the code on the control panel that causes his airplane's ID to appear on the radarscope. Sometimes a controller may need to verify the ID, in which case he asks the pilot to "Ident." The pilot responds by pressing an ID button on the transponder, which causes his target on the ground radar to "bloom," creating a circle of light that clearly indicates the location of the airplane.

The word "squawk" goes back to World War II when the British, to keep their new transponder secret,

Control-Display Unit

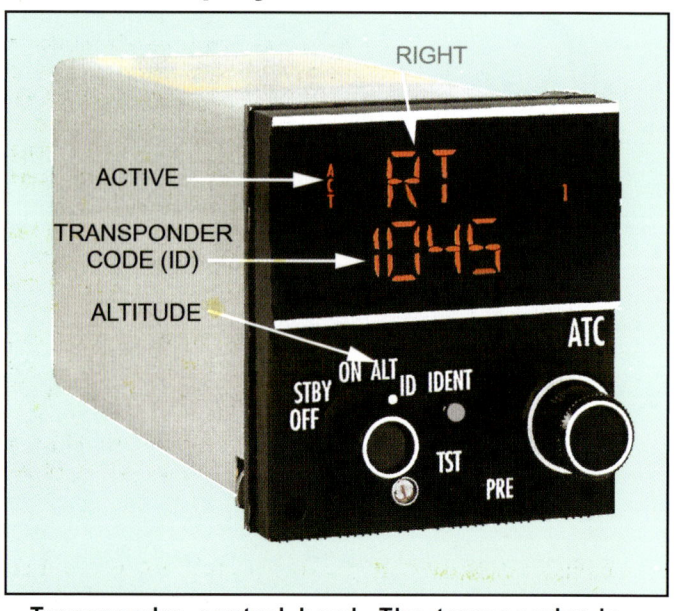

Transponder control head. The transponder is "squawking" an identification code of "1045," which is the Mode A function. Also transmitted is altitude, the Mode C function. The ident button is pressed only when requested by the controller. The standby (STBY) position keeps the transponder hot, but prevents replies. This reduces clutter on the radar screen while the airplane taxis on the ground at an airport. As the airplane lifts off, the pilot turns the knob to Mode C (which activates ID and altitude replies).

Note "ATC" at the upper right. This term is used for the transponder in large and airline aircraft.

Transponder Interrogator

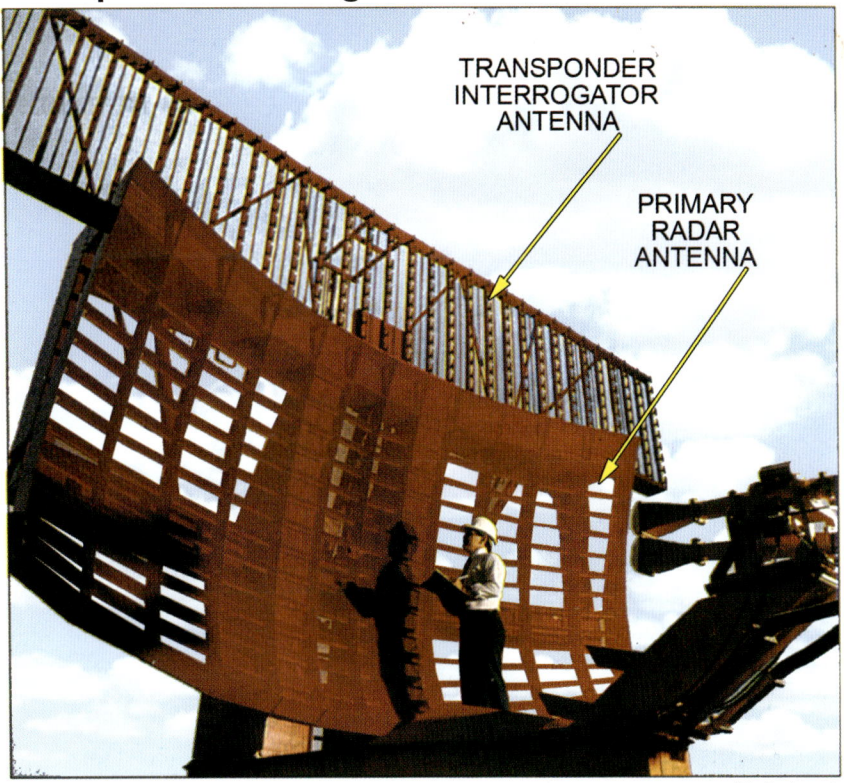

(Westinghouse)

The aircraft transponder sends to, and receives from, the top section of a surveillance radar known as a "beacon interrogator."

The larger antenna below it is the older primary radar, which sends out a pulse and picks up the signal reflected from the skin of the aircraft. Because skin returns are weak, difficult to see and carry no information other than the range and bearing of the aircraft, they are used only as a back-up.

The beacon interrogator on top, on the other hand, picks up a signal that's strengthened thousands of times by the aircraft transponder. Besides a bright display, the image on the radar screen carries data such as aircraft ID, transponder code and ground speed.

The surveillance radar shown above is an ASR---airport surveillance radar---that covers up to about 60 miles from the airport. To the pilot, this is "approach control" or "Tracon" (terminal radar approach control).

During cross-country flight, airplanes receive longer range coverage from "en route" radars of larger size and power.

---

called it "Parrot." It survives to this day in the military; when a (British) controller wants a pilot to turn off his transponder, he says, "Strangle your parrot!"

The word "parrot" also explains why controllers today ask the pilot to "Squawk" a transponder code.

Grand Canyon. Transponders came into widespread use after a mid-air collision between two airliners over the Grand Canyon in 1956. A DC-7 and a Constellation requested permission from ATC to fly off course so passengers could enjoy the view. Flying outside controlled airspace (on a sunny day) the airplanes collided with the loss of 128 lives. The disaster began an overhaul of the ATC system (and created the FAA). With an ability to put strong targets and flight data on the radar screen, transponders became a key component in the air traffic system.

Two Systems: ATCRBS and Mode S

ATCRBS. The transponder improved air traffic control for a half-century, operating under the name, ATCRBS, for Air Traffic Control Radiobeacon System." But it began showing its age as the aircraft popu-

95

Panel-Mount Transponder (Mode S)

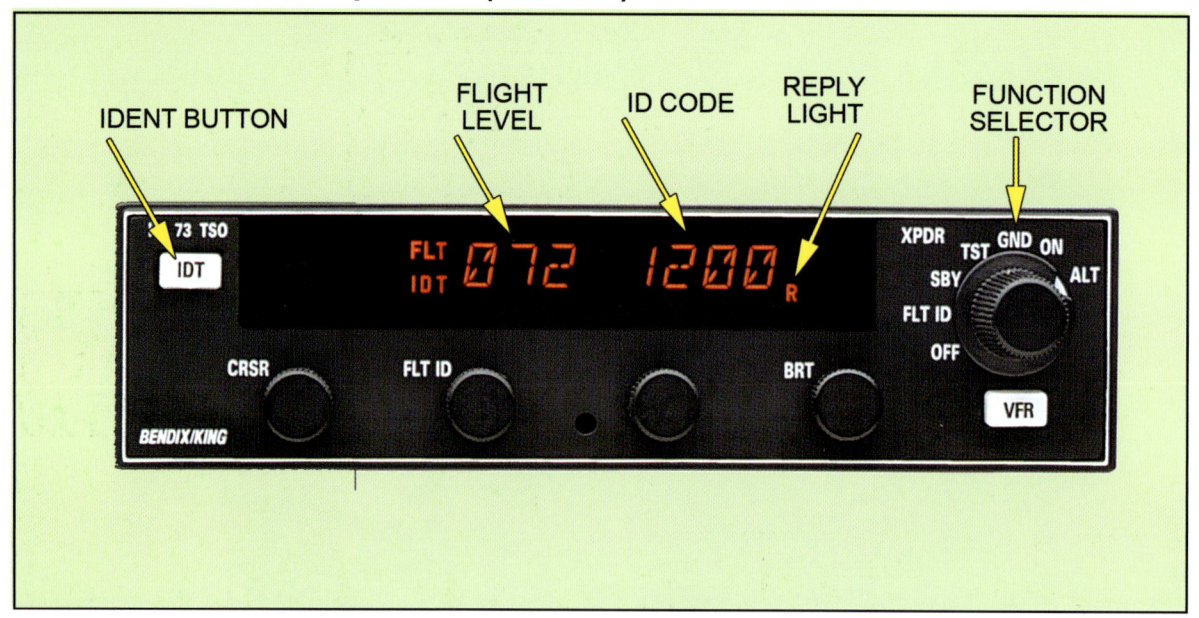

A Mode S transponder for General Aviation, the Bendix/King KT-73. Its controls and displays include:

IDENT BUTTON. The pilot presses the button when air traffic control requests "Ident." The reply light (R) illuminates for several seconds while the reply transmits.

The same Reply Light also blinks when the transponder answers interrogations from the ground.

FLIGHT LEVEL. The altitude of the airplane, as reported by the transponder in hundreds of feet. Thus, "072" on the display is 7200 feet (add two zeroes).

ID CODE. The squawk code assigned by ATC under the older transponder system (ATCRBS). It is dialed in by four knobs along the bottom of the panel.

FUNCTION SELECTOR. This turns the transponder on, displays the flight ID code and tests all lighted segments of the display. The Ground position disables most transponder functions because they are not needed on the ground. A large number of airplanes on the airport surface would clutter radar displays. The pilot switches, shortly after take-off to Alt (altitude) to resume normal transponder operation. The Alt position reports ID and altitude.

The "VFR" button at the bottom right automatically sets the transponder to 1200. This code is selected when the airplane is not on an instrument flight plan and is flying under visual flight rules (VFR).

---

lation grew and more airplanes operated out of fewer airports.

There are several limitations on ATCRBS. First, it wastes space in the radio spectrum. When the radar antenna on the ground sweeps around, it interrogates all airplanes within range---and all airplanes reply. The controller cannot obtain a reply only from the airplane it needs to contact.

Also, when radar sends a beam, it sweeps across the airplane, making 20 or more interrogations in one pass. Only one interrogation and reply are required; extra replies limit the capacity of the system.

Another shortcoming is "synchronous garble," which is two replies happening at the same time. Let's say two airplanes are on the same line north of the radar site; one at 20 miles, the other at 30 miles. Because they are both struck by the same radar interrogation, they reply nearly at the same time. The two replies move along the same line back to the radar antenna and interfere with each other. To the controller, the targets appear confused or "garbled."

Another limit for the ATCRBS transponder is in the collision avoidance system (TCAS). As described in the chapter on TCAS, two aircraft approaching each other must fly an escape maneuver that keeps them apart; for example, to avoid a collision, one flies up, the other flies down. That maneuver must be coordinated by the transponder, but ATCRBS cannot provide this function.

Mode S. The answer to these problems came as a completely new transponder known as Mode S (S for Select). It means "selective addressing," which enables a controller to request a specific airplane to reply, not the whole fleet. This greatly reduces the number of unnecessary signals filling the air.

A requirement of Mode S is that it does not rapidly obsolete the ATCRBS transponder. The two must be able to exist side by side during a long transition

Transponder

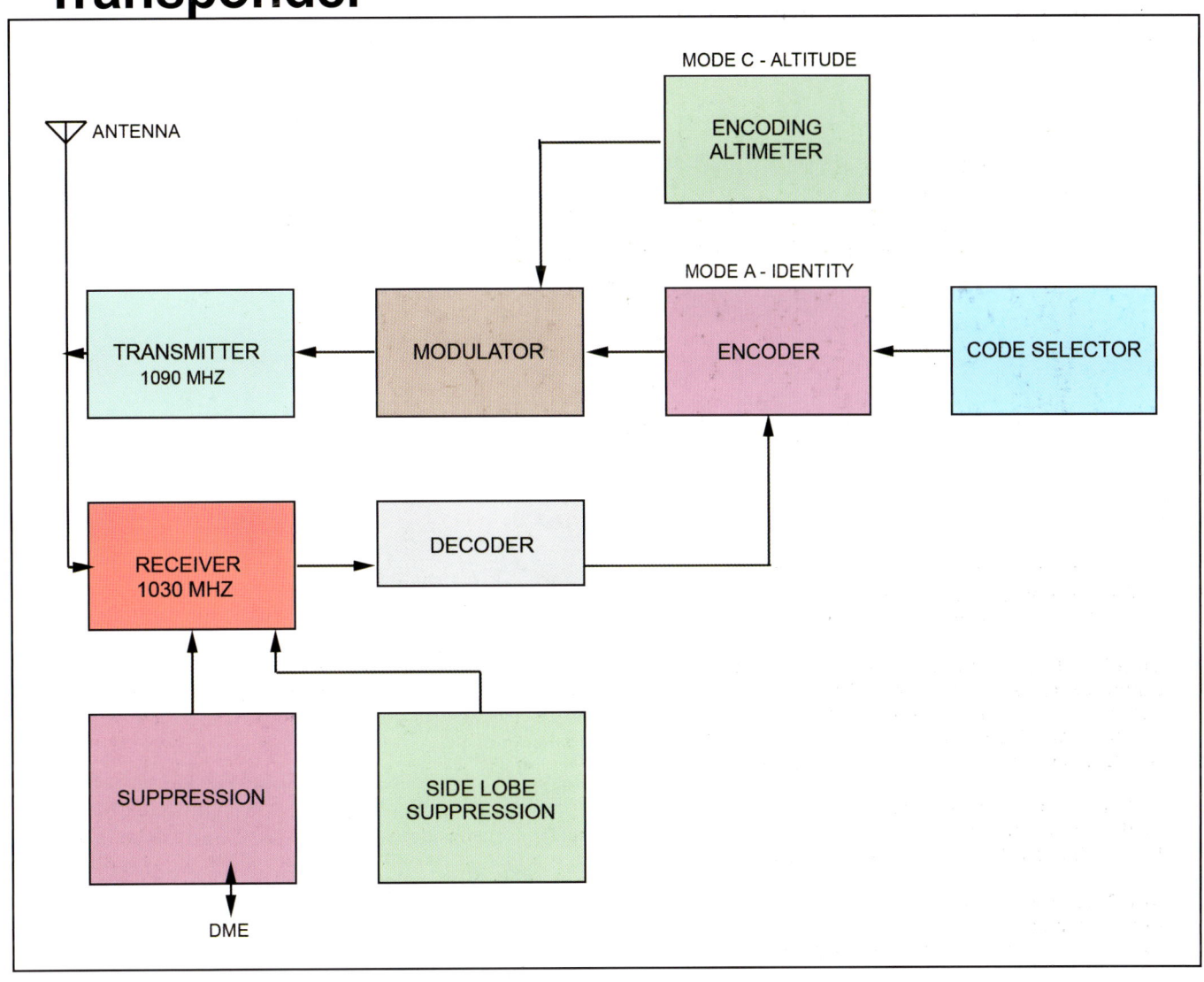

The transponder is a transmitter-receiver with these major building blocks:

RECEIVER
Interrogations from the ground station are picked up by the antenna on a frequency of 1030 MHz. The pulses are applied to the decoder.

DECODER
Measuring incoming pulses, the decoder identifies the type of interrogation. If they are recognized they are passed on to the encoder.

ENCODER
The encoder creates the pulse train which contains the reply.

ENCODING ALTIMETER
After converting barometric pressure (based on 29.92 inches of mercury) to electrical signals, the encoding altimeter sends altitude information to the encoder for the Mode C reply.

CODE SELECTOR
The pilot dials in the 4-digit transponder code, which is sent to the encoder.

MODULATOR
Pulses that form the reply are amplified in the modulator and applied to the TRANSMITTER for transmission on 1090 MHz.

SIDE LOBE SUPPRESSION
The radar signal from the ground contains a main lobe and several side lobes. If the transponder replies to a side lobe, the radar operator will see the airplane at the wrong position. The side lobe suppression circuit prevents the transponder from replying if it senses reception of a side lobe.

SUPPRESSION
There is a chance that other transmitters aboard the airplane might interfere with the transponder. This usually caused by the DME, which operates close in frequency. To avoid interaction, the transponder receiver is suppressed when the DME transmits. The DME receiver is also suppressed when the transponder is transmitting.

Transponder Control-Display (Airline)

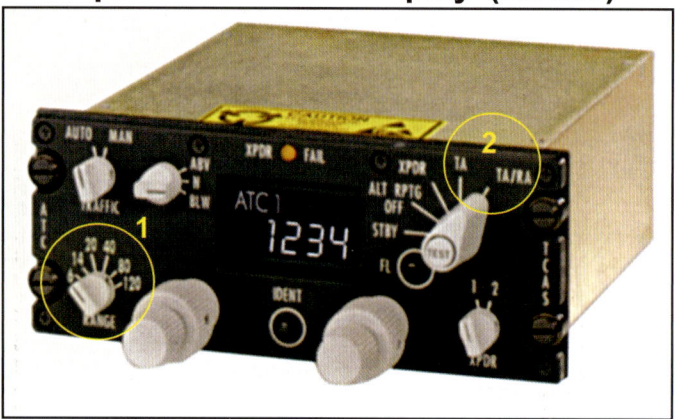

A Mode S transponder installed in the B-777 and numerous other large aircraft. Because it works so closely with TCAS (anti-collision system), several TCAS controls appear with the transponder knobs. For example, in the circle marked "1," is the mileage range for TCAS surveillance. In "2" are types of collision warnings selected for display.

Line Replaceable Unit (LRU)

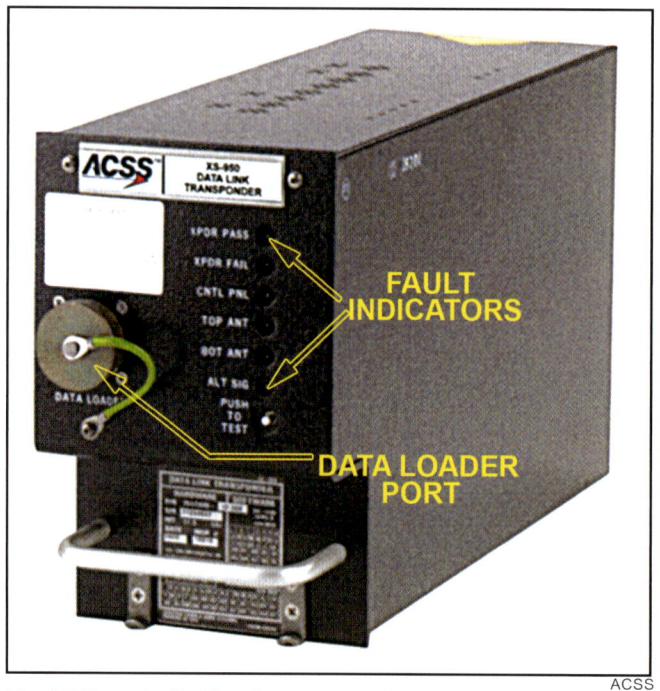

The LRU controlled by the transponder panel unit. Troubleshooting is helped by fault indicator lights that monitor the transponder, antennas and control panel.

When software upgrades occur, they are loaded through the data loader port.

period. The Mode S system, therefore, is "backwardly compatible" in that ATCRBS and Mode S transponders work within all air traffic control systems. Surveillance radars for both types are now in service.

The FAA had issued an end date for the manufacture of ATCRBS transponders, expecting Mode S to gradually take over as old transponders wore out. Announcing the end of ATCRBS, however, raised protests from the General Aviation community. The Mode S transponder was more expensive and didn't offer any advantages to owners of light aircraft. The only buyers of Mode S were airlines because it was a requirement for the anti-collision system (described in the chapter on TCAS).

But a new development changed opposition to Mode S. Sales of the new transponder surged in General Aviation after the year 2000 because valuable new pilot services were added to Mode S. The FAA introduced the Traffic Information Service (TIS) which uplinks, through the Mode S transponder, images of air traffic throughout the U.S. It provides an option for light aircraft to have an anti-collision system at relatively low cost.

Aircraft Address. For ATC to single out one airplane each Mode S aircraft has its own "Aircraft Address." It's 24 bits long and obtained through the aviation authority of each country. The address is programmed into the transponder during installation, along with the aircraft's maximum speed.

A caution about addresses was issued by Eurocontrol, the air traffic agency for Europe. It reported instances of errors by technicians in entering the address during installation, or when changing the country of registration. Eurocontrol says that such errors can disable an Airborne Collision Avoidance System (ACAS), which is the same as TCAS in the U.S.

Mode A Interrogation

P1 INTERROGATION — **P2 SIDE LOBE SUPPRESSION** — **P3 INTERROGATION**

8 microseconds (between P1 and P3)

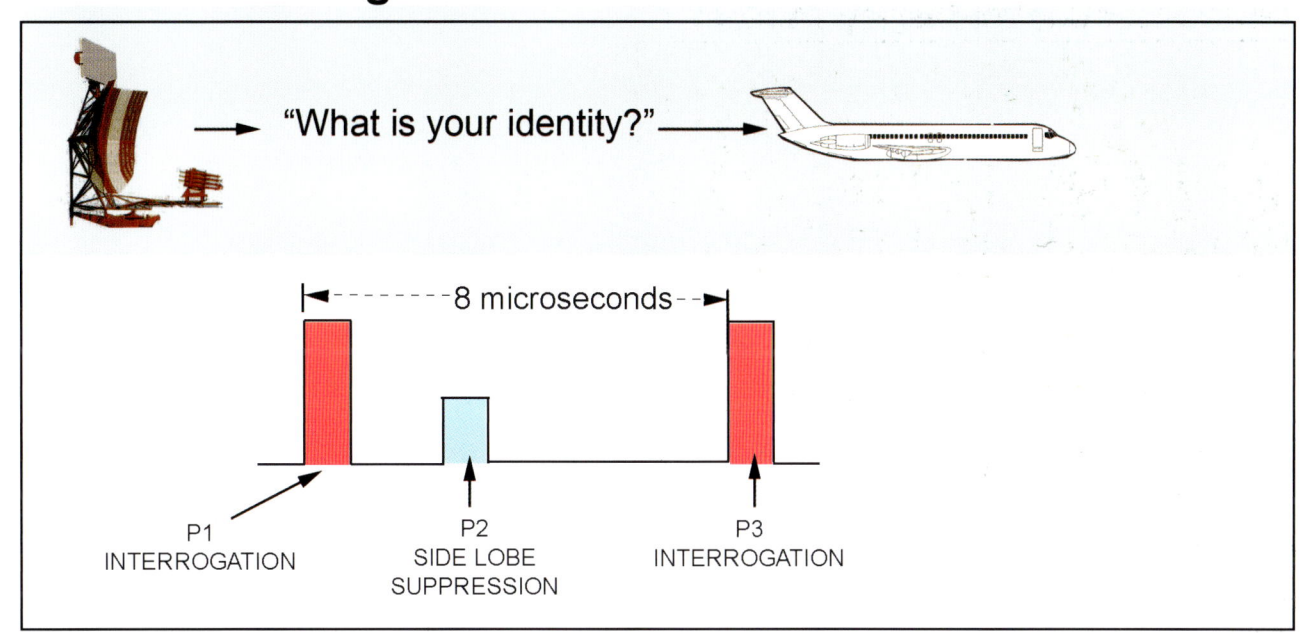

"What is your identity?"

Surveillance radar on the ground sends out an interrogating pulse to learn the aircraft ID (the code selected by the pilot on the transponder). The upgoing radar signal consists of three pulses, as shown above; P1, P2 and P3.

Consider P1 and P3, which tell the airplane this is a Mode A interrogation. It's done by the 8 microsecond time period between the two pulses. On measuring this interval, the transponder recognizes it as a Mode A interrogation and sends a reply containing the aircraft ID.

P2 overcomes a problem with the radar (and any other) directional antenna known as "sidelobes." These are unavoidable loops of radio energy that lie on either side of the main beam of the radar antenna. The problem is that an aircraft transponder may reply to a sidelobe, which places the airplane in the wrong location on the radar screen. Pulse P2 eliminates the problem. If P2 remains *below* P1 in strength, it means the transponder is receiving the main beam of the radar. This triggers a reply. However, if P2 is *higher* than P1, it means a sidelobe is being received; now the transponder will not reply. When the airplane is correctly illuminated by the main beam, P1 remains higher than P2 and the transponder replies.

Mode C Interrogation

"What is your altitude?"

P1 INTERROGATION — **P2 SIDE LOBE SUPPRESSION** — **P3 INTERROGATION**

21 microseconds (between P1 and P3)

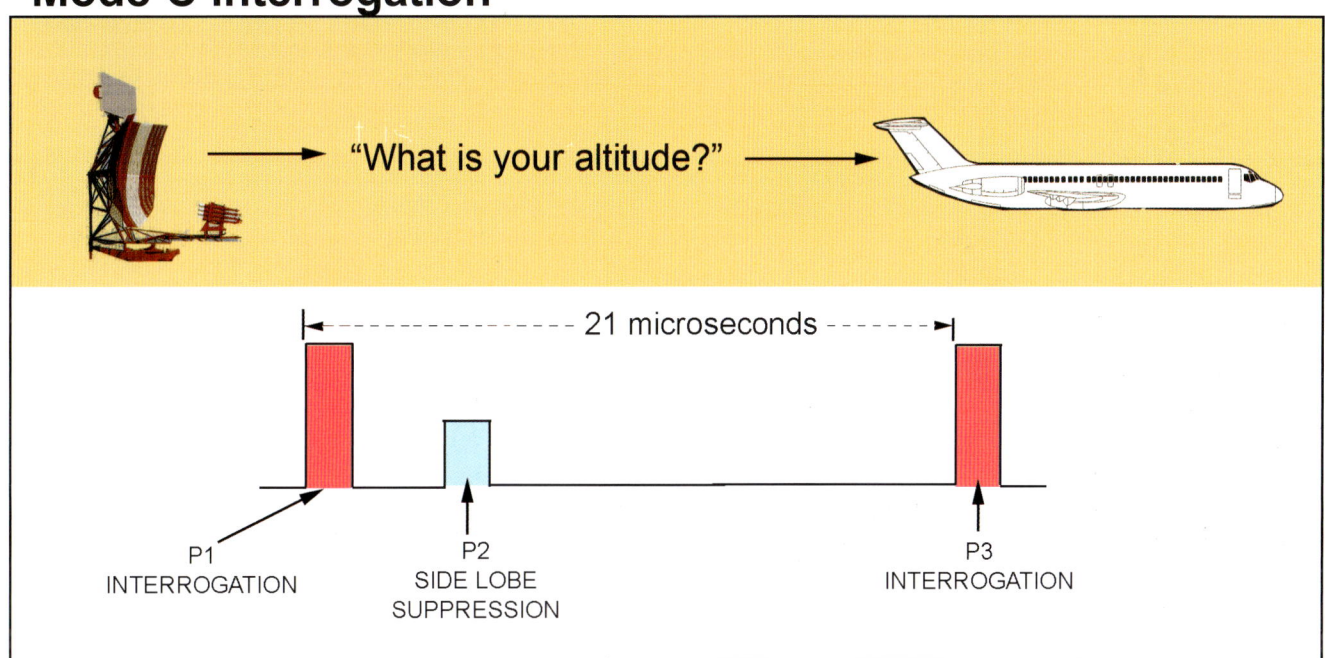

Next, the surveillance radar sends out a set of pulses to learn the aircraft's altitude (Mode C). The aircraft transponder recognizes it by the spacing between P1 and P3, which is now 21 microseconds long (more than 2 1/2 times longer than for a Mode A interrogation). The purpose of P2 is the same as already described for Mode C.

As the ATC radar antenna makes one full rotation, it transmits a Mode A interrogation, followed by a Mode C interrogation on the next rotation.

Altitude Reporting; Mode C

ENCODER

TRANSPONDER

ANTENNA

CODE LINES

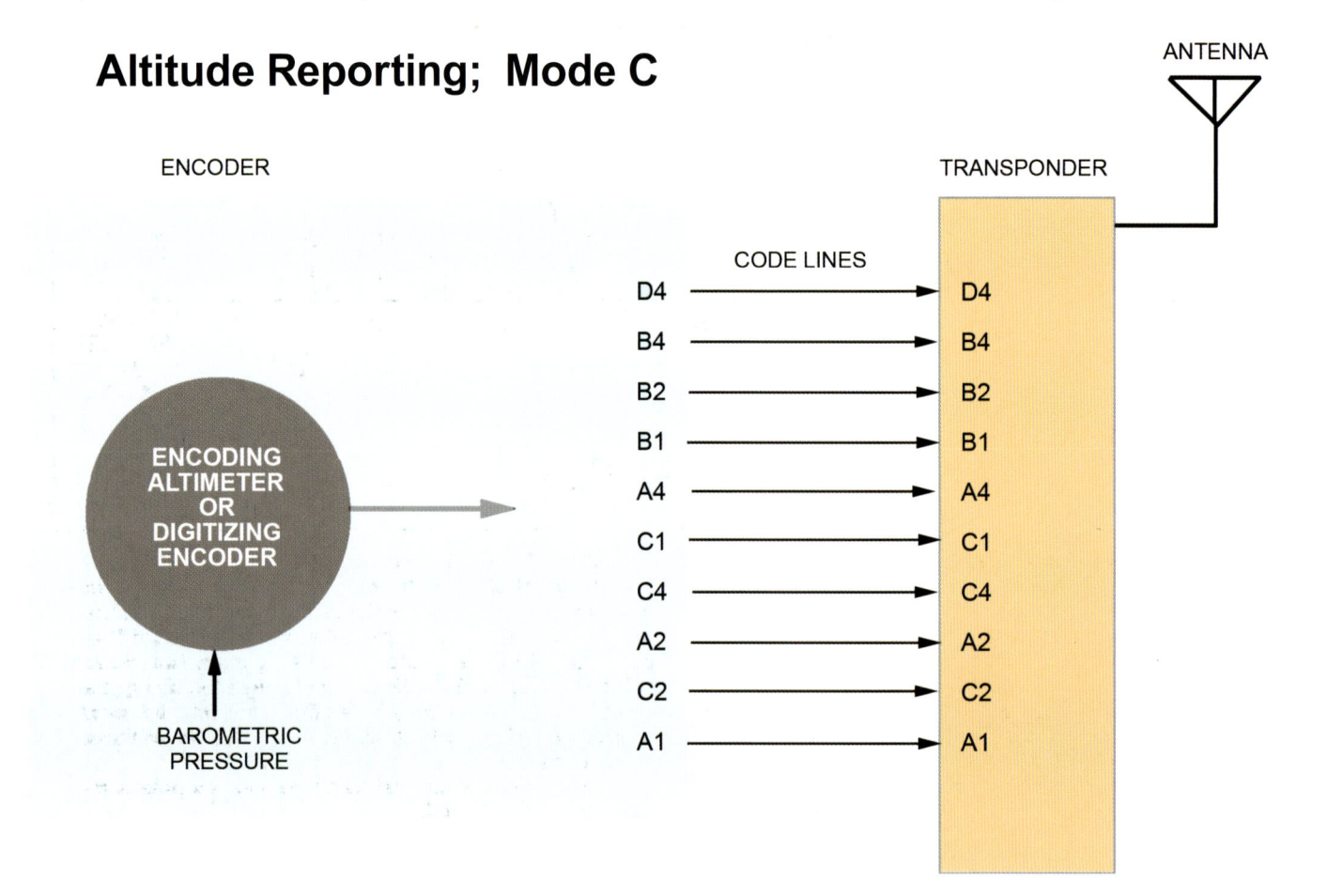

ENCODING ALTIMETER OR DIGITIZING ENCODER

BAROMETRIC PRESSURE

D4 → D4
B4 → B4
B2 → B2
B1 → B1
A4 → A4
C1 → C1
C4 → C4
A2 → A2
C2 → C2
A1 → A1

The transponder reports an airplane's altitude when replying to a Mode C ground interrogation. It begins by measuring air pressure surrounding the airplane, often done with an aneroid sensor, a capsule which is mechanically squeezed by pressure. This movement is converted to an electrical signal representing altitude and is connected to the transponder.

There are two possible locations for an aneroid sensor. When built into the altimeter, the instrument is called an "encoding altimeter." When mounted separately, it is known as a "blind encoder" (because it has no dial and is hidden from view).

Altimeters have a knob for setting local air pressure because weather is always changing---as high and low pressure systems move across the country. During a flight the pilot resets the altimeter to maintain an accurate reading above sea level.

An important feature of transponder operation is that the aneroid sensor is *never* adjusted by the pilot. It is permanently fixed to 29.92 inches of mercury, standard sea level pressure, and is known as "pressure altitude." In the metric system it is 1013 millibars. A pilot may make many adjustments to correct his altimeter reading in different pressure areas, but this does not change the pressure sensor. The sensor is preset at the factory for 29.92 and recalibrated every two years when the transponder is recertified by a technician.

There are several reasons for reporting altitude based on a standard pressure. One is that pilot error could transmit the wrong altitude to air traffic control. It could also send incorrect information to TCAS (anti-collision) systems aboard nearby aircraft, which also interrogate the transponder for altitude.

If all transponders report altitude based on a pressure of 29.92, the altitude sent to ground radar will contain the error caused by changes in the air mass. This is corrected by the air traffic facility when it receives the Mode C reply; it corrects to 29.92 against local air pressure at sea level.

As seen in the illustration above, the encoder sends information via code lines to the transponder. The letters represent pulse positions on the transponder reply signal, which form binary words that encode altitude every 100 feet. Known as the "Gray" or "Gilham" code, it can transmit altitudes from -1000 feet to 126,700 feet.

During most of aviation's history, altitude has been based on instruments driven by air pressure. In the future, this function will be increasingly provided by GPS or other satnav system. GPS can already fix the vertical position of an airplane to within a few centimeters, with no reference to air pressure.

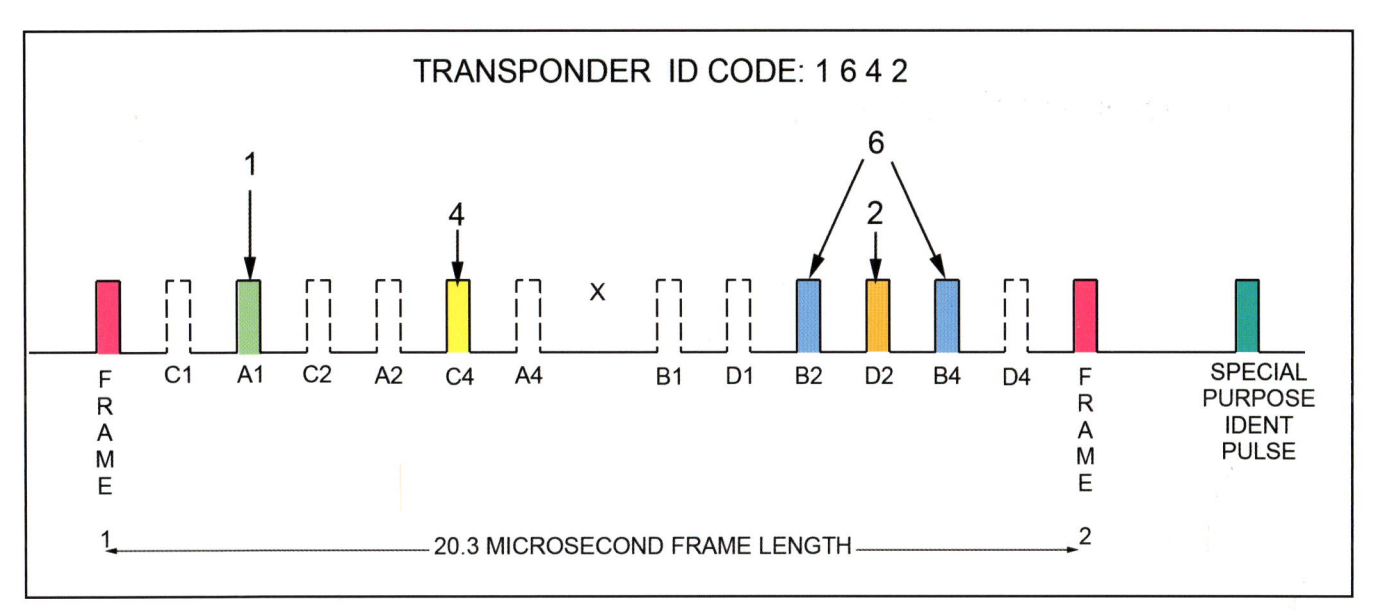

TRANSPONDER ID CODE: 1 6 4 2

FRAME | C1 | A1 | C2 | A2 | C4 | A4 | X | B1 | D1 | B2 | D2 | B4 | D4 | FRAME | SPECIAL PURPOSE IDENT PULSE

1 ← 20.3 MICROSECOND FRAME LENGTH → 2

The transponder sends its ID by selecting various pulse positions (time slots) spread over a "frame." Shown above are positions selected for an ID of "1642". Although the coding uses digital signals (on-off pulses), it was developed over 50 years ago and has only 4096 codes. This will eventually be replaced by the Mode S transponder, with far more sophisticated coding and greater capacity.

At the far right is the Special Purpose Identification Pulse. In normal air traffic operations, the controller easily identifies each aircraft by examining its "data block," an area on the radar screen next to the target showing aircraft ID, altitude and ground speed. Occasionally, however, the controller wants to verify that he is looking at the correct target and asks the pilot to "Ident." When the pilot presses the Ident button, it sends the Special Purpose Identification Pulse. The target on the radar "blooms" in a circle of light, positively identifying the aircraft from all other targets. At the same time, the pilot sees the reply light on his transponder remain on for about 15 seconds.

The Ident button should never be pressed unless requested by the controller.

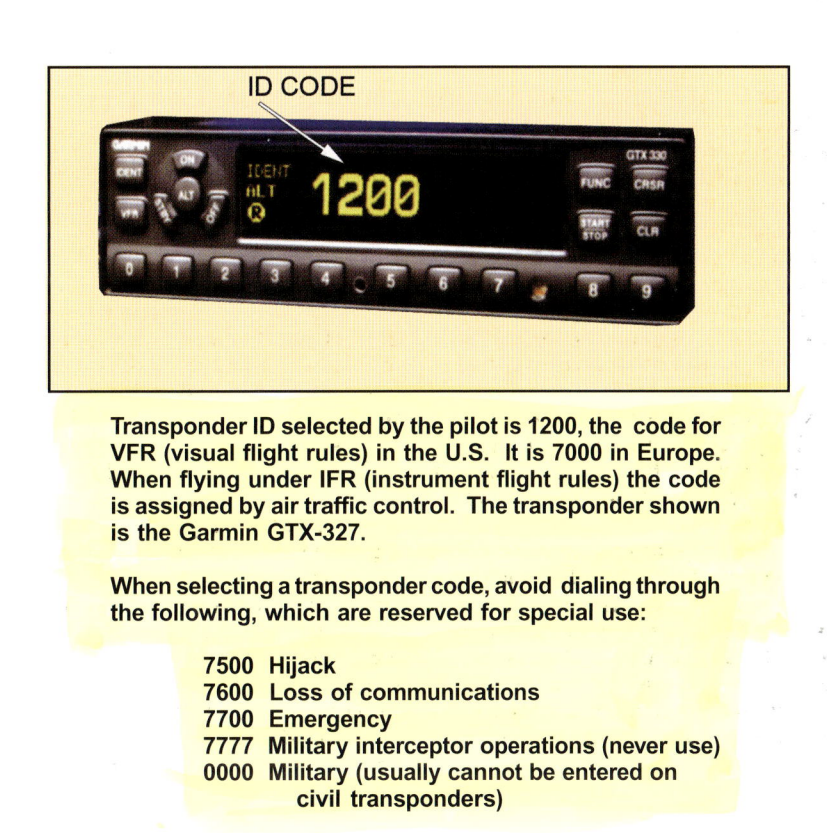

ID CODE — 1200

Transponder ID selected by the pilot is 1200, the code for VFR (visual flight rules) in the U.S. It is 7000 in Europe. When flying under IFR (instrument flight rules) the code is assigned by air traffic control. The transponder shown is the Garmin GTX-327.

When selecting a transponder code, avoid dialing through the following, which are reserved for special use:

- 7500 Hijack
- 7600 Loss of communications
- 7700 Emergency
- 7777 Military interceptor operations (never use)
- 0000 Military (usually cannot be entered on civil transponders)

Mode S: Interrogations and Replies

1. Mode S - ATCRBS All Call
In a mix of traffic carrying ATCRBS and Mode S transponders, the radar sends out an "all call" interrogation. All transponders reply with ID and altitude. Each Mode S transponder replies with its own 24-bit address. As each aircraft enters the radar coverage area, it responds to the "all call" interrogation.

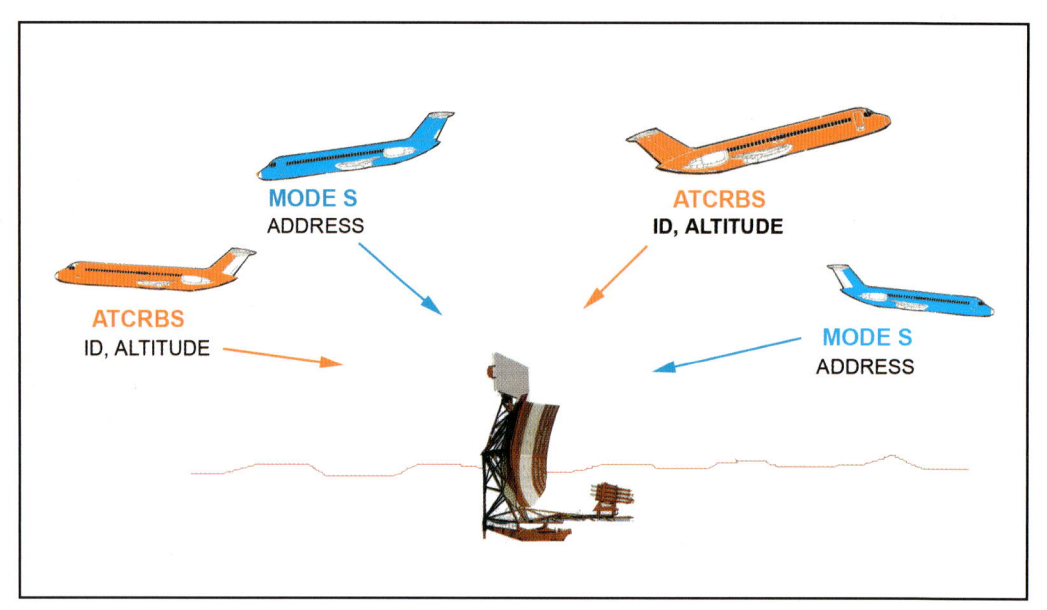

2. Mode S Discrete All Call
All Mode S transponders reply with their 24-bit address. ATCRBS aircraft do not reply.

When the radar gets a Mode S address, it locks out the transponder from replying to further all call interrogations. Until the airplane leaves the area, it replies only when radar interrogates it selectively. The airplane, however, is still tracked on the radarscope but with many fewer interrogations.

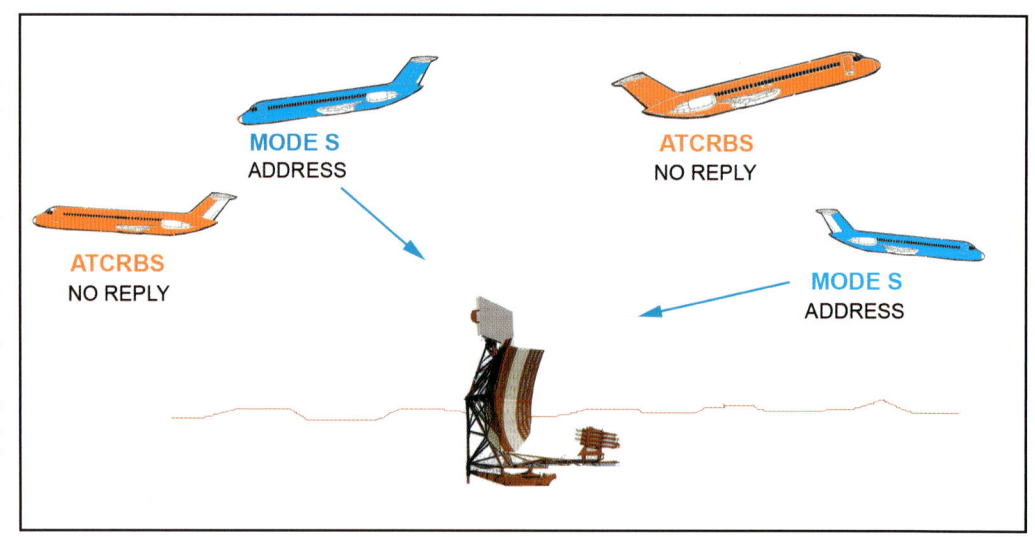

3. Mode S Selective Address
If a controller needs to communicate with only one aircraft, it transmits the Mode S address. Only that airplane replies.

This greatly increases capacity of the air traffic system. It also eliminates "synchronous garble"—where two airplanes reply to the same interrogation. This can happen when airplanes are on the same line to the radar antenna or are closely spaced.

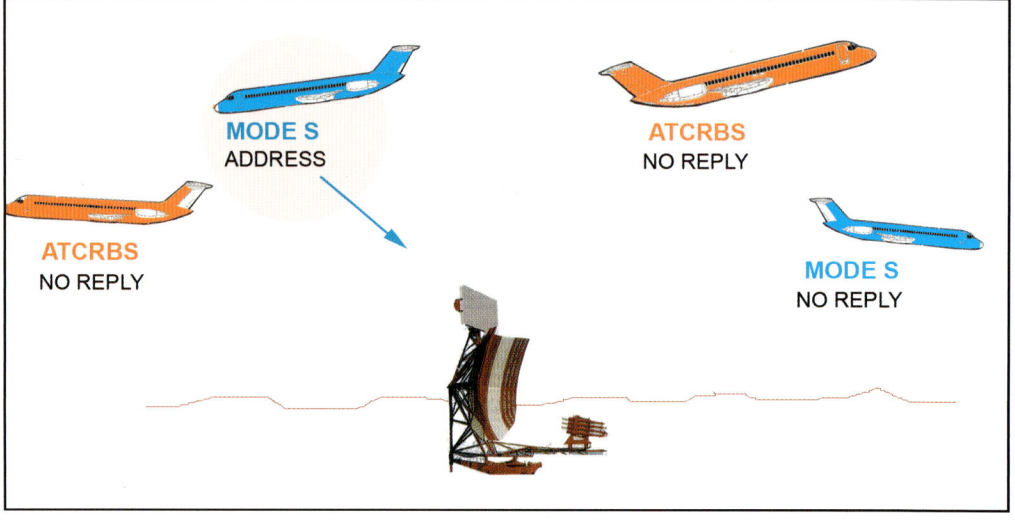

Review Questions
Chapter 14 Transponder

14.1 A transponder receives an _____ from a ground station and transmits a _____.

14.2 The transponder operates in a system known as "SSR". What do the letters mean?

14.3 How is an airline transponder labelled?

14.4 How many digits are in a transponder ID code?

14.5. What are two advantages of transponder signals over primary radar, or "skin" return?

14.6 The first secondary surveillance system is known as _____. The improved system is called _____.

14.7 What is the main benefit of Mode S?

14.8 What transponder information is carried by Mode A? Mode C?

14.9 When a radar interrogator wants every aircraft in range to reply (ATCRBS and Mode S), it transmits _____.

Chapter 15

Radar Altimeter

The radar altimeter is required for aircraft operating during the very low ceilings and visibilities of Category II and III instrument approaches. The conventional barometric altimeter can only read altitude above sea level and needs frequent readjustment as weather brings changes in pressure. Under the best of conditions, a conventional altimeter is usually accurate to 60 or 70 feet. That's not adequate when landing an airplane in near-zero visibility, where errors of more than about six feet may cause a hard landing or worse. Radar altimeters, on the other hand, give altitude with errors as small as two feet. Although the fully "blind" landing is not regularly flown, most widebody transports are equipped for "autoland"---which can fly the airplane through descent, decrab (nose straight), flare, touchdown and rollout. The radar altimeter is essential for such operations so close to the ground.

A radar altimeter makes it possible by measuring *absolute* altitude. Instead of deriving altitude from air pressure, it transmits a radio wave, listens for the echo from the ground, then computes altitude. The result is a reading in feet "AGL", above ground level. The computation is done by knowing the speed of the radio wave (speed of light) and amount of time that

Radar Altimeter Display

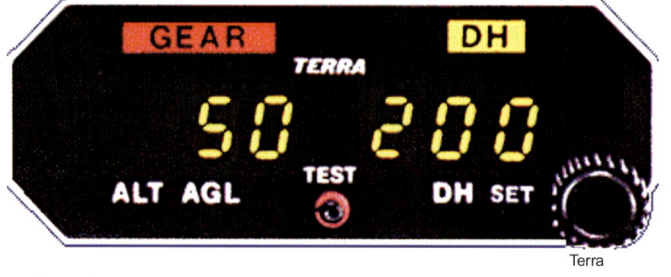

Panel-mounted radar altimeter indicator. The transmitter-receiver is located remotely. The LED display is indicating ALT AGL (altitude above ground level) of 50 feet. The pilot preselected decision height (DH) to 200 feet, a typical ceiling minimum for a Category I instrument landing. When the airplane descends below DH, the pilot hears an audible warning (1 kHz tone).

"GEAR" also illuminates and sounds the tone when the airplane descends below 100 feet and the landing gear is not down and locked.

Alerts can also give warnings every 100 feet, starting at 800 feet, to help the pilot find the ground.

When the TEST button is depressed, the display should read a 40-foot altitude and sound an audible alert.

Radar Altimeter Antennas

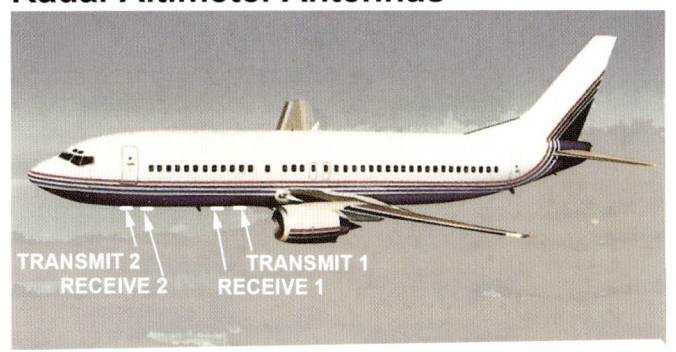

Two independent radar altimeters are installed for some instrument operations requiring separate antenna installations.

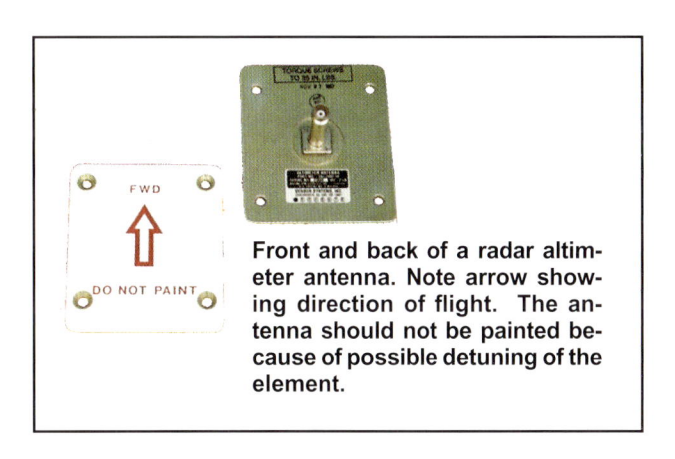

Front and back of a radar altimeter antenna. Note arrow showing direction of flight. The antenna should not be painted because of possible detuning of the element.

elapses for the wave to travel from airplane to ground and back. This is similar to radar, which sends out a pulse and measures the returning echo, but most radar altimeters do not use pulses. They operate on the principle of FMCW; for "frequency modulation continuous wave".

The radar altimeter is sometimes called by its older name, "radio altimeter." On the instrument panel is it often abbreviated as "RAD ALT."

The main components are a transmitter and receiver operating on a center frequency of 4300 MHz. The transmitter sends out low power of about 350 milliwatts. The receiver and transmitter each has its own antenna.

As shown in the illustration, the cycle begins with a transmitted wave Although it is basically on 4300 MHz, it continuously shifts in frequency (thus the name, frequency modulation). Let's say when the wave hits the ground, it is slightly *higher* than the resting frequency of 4300 MHz. When the echo reflects back to the airplane the round trip takes a certain amount of time. When the echo reaches the airplane, the passage of time has allowed the transmitter to shift to an even higher frequency.

Now the receiver has the information it needs; two frequencies that mark the beginning and end of the signal's round trip. Because the receiver knows how long it takes for the transmitter to shift frequency, it uses this value (time) in a calculation: travel speed of the signal multiplied by time equals distance. The rate of travel is constant; radio waves move like light (6.18 microseconds per mile). Because the signal travels down and up, the answer is divided in half before it is displayed to the pilot as altitude.

Radar altimeters are used only at low altitude, usually from -20 feet to 2500 feet. They are intended mostly for "altitude awareness" when the airplane is low to the ground in low visibility, and as part of the automatic flight control system. When the airplane cruises at altitude well above obstacles, or during a landing in good visibility, the regular barometric altimeter is sufficiently accurate.

The key functions of the radar altimeter are:

Decision Height (DH). During an instrument approach, a pilot makes a decision to: continue to land, abort the approach, go around and try again, or fly to better weather at an alternate airport. The radar altim-

Radar Altimeter: Operation

1. **The radar altimeter** produces a radio frequency carrier with a resting frequency of 4300 MHz. As it transmits, the frequency shifts 50 times per second off the resting frequency. The carrier increases to about 4350 MHz and down to 4250 MHz. This is shown at points A (Low Freq) and B (High Freq). From this process, the system derives its name; FMCW, for Frequency Modulation Continuous Wave.

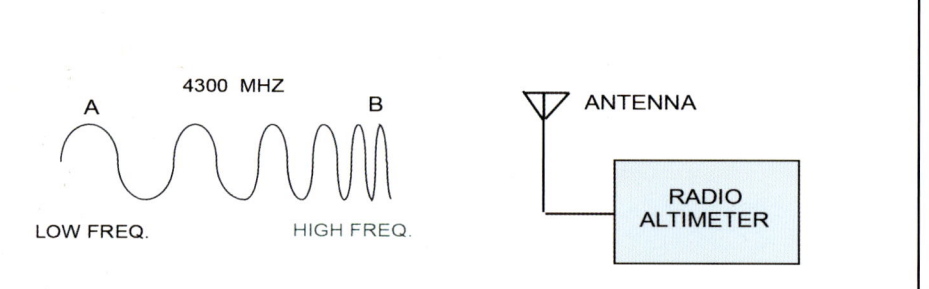

2. The airplane is shown over a point on the ground. At this instant, the radar altimeter is transmitting the carrier at its lowest frequency (A). The signal travels to the ground and is reflected back to the airplane at the same frequency (A).

The returning signal reaches the airplane after an interval of time, the number of microseconds it takes for the round trip.

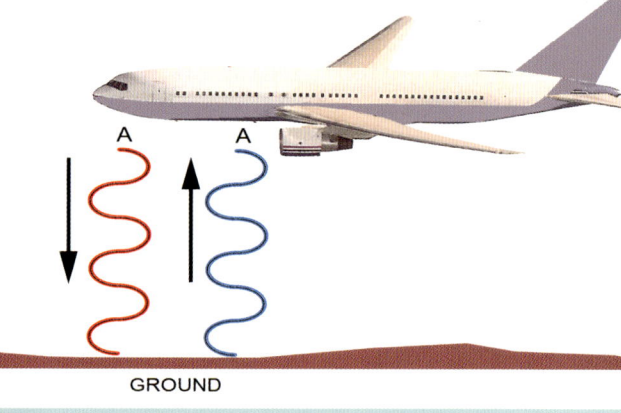

3. Because time has passed, the radar altimeter is now transmitting on a higher frequency (B). The system has two frequencies to process; A, the reflection from the ground, and B, the higher frequency. Three quantities are now known: the amount the carrier frequency is changing, elapsed time for the signal to make the round trip and the speed of the signal (speed of light). They are computed as altitude above ground in feet.

The shift in the carrier during frequency modulation is 50 times per second (50 Hz), enabling the radar altimeter to rapidly update. This is critical when an aircraft is descending through the last few hundred feet above ground.

eter is required for performing a Category II or III instrument approach, where decision height is 50 feet or lower.

Altitude Trips. In this function, the radar altimeter sounds a warning as the airplane descends through preselected altitudes. It gives the pilot a clear indication of when Decision Height will occur.

Gear Warning. If the radar altimeter senses the aircraft is close to touchdown, and wheels are not down and locked, it sounds a warning.

Other Applications. Radar altimeters work with other avionics systems aboard the airplane. A ground proximity warning system needs several inputs, including the radar altimeter, for determining whether to warn the pilot of a ground collision. It also operates the "rising runway," a symbol that appears at the bottom of the attitude director display for a Category II or IIIA instrument landing. It helps the pilot keep the airplane aligned with the runway during touchdown and roll-out. Other outputs of the radar altimeter go to the autothrottle, flight control system and flight data recorder.

Terra
A round analog display indicating 160 feet above ground level.

Although not a requirement for helicopter flight, radar altimeters are useful to pilots making a vertical descent in darkness. The pilot sets Decision Height at around 100 feet and, upon reaching that altitude, switches on a floodlight to see the landing spot below.

Review Questions
Chapter 15 Radar Altimeter

15.1 What is the main application of the radar altimeter?

15.2 The radar altimeter measures _____ altitude.

15.3 The altitude indicated by a radar altimeter is AGL. What do the letters mean?

15.4 What are three major components of a radar altimeter?

15.5 What is the resting carrier frequency of a radar altimeter?

15.6 What categories of approach require a radar altimeter?

15.7 The carrier of a radar altimeter is frequency modulated, meaning it moves _____ and _____ in frequency.

15.8 One factor in measuring radio altitude is the difference in frequency between the _____ and _____ wave.

15.9 When are radar altimeters useful in helicopter operations?

Ch 16

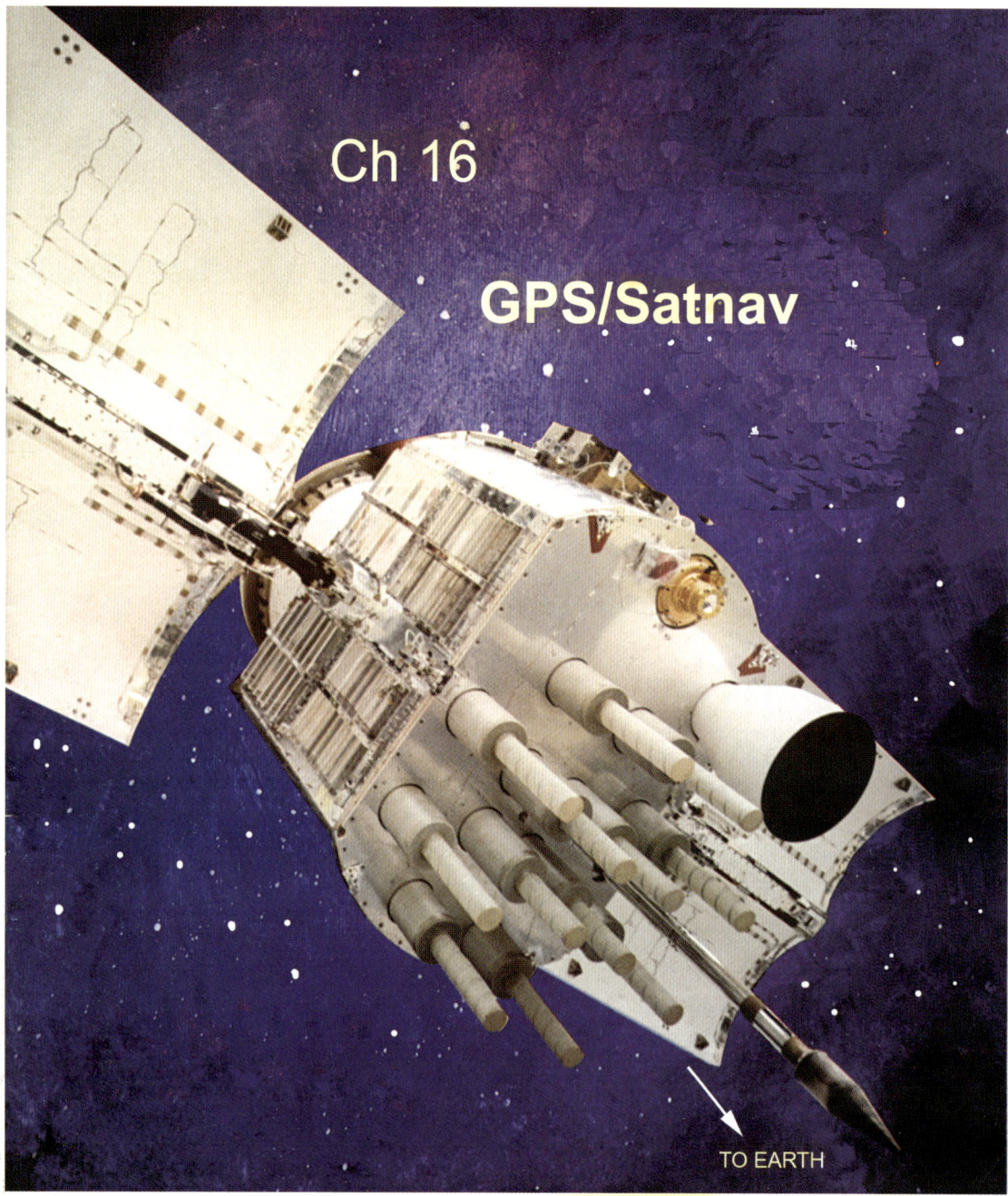

GPS/Satnav

TO EARTH

Rockwell

Satnav---satellite navigation---will replace nearly every other form of radionavigation. It meets or exceeds the accuracy, reliability and global coverage of land-based systems. Satnav is also supported by world aviation agencies to eliminate the high cost of servicing thousands of VOR, ILS, NDB and other ground stations.

The benefits of satnav will multiply as GPS, a U.S. system, is joined in coming years by Europe's Galileo. To avoid obsoleting the huge avionics investment in old aircraft, however, today's land-based stations will continue to operate well into the 21st Century.

Satnav. In the 1980's, the captain of a trans-atlantic flight between Sweden and New York walked back to the passenger cabin. He was approached by a man who said, "Captain, I believe we are 10 miles off course."

A typical GPS satellite weighs nearly 1800 pounds and is powered by a combination of solar panels (in sunlight) and rechargeable batteries (in the dark). Because precise time is essential, each satellite carries four atomic clocks that keep time accurate to one second in 80,000 years. The overall width of the satellite, including solar panels, is about 19 feet.

The satellite continuously broadcasts the GPS signal through 12 rod-like (helical) antennas, shown in the illustration as pointing "To Earth."

The captain looked at him and replied, "You're probably right." He had noticed the man holding a portable GPS. The airliner was navigating by the most advanced system known; the ring laser gyro. (Three of them, in fact, for redundancy.) The laser gyro is guaranteed not to exceed an error of 1 mile per hour---meaning that after a 4-hour flight across an ocean, the airplane would be no more than 4 miles off either side of its assigned track. That passenger's pocket

GPS Constellation: "Space Segment"

Number of satellites in orbit..........24
(21 active and 3 spares)
Height above earth...........11,000 n miles
Orbital planes....................................6
Time to complete one orbit.......12 hours
(circles the earth twice a day)

There are six paths, or "planes," with up to five satellites each. To an observer on earth, they appear to rise at the horizon, cross the equator, then descend below the opposite horizon. Over most of the earth, at least five or more satellites are available for navigation at any time.

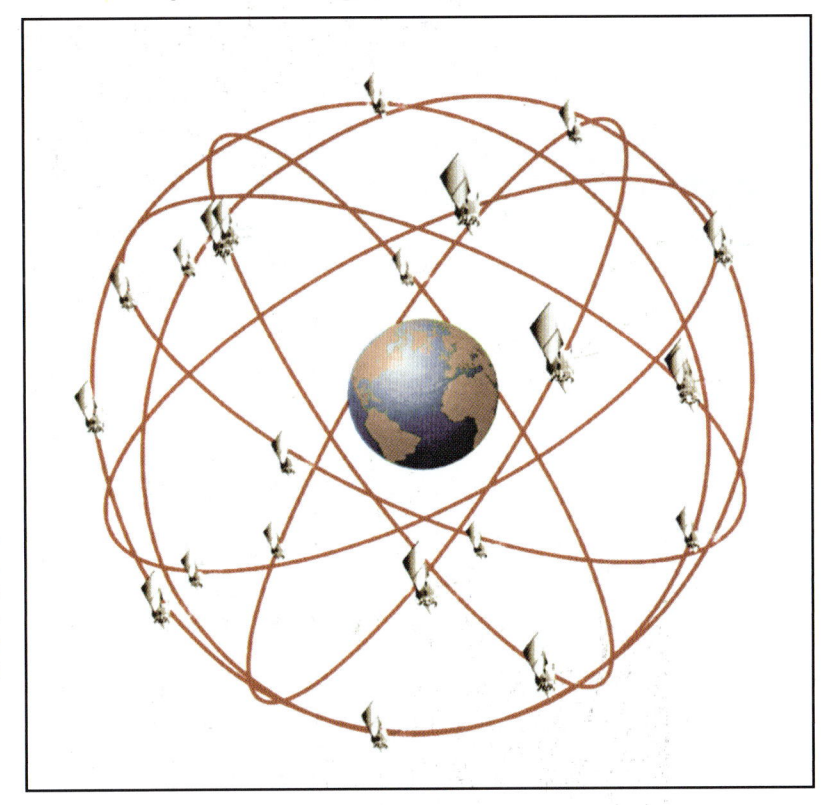

GPS not only knew the aircraft position within less than 100 meters, but held that accuracy throughout the trip (updating once per second).

Radionavigation for nearly 100 years sent out signals to be detected by an airborne receiver. But by the 1960's the microcomputer introduced digital signal processing. Instead of transmitting simple tones or timing pulses, a satellite could encode large amounts of data and broadcast it over the earth's surface. This data could then control the accuracy and performance of even the most inexpensive GPS receiver.

This places the costliest components---like precision timing generators---aboard the satellites, not in the receiver. Each GPS satellite carries at least two atomic clocks, which measure time by the motion of atomic particles of the elements rubidium or cesium. When the Galileo satellites appear, they will add yet another technology; the "maser," a cousin to the laser. In the maser, radio waves are driven to a high state of energy, then allowed to fall to a lower level. During the fallback, they produce oscillations which are extremely accurate timing references.

The GPS clock aboard the airplane, one the other hand, is a simple quartz type, like the one in wrist-

Launch Vehicle

GPS satellite is launched by Delta II rocket at Cape Kennedy. Although satellites are usually rated for a life of 7.5 years, many have operated 11 years.

109

watches. Its accuracy is controlled by precision data streaming down from the satellite.

Clocks are the key to fixing the position of the airplane. They measure the time for a signal to leave the satellite and arrive at the airborne receiver. By multiplying travel time by speed of the signal (which is the speed of light), the answer is the number of miles to the satellite. By solving that for several satellites, the receiver fixes the position of the airplane.

As shown in the illustrations, a satellite transmits its position and health, plus the position and health of *every* other satellite in the GPS constellation. Data is received continuously and stored in the receiver database. This enables the receiver to select a satellite and perform a "correlation". The receiver tries to locate and match the signal pattern transmitted by the satellite with patterns stored in its database. Once a match is discovered, the receiver can measure the travel time of the signal. The receiver knows when the signal left the satellite because the satellite transmitted a "navigation" message which tells when the signal was broadcast and the satellite's location in orbit. The receiver then compares the time with its own internal ("local") clock time. The difference between them is the amount of time the signal travelled. The receiver doesn't require an expensive atomic clock because its time is kept accurate by reference to the satellite clock.

The ability to correlate signals between satellite and receiver provides another benefit. Satellites carry a limited payload and generate radio carriers of very low power. So low, in fact, that when the signal arrives at the receiver it is below the "noise" (natural and man-made). In most radio communications, it is nearly impossible to pick out a signal when it is below the noise level. The GPS receiver, however, has stored in its database an exact pattern of the signal and can narrow its response to that coding. By acting so selectively, the receiver rejects much of the noise.

Early GPS receivers acquired only one satellite at a time, which is too slow for aviation. Today's GPS receivers are often "12-channel parallel," meaning they can process up to 12 satellites simultaneously. If a GPS receiver is taken thousands of miles from its home base with the power turned off, it "finds itself" a minute or two after being turned on. With a 12-channel parallel system, the receiver performs a rapid "sky search".

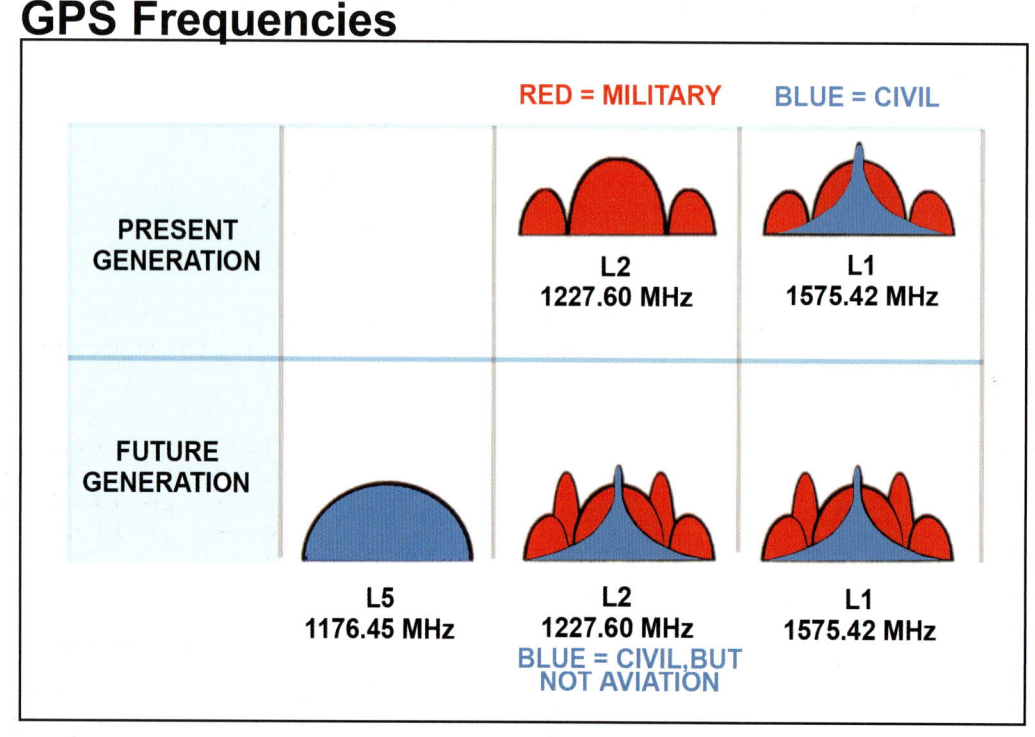

GPS Frequencies

RED = MILITARY **BLUE = CIVIL**

PRESENT GENERATION		L2 1227.60 MHz	L1 1575.42 MHz
FUTURE GENERATION	L5 1176.45 MHz	L2 1227.60 MHz	L1 1575.42 MHz

BLUE = CIVIL, BUT NOT AVIATION

In early GPS generations, civil aviation used L1 and shared it with the military. L2 was exclusively for the military who used it with L1 to achieve high accuracy. In the coming generation, civil users get a second frequency (L5) for increased accuracy. Error will drop to 3 to 10 meters. A major advantage of two frequencies is the ability of the receiver to measure and correct propagation error caused by differences in signal speed through the unstable ionosphere.

Satnav Terms and Service

Some major terms and acronyms to describe satnav systems:

GNSS: Global Navigation Satellite System. The international term to describe navigation based on satellites. When GNSS describes a precision instrument approach, it is called **GLS** (the LS meaning Landing System).

SA: Selective Availability. Because GPS began as a system for the U.S. Dept. of Defense, the highest accuracy was reserved for the military. The signal available to civil users was degraded by "SA," which limited receivers to about 100-meter accuracy.

In the year 2000, Selective Availability was turned off and accuracy improved by about five times.

SPS: Standard Positioning Service. This service is intended for civil users and is located on L1, one of two existing GPS channels. More channels will be added in the future.

Panel-Mount GPS Receiver

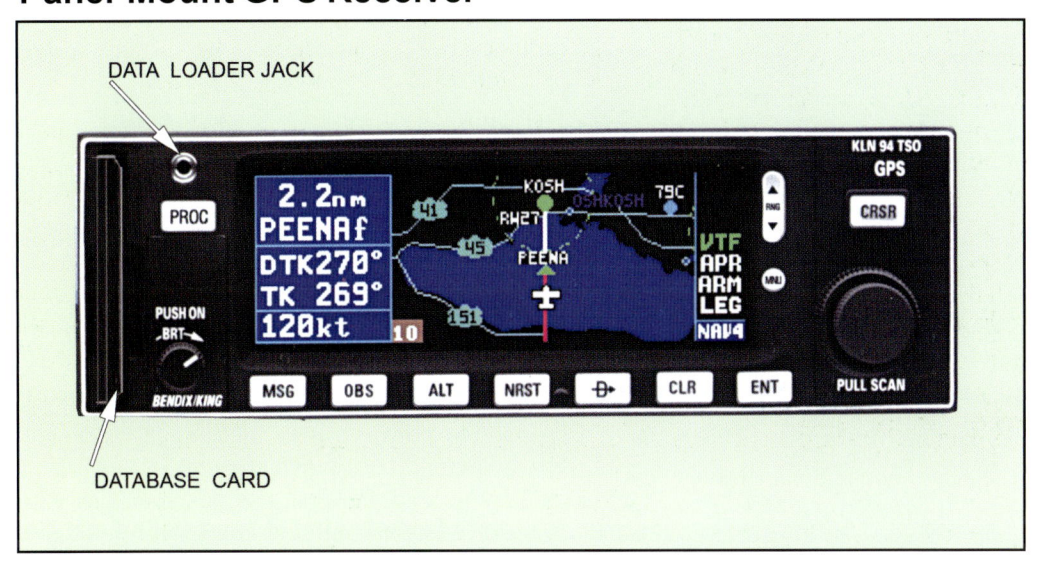

A GPS receiver used in IFR (instrument flight rules), like this Bendix-King KLN-94, must update its navigation database regularly. This can be done in two ways. A database card is inserted into the slot, or the dataloader jack connected to a PC. Updates may be obtained over the Internet.

Airplane Measures Time to Compute Distance to Satellite

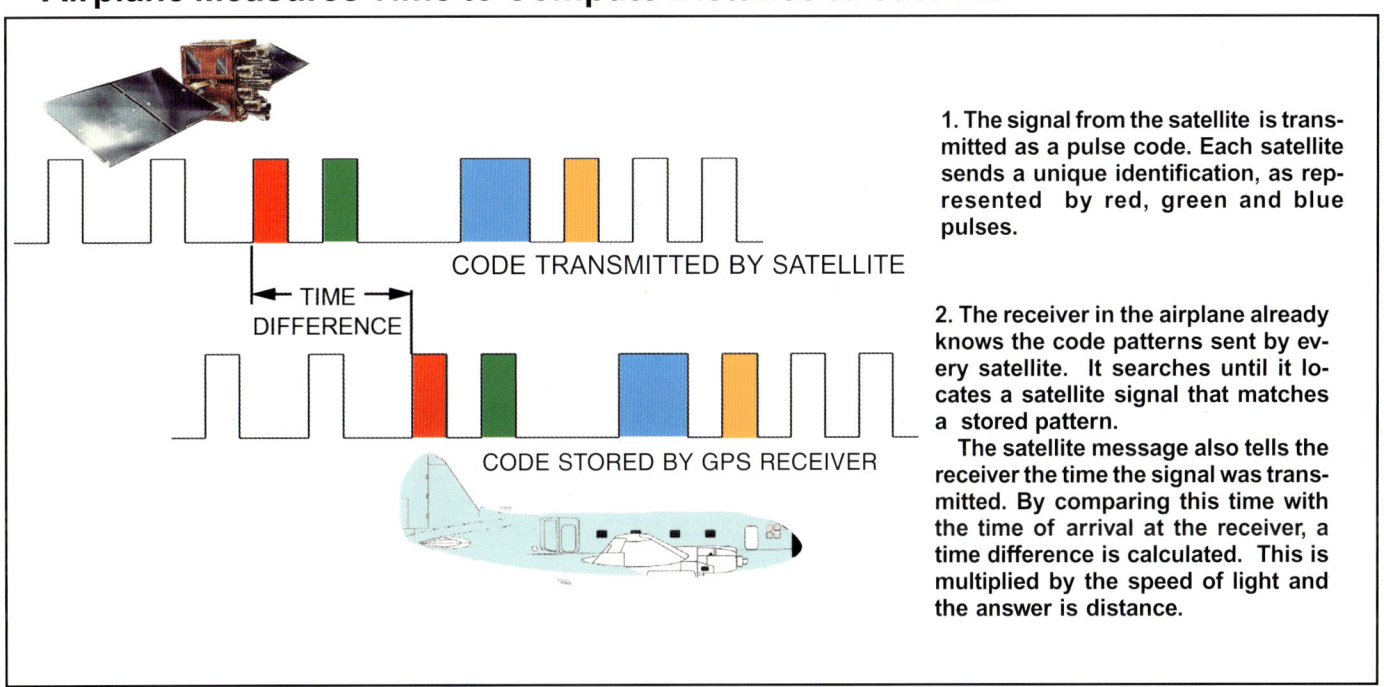

1. The signal from the satellite is transmitted as a pulse code. Each satellite sends a unique identification, as represented by red, green and blue pulses.

2. The receiver in the airplane already knows the code patterns sent by every satellite. It searches until it locates a satellite signal that matches a stored pattern.

The satellite message also tells the receiver the time the signal was transmitted. By comparing this time with the time of arrival at the receiver, a time difference is calculated. This is multiplied by the speed of light and the answer is distance.

111

Finding Position

When only one signal is received, the airplane may be located anywhere on the surface of a sphere (or "bubble"), with the satellite (SV1) at its center. After receiving a second satellite (SV2) the spheres intersect and narrow the position. With SV3, the position is further refined. It takes a fourth satellite to obtain latitude, longitude and altitude, which is a 3-dimensional fix.

Receiving a fourth satellite is required for correcting the clock in the GPS receiver. That enables a low-cost clock to keep sufficiently accurate time for the distance-solving problem.

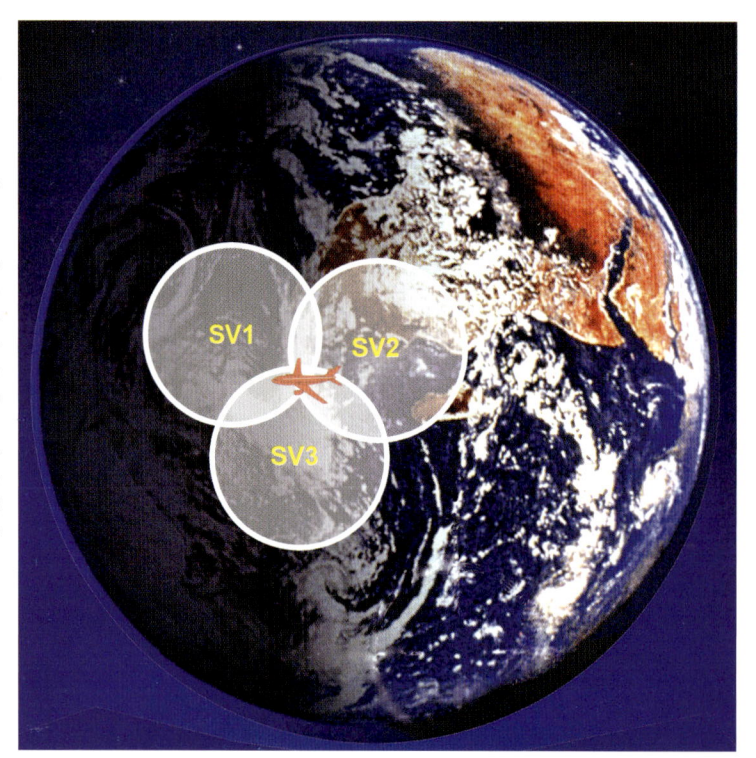

C/A Code: Coarse/Acquisition Code. This is the code transmitted on the civil channel, L1. It is "coarse" because it has the least accuracy.

PPS: Precise Positioning Service. Provides the greatest accuracy for military users and operates on the two existing GPS channels, L1 (shared with civil) and L2 (military only). It uses the "P Code," which is encrypted and available only to qualified users.

Propagation Corrections. The advantage of two channels, L1 and L2, is reducing propagation error. As satellite signals move toward earth, they pass through the ionosphere, which lies from about 60 miles to 200 miles above earth. Although radio signals move at close to the speed of light, higher frequencies move faster through the ionosphere, which introduces error. By measuring different GPS frequencies, L1 and L2, the receiver computes the "propagation" error and removes it.

The military removes the error because it is able to receive both L1 and L2. However, future satellites will add a civil code to L2, enabling any civil receiver to solve the error. Later, civil aviation will receive a third civil frequency, L5.

PRN Code

Each satellite transmits a unique identity known as a "pseudorandom" code. The term "pseudo" usually means "false" but in GPS it means "uncorrected". It is called "random" because the GPS signal resembles random noise.

The atomic clocks aboard satellites are extremely precise, however, they do drift in time. The error is measured by ground stations, but this is not used to correct satellite time. Atomic clocks are not easily reset while in orbit. It is more practical to develop a time correction factor based on the error and send it to the GPS receiver. The receiver stores the correction factors for all satellites in orbit and applies them while developing a position fix. For maximum timing accuracy, at least four satellites should be received.

Position-fixing. Determining aircraft position is similar to "triangulation." The receiver draws lines of position from several satellites, and locates the airplane where they intersect. One way to visualize this is to imagine each satellite surrounded by a large bubble. When the airplane measures distance from a satellite, the receiver places the airplane somewhere on the *surface* of the bubble. For example, if the range is 15,000 miles, the airplane may be anywhere on the bubble and still be at 15,000 miles.

Next, assume the receiver measures 18,000 miles to another satellite, placing the airplane on the surface of the bubble surrounding that satellite. As more satellites are measured, intersections among the bubbles grow smaller until the receiver has a position accurate to several feet. It is not unusual to receive eight or more satellites at once.

The Satellite Signal

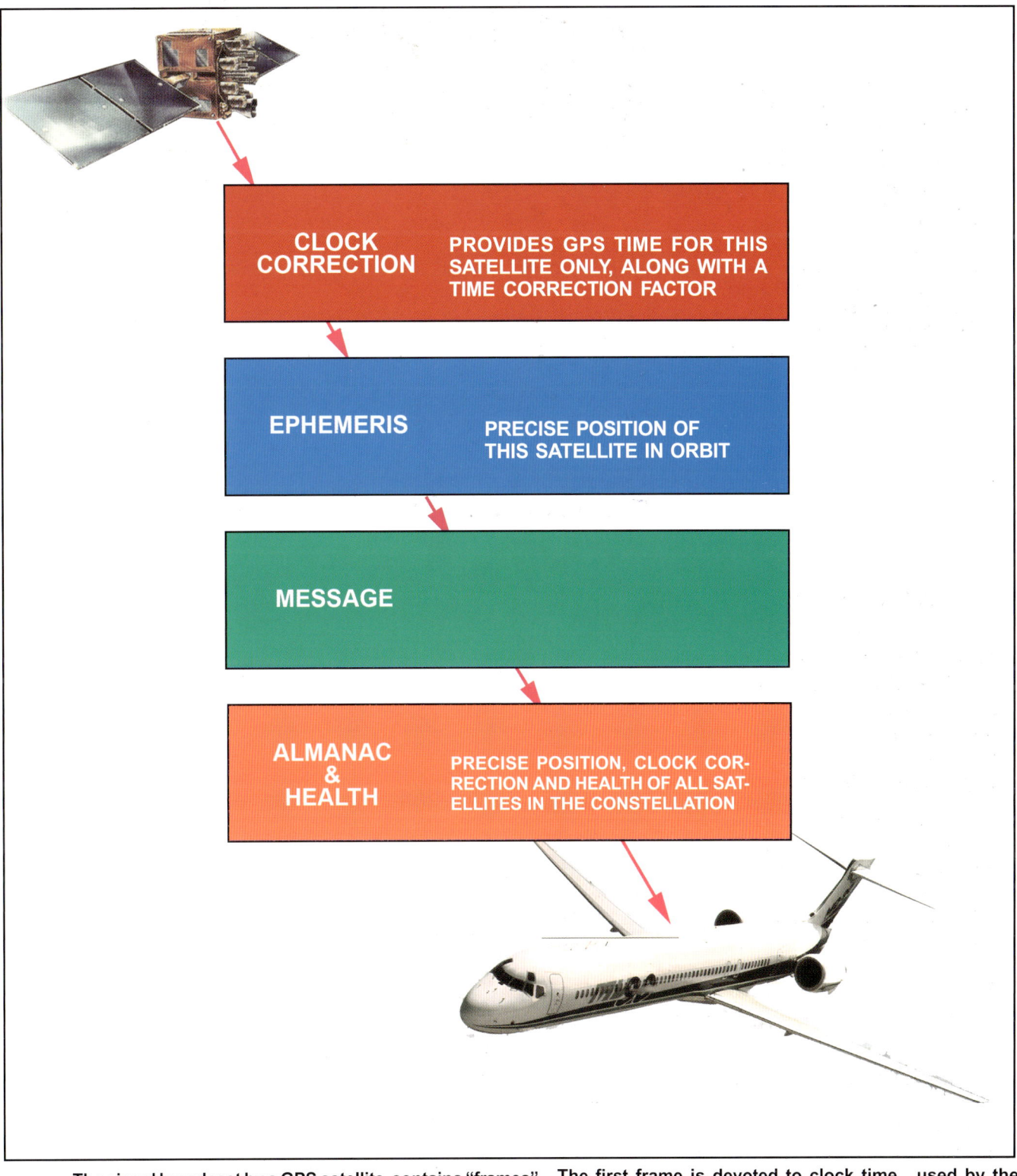

CLOCK CORRECTION — PROVIDES GPS TIME FOR THIS SATELLITE ONLY, ALONG WITH A TIME CORRECTION FACTOR

EPHEMERIS — PRECISE POSITION OF THIS SATELLITE IN ORBIT

MESSAGE

ALMANAC & HEALTH — PRECISE POSITION, CLOCK CORRECTION AND HEALTH OF ALL SATELLITES IN THE CONSTELLATION

The signal broadcast by a GPS satellite contains "frames" of information. They not only describe the position and health of the satellite, but data for every other satellite in the system.

The first frame is devoted to clock time, used by the aircraft receiver for measuring travel time of the signal (and thus distance).

GPS Segments

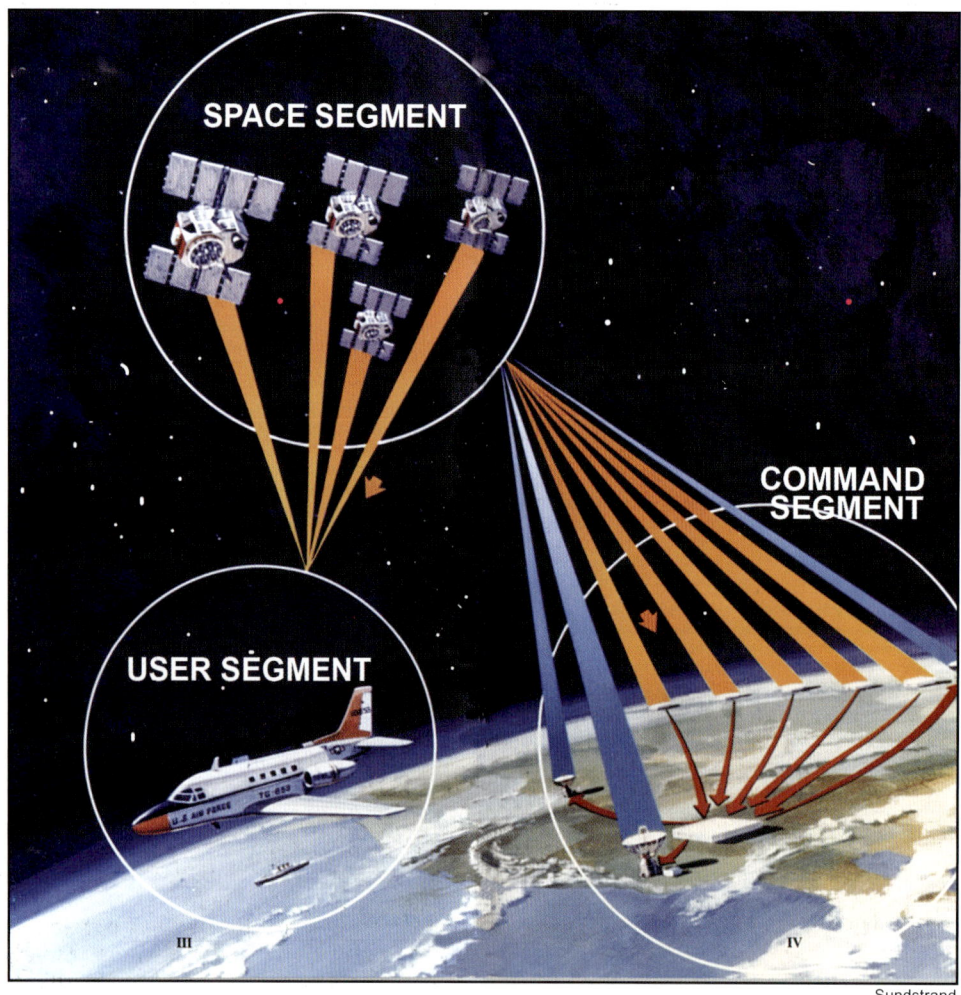

The three major components of the GPS system:

Space Segment: The satellites, also known as "SV" (satellite vehicle), circle the earth every 12 hours at an altitude of 11,000 miles. At any point on earth, up to ten satellites are in view.

Control Segment: A master control station in Colorado Springs connects to five or more ground stations around the world. They track satellites for performance and health. Orbit information and clock corrections are uplinked to the satellites several times a day. This data is broadcast on the navigational signal to users.

User Segment: Consists of airplanes, ships, other vehicles and portable GPS.

For an airplane flying over the earth, three satellites provide a 2-dimensional fix; which is latitude and longitude. Receiving four satellites provides a 3-dimensional fix; longitude, latitude and altitude.

WAAS: Wide Area Augmentation System

As GPS spread through aviation, it proved highly successful for cross-country travel between cities. But its full value would never be realized unless it replaced the instrument landing system (ILS), the key to on-schedule operations in almost any weather. GPS would have to equal the performance of a Category I ILS (200-ft ceiling and 1/2-mile visibility). Although GPS accuracy increased when Selective Availability (SA) was turned off in 2000, it was inadequate for Category I.

The solution in the U.S. is WAAS, or Wide Area Augmentation System. A network of 25 ground reference stations are placed around the country. The location of each station is surveyed with great accuracy. A GPS receiver at the station picks up the satellite signal and compares it with the known geographic position. The difference is the error (mostly due to propagation effects in the ionosphere). The ground stations relay that error to a geostationary satellite, which rebroadcasts it as a GPS signal. Airplanes in the general area of the ground station, therefore, can use the error to correct their GPS position. It is estimated that WAAS

Wide Area Augmentation System (WAAS)

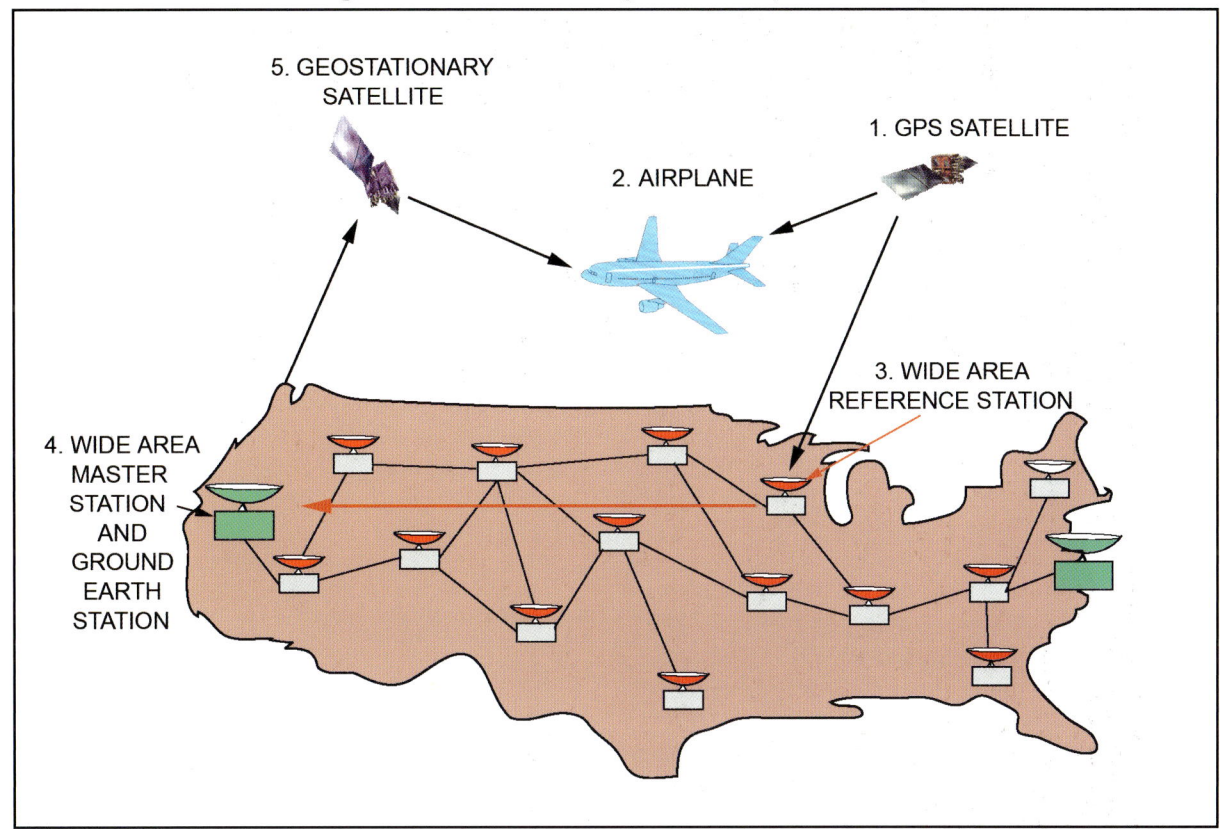

WAAS is a form of differential GPS designed to bring the equivalent of an Instrument Landing System (Category I ILS) to almost any airport. The sequence of events:

1. GPS satellite transmits a navigation signal.
2. The aircraft receives the signal, which has accumulated errors, mainly because of delays through the ionosphere.
3. Wide Area Reference Station. About 25 such ground stations cover continental U.S., spaced several hundred miles apart. They pick up the same signal as the airplane. A ground station, however, knows the error because its location was precisely surveyed. The surveyed location is compared with the position given by the GPS receiver at the ground station---and the error is determined.
4, Ground Earth Station-Wide Area Master Station. The error is sent cross-country through a network to one of two Wide Area Master Stations (on east and west coasts). A correction is developed.
5. Geostationary Satellite. The correction is sent to a geostationary satellite which appears fixed overhead. The satellite then broadcasts corrections (from every location in the country) to the airplane. This removes the "differential" error and raises accuracy sufficiently for precision landings (vertical and horizontal guidance).

The geostationary satellite transmits the error in a GPS-like format, which can be received as if it were another GPS satellite. Thus, WAAS for an airplane needs no separate receiver.

Several countries have WAAS systems, but call them different names. All systems, however, are compatible and can be used by any satnav receiver.

could add precision instrument approaches to over 5000 airports in the U.S.

Only one WAAS ground station is needed to cover many airports within its coverage area. Continental U.S. can be serviced by only 25 stations. They provide accuracy down to about 1 meter, which satisfies the most critical requirement of Category I, vertical guidance for the glideslope.

Another benefit involves the aircraft receiver. Because the WAAS correction arrives as a GPS signal, a separate receiver is not required. Also, no ground station is required at each airport and an airplane can use WAAS to approach any runway at the airport (once the approach has been designed and certified by a government agency).

GPS approaches use terminology that is different from that of ILS components, such as localizer and glideslope:

LNAV/VNAV. This refers to "lateral and vertical navigation". "Lateral" is equivalent to the localizer; "vertical" is the glideslope component. LNAV/VNAV provides GPS guidance down to a 350-ft ceiling and 1.5 mile visibility, which is not as good as a Cat I ILS, but offers thousands of airports their first precision (or any) instrument approach.

LPV: Localizer Performance Approach with Vertical Guidance. This improvement reduces the weather minimums to 250-ft ceiling and 1.5-mile visibility. It's still not as good as an ILS Category 1 approach but close to it.

SBAS: Space Based Augmentation System. The WAAS system is used in the U.S., but is also part of an international system known as "SBAS,' for space-based augmentation system. It is so-called because it uses geostationary satellites for relaying correction signals. Other countries use different names:

EGNOS: European Geostationary Navigation Overlay Service.

MSAS: Multifunction Transport Satellite System (Japan).

LAAS: Local Area Augmentation System

The first "blind" landing was demonstrated by Jimmy Doolittle in 1927. It is surprising that well into the 21st Century, the complete blind landing was still not achieved. It is true that widebody airliners are often equipped for "autoland," which provides a hands-off landing and roll-out. But civil air regulations in most countries do not permit it in actual instrument conditions, where the pilot sees nothing beyond the windshield.

This is the "zero-zero" landing---no ceiling and no visibility (or Runway Visual Range). It is known as "Category IIIc." Even Category II requires special equipment aboard the airplane and tighter specifications for the ILS transmitter on the ground. As a result there are few Category III operations anywhere in the world. Another obstacle is that autoland systems can put the airplane on the runway, but there is no guidance for taxiing to the terminal when dense fog blocks any view outside. (This should be solved with emerging "synthetic" and "enhanced" vision systems, which see the runway with infrared light, or construct an image from a mapping database.)

LAAS. A system known as LAAS, for Local Area Augmentation System, raises the accuracy of the GPS landing system above Category I. It is similar to the

Set up for a GLS Approach

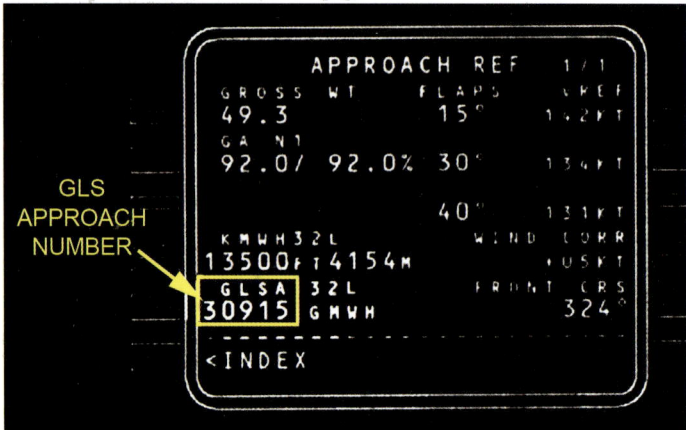

Every GLS approach has a unique 5-digit channel number, in this example, "30915." Above it is "GLSA" to identify it as a GLS approach. The display is a page in a Flight Management System typical of a large transport. Once the GLS channel is selected, the pilot (or flight control system) is guided through courses and altitudes to make the approach to the airport. This example is the approach to Runway 32L (left) at Grant County International Airport in Washington State.

Block IIR: Second Frequency for Civil Aviation

GPS continuously improves with each generation. Shown here is a Block IIR satellite (built by Lockheed Martin). The major enhancement is a second GPS signal for civil aircraft to improve accuracy. Earlier satellites provided only one civil signal, which picks up errors when it changes speed through the ionosphere. With two signals, their speeds are compared and used to reduce the error.

The new satellites also provide the military with an "M code," which operates at higher power to improve resistance to jamming.

wide area system (WAAS) described above, except that the ground monitoring station for correcting error is located on the airport. Because GPS error is sensed within hundreds of feet (not miles) from runways, the correction provides high accuracy in the airborne receiver. As shown in the illustration, corrections are not relayed via an orbiting satellite, but through a VHF station whose signal is picked up by the airplane. Now the accuracy of navigation is down to 1 meter, and satisfies the demands of vertical descent. It can also provide a moving map of runway and taxiways for the pilot to find his way to the terminal in the densest fog.

The international term for the LAAS system is GBAS, for "ground-based augmentation system". It refers to the VHF ground stations that transmit the correction signal.

Multimode Receivers. The aviation world is in a long-term changeover from ILS to GLS, and both systems will exist side by side for many years. For this reason, many new long-range aircraft are equipped with MMR---multimode receivers--- that operate on all systems, including Microwave Landing System (MLS), VOR, ILS and GLS. No matter what system is in operation at an airport, the pilot sees the same steering commands on his instruments and displays.

RAIM: Receiver Autonomous Integrity Monitoring

During the transition to satnav, GPS will not be certified for instrument approaches unless there is a back-up system on the airplane in the event of failure. The backup may be another form of navigation, such as ILS (instrument landing system). The pilot must not only monitor the GPS, but the other source, as well. This is hardly practical and places a heavy workload on the pilot.

To avoid this problem, the RAIM system was developed, which enables the GPS receiver to check itself. Meaning "Receiver Autonomous Integrity Monitoring," RAIM checks whether there are enough satellites for the approach, whether they are healthy and if their geometry is adequate. The last factor, geometry, is the layout of satellites during reception; their signals must arrive from widely different angles for the GPS receiver to develop an accurate position.

It usually takes four satellites to determine longitude, latitude, altitude and a correction for the receiver clock. RAIM is accomplished by receiving at least five satellites, which should be possible anywhere in the world. The receiver searches for the best satellite combination and, when satellites drop below the horizon, new ones are acquired as they rise into view. Five satellites assure that integrity is sufficient to fly the instrument approach. The GPS receiver will "look ahead" and determine if RAIM will be acceptable before the airplane commences the instrument approach. If it's inadequate a warning appears; "No GPS RAIM."

LAAS for High Accuracy

A Local Area Augmentation System (LAAS) installed at an airport. The GPS ground station receives the same satellites as the airplane. The station location, however, was surveyed and its precise position known. If it detects differences between its known geographic location and GPS position, a correction is transmitted to the airplane.

The technique is called "Differential GPS" (DGPS) because it detects error as a "difference" signal. Internationally, it's called "GBAS," for ground-based augmentation system.

117

Galileo Constellation

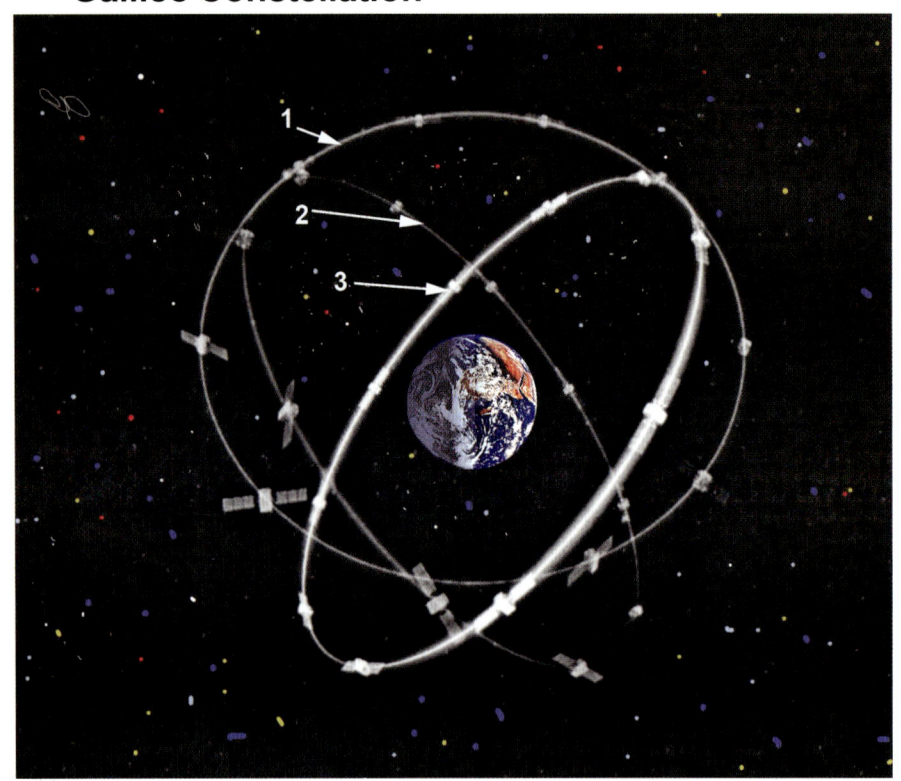

The European system, Galileo, inserts 30 satellites into orbit, with three standing by as spares. Unlike GPS, which has six orbits, Galileo has three (shown in the illustration).

The constellation, at an altitude of 14,600 miles (23,616 km), is slightly higher than GPS (12,500 miles).

Two ground stations in Europe gather data from 20 sensor stations around the world and uplink data to synchronize clocks and maintain orbits.

The source of Galileo's timing accuracy are two clocks carried aboard each satellite. One is a rubidium atomic clock (as in GPS), the other is a newer type, "passive hydrogen maser."

Galileo is interoperable with GPS; a receiver aboard the airplane processes both signals.

Galileo Satellite

Solar panels, which generate 1500 watts of primary power, rotate with the satellite around an earth-pointing axis. The panels are kept facing the sun, while navigation antennas point toward earth. Each satellite weighs 1400 pounds (650 kg).

A goal of Galileo is to achieve navigational accuracy to within 5 meters without augmentation systems on the ground (see WAAS and LAAS).

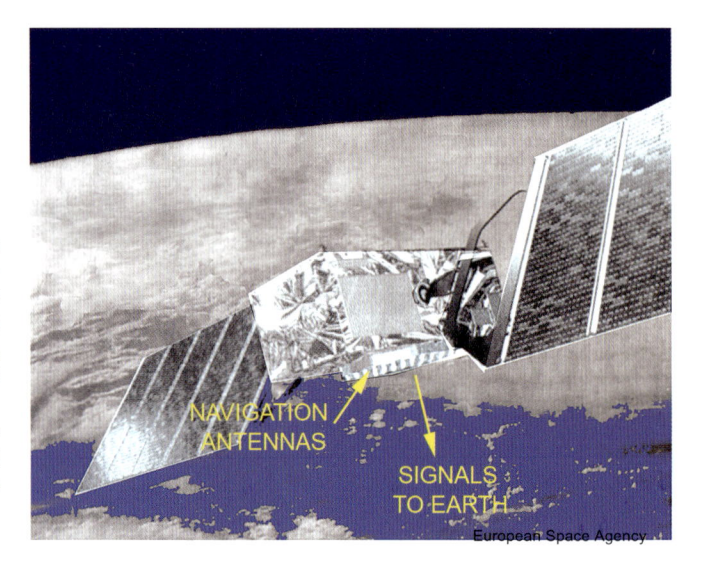

Review Questions
Chapter 16 GPS/Satnav

16.1 How many satellites are there in a GPS constellation?

16.2 How many GPS satellites are active? How many are spares?

16.3 What is the European equivalent of GPS?

16.4 Why don't GPS receivers in airplanes require expensive atomic clocks (like those in satellites) to measure time with high accuracy?

16.5 How does a GPS receiver identify a satellite?

16.6 What is the term for a satellite's identity?

16.7 How does a GPS receiver measure the time for the signal to travel from satellite to receiver?

16.8 How is distance determined between the GPS receiver and the satellite?

16.9 GPS frequencies, or channels, are designated by the letter _____

16.10 How many satellites are required for a three-dimensional fix (latitude, longitude and altitude)?

16.11 How many satellite frequencies are required to perform propagation corrections?

16.12 What part of the satellite signal carries the satellite's precise position in orbit?

16.13 Name the three segments of the GPS system.

16.14 The Wide Area Augmentation System (WAAS) uses ground stations and satellites to _____

16.15 What is the advantage of LAAS (Local Area Augmentation System) over WAAS?

16.16 What is the purpose of RAIM (Receiver Autonomous Integrity Monitoring)?

Chapter 17

EFIS

Electronic Flight Instrument System

Along with digital electronics and GPS navigation, the Electronic Flight Instrument System changed the face of the flight deck. The term EFIS originally described an airline system (that first rolled out with the Boeing 767 in 1981) but today it identifies electronic instruments for aircraft of all sizes.

EFIS is often called the "glass cockpit" because TV screens replace mechanical and electromechanical instruments. Dozens of old "steam gauges" are now replaced by an EFIS display that is rapidly changing from about a half-dozen separate screens to "wall-to-wall" glass.

Six large EFIS screens span across the main instrument panel. This is the flight deck of a B-757, first airliner to adopt electronic flight instruments. Although these displays are CRT's, newer aircraft use LCD flat panels

Replacing Old Instruments

Electronic instruments are also designed as direct replacements for old electromechanical equivalents. In these two examples, the instruments are LCD that provide both analog and digital read-outs. Reliability of these solid-state devices is much greater than that of mechanical types.

The instrument on the left is a torque indicator used in military helicopters. The one on the right is a vertical torque indicator for commercial helicopters

Aerospace Display Systems

Electronic instruments bring many benefits. They eliminate hundreds of gears, bearings, pointers, rotating drums and other fragile mechanical components. Any instrument is easily duplicated on the screen by programming its image.

EFIS Pictorial Display

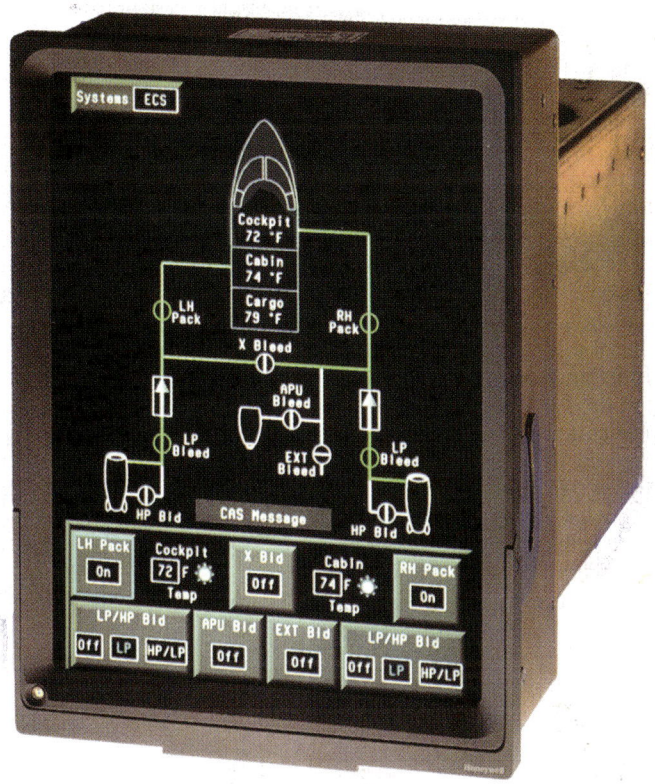

Instead of spreading information over different instruments on the panel, EFIS overlays them into a single, easy-to-understand image, for example; a map display can also show thunderstorms, high terrain and nearby aircraft.

An EFIS display may be decluttered to show only information required for that phase of the flight. If there's an "exceedance," meaning a system is developing a fault, it automatically appears to warn the pilot.

Because there is almost no limit to what can be shown as an image on a screen, EFIS brought in new generations of symbols that are pilot-friendly. The first systems simply created pictures of instruments they replaced, but it became apparent there were better images. For example, pilots fly an ILS (instrument landing system) by keeping two needles centered; one for

An example of how EFIS presents an easy-to-read pictorial view. This environmental control system was once a collection of knobs and gauges on an overhead panel. Now it's in the instrument panel. At a glance, the pilot sees how bleed air flows from the engines (lower left and right) and is distributed for controlling temperature in the cockpit and passenger cabin, including position of valves.

Also shown is bleed air from the APU (auxiliary power unit) and from an external source (just below center of screen).

This presentation is more useful for troubleshooting the system in flight than consulting a paper manual.

Honeywell: Primus Epic

121

the localizer to remain on the runway centerline, the other a glideslope for vertical guidance. In a series of experiments by NASA, the "highway in the sky" was developed. The pilot aims the airplane at a rectangle (or hoop) on the screen and flies through it. Additional hoops appear in the distance; if the pilot keeps flying through them (like threading a needle), the airplane remains on the localizer, glideslope or other 3-dimensional path.

The original technology for EFIS was the heavy glass cathode ray tube. Flat panel LCD's of the 1970's were not ready for aviation; they were monochrome, had narrow viewing angles and low resolution. Driven by the large market in portable PC's, the technology advanced rapidly and all new EFIS systems are flat panel.

There is much retrofitting of old aircraft to replace their electromechanical instruments with EFIS. It's happened in most major transport aircraft in "derivative" models, usually shown by a "dash number;" for example; the Boeing 737-100 first rolled out in 1967 with conventional instruments. It is now up to 737-900, with recent generations equipped with EFIS.

Transition from Electromechanical to EFIS

This EFIS screen is a "Primary Flight Display" and combines many early instruments into a single screen. The top half was once the artificial horizon. One of the first improvements was the addition of "command" bars---the V-shape near the middle of the screen. Driven by the flight control system (autopilot), the bars helped guide the pilot fly manually or enabled him to observe commands of the autopilot. At this stage, the instrument was called an "ADI," for attitude director indicator.

When EFIS appeared, all the same functions were pictured on a video screen. This called for a new name, EADI, for electronic attitude director indicator. At the same time, several other instruments were added to the image; air speed indicator, altimeter, vertical speed indicator and others.

The lower half of the screen was once the horizontal situation indicator (HSI). When the electromechanical instrument is shown on an EFIS screen it is known as an EHSI, the "E" is for electronic.

When the two major flight instruments---ADI and HSI---are placed one above the other and connected to the autopilot, they are known as a "flight director."

The trend in EFIS, however, is to combine those instruments onto one screen, as shown here, and call it a "Primary Flight Display."

The system in the illustration is the Honeywell Primus EPIC, a flat panel measuring 8-in by 10-in.

122

Three-Screen EFIS

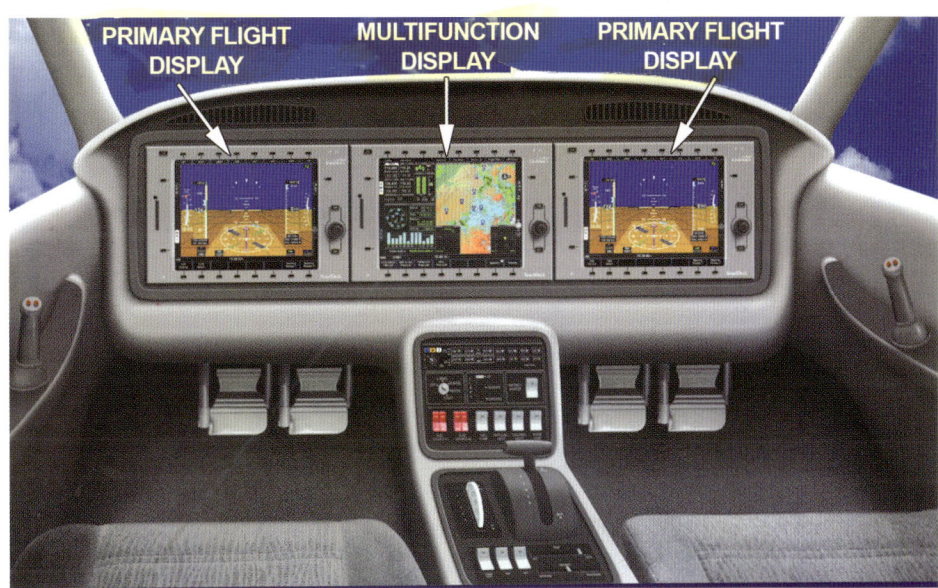

PRIMARY FLIGHT DISPLAY — MULTIFUNCTION DISPLAY — PRIMARY FLIGHT DISPLAY

L3 Communications

The future of instrument panels is shown in this "Smartdeck" by L3 Communications. It is "wall-to-wall" glass, with three 10.5-inch panels that display information once required by three separate instruments.

Panels like these are usually interchangeable, with their function determined by how their software is programmed. This permits "reversionary modes," meaning that any display on one panel can be switched over to another.

The panels are arranged as Primary Flight Displays for captain (left) and first officer (right). In the center is the multifunction display. Because the multifunction display typically displays engine instruments and warnings it is also called EICAS, for engine instrument and crew advisory system.

The Primary Flight Display is mainly for controlling the attitude of the airplane and for navigating.

Primary Flight Display

Attitude	Engine Power
Heading	Selected Heading
Altitude	Selected Course
Airspeed	Autopilot/Flight Directo
Vertical Speed	Navigation
Lateral/Vertical Path	Timer

Multifunction Display

Radio Management	Charts
Aircraft Systems	Runway Diagrams
Engine Instruments	Wind Direction/Speed
Checklists	Ground Track
Moving Map	Caution/Warning
Flight Plan	Geographic Overlays
Terrain	Lightning/Weather
Datalink (Traffic/Weather)	Traffic Information

EFIS Architecture

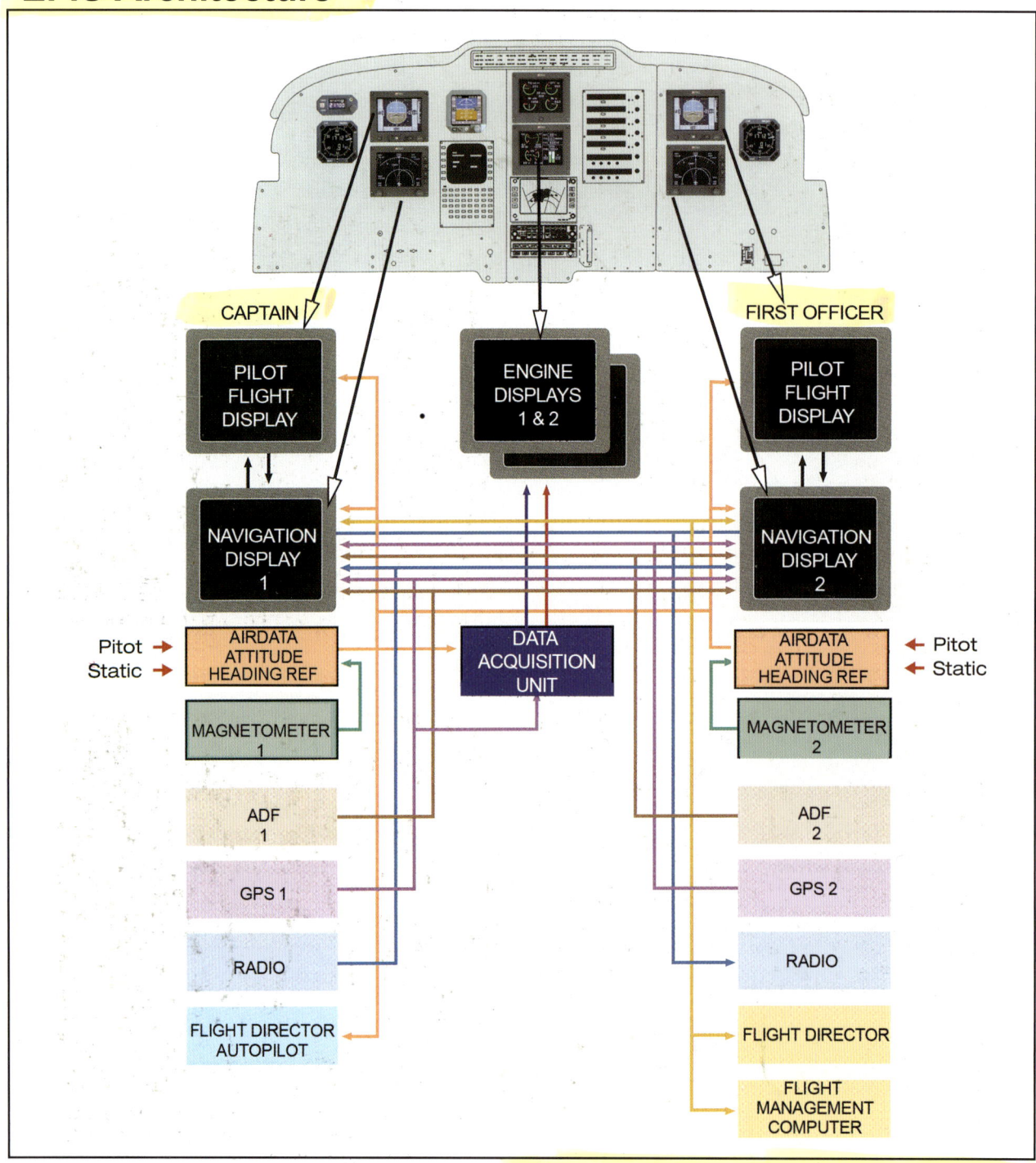

CAPTAIN — **FIRST OFFICER**

- PILOT FLIGHT DISPLAY
- ENGINE DISPLAYS 1 & 2
- PILOT FLIGHT DISPLAY
- NAVIGATION DISPLAY 1
- NAVIGATION DISPLAY 2
- AIRDATA ATTITUDE HEADING REF (Pitot, Static)
- DATA ACQUISITION UNIT
- AIRDATA ATTITUDE HEADING REF (Pitot, Static)
- MAGNETOMETER 1
- MAGNETOMETER 2
- ADF 1
- ADF 2
- GPS 1
- GPS 2
- RADIO
- RADIO
- FLIGHT DIRECTOR AUTOPILOT
- FLIGHT DIRECTOR
- FLIGHT MANAGEMENT COMPUTER

An EFIS system requires inputs from various sources, as shown in this system known as "MAGIC," for Meggitt Avionics new Generation Integrated Cockpit."

Because the electronics are digital, any analog signals from outside must go through the Data Acquisition Unit (DAU). Signals from engine sensors and fuel probes, for example, are converted to digital format. The DAU can also store data for engine trend monitoring, which can detect faults before they cause a failure.

The Air Data Attitude Heading Reference System (ADAHRS) replaces conventional sensors for measuring temperature, pressure, altitude, airspeed and others. It also eliminates "spinning iron" gyro's for aircraft attitude and heading. It's done with solid-state devices containing almost no moving parts.

Note how the instrument panel is divided into to nearly identical halves; for captain and first officer (co-pilot). This provides the safety of redundant systems, which are powered from different sources. In a typical EFIS a display on one side of the panel can be switched and viewed on the other side.

Multifunction Display: MFD
One MFD, like this Apollo MX-20 displays a wide variety of navigation, weather and traffic information.

The 360-degree compass rose is a horizontal situation indicator. The red, yellow and green areas are terrain warnings.

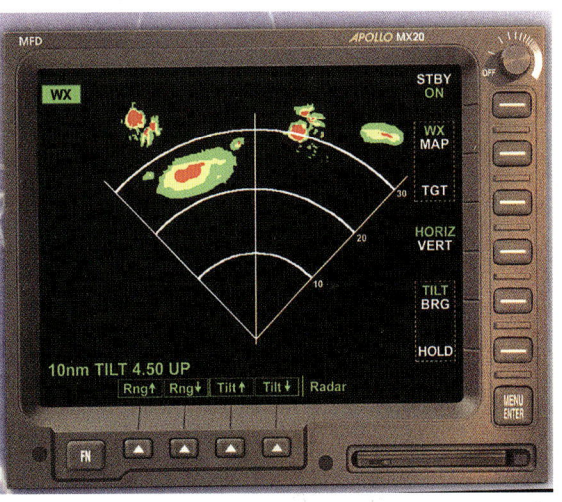

This screen shows weather radar images and works with several makes of radar sets.

Weather shown here is not from aircraft radar, but signals from NEXRAD, a nation-wide system of government ground radars. The pilot may select weather images from any area of the country.

Horizontal Situation Indicator with waypoints along the route. Also shown is nearby traffic; targets appear as small blue arrowheads.

Navigational charts for enroute and approach phases of flight. When the airplane lands, the chart changes to a taxi diagram.

125

EFIS on the B-747-400

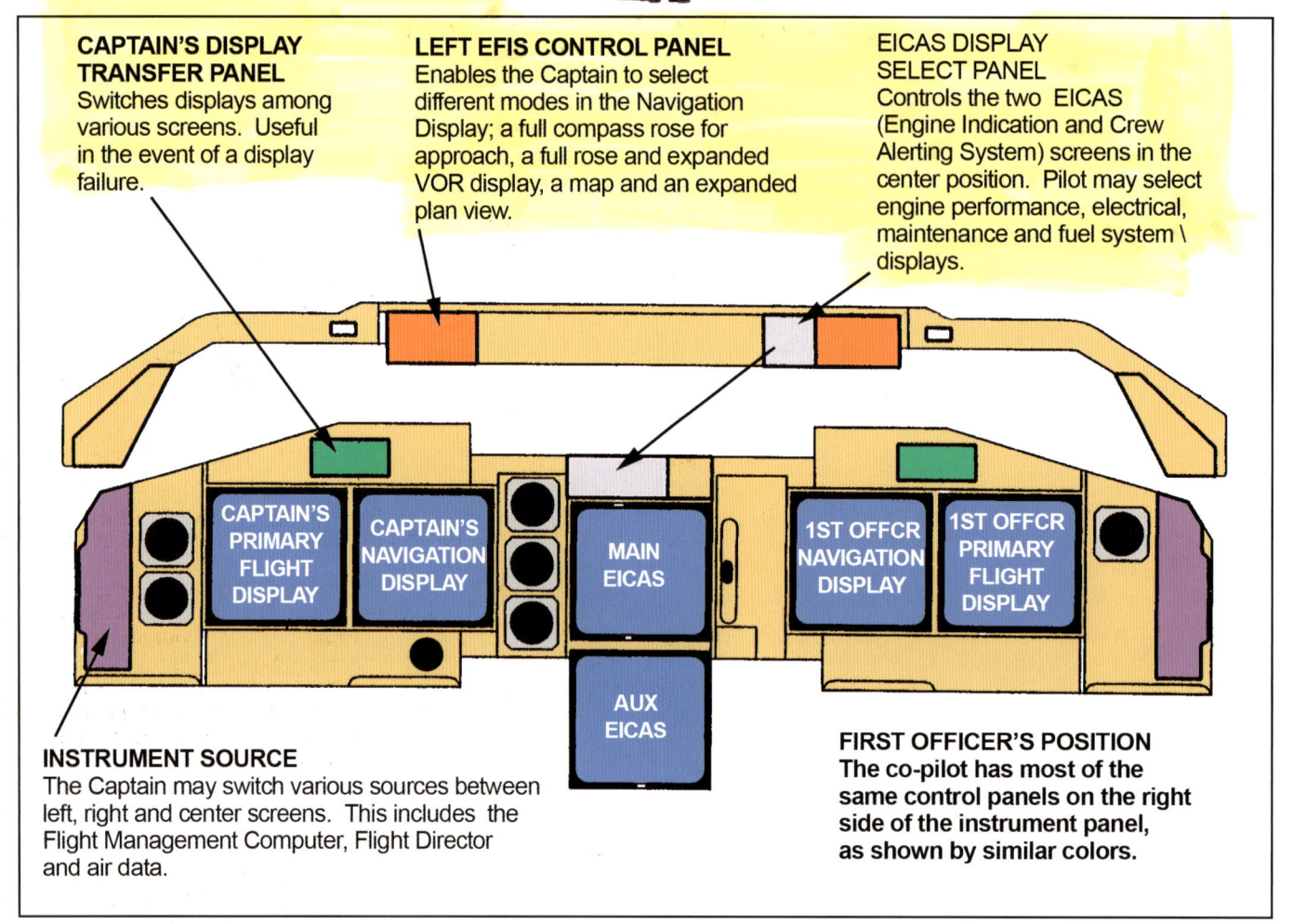

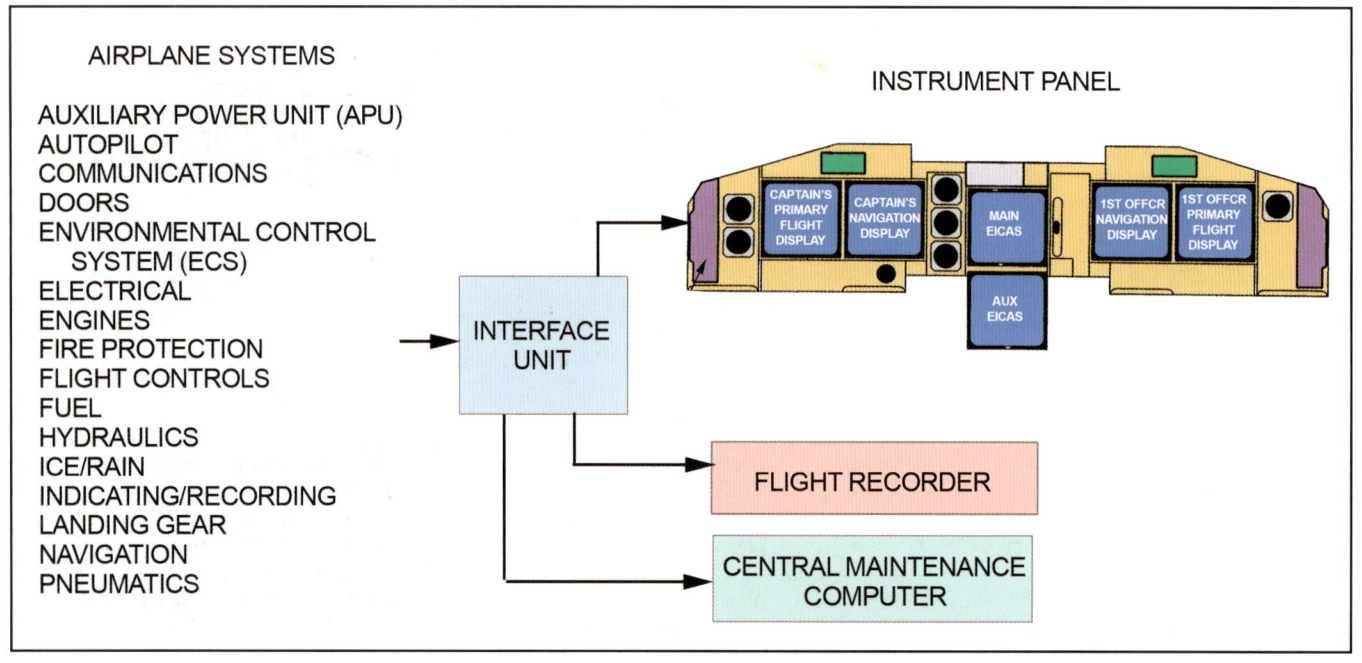

CAPTAIN'S DISPLAY TRANSFER PANEL
Switches displays among various screens. Useful in the event of a display failure.

LEFT EFIS CONTROL PANEL
Enables the Captain to select different modes in the Navigation Display; a full compass rose for approach, a full rose and expanded VOR display, a map and an expanded plan view.

EICAS DISPLAY SELECT PANEL
Controls the two EICAS (Engine Indication and Crew Alerting System) screens in the center position. Pilot may select engine performance, electrical, maintenance and fuel system displays.

INSTRUMENT SOURCE
The Captain may switch various sources between left, right and center screens. This includes the Flight Management Computer, Flight Director and air data.

FIRST OFFICER'S POSITION
The co-pilot has most of the same control panels on the right side of the instrument panel, as shown by similar colors.

AIRPLANE SYSTEMS

- AUXILIARY POWER UNIT (APU)
- AUTOPILOT
- COMMUNICATIONS
- DOORS
- ENVIRONMENTAL CONTROL SYSTEM (ECS)
- ELECTRICAL
- ENGINES
- FIRE PROTECTION
- FLIGHT CONTROLS
- FUEL
- HYDRAULICS
- ICE/RAIN
- INDICATING/RECORDING
- LANDING GEAR
- NAVIGATION
- PNEUMATICS

Information from airplane systems is applied to an interface unit. The data is digitized and symbols generated for displaying images on the EFIS and EICAS screens. The interface also sends some of that data to the flight data recorder and the central maintenance computer for storage.

Airbus A-320 Flight Deck

The A-320 began flying in 1988 as a twin medium-range transport. Because EFIS panels are interchangeable, fewer spares are required for maintenance.

A feature of the A-320 is the absence of control yokes for captain and first officer. Yokes are replaced by two sidestick controllers, as found in fighter aircraft. This gives a wide, unobstructed view of the instrument panel.

It is also "fly-by-wire," where the sidesticks drive computers that, in turn, control actuators for rudder, ailerons, elevator and spoilers. Safety is assured by operating each sidestick through five computers, each with different software, microprocessors and manufacturers.

The advantages of fly-by-wire: large mechanical linkages and cables are eliminated, less weight, built-in test and flight envelope protection (which prevents excessive control inputs). It also provides "gust load alleviation," which senses turbulence, then operates aileron and spoiler to relieve strain on the wingtips. This enables a lighter, longer wing for better fuel economy.

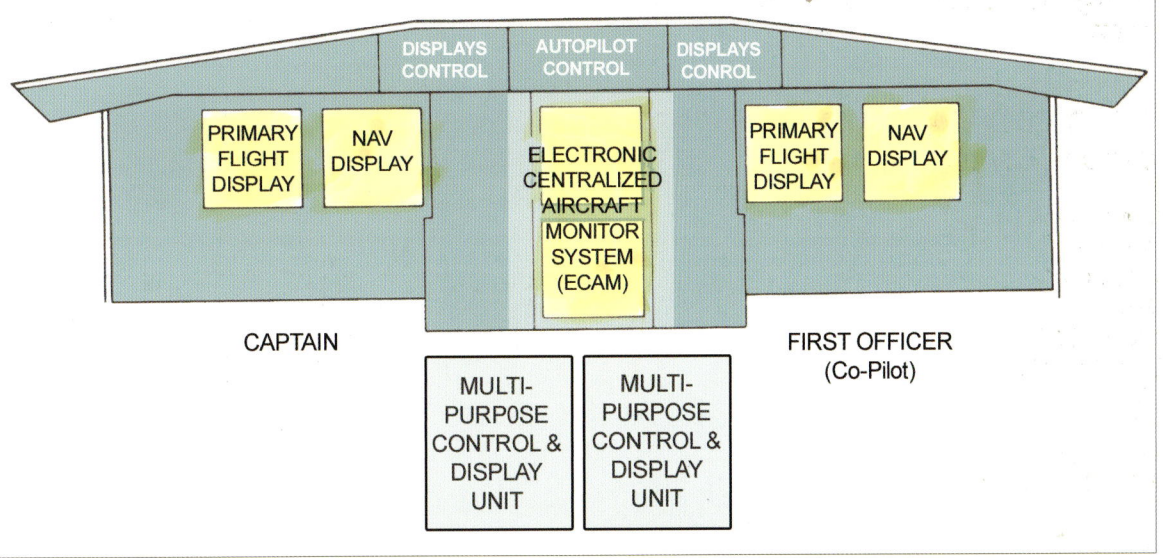

The instrument panel of the A-320 has six main CRT displays, all physically interchangeable. This eliminates 75% of conventional instruments.

The two screens in the center (ECAM) monitor engines, flap and other settings, and system malfunctions.

The two multipurpose displays at the bottom have built-in test equipment (BITE) that show malfunctions, diagnostic data and failed components. It reduces the problem of returning a unit to the shop and finding nothing wrong (a major cost item for the airlines).

The engine thrust levers are controlled by FADEC (Full Authority Digital Engine Control). It adjusts fuel and power setting for best efficiency. Weighing less than the conventional (hydro-mechanical) system, FADEC also provides engine protection (from exceedances) and health monitoring.

Review Questions
Chapter 17 EFIS

17.1 What are two benefits EFIS?

17.2 The EFIS screen directly in front of the pilot, which shows attitude instruments, is called a _____.

17.3 The EFIS screen usually in the center of the instrument panel is the _____.

17.4 The center screen of a typical airline displays EICAS, which means _____.

17.5 An EFIS system can display BITE, which stands for _____.

17.6 The control yokes on recent Airbus aircraft are replaced by sidestick controllers. Why?

17.7 First-generation EFIS was based on cathode ray tubes. What replaced them?

17.8 Why are fewer spares required to maintain an EFIS system?

17.9 What can a pilot do if images fail to appear on his EFIS screen?

17.10 Why is it easier to troubleshoot a problem in flight with an EFIS-equipped airplane?

Chapter 18

Cockpit Voice and Flight Data Recorders

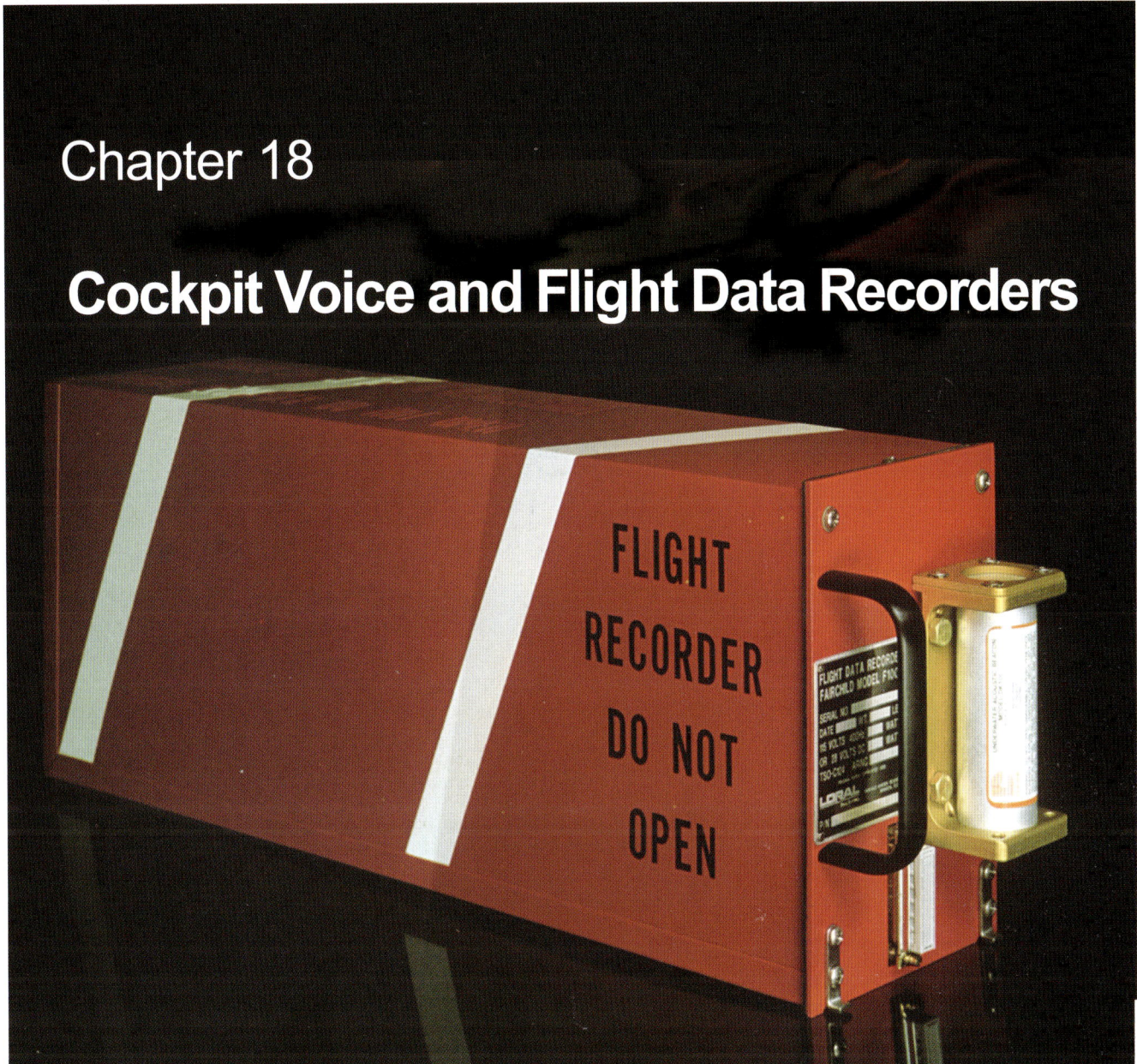

Fairchild

Recording information in flight is among the most valuable methods of determining the cause of airplane accidents. There are several types of recorders, some required by law, others installed because they reduce the cost of maintenance. In the newest airliners hundreds of points are measured on engine, airframe, hydraulic, pneumatic and other systems. When downloaded later on the ground, the data often warns of trouble well in advance of a full-blown inflight problem. Another trend, now possible with worldwide satellite communications, is to transmit flight data as it is collected, and downlinking it to a maintenance facility even before the airplane lands.

Two devices required in airliners and other high-performance aircraft are the CVR (cockpit voice recorder) and FDR (flight data recorder). These are the "black boxes" mentioned by news reporters following an air disaster. As seen above, they are not black but a bright "international orange" used on emergency equipment for high visibility.

An improvement in flight recorders is the transition from recording on tape to storing data on solid-state memories. Not only does it improve reliability, but stores far more data. Early recorders required high maintenance, and tapes often fouled in the mechanism, losing valuable accident information.

CVR Basics

A typical CVR is required (by U.S. law) to record for at least 30 minutes, then start again, while erasing the previous 30 minutes. In other countries the requirement is 120 minutes. After the airplane lands safely, the pilot may bulk-erase the tape. Erasing is not possible in flight because the erase circuit is disabled unless the system senses the airplane is on the ground. This is usually done by a weight-on-wheels, or "squat switch."

The new CVR's are easier to download than early models. Instant playback is possible with a portable device. Any place on the recording is quickly located by forward, reverse and stop commands.

The power source can be either 115 volts 400 Hz or 28 VDC. With so few moving parts, the solid-state CVR requires no periodic maintenance or scheduled overhaul.

Inertial Switch. If a CVR continues to receive aircraft power after a crash, the recorded audio is wiped clean and lost. This is prevented by an inertial switch. It responds to high G forces of a crash by interrupting power to the voice recorder.

Audio Channels. The CVR provides four audio channels into the recorder:

Captain: Any microphone used by the captain, such as the normal boom mike, as well as the mike in an oxygen mask or hand mike. This assures a recording of radio communications.

Co-Pilot (First Officer) The same as for the captain.

Public Address (PA) This channel picks up announcements by the crew to passengers in the cabin.

Cockpit Area Mike. This is designed to pick up crew member voices and other sounds in the cockpit. There have been problems with cockpit area mikes. After a crash, safety investigators often complain that

Manufacturers of flight data recorders must comply with standards for survivability in a crash. A recorder should withstand a temperature of 1100 degrees C for 30 minutes, as shown in this test. In another test, a 500-pound weight is dropped on the recorder from 10 feet.

audio from the cockpit area mike is impossible to hear because it's drowned out by nearby loudspeaker audio. Not only does this eliminate important conversation between pilots, but sounds which can point to problems---sounds such as changes in engine speed, switch clicks and flap motors. A technician must follow the manufacturer's installation instructions carefully for good cockpit area pickup. In airline installation, the airframe manufacturer will have determined all locations. In General Aviation, where there is a choice for locating the area microphone, typical techniques include using a directional microphone facing the crew and one that is noise-cancelling.

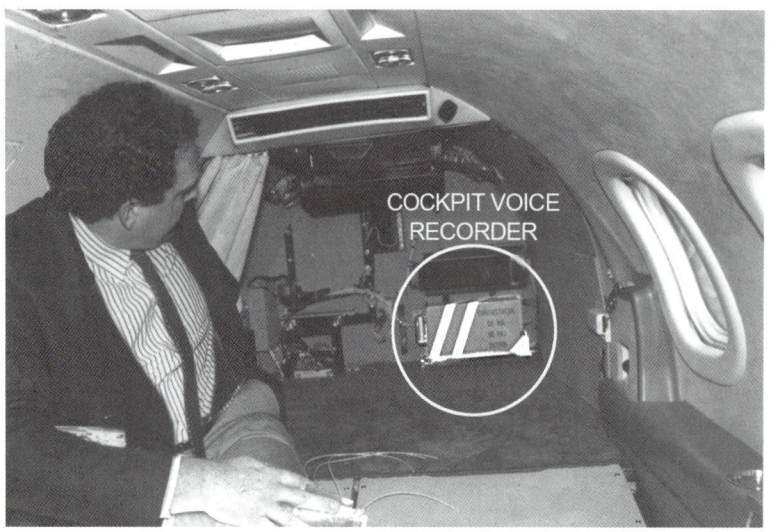

Line replaceable unit (LRU) for a cockpit voice recorder shown located in the aft fuselage of a Learjet. It is usually on the pressurized, or cabin, side.

Cockpit area microphone picks up conversation between pilots and other sounds that provide clues for accident investigators. This mike is located atop the glareshield on a small corporate jet. In airliners, the mike is usually above, on the overhead panel.

Underwater Locating Device

Both cockpit voice and flight data recorders are required to be fitted with a ULD, or underwater locating device. They are also known as ULB, for underwater locator beacon. Each recorder usually has one, but when both CVR and FDR are located next to each other and are not likely to become separated during a crash, a single ULD may, in some cases, be used.

Most ULD's are "pingers," sending out an ultrasonic tone on 37.5 kHz, which is too high for human's to hear. (The high frequency is more effective for homing on with a listening device.)

The ULD is triggered when moistened by water (salt or fresh). It must start pinging no more than four hours after the airplane goes underwater, then continue to broadcast for at least 30 days. It is rated to perform at depths up to 20,000 feet. (The average depth of the world's major oceans is 13,000 feet.)

Flight Data Recorders (FDR)

The second "black box" needed by crash investigators is the Flight Data Recorder. Early FDRs used a sharp-pointed stylus to scratch lines into a band of thin steel. Although the steel "memory" resisted heat and flames, it had low capacity for storing information. Like the cockpit voice recorder, the FDR is always a bright orange.

Under pressure from accident investigators for more parameters (measuring points), the FAA required

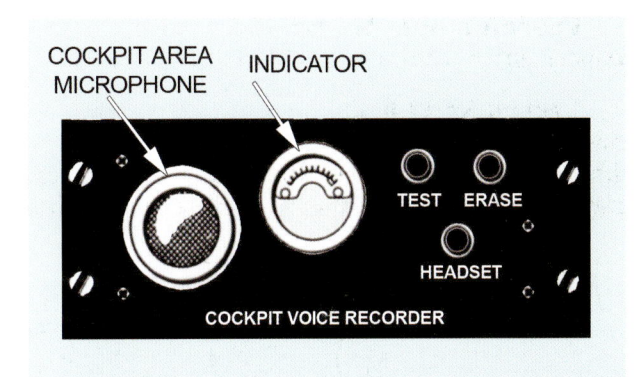

Controller for a cockpit voice recorder (early type). The cockpit area microphone picks up sound of pilot conversation, airplane, engine mechanical noises and warning tones.

To check the system, the test switch is held down for five seconds; if OK, a green light illuminates. If it doesn't light in six seconds, the CVR must be removed for service

The erase switch works only when the airplane is on the ground (either a cabin door must open, a squat switch energized or other interlock). Erasing is indicated by an audio tone.

The headset jack enables pilot or technician to plug in and hear if audio is distorted; he speaks into the cockpit area mike and listens to the playback quality on headphones.

Cockpit Voice Recorder: Interconnect

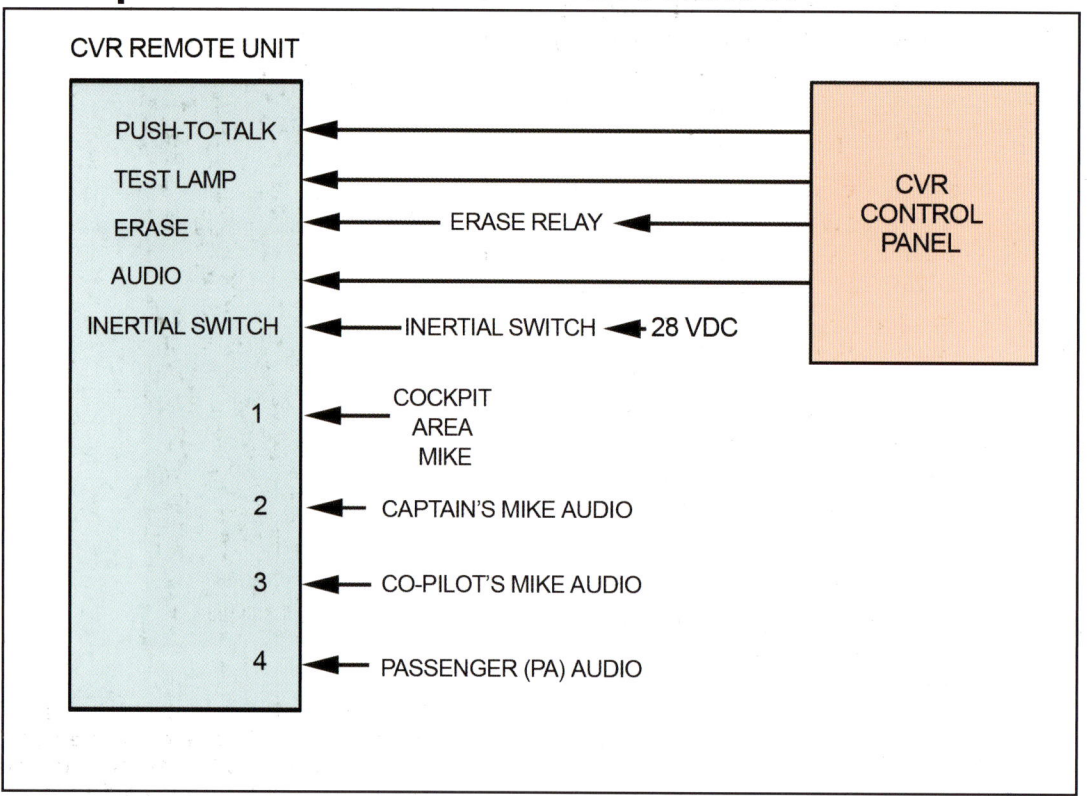

```
CVR REMOTE UNIT
  PUSH-TO-TALK   ←──────────────────────┐
  TEST LAMP      ←──────────────────────┤   CVR
  ERASE          ←──── ERASE RELAY ←────┤  CONTROL
  AUDIO          ←──────────────────────┤   PANEL
  INERTIAL SWITCH ←── INERTIAL SWITCH ←─ 28 VDC
  1 ← COCKPIT AREA MIKE
  2 ← CAPTAIN'S MIKE AUDIO
  3 ← CO-PILOT'S MIKE AUDIO
  4 ← PASSENGER (PA) AUDIO
```

Line Replaceable Unit, at left, is mounted in the aft fuselage in a crash-hardened housing. It records four audio channels; captain's mike, co-pilot's mike, cockpit area mike and public address. Test and erase functions are done at the controller on the instrument panel (right).

Erasing the tape can only happen after the airplane is safely on the ground. This function is protected by an external "squat," or weight-on-wheels switch.

The inertial switch reacts to forces of a crash and shuts off power to prevent the tape from running and erasing the last 30 minutes.

Although the CVR tape is rated for 30 minutes of recording in the U.S., other countries require 120 minutes.

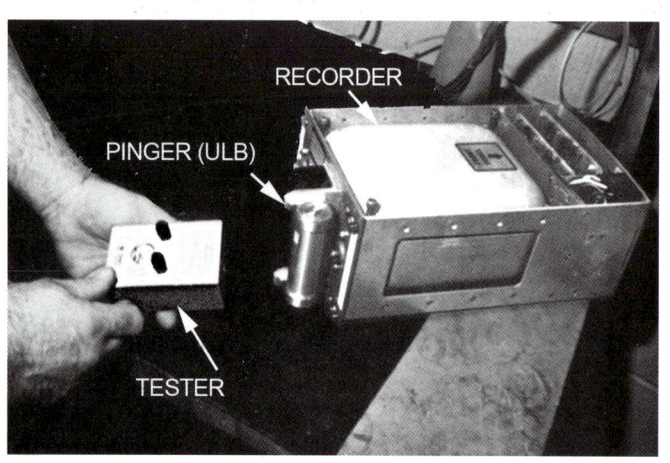

When testing the underwater locator beacon (the "pinger") the tone is ultrasonic and cannot be heard by the ear. The tester receives the tone and converts it down to the range of human hearing. To start the pinger operating, one end is moistened to simulate an underwater condition.

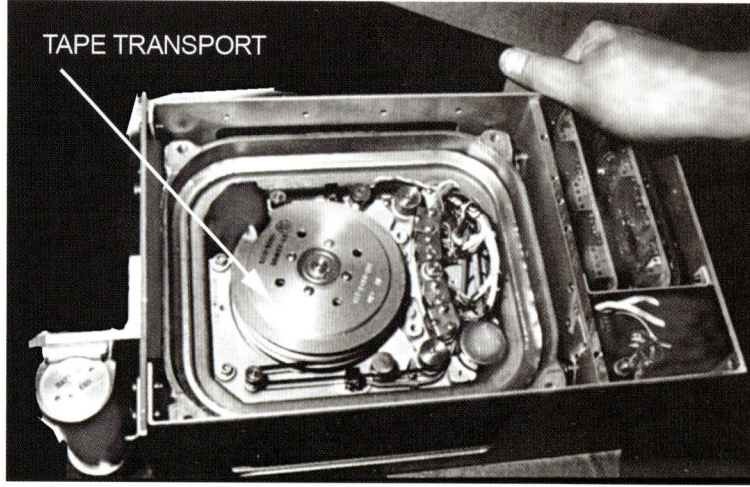

Early cockpit voice recorders use a tape and mechanical drive. Next-generation recorders eliminate tape with more reliable solid-state memory. New recorders meet tougher requirements for heat and G-forces and need less maintenance.

large aircraft be equipped with digital flight data recorders of greater capacity and reliability. Depending on date of manufacture all such airplanes had to retrofit anywhere from 22 up to 57 parameters. Aircraft manufactured after 2002 require 88 parameters.

FDR's grew even more important with the arrival of electronic instruments. In airplanes with mechanical gauges, accident investigators could look at an airspeed needle pinned in place by the crash and obtain valuable information (such as airspeed when the airplane struck the ground). They could tell if warning lights were on or off at the time of impact by looking at condition of filaments in the bulbs. But as this information went from instruments, switches and lamps to electronic displays, it disappeared when the screen went dark. Thus the urgency of storing data on a flight recorder.

Many in the aviation industry want to add to the present generation of flight data recorders. One idea is to equip large aircraft with two recording systems; forward and aft, to assure sufficient data. There is also a move to equip the cockpit with a video camera. Video images stored in the FDR could yield valuable information about what happened just before the crash.

The digital FDR (DFDR) takes analog signals (heading, altitude, airspeed, etc.)--which usually vary in a smooth, continuous fashion and converts them to digital format for storage in a solid-state memory. Some signals are "synchro," meaning signals from electro-mechanical instruments. Yet another type of input is from the aircraft databus, such as ARINC 429, which is a stream of data from many aircraft systems.

Unlike the old, mechanical recorder, there is no scheduled overhaul and little maintenance for digital models. Reliability extends to 20,000 hours (on average) before failure and data is easily recovered with a portable unit. A typical flight data recorder stores 25 hours of information before starting over again.

Cable Assemblies

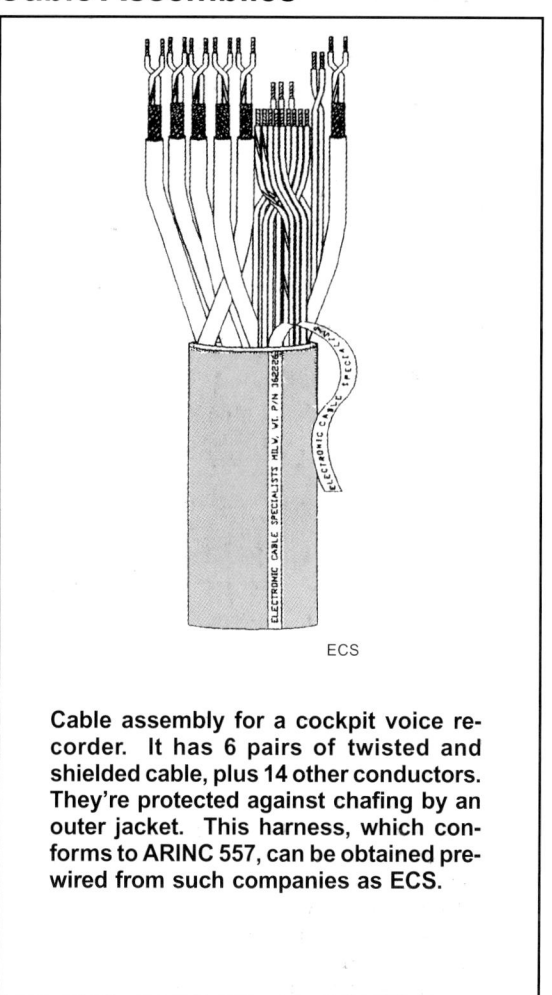

Cable assembly for a cockpit voice recorder. It has 6 pairs of twisted and shielded cable, plus 14 other conductors. They're protected against chafing by an outer jacket. This harness, which conforms to ARINC 557, can be obtained pre-wired from such companies as ECS.

Flight Data Recorder: Solid-State

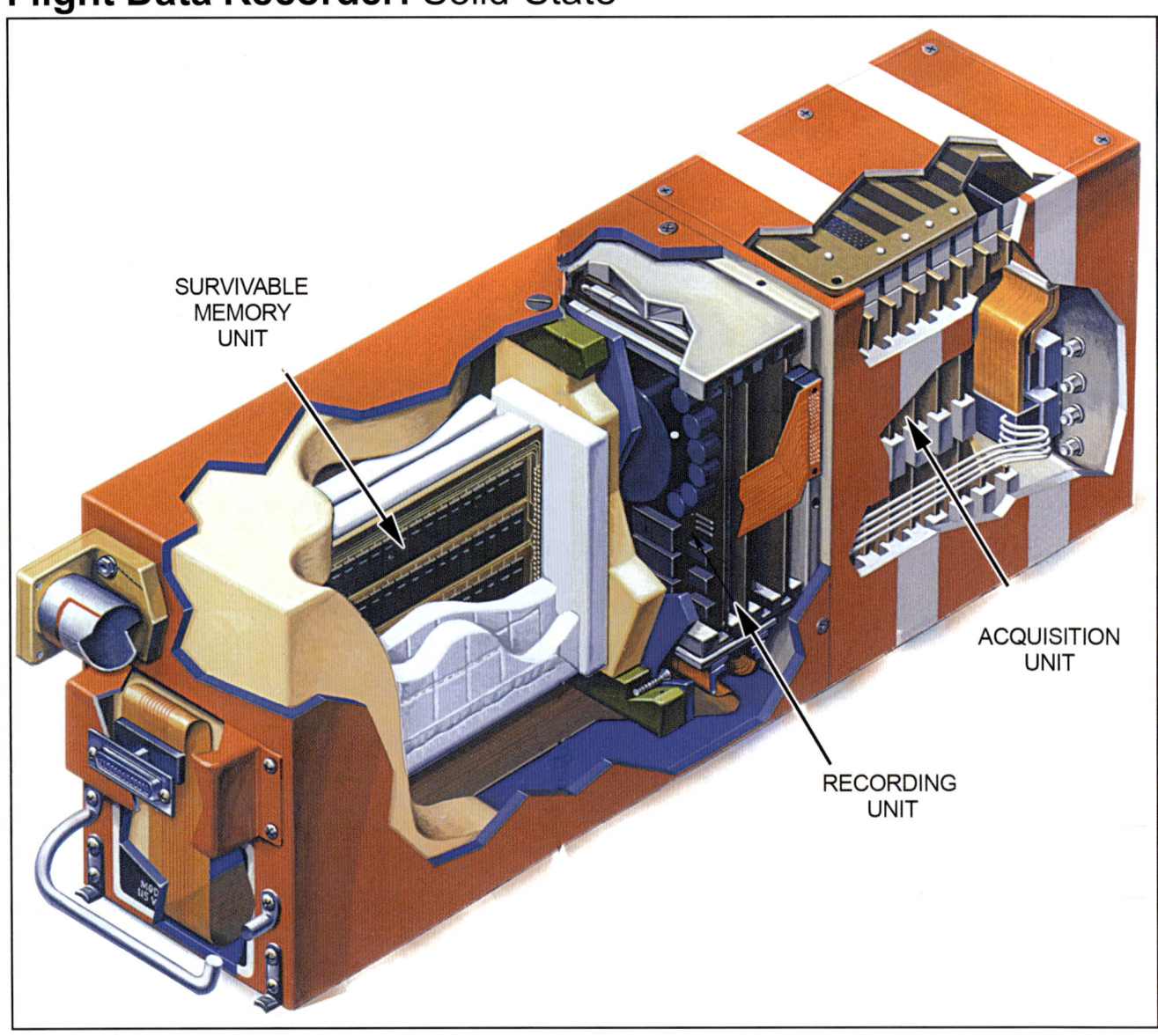

Labels on main diagram: SURVIVABLE MEMORY UNIT, RECORDING UNIT, ACQUISITION UNIT

Solid-State Flight Data Recorder (SSFDR) by Lockheed eliminates tape storage with a survivable solid-state memory. It is interchangeable with earlier-generation recorders without wiring changes.

The system uses direct recording, which eliminates data compression. This permits the memory (non-volatile flash) to be downloaded without a delay of 8-10 hours. The unit doesn't have to be removed from the airplane to retrieve stored data and is done with a PC. Unlike early recorders this one has much greater MTBF (mean time between failures) of 20,000 hours and requires no scheduled overhaul.

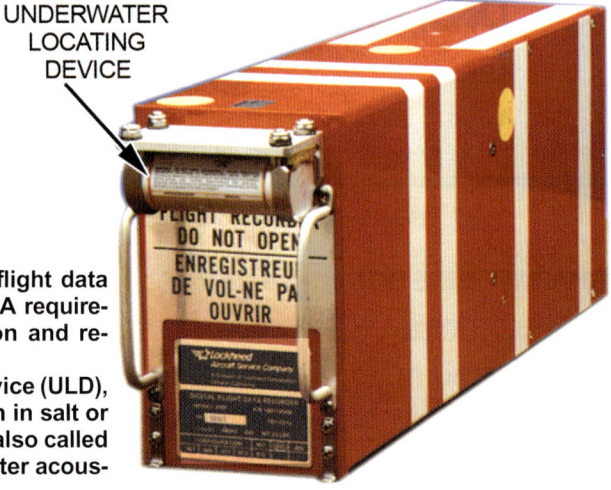

Another recorder type (at right) is the digital flight data recorder (DFDR). It is designed to meet an FAA requirement for an expandable flight data acquisition and recording system (EFDARS).

All recorders have an underwater locating device (ULD), seen on this model. It is triggered after a crash in salt or fresh water and emits an ultrasonic tone. It is also called a ULB (underwater locator beacon) or underwater acoustic beacon.

Label on second image: UNDERWATER LOCATING DEVICE

Information Stored by Digital Flight Data Recorder (36 Parameters)

1. Begin recording prior to takeoff:
a. Record time of flight control check (hold flight controls at full travel for 2 to 5 seconds, each position).
b. Takeoff flap Setting.
c. Takeoff thrust setting.
d. Brake release time.
e. Rotation Speed (VR) and time of rotation.
f. Aircraft attitude after rotation.

2. During stabilized climb (wings level) after takeoff record:
a. Altitude and time at which climb stabilized.
b. Airspeed.
c. Vertical speed.
d. Pitch attitude.
e. Displayed angle of attack.
f. Heading (note true or magnetic).

3. During level flight (wings level) at maximum operating limit speed (VMO./MMO) or at VMAX record:
a. Altitude and time at start of level flight.
b. Airspeed.
c. Ground speed and time at which recorded (three times).
d. Outside or total air temperature.
e. Automatic Flight Control System (AFCS) Mode and engagement status including autothrottle.
f. Pitch attitude.
g. Displayed angle of attack.
h. Heading (note true or magnetic).
i. Drift angle and time at which recorded (three times).
j. All displayed engine performance parameters for each engine.
k. Altitude and time at end of level flight.

4. During a banked turn (90° to 180° heading change) record:
a. Altitude, heading and time at beginning of turn.
b. Stabilized roll attitude (bank angle).
c. Altitude, heading and time at end of turn.

5. During stabilized (wings level) descent, record:
a. Altitude and time at which descent initiated.
b. Airspeed.
c. Pitch attitude.
d. Displayed angle of attack.
e. Heading (note true or magnetic).
f. Altitude and time at which leveled off.

6. During approach at level flight (wings level) deploy flaps throughout the flap operating range in all available settings (or at 5° increments) and hold for 5 seconds at each setting. Record:
a. Altitude and time at beginning of flap deployment sequence.
b. Flap setting and time when each setting is reached.
c. Altitude and time at end of flap deployment sequence.

7. During final approach, record:
a. Altitude and time at beginning of final.
b. Radio altitude and time at which recorded (three points).
c. Localizer deviation and time at which recorded (three times).
d. Glide slope deviation and time at which recorded.
e. Time of outer marker passage.
f. Time of landing gear deployment.
g. Final flap setting.
h. Time of inner marker passage.

8. During landing and rollout, record:
a. Time when thrust reversers deployment sequence was initiated.
b. Ground spoiler or speed brake setting and time ground spoiler deployed.

9. During all flight phases, record:
a. Time of any three radio transmissions from each flightcrew position.
b. Any warning or caution lights that illuminated and the time at which they illuminated.

Review Questions
Chapter 18 CVR and FDR

18.1   A CVR is required to record for _____ minutes before erasing and recording again.

18.2   What is the purpose of an inertial switch?

18.3   Name the four audio channels into a CVR.

18.4   What is the purpose of the cockpit area mike?

18.5   The erase switch on a CVR works only _____.

18.6   How long must an Underwater Locating Device send signals after an aircraft ditches in the water?

18.7   FDR's for aircraft manufactured after 2002 must record up to ____ parameters.

18.8   The solid-state FDR replaces tape storage with _____.

Chapter 19

Weather Detection

The thunderstorm cell shown above is producing a "microburst," a powerful downdraft and outflow from its central core. Once the cause of many airline accidents, it is no longer a major problem. Windshear devices that give warning are aboard all commercial airliners.

The earth is a weather factory generating many hazards to flight; thunderstorms, lightning, fog, turbulence, haze, hail, rain, blowing snow and windshear. Nevertheless, airliners complete their scheduled flights 98.7 percent of the time. Much of this success is owed to a network of weather-reporting stations on the ground which deliver timely information to the pilot. Just as important is weather-detecting equipment aboard the airplane to sense dangerous conditions ahead and help the pilot plan an escape route.

One of the greatest weather threats to aircraft is the thunderstorm. Few aircraft have the performance or structural strength to withstand turbulence generated inside storm clouds. It's proven especially deadly when the airplane is arriving or departing the airport and is low to the ground, where it is known as "wind shear."

Clear Air Turbulence

Another hazard is CAT, for clear air turbulence. It occurs at high altitudes of the jet stream between fast-moving currents of air. Because the air masses

Color-Coding the Radar Display

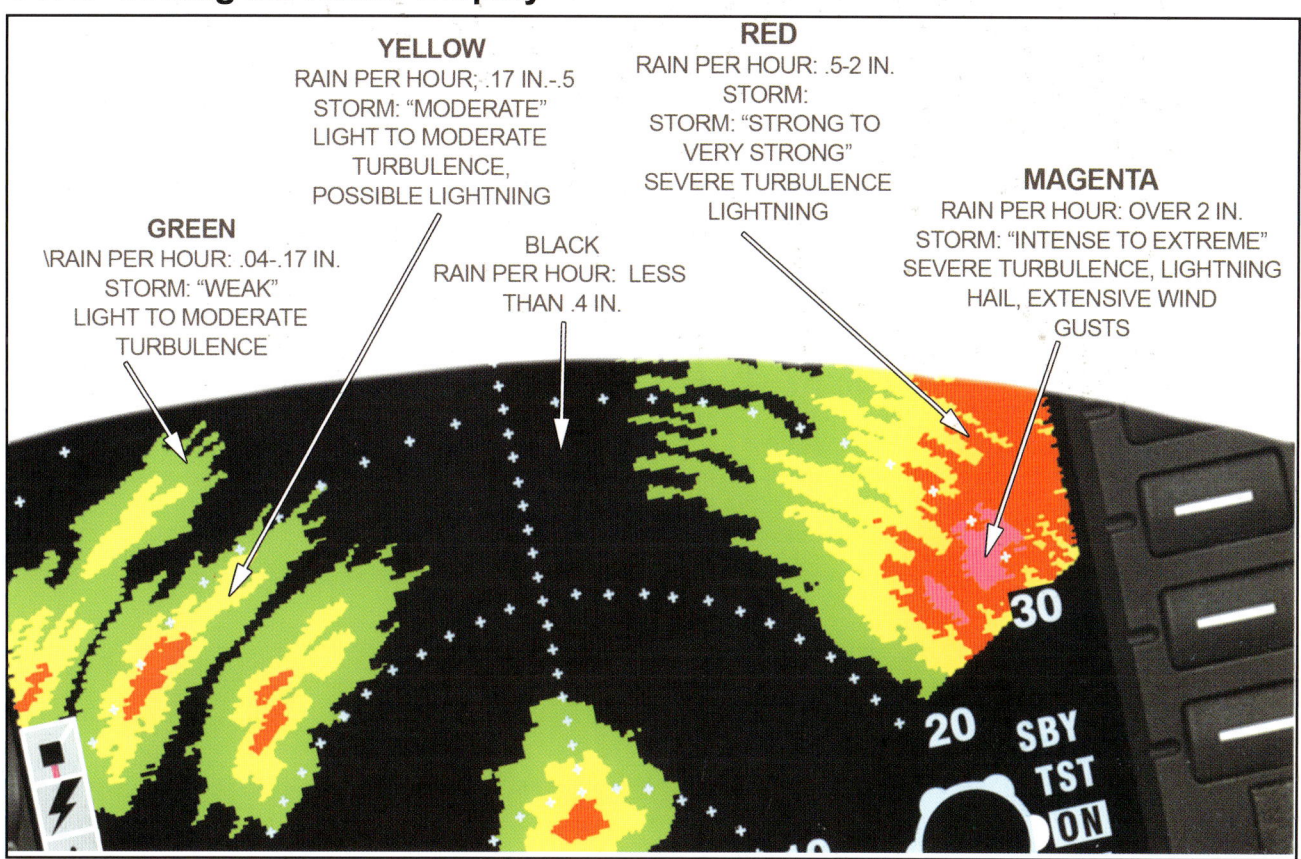

YELLOW
RAIN PER HOUR: .17 IN.-.5
STORM: "MODERATE"
LIGHT TO MODERATE TURBULENCE,
POSSIBLE LIGHTNING

RED
RAIN PER HOUR: .5-2 IN.
STORM:
STORM: "STRONG TO VERY STRONG"
SEVERE TURBULENCE
LIGHTNING

GREEN
\RAIN PER HOUR: .04-.17 IN.
STORM: "WEAK"
LIGHT TO MODERATE TURBULENCE

BLACK
RAIN PER HOUR: LESS THAN .4 IN.

MAGENTA
RAIN PER HOUR: OVER 2 IN.
STORM: "INTENSE TO EXTREME"
SEVERE TURBULENCE, LIGHTNING
HAIL, EXTENSIVE WIND GUSTS

Radar image: Bendix-King

The radar image uses five colors to indicate severity of weather; black, green, yellow, red and magenta (a purple-red). The colors are based on the rate of rainfall in inches per hour. Rainfall is also a guide to turbulence in clouds. A pilot may enter the green region, he tries to avoid the yellow, and carefully flies around red and magenta.

Weather on Multifunction Display

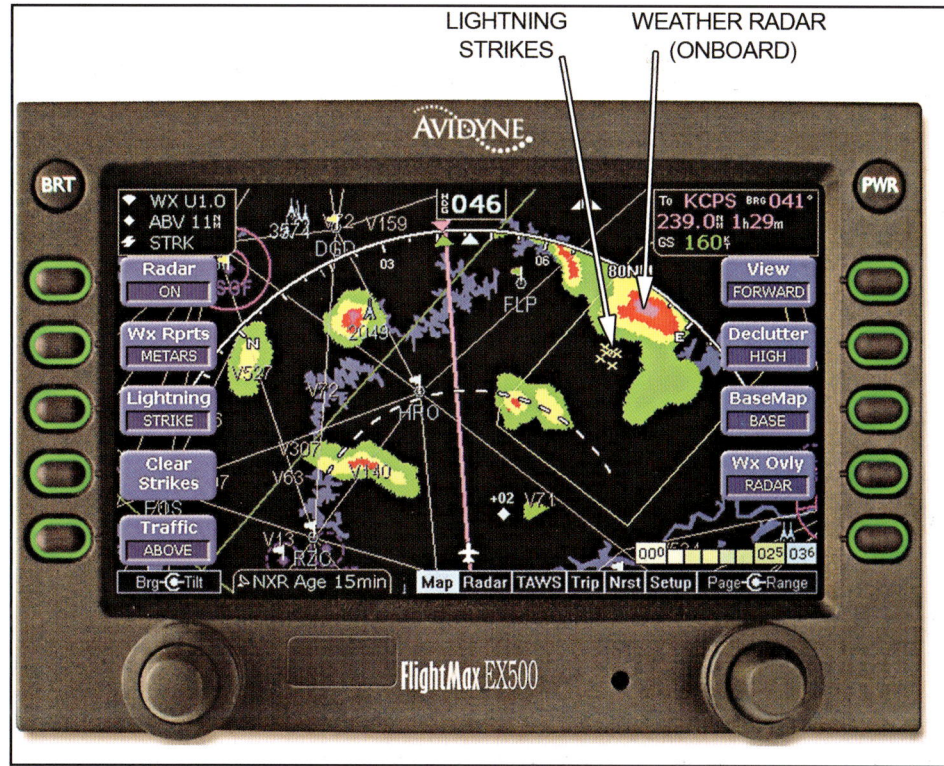

LIGHTNING STRIKES
WEATHER RADAR (ONBOARD)

More aircraft are now outfitted with multifunction displays which overlay several sources on a basic moving map. This includes weather (rain or other precipitation) and lightning strikes. The information may be picked up by satellite or from an onboard radar or Stormscope.

Besides weather, the displays show traffic and terrain hazards.

Avidyne Flight Max

137

move in different directions, an airplane hits heavy turbulence when it enters the boundary between them. The damage is usually not to the airplane but to passengers. They are tossed about and injured in the cabin (thus the request to keep the seat belt buckled.)

Thunderstorms

Because thunderstorms are accompanied by lightning, the earliest attempt at detection was the ADF, or automatic direction finder, already aboard many aircraft. Lightning is an electrical discharge that generates not only flashes of visible light but radio frequencies in the low- and medium-frequency bands. The ADF receiver, therefore, responds to this energy. With each lightning discharge, the ADF needle dips away from its rest position. According to the folklore of aviation (which many pilots believe) an ADF needle points toward the storm. This is dangerous because the needle and its mechanism do try to point to the storm, but swing too slowly. As the lightning discharges in different directions, the needle lags behind, becoming confused and erratic. But as we'll see, lightning can provide valuable information about storm location.

Types of Detection

Weather Radar. The leading airborne weather-detecting device, first put aboard a DC-4 airliner in 1946, is weather radar. Adapted from military models of World War II, it proved so effective it became required equipment aboard all commercial flights. The radar system operates on the principle of reflectivity; a pulse emitted from the radar antenna strikes water droplets in a cloud and reflects back as an echo. By plotting the strength and direction of the echoes, areas of heavy rain are "painted" on a graphic display, and form an outline of thunderstorm cells. The first radars were monochrome, showing rain intensity in shades of gray. Present-day radars give a much clearer presentation in color.

Weather radar has been much improved in recent years. It is less prone to an early problem, known as "attenuation," where an area of moderate rain blocks echoes from a more turbulent cell behind it. Newer radars are less responsive to "ground clutter," where radio energy strikes the ground and interferes with the image. Modern radars can present a vertical, or profile, view of the storm, showing the height of the clouds, which is a good clue to storm intensity.

The most significant development for weather radar in recent years is turbulence detection. The first radars could only sense rain as it fell in the vertical direction. By the 1980's, however, designers could build radar sets which also measure the *horizontal* movement of rain. Although very heavy rain is usually a good indicator of turbulence, rain that moves horizontally is a sure sign of powerful, dangerous winds.

The new radars detect this with a turbulence detection system based on the doppler shift. If a radar pulse hits a rain drop moving horizontally away from the airplane, the returning pulse is slightly reduced in frequency (the doppler shift). When the rain drop moves toward the airplane, the echo frequency rises in frequency. (It's the same doppler shift that causes a train whistle sound to higher in pitch as it approaches; the waves are squeezed together and you hear a rising tone. After the train passes, the waves stretch out, causing a lower frequency.)

Sensors for Multifunction Display

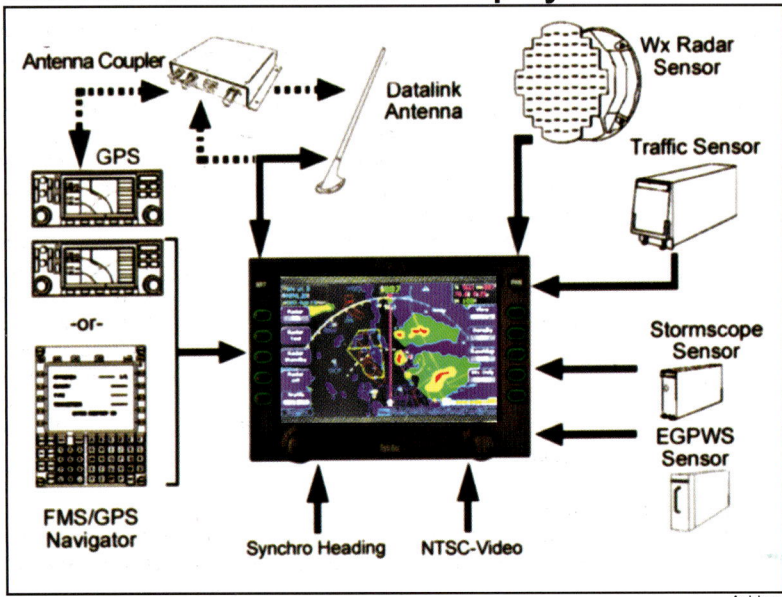

Many sensor and data inputs are required to drive the MFD. The system has standard industry interfaces (such as ARINC 429 and IEEE RS-232), as well as interfaces to accept signals from other manufacturers' products for radar, lightning, terrain and traffic information.

Avidyne

Weather Radar Transmitter-Receiver

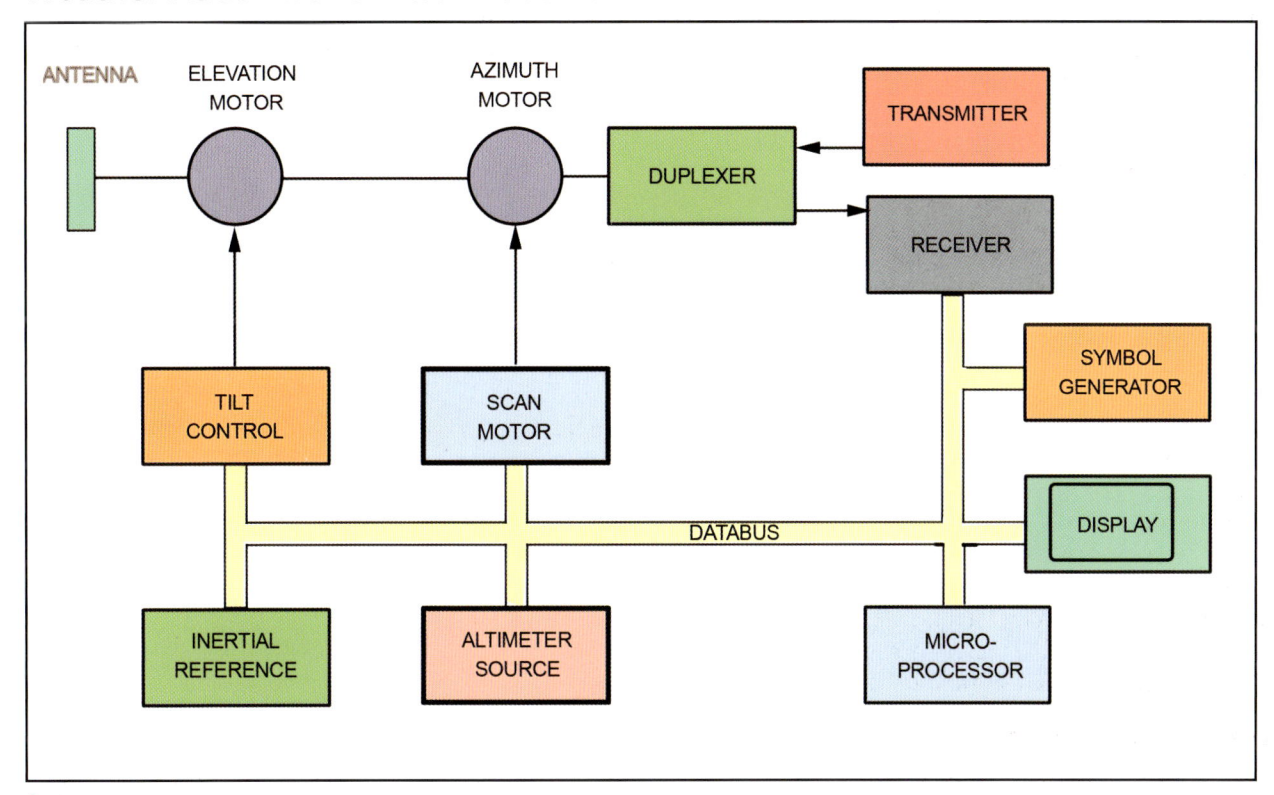

Antenna
A single antenna located in the nose transmits radar pulses and receives echoes from rain and other precipitation.

Elevation Motor
Raises and lowers the antenna vertically to keep it stabilized on the same area of the sky, even as the airplane nose moves up or down.

The elevation motor also enables the pilot to "tilt" the antenna to keep it just above the horizon and avoid receiving echoes from the ground. The radar, however, may operate in a "mapping" mode, which provides an image of the ground, if desired. (One example; when approaching a coastline.)

The latest application for elevation is "vertical profile radar." The motor sweeps the antenna vertically to show the height of the clouds, an indication of the storm's strength. The typical tilt range in a weather radar is plus and minus 15 degrees.

Azimuth Motor
This sweeps the antenna from side to side in a scanning motion.

Inertial Reference
The inertial reference senses aircraft pitch and roll and provides information required by elevation and azimuth motors to stabilize the antenna. The inertial source may be laser gyros or electromechanical gyros which also operate the airplane's flight attitude instruments.

Duplexer
In order for one antenna to serve both transmitter and receiver, a "duplexer" is used. It directs radio energy from the transmitter to the antenna, and connects the receiver to the antenna for receiving returning echoes.

Microprocessor
This microcomputer converts switch positions selected by the pilot into digital words and applies them to one or more databuses. It also computes the azimuth and elevation of the antenna to keep it stabilized.

Transmitter
The transmitter sends out pulses of radio energy, usually on 9.333.8 MHz. The receiver then listens for echoes between pulses.

Receiver
The strength of echoes varies according to the rainfall rate and they are divided into colors for the display; black, green, yellow and red.

The most recent is the color magenta, for turbulence. In this mode, the receiver measures *horizontal* movement of rain, which is a measure of turbulence. Under the Doppler effect, the returning echo rises or falls in frequency, depending on the direction of the rain drop.

When the radar is set to the turbulence mode, the number of pulses transmitted per second increases from several hundred per second to over 1000 per second. This is because stronger echoes are required to measure the very small frequency change. Also, the turbulence mode has a range of less than 50 miles. This limit occurs because a high pulse rate allows little time for the echo to return to the airplane before the next pulse is transmitted.

Symbol Generator
This section converts weather information from a digital form into graphics that can be displayed for the pilot.

Early radar did not have circuits which could measure doppler shift. They were too unstable to measure small frequency changes. Today's radars use solid-state devices that generate precise frequencies and have the stability to measure frequency shifts in the returning echo. As seen in the illustration, turbulence is shown on the radar screen by the color magenta.

Single Engine Radar. Mounting radar in a light aircraft has been a problem because the antenna interferes with the propellor and engine. To avoid this area, the radar antenna is slung under a wing or built into the wing's leading edge. Small antenna size, however, limits operating range of these single-engine installations.

Lightning Detection

Weather research shows that thunderstorms create lightning in strong up and down drafts. Particles of dust, ice crystals and water rub against each other and build static electricity. When voltage rises sufficiently, an electrical discharge jumps between clouds (most of the time), while some charges move from cloud to earth. As heavy electrical currents heat and expand the air, they produce the sound of thunder---plus a wide spectrum of radio energy that travels hundreds of miles. You hear it as static on an AM radio during a storm. That energy is also an indicator of where turbulence is located.

Stormscope appeared in the 1970's as the first practical lightning detection system for aircraft. It became successful in single-engine airplanes because it doesn't need a radar antenna on the nose; just a small receiving antenna on the belly of the airplane.

The Stormscope is tuned to a region where radio energy of lightning is concentrated; the very low frequency of 50 kHz. The display is electronic which means there are no mechanically moving parts to lag behind, as in the case of an ADF needle.

The display also maps the storm. When a lightning stroke is sensed, a dot is placed on the screen that shows the direction and distance of the stroke. The dot is held on the screen and joined by the next dot. Storing these signals, therefore, builds a graphic image of thunderstorm cells and places them in the proper position relative to the nose of the airplane.

Weather Radar Control Panel

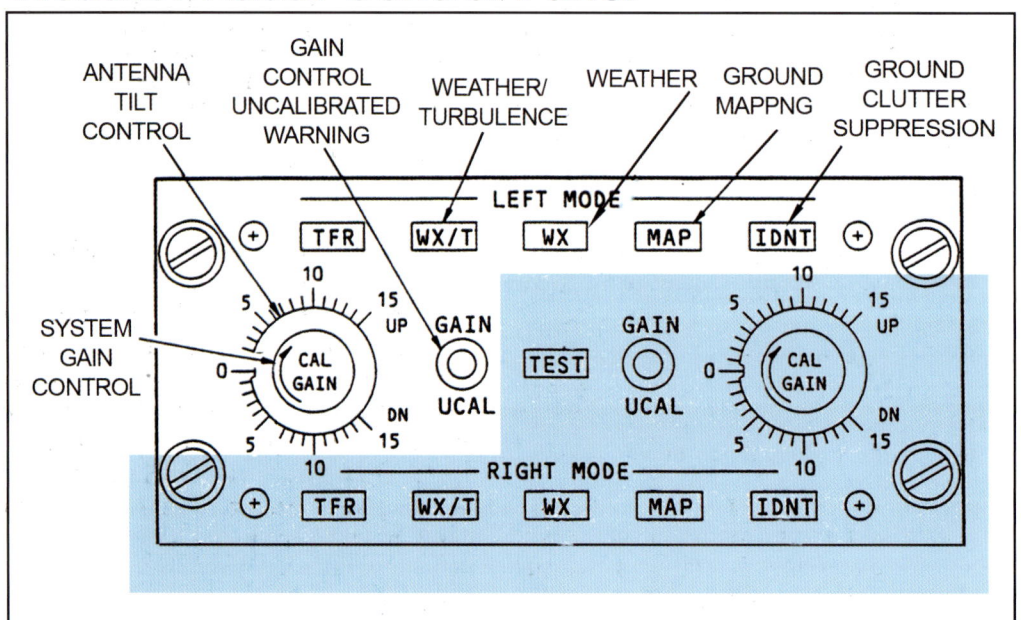

Both the captain and first officer operate the weather radar from the same control panel in this ARINC-type unit. Nearly all controls are duplicated; the captain's side is the white area, the co-pilot's is shown with a blue background. They are grouped as ""Left Mode" and "Right Mode" to indicate that the controls affect left and right sides of the instrument panel.

The panel, however, controls just one radar set and antenna. If captain and co-pilot choose different modes or ranges, they will see these selections on their displays. This is done by "time sharing" the radar scan. When the antenna swings from left to right, it obeys the captain's settings. When it scans from right to left, it reconfigures and responds to the co-pilot's switch settings. Thus, the two pilots may be viewing different weather situations on their displays---all from the same radar at nearly the same time.

Radar Antenna

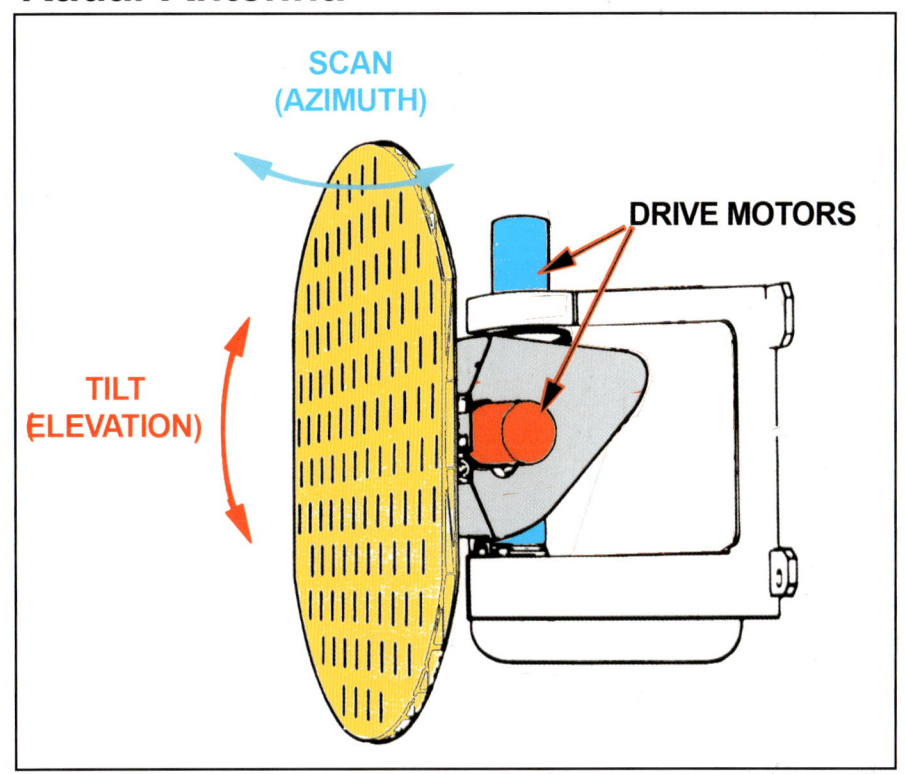

Mounted in the nose of the aircraft, the weather radar antenna sends and receives up to about 300 miles. Scanning motion (side to side) covers an arc of about 120 degrees ahead of the aircraft. The tilt motor keeps the antenna pointed high enough to avoid receiving returns from the earth's surface and cluttering the display. If the antenna is tilted down for the mapping mode, the pilot sees large geographical features such as lakes and coastlines.

Early antennas followed the "dish" design (parabolic reflector), but later aircraft use the "flat plate" design shown above.

A little-understood function of the Stormscope is how it determines the distance to the storm. It's done by measuring the strength of the incoming signal and converting it to miles. This sounds plausible until the question arises; how does the Stormscope know if the storm is small and close by, or large and far away? Each condition would seem to produce the same strength.

Stormscope determines the difference because large storms don't produce more energy per stroke, but more strokes per second. The reason is, lightning is created when voltage between two air masses reaches a breakdown, or flashover, point. Let's assume a small cell discharges at 100 million volts and contains electrical energy of 500 megajoules. After the cloud charges again to 100 million volts, another stroke occurs. Next, consider a larger cell at the same distance. It also flashes over at 100 million volts and 500 megajoules of energy. The difference, however, is that a large cell has a greater source of energy (more area) and generates the next stroke in less time. Thus, all single strokes (from large and small cells) generate about the same amount radio energy. Using this reference, the Stormscope can determine the distance to any stroke by measuring its strength at the time of arrival.

Lightning detectors of this type are based on "sferics," derived from the world "atmospherics." They are not as accurate in range as weather radar and may show "radial spread," where dots appear closer than they actually are, especially during strong thunderstorm activity. Because the error makes a dot appear closer (and thus give an earlier warning) it is not considered a major flaw in the instrument.

Datalink

A recent addition to weather detection not only solves the single-engine radar problem but extends new services to aircraft of all sizes. It is datalink; sending weather images from National Weather Service radar sites to aircraft. The link is done via satellite and requires only a receiver and display.

The images are the same ones seen on TV weather broadcasts. The system is Nexrad (Next Generation Radar), a network of high-power ground radar stations. Because of their megawatt power and large antenna

(continued page 144)

Radomes

The radome, which appears as a nose cone, protects the radar antenna from high speed impacts of rain, freezing moisture, hail and abrasive dust. Radomes must not only be structurally strong, but avoid reducing radar power by more than about 10 per cent. As the radome ages, it develops cracks and damage which eventually reduce the range and accuracy of the radar image. Frequent inspection and maintenance prevent this.

SIMULATED HAIL IMPACT

QUARTZ PANEL   KEVLAR PANEL   FIBERGLASS PANEL

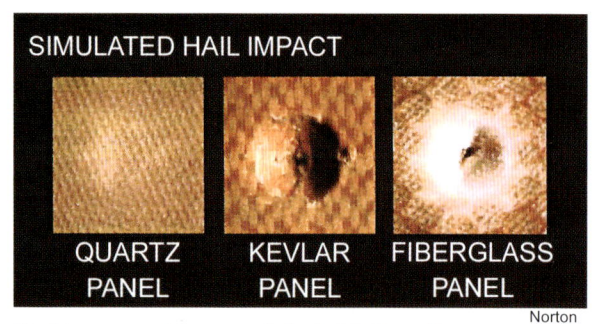

Norton

Radomes erode, especially in high-performance aircraft. The test shown above illustrates a radome constructed of quartz, which has proven light and strong compared to other materials.

Radome Boot

(Norton, and below)

3M

A pre-formed boot made of polyurethane may be applied over the radome for added protection. It reduces the effects of rain, snow, sleet, insects, sand and ultraviolet light

Quartz radome used on the Airbus A-320.

UNDERWING RADAR POD

Radar antenna for a single-engine airplane cannot mount in the nose because of the propellor. Instead, it is located under the wing, as shown above, or built into the leading edge of wing.

142

Windshear

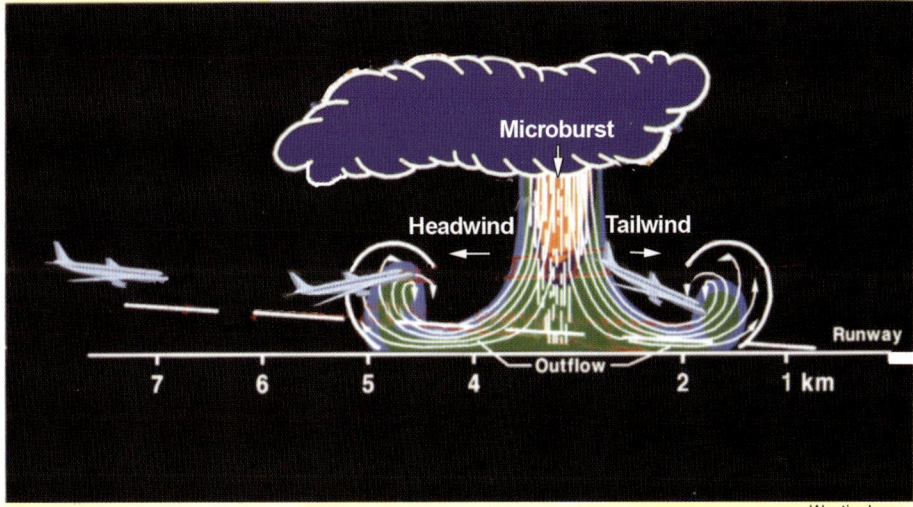

Westinghouse

Wind shear---a sudden change in wind direction or speed---is most hazardous when the aircraft is close to the ground, as during an approach. Windshear mostly affects pure jet aircraft because of slow turbine "spool-up" time; a delay of about 4 to 7 seconds after the pilot calls for full power.

The discovery of the "microburst" (pictured above) shows what happens. A small thunderstorm cell across the approach path is sending down a column of air from its core. As the wind strikes the ground it spreads in all directions. The airplane at the left is stabilized on the glideslope. When it reaches point "5" it enters a headwind at the edge of the microburst. This lifts the airplane above the glideslope, causing the pilot to reduce power or lower the nose to get back on. Next, the airplane reaches the strong downdraft from the center of the microburst and the airplane sinks further. The final phase is entering the tailwind portion of the microburst ("2"), causing further sinking and loss of performance. The complete windshear encounter may take less than a minute, hardly enough time to recover---and the airplane crashes short of the runway.

Because so many landing accidents were caused by windshear, protection systems are now required aboard airline aircraft. They not only give advance warning, but help the pilot fly the correct attitude for maximum climb out of the windshear condition.

Lightning Detection

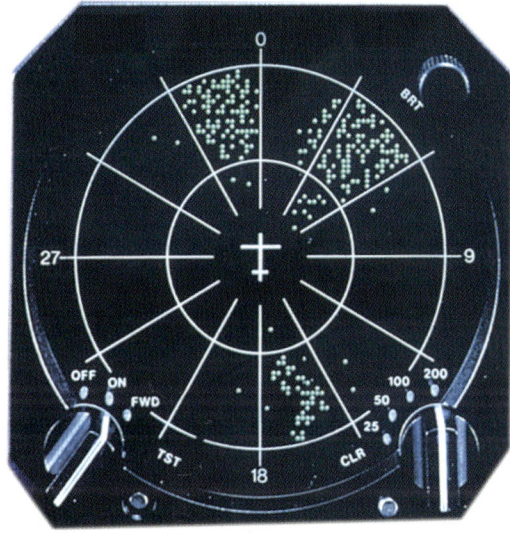

L3 Communications

The Stormscope shows each lightning stroke as a green dot. The three large dot clusters are groups of thunderstorm cells. The display is 360 degrees, with the airplane in the center, thus showing activity behind the airplane. If the pilot wants to avoid storms ahead, he'll know not to make a 180-degree turn and fly into more storms.

If the Stormscope is connected to a magnetic heading source, it keeps the dots correctly oriented to the nose of the airplane. Without this connection, the pilot must manually clear the display after the airplane turns and await the buildup of more dots.

When there is little storm activity along the route, the pilot may choose the 200-mile range to see the "big picture." If dots start to appear, he shortens the range for greater accuracy; down to 100, 50 and 25 miles.

Because Stormscopes are sensitive to electrical discharges, the installation must be done carefully to avoid false dots due to strobe lights, magnetos and other electrical equipment.

Windshear Computer

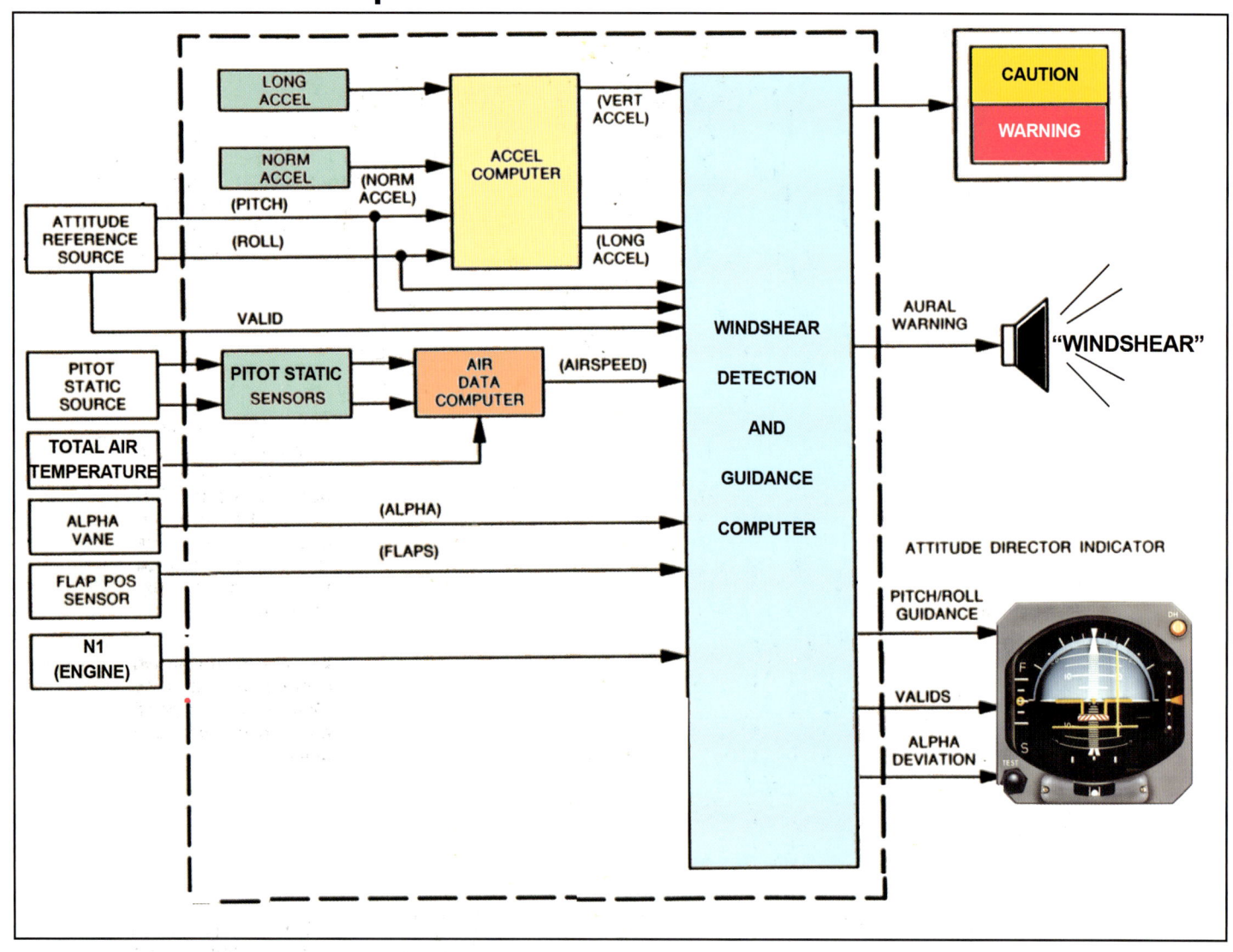

The system detects windshear before the pilot sees it on his instruments or senses any danger.

The basic principle is to measure air speed, ground speed and inertial forces. If they start to differ at an excessive rate, it's caused by wind shear. For example, the pitot tube, which measures airspeed, is compared with an inertial sensor aboard the airplane which measures changes in acceleration of the airplane.

As seen in the diagram above, various sensors provide other information such as angle of attack (alpha) and temperature. Outside air temperature is also monitored because windshear is often accompanied by rapid temperature change.

Two alerts are developed: "Caution," which indicates the airplane is encountering a headwind and updraft. This is considered an increase in airplane performance. The second alert is "Warning," for a tailwind and downdraft (or a decrease in performance). Now the voice says "Windshear."

Guiding the pilot out of the windshear condition follows the warning. Without guidance, the pilot may simply add full power and raise the nose, which could stall the airplane. To avoid this, the windshear computer indicates the ideal flight path on the instruments (done by pitch cues on the attitude indicator).

dishes, they produce images of high quality. When severe weather is in an area, a Nexrad site repeatedly sweeps the sky for five minutes, mapping precipitation horizontally (for a conventional radar image) and sampling 14 different elevations (for a profile view), up to 140 miles away.

A feature unique of the Nexrad system is that a pilot may "look ahead" and see current weather any where in the country.

Lightning

The most hazardous weather to aircraft is turbulence in thunderstorm clouds. Turbulence also generates lightning (due to friction between particles), which may reach 100 million volts and 200,000 amperes. Detecting lightning is a good indication of turbulence. This information is available from ground stations or with a lightning detector aboard the aircraft. Radar, on the other hand, reads rainfall rate, also a good indicator of turbulence. The newest radars also measure turbulence by the horizontal motion of raindrops.

On average, an airliner is struck once a year by lightning, but with minor damage. Only one or two air crashes have ever been suspected of resulting from a lightning strike. Turbulence is the danger.

Datalink

Images and text are transmitted by datalink direct to the airplane from a geosynchronous satellite. The datalink service shown is by XM Satellite Radio.

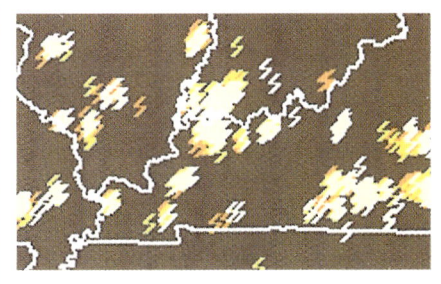

Lightning strikes measured from the ground by the National Lightning Detection Network are updated every five minutes and sent to the aircraft by datalink.

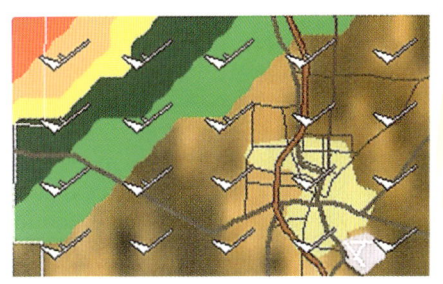

Winds aloft are shown every 3000 feet up to 42,000 feet. Speed and direction are given.

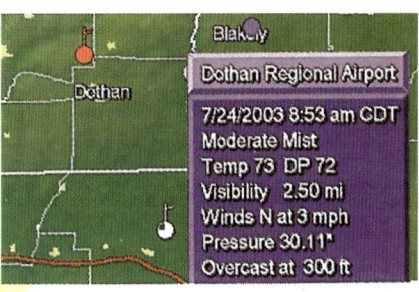

METARS (aviation weather reports) are transmitted from the National Weather Service every 15 minutes.

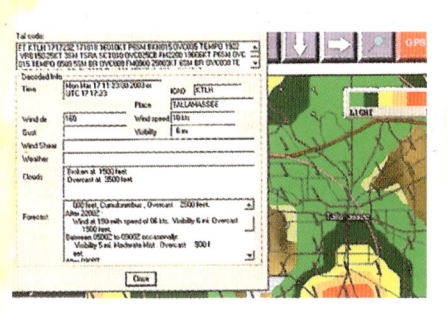

TAFS (aviation forecasts) are also available from the National Weather Service.

145

Review Questions
Chapter 19 Weather Detection

19.1 What is the greatest threat of a thunderstorm to an aircraft?

19.2 On a weather radar display, what color indicates maximum hazard to an aircraft?

19.3 Weather radar detects storms by transmitting _____ of radio energy and measuring their echoes from water droplets.

19.4 Detecting turbulence in a storm is done by measuring echoes from the _____ movement of water droplets.

19.5 What is the normal use of the tilt control?

19.6 What raises and lowers the radar antenna in a vertical direction for tilt control?

19.7 What causes the radar antenna to scan left and right (horizontal motion)?

19.8 How is the radar antenna stabilized as the airplane maneuvers through pitch and roll?

19.9 What is a typical frequency for an airborne weather radar?

19.10 Lightning detection systems are usually tuned to a frequency of _____.

19.11 What is the most recent method for delivering weather images to the cockpit?

19.12 What is the purpose of a radome?

19.13. Radomes must reduce radar power by no more than about _____.

19.14 Wind shear is a sudden change in wind _____ and is most dangerous near the _____.

19.15 A dangerous form of windshear, which occurs over a small area, is known as a _____.

19.16 Windshear detection systems warn the pilot and also provide _____.

Chapter 20

TCAS
Traffic Alert and Collision Avoidance System

Honeywell

Keeping aircraft safely separated had been the task of air traffic control since the 1930's when pilots radio'ed position reports by voice. This was followed primary surveillance based on radar "skin returns," then secondary surveillance using transponder interrogation and reply. But as airplanes began cruising near Mach 1 and air traffic multiplied, so did the threat of the "mid-air."

The search for a workable anti-collision system persisted for 50 years. Early experimental systems required costly atomic clocks, complex antennas and techniques borrowed from electronic warfare. Progress was slow until, in 1956, two airliners collided over the Grand Canyon on a sunny day. Closing at about 900 miles per hour, the pilots would have to see the other airplane at four miles, decide on the correct response, then maneuver off the collision course. All this would have to happen in 15 seconds. As a result of the accident, the U.S. Congress brought pressure on the FAA to develop an anti-collision system, and for airlines to install it at an early date.

During the 1960's, the transponder was spreading through aviation and researchers decided to abandon earlier technology and adopt the transponder as a building block in a new anti-collision system. After trying several variations, TCAS (Traffic Alert and Collision Avoidance System) was chosen as a world standard and it's now in widespread use everywhere, with scaled-down versions for business and light aircraft. In Europe the system is known as ACAS, for Airborne Collision Avoidance System, but all systems follow the standard adopted the International Civil Aviation Organization (ICAO).

While the transponder is a major component, the foundation is the TCAS processor. It performs one of the most intensive and rapid computations aboard the aircraft, executing software for collision logic. It must acquire, track and evaluate dozens of aircraft up to about 40 miles away---then issue commands on how to avoid a collision---all within seconds.

The road to TCAS was not entirely smooth. As

TCAS Symbols on a Radar Display

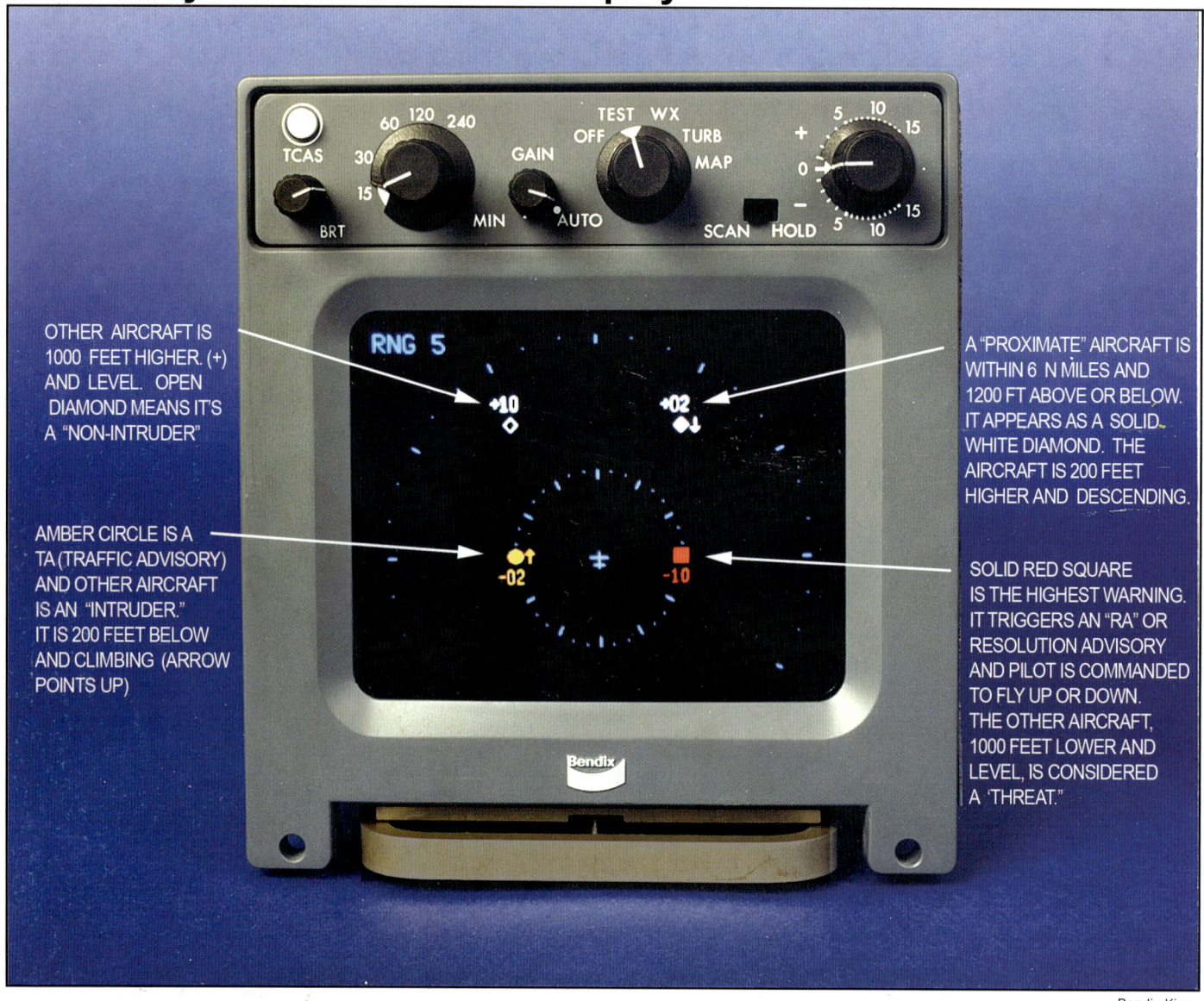

OTHER AIRCRAFT IS 1000 FEET HIGHER. (+) AND LEVEL. OPEN DIAMOND MEANS IT'S A "NON-INTRUDER"

AMBER CIRCLE IS A TA (TRAFFIC ADVISORY) AND OTHER AIRCRAFT IS AN "INTRUDER." IT IS 200 FEET BELOW AND CLIMBING (ARROW POINTS UP)

A "PROXIMATE" AIRCRAFT IS WITHIN 6 N MILES AND 1200 FT ABOVE OR BELOW. IT APPEARS AS A SOLID WHITE DIAMOND. THE AIRCRAFT IS 200 FEET HIGHER AND DESCENDING.

SOLID RED SQUARE IS THE HIGHEST WARNING. IT TRIGGERS AN "RA" OR RESOLUTION ADVISORY AND PILOT IS COMMANDED TO FLY UP OR DOWN. THE OTHER AIRCRAFT, 1000 FEET LOWER AND LEVEL, IS CONSIDERED A "THREAT."

Bendix-King

If an airplane has an EFIS or radar display, it can show TCAS information. The weather radar control panel is at the top, with a button at top left for activating the TCAS display.

Besides TCAS symbols on the display, there are voice announcements. If a threat advisory (TA) appears on the display, the voice says, "Traffic, Traffic." If it turns into a resolution advisory (RA), the voice gives a command to climb or descend.

the first systems were fitted to aircraft, pilots complained about false alarms (and shut them off). It mostly happened near crowded terminals and at low altitude. The technical committee responsible for TCAS responded with software upgrades ("Changes") that address each complaint. The performance of TCAS is now so effective, the FAA ruled that if a pilot receives a clearance from a controller that conflicts with TCAS, the pilot must obey the TCAS. In 2002 a pilot ignored that procedure and caused a mid-air collision 35,000 feet over Europe between an airliner and a cargo plane. Air traffic control had instructed the pilot to descend, while TCAS advised him to climb. All 69 people perished in the collision. Both aircraft had fully functioning TCAS.

Basic Operation

Once every second, the transponder of a TCAS airplane automatically transmits an interrogation. This is similar to the interrogations sent out by air traffic surveillance radar and the frequencies are the same.

If another airplane is within range, its transponder replies to the interrogation. The first airplane measures the time between interrogation and reply to determine the distance (range) to the other aircraft. Also received is the altitude of the other aircraft, which is encoded in the transponder reply (mode C). If the other aircraft has a Mode S transponder, its address is also sent. Directional antennas aboard the interrogating airplane determine the bearing (direction) to the threat aircraft.

Because TCAS exchanges data between airplanes, it does not require ground stations. Thus, it can operate where there is no radar coverage, such as oceanic flight and over remote areas.

Once the TCAS processor acquires information about the other aircraft, it looks at the potential for a collision. A major factor is "range rate," which tells the rate at which distance is changing between the two aircraft. If that change is *constant*, the two aircraft are on a collision course. This is similar to what happens if a pilot looks out and sees another airplane that appears stationary in the sky. It means the two airplanes are converging. TCAS detects such threats long before they are visible to the pilot.

Tau

Airplanes differ greatly in speed and performance and TCAS must work with them all. This is done through a concept known as "tau" (the Greek letter) to adjust warnings to the actual situation. By measuring distance and closing rate to the target, TCAS might issue the first warning 40 seconds before a potential collision and a second one 25 seconds before. TCAS adjusts warning times according to aircraft speeds.

Vertical Speed Indicator Adapted for TCAS

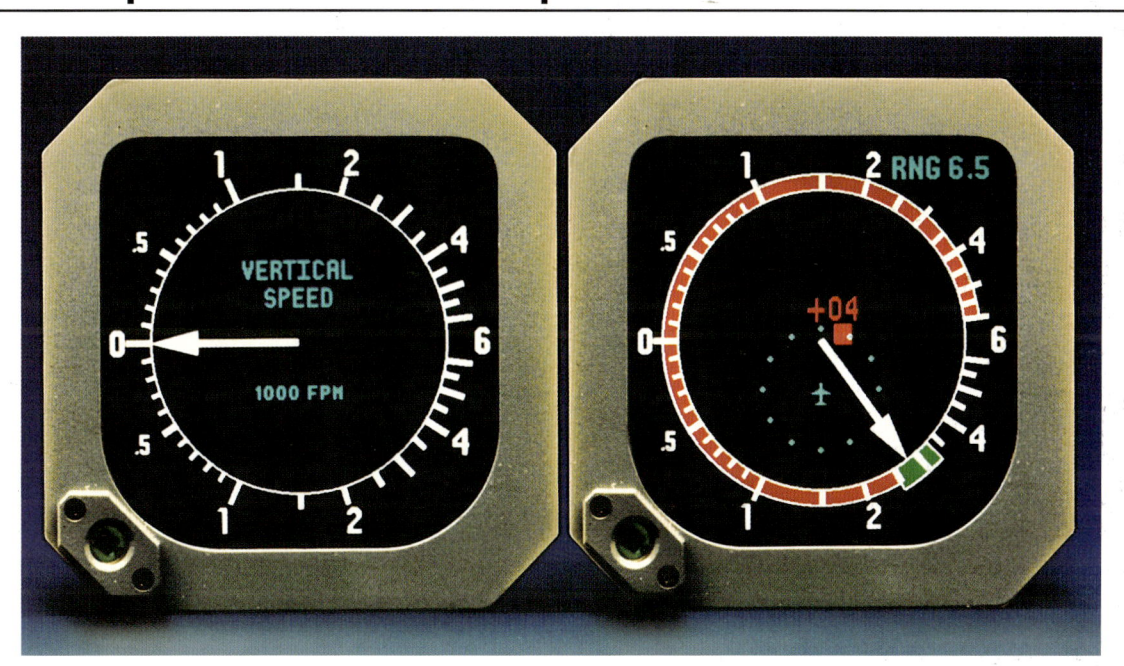

Aircraft without electronic flight instruments (EFIS) may add a TCAS display by replacing the conventional VSI (vertical speed indicator) shown at the left.

The new instrument (right) still functions as a vertical speed indicator but adds TCAS symbols. In this example, the airplane (green symbol) is encountering a threat (red square) 6.5 miles ahead at 1 o'clock. The "+04" means the threat is 400 feet higher and remaining at that altitude

Because two airliners typically close at about 1000 nm/hour, they could be less than 30 seconds from a collision.

The TCAS system is issuing an "RA," or resolution advisory. This is a command for the pilot to make a rapid descent, as shown by the green area at lower right. The pilot is complying by flying toward that area, as shown by the vertical speed needle. The airplane is descending vertically about 3000 feet per minute.

Note the large circle of red around the instrument. It is warning the pilot not to climb or descend in this region, but to go for the green.

In this example, TCAS logic instructed the pilot to descend. Because TCAS in both aircraft are communicating by datalink, the other aircraft is commanded to climb. This is a "cooperative maneuver," and produces maximum separation between aircraft.

TCAS System

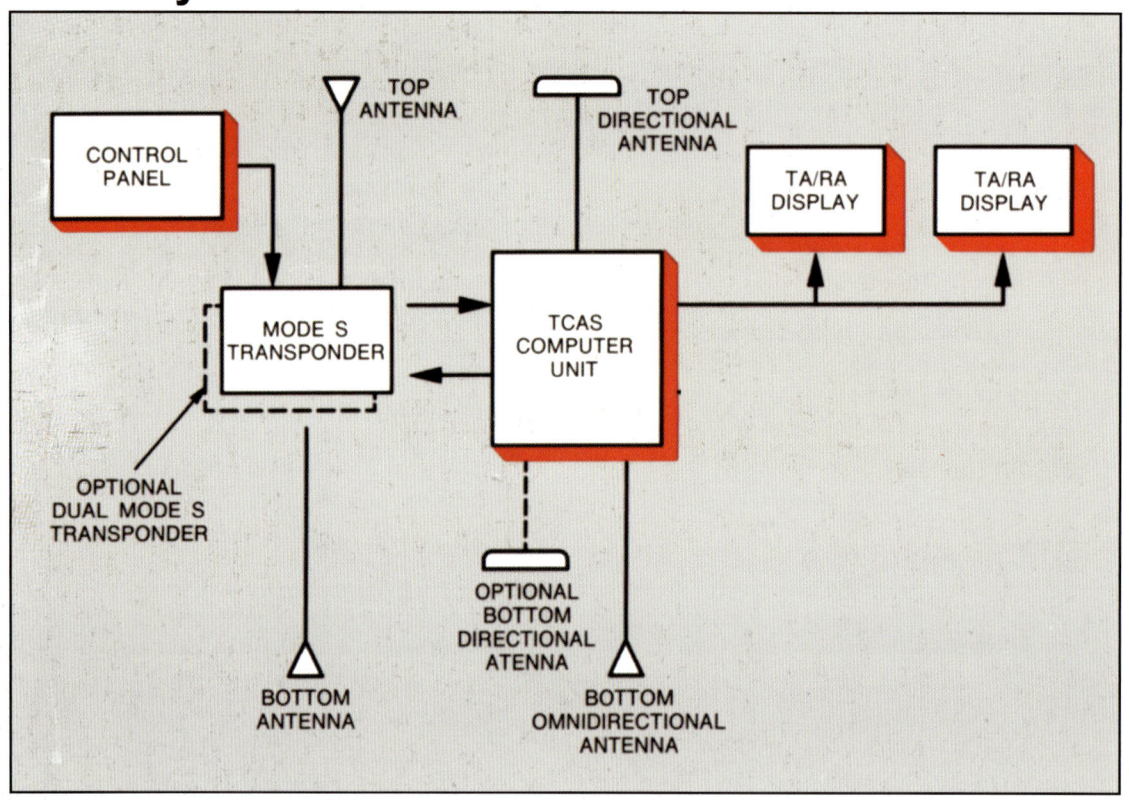

Major functions of a TCAS II system. It requires a Mode S transponder to enable two closing aircraft to communicate and determine which direction to fly (up or down) to avoid a collision. The transponder often uses a top and bottom antenna on the aircraft to assure full coverage above and below.

The computer processes large amounts of information; transponder replies of other aircraft, target tracking, threat assessment, visual and aural advisories, escape maneuvers and coordinating maneuvers between closing aircraft.

Traffic and Resolution Advisories

If a collision is possible, TCAS delivers two kinds of warnings:

•**Threat Advisory (TA)**. This is the less serious of the two. It means another aircraft might be 45 seconds from the closest point of approach (CPA). The pilot sees the TA on a display (shown in the illustration) and becomes aware of the threat.

•**Resolution Advisory (RA)** With this warning the conflict is rapidly growing more serious. The threat aircraft could now be 30 seconds from closest point of approach. TCAS issues a Resolution Advisory, which commands the pilot to climb, descend, remain level or observe a vertical restriction, as shown.

TCAS I and TCAS II

There are two versions of TCAS, for large and small aircraft. The full system, TCAS II, is required aboard airliners and large transports with 31 or more seats. In TCAS II, the full collision logic is provided to generate the two types of warnings; TA (threat advisory) and RA (resolution advisory).

TCAS I is a scaled-down system that issues only TA's (threat advisories). Otherwise, everything is much the same as TCAS II; the symbols, warnings and displays. Lower in cost, TCAS I is designed for corporate, business and light aircraft.

The added complexity of TCAS II is in the collision logic for developing the evasive commands, a more elaborate antenna system, the need for a Mode S transponder and a method of air-to-air communication known as "datalink."

Coordinating Climb and Descend

When TCAS issues a Resolution Advisory (RA) it instructs the pilot how to avoid a collision by flying up or down. Obviously, if both aircraft fly toward each other and perform the same escape maneuver (both fly up, for example) they would collide. This is prevented by "coordination interrogations" transmitted by each aircraft once per second. These are regular transponder signals on 1030 and 1090 MHz, but now used as a datalink to exchange information between aircraft.

TCAS Components

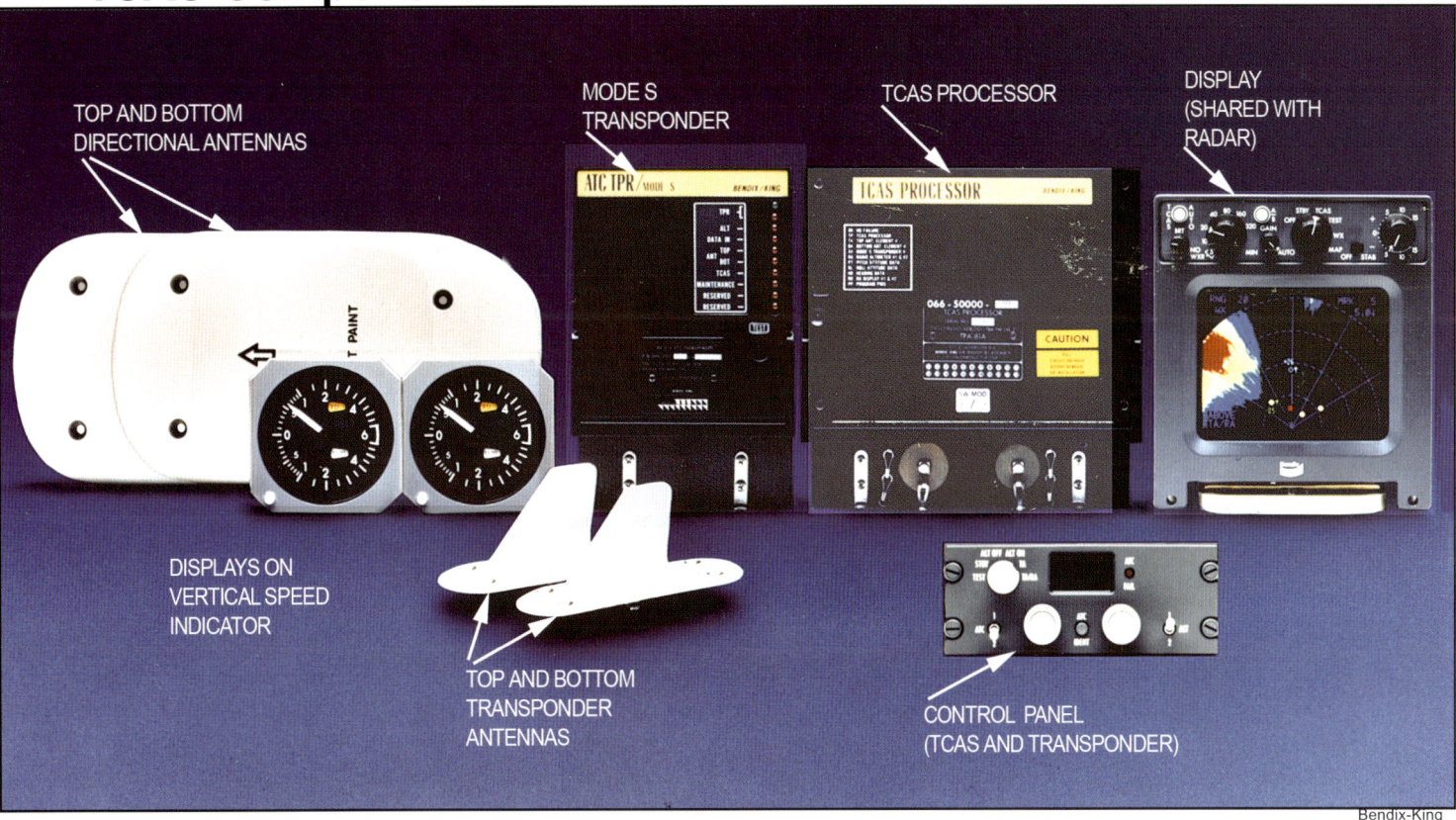

The control panel at lower right selects TCAS and transponder functions. Two antennas are used for the transponder---placed at top and bottom of the fuselage---to assure complete signal coverage around the airplane. Two directional antennas (left) also determine the bearing of a threat aircraft above and below the airplane.

Let's say the TCAS of one aircraft decides on a "fly up" maneuver. This is considered an *intention* and, in this example, is an "upward sense" (climb). The intention is transmitted to the other aircraft. This causes the TCAS of the second aircraft to select a "downward sense" (descend). Thus, when one aircraft receives the other's intention, it selects the opposite sense---so one flies up, the other flies down.

There is a possibility that both aircraft will see each other as a threat at the same instant and both select the same sense. If this happens TCAS logic goes to another source to break the conflict; the transponder address. (All Mode S transponders have a permanent address.) The aircraft with the *higher* address will reverse its sense.

Whisper-Shout

During the design of TCAS there was concern the system would overload because of too many replies, especially as airplanes converged on a busy airport. This was solved by the "whisper-shout" technique. As the airplane cruises, it transmits an interrogation at low transponder power, say 2 watts. Only the closest aircraft and those with the highest transponder sensitivity can hear it and reply. This is the "whisper"---which limits the replies to the closest airplanes. TCAS processes these replies, which are a small portion of the total number of targets.

Next, the transponder increases power slightly to trigger replies from aircraft slightly farther away. At the same time, however, the transponder also sends a "suppression" pulse which silences the first set of transponders and prevents their replies. In rapid steps, the interrogations increase in power, until they're "shouting" at 250 watts. These high-level signals now reach aircraft at the outer edge of coverage. It's important to note that each time the power ramps up it is followed by a suppression pulse that silences all transponders that replied earlier.

A complete whisper-shout cycle repeats once per second, effectively placing replies into small groups that are processed in sequence. This reduces clutter and overload.

Directional Interrogation

Besides whisper-shout, another technique reduces the number of replies received each second. The inter-

151

rogations are transmitted through a directional antenna which electronically rotates 90 degrees at a time. This covers a full circle in four quadrants and limits replies to the active quadrant.

Non-TCAS Airplanes

The system can also recognize aircraft that are not carrying TCAS or Mode S transponders. Such aircraft typically have the earlier ATCRBS transponder. A TCAS-equipped aircraft, however, interrogates these aircraft and computes information required to display a threat advisory (TA). There can be no cooperative maneuvering because this requires Mode S transponders on both aircraft, as well as a TCAS system.

TCAS III

TCAS II commands the pilot only in the vertical direction, which is sufficient to avoid a collision. The industry had started work on TCAS III, to add commands in the horizontal direction (fly left, fly right) but it never was completed. The problems of issuing both vertical and horizontal maneuvers proved extremely difficult. Maneuvering in two dimensions simultaneously multiplies the chances for aircraft to create new collision courses with second and third airplanes as they avoid the first one. Before these problems were solved, TCAS III was abandoned as new systems began to examine the collision threat.

A new global air traffic system is emerging with collision avoidance based on GPS and satellites. It is ADS-B---automatic dependent surveillance-broadcast. As aircraft cruise they "squitter" (automatically transmit) their position based on GPS. That information is picked up by nearby aircraft for collision avoidance and also relayed via satellite to air traffic control for managing traffic.

Yet another system began during 2004. Known as TIS, Traffic Information Service, it broadcasts the targets shown on all surveillance radars on the ground. The images are downlinked via satellite to aircraft, which display traffic, as done with TCAS.

TCAS, however, will be operational for many generations. It is still unequalled as the tactical collision avoidance system anywhere on earth.

TCAS Voice Warnings

1. Traffic Advisory (TA): "TRAFFIC, TRAFFIC"

2. Resolution Advisories (RA):

Preventive:
"MONITOR VERTICAL SPEED, MONITOR VERTICAL SPEED"
The pilot keeps the VSI needle out of the lighted segments.

Corrective:
"CLIMB, CLIMB, CLIMB"
Climb at the rate shown on the RA indicator; nominally 1500 fpm.
"CLIMB, CROSSING CLIMB, CLIMB, CROSSING CLIMB".
As above, except that it further indicates that own flight path will cross through that of the threat.
"DESCEND, DESCEND, DESCEND".
Descend at the rate shown on the RA indicator; nominally 1500 fpm.
"DESCEND, CROSSING DESCEND, DESCEND, CROSSING DESCEND"
As above except that it further indicates that own flight path will cross through that of the threat.
"REDUCE CLIMB, REDUCE CLIMB"
Reduce vertical speed to that shown on the RA indicator.
"REDUCE DESCENT, REDUCE DESCENT"
Reduce vertical speed to that shown on the RA indicator.
"INCREASE CLIMB INCREASE CLIMB"
Follows a "Climb" advisory. The vertical speed of the climb should be increased to that shown on the RA indicator, nominally 2500 fpm.
"INCREASE DESCENT, INCREASE DESCENT"
Follows a "Descend" advisory. The vertical speed of the descent should be increased to that shown on the RA indicator, nominally 2500 fpm.
"CLIMB, CLIMB NOW, CLIMB, CLIMB NOW".
Follows a "Descend" advisory when it has been determined that a reversal of vertical speed is needed to provide adequate separation.
"DESCEND, DESCEND NOW DESCEND, DESCEND NOW"
Follows a "Climb" advisory when it has been determined that a reversal of vertical speed is needed to provide adequate separation.

Review Questions
Chapter 20 TCAS (Traffic Alert and Collision Avoidance System)

20.1 A TCAS aircraft transmits an interrogation once per _____.

20.2 How does an intruder aircraft with an ATCRBS (early type) transponder reply to TCAS interrogations?

20.3 How does an intruder aircraft with a Mode S transponder reply to TCAS interrogations?

20.4 How does TCAS determine the direction of a threat?

20.5 How does TCAS determine the distance of a threat?

20.6 How does TCAS determine whether the other aircraft is a threat?

20.7 What is the concept of "Tau".

20.8 Name the two kinds of warnings issued by TCAS.

20.9 Does a Threat Advisory (TA) command the pilot to maneuver out of the way?

20.10 What does an Resolution Authority (RA) do?

20.11 If two TCAS aircraft are closing, what prevents them from climbing, and flying into each other?

20.12 What is the technique of "whisper-shout"?

20.13 How does the directional antenna reduce the number of replies for each interrogation?

Chapter 21

VFR — Visual Flight Rules

Planning the Installation

Installations vary, from wiring a headset jack to rebuilding an instrument panel. No matter how extensive, it must follow rules of "airworthiness"---guidance by a civil aviation authority such the FAA in the US or a CAA in other countries.

Observe the TC. For major rebuilding of an instrument panel, there is an overriding rule about where you can place equipment. *Certified* airplanes---those built in a factory and sold ready to fly---must obtain a TC, or Type Certificate. The TC shows all equipment delivered with the airplane. Such equipment may not be moved to other locations on the panel without violating the TC. They may be replaced with equivalent units, but not shifted around. This does not prevent adding new equipment to the panel, or minor relocation of radios in a center stack, for example. These alterations will be noted on forms submitted for approval to the government agency.

The pilot/owner handbook or flight manual typically lists the equipment installed under the Type Certificate.

STC. When adding systems to a factory-built airplane, using equipment critical to flight, this is usually done under an STC, or *Supplemental Type Certificate*. The manufacturer of the new system proved its airworthiness to obtain the STC. Examples include autopilots, displays and fuel management systems. For such installation, you will work from drawings prepared by the STC-holder showing precisely where and how components mount.

STC's can be compared to a patent; they are owned exclusively by the designer and protected by law. Often, the STC is offered for sale to avionics shops, along with the system and an installation kit. In cases where a manufacturer is selling a major system, such as an autopilot, he often allows the buyer to use the STC at no extra cost.

For large aircraft, expect more support from the avionics manufacturer. If a fleet of 30 air transports will be upgraded with a collision avoidance system, chances are a field representative from the manufacturer will assist in early installations.

Non-certified airplanes. There is a wide range of aircraft operating in the "Experimental" category, which includes kit-built, built-from-plans, antiques, warbirds

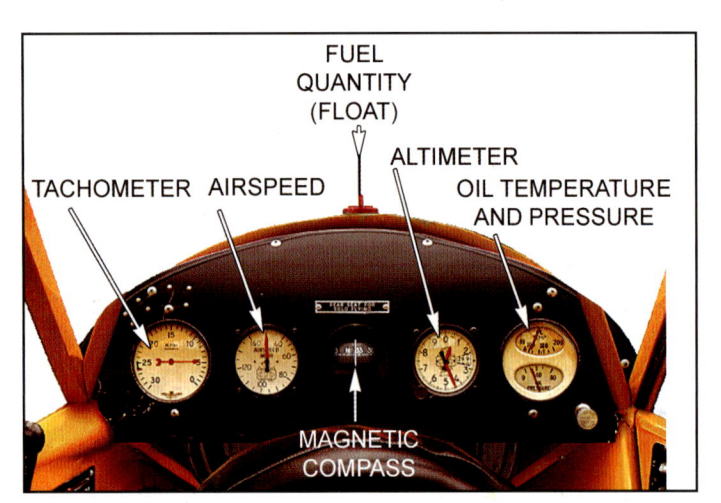

Instrument panel of the original Piper Cub, which received its Type Certificate in 1931. Today, the same instruments are still required for "day VFR" flying. More airplanes, however, show the same information on an electronic display, as seen on the next page.

One EFIS Screen Replaces Ten "Steam Gauges"

Tachometer from a 1931 Piper Cub is a 3-inch instrument that shows one function: RPM.

The future of instrument panels is EFIS (Electronic Flight Instrument System). In this comparison, the Piper Cub tachometer is only slightly smaller than the EFIS screen below, which displays ten or more instruments.

Nevertheless, the technician will see "steam gauge" instruments for generations to come. It will take that long for over 100,000 airplanes in the U.S. alone to fully change over to the new technology.

The Dynon system below is the first of the simple, low-cost EFIS screens. As this and other systems gain certification for production aircraft, they will gradually be installed as an upgrade to existing instrument panels. By 2005 nearly all airframe manufacturers announced they will install EFIS in their new airplanes.

The airlines have been flying with EFIS since 1982, beginning with the Boeing 757 and 767. They use cathode ray (TV) tubes, while new EFIS has flat-panel LCD displays.

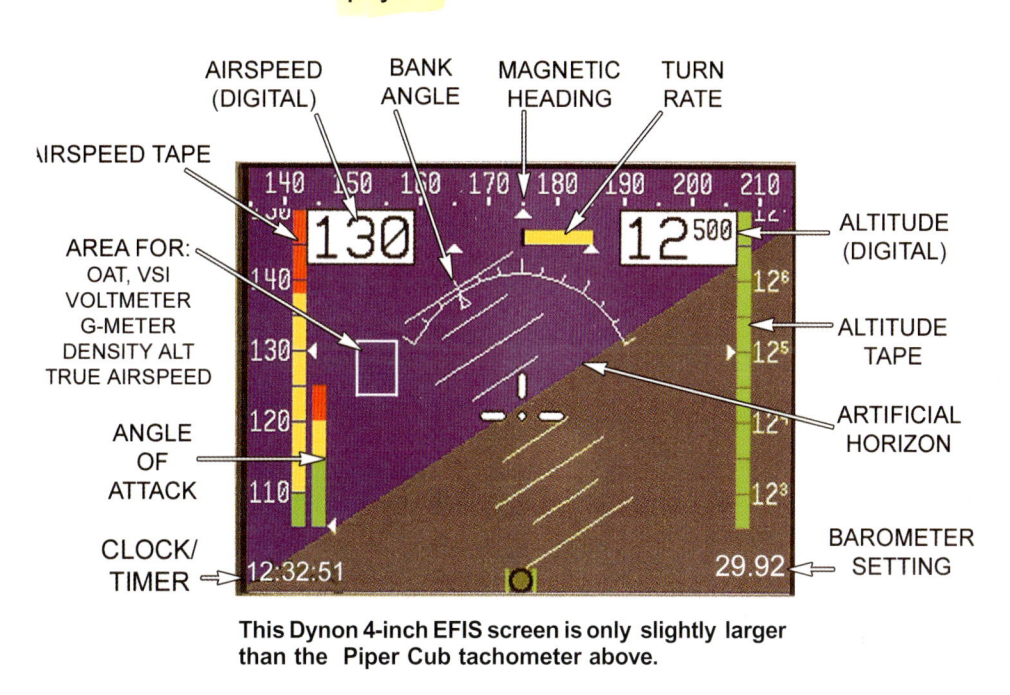

Labels on EFIS screen:
- AIRSPEED (DIGITAL)
- BANK ANGLE
- MAGNETIC HEADING
- TURN RATE
- AIRSPEED TAPE
- AREA FOR: OAT, VSI, VOLTMETER, G-METER, DENSITY ALT, TRUE AIRSPEED
- ANGLE OF ATTACK
- CLOCK/TIMER
- ALTITUDE (DIGITAL)
- ALTITUDE TAPE
- ARTIFICIAL HORIZON
- BAROMETER SETTING

This Dynon 4-inch EFIS screen is only slightly larger than the Piper Cub tachometer above.

and space vehicles. If they are registered as experimental, the avionics installation does not have to follow the same rules of certified aircraft. Instrument and radio placement may be designed by the builder. Many kit-built aircraft are capable of speeds greater than production aircraft, fly at higher altitudes and with such advanced systems as integrated displays, pressurization and turboprop powerplants. For safety's sake, these installations should also follow recommendations for airworthiness that apply to certified airplanes. Experimenters are encouraged (the Wright brothers began as bicycle mechanics), but home-built aircraft are inspected and an FAA representative may not accept something which appears unsafe.

Type of Flying

An airplane is typically outfitted according to type of flying, which informally divides as follows:

Day VFR. The airplane flies during daylight hours and under VFR (visual flight rules). Besides required instruments (see table) the pilot may want nothing more than a handie-talkie for communication and a portable GPS for navigation. This is often a solution when the airplane has no electrical system (battery and generator).

Night VFR. Even on clear, moonlit nights, flight after sundown should have avionics redundancy; a second com and second means of navigation. Flying VFR after dark is not only ruled out in every country out-

155

side the U.S., but the accident rate is ten times higher on dark, moonless nights. The pilot should be able to call for help if he inadvertently flies into a cloud at night or is lost with no backup navigation.

Light IFR. Many pilots obtain a rating to fly under IFR (instrument flight rules), but rarely use it. But it is a great timesaver when the obstacle is a low cloud layer only in the vicinity of the airport. The IFR rating is used only to fly for the few minutes it takes to climb above, or descend through, thin layers.

Low IFR. This is for the serious pilot who needs to get through widespread areas of low visibility, then shoot an instrument landing to a runway under a low ceiling. This aircraft needs reliable, redundant avionics. Safety will greatly improve with a terrain avoidance advisory system, weather detection and a satellite datalink that delivers the images of Nexrad, the ground weather radar network. Although not a requirement for private pilots, an autopilot is essential to safe single-pilot IFR operations.

Aircraft, flying under any condition---day, night or on instruments---benefit from some type of collision avoidance. The chance of a mid-air is the opposite of what is generally believed. Virtually no collisions occur inside clouds or at night. Most happen on a bright VFR day in the vicinity of an airport when airplanes converge for landing. As the chapter on collision avoidance describes, there are anti-collision systems to fit any size airplane.

Instruments and Radios

Applies to powered civil aircraft with a standard airworthiness certificate operating under FAA Part 91 (mainly private and corporate aircraft). For more specific requirements, and air transport requirements, check Federal Air Regulations.

Day VFR
1. Airspeed
2. Altimeter
3. Magnetic direction indicator (compass)
4. Tachometer for each engine
5. Oil pressure gauge for each engine using pressure system.
6. Temperature gauge for each liquid-cooled engine.
7. Oil temperature gauge for each air-cooled engine.
8. Manifold pressure gauge for each altitude engine (usually applies to aircraft with controllable pitch propellers).
9. Fuel gauge showing quantity in each tank.
10. Landing gear position indicator (
11. Anti-collision light
12. Emergency locator transmitter (ELT)
13. Transponder, with Mode A and C (when operating in high-traffic areas and within 30 miles of large airports.

Night VFR
1. All instruments for day VFR.
2. Position lights
3. Anti-collision light
4. Landing light (if operating for hire)
5. Adequate source of electrical energy for electrical and radio equipment.
6. Spare set of fuses or three spare fuses of each kind required, available to pilot in flight.

Instrument Flight Rules (IFR)
1. All instruments for day and night VFR
2. Two-way radio and navigation equipment appropriate to the ground facilities used.
3. Gyroscope rate of turn indicator (except where aircraft has a third attitude instrument.
4. Slip-skid indicator
5. Sensitive altimeter with setting for barometric pressure.
6. Clock with hours, minutes, seconds with sweep-second pointer or digital display.
7. Artificial horizon (gyroscopic pitch and bank)
8. Directional gyro
9. Flight at or above 24,000 ft MSL. If VOR navigation is used, DME is required.

Other Requirements
1. Altitude alerting system for turbojets
2. Large and turbine-powered multiengine air planes: flying over water (more than 30 minutes' flying time or 100 nautical miles from shore).
   Two transmitters
   Two microphones
   Two headsets or one headset and one speaker
   Two independent receivers
   Two independent electronic navigation units (appropriate to the air space flown)
   HF communications, if necessary to the flight.

Basic T Instrument Layout

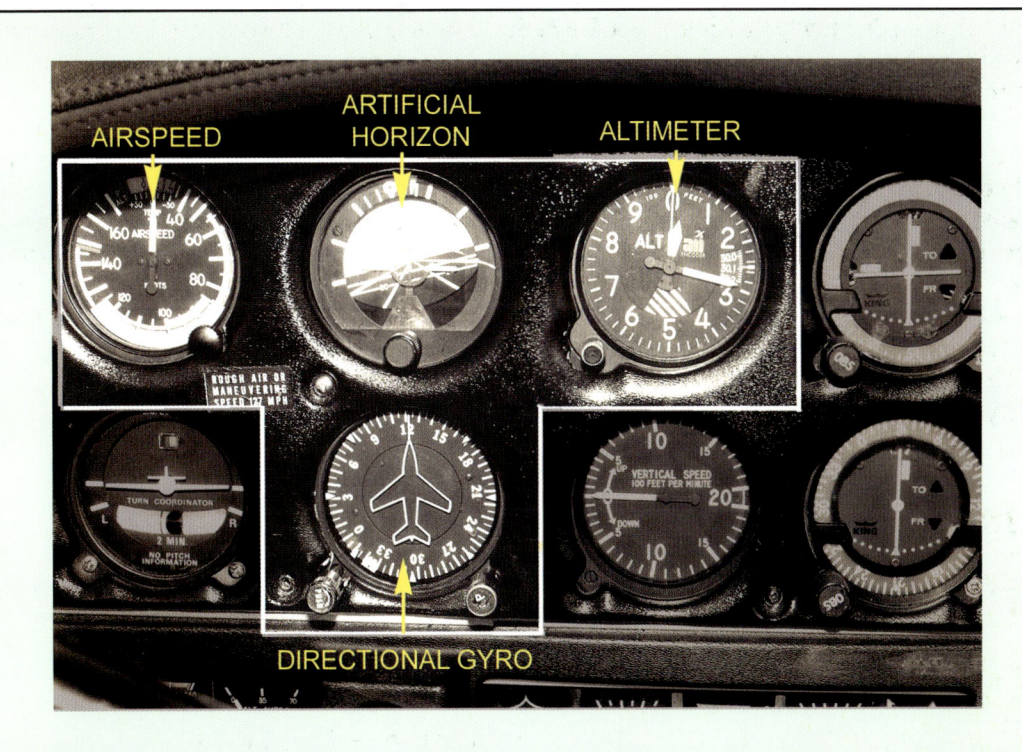

Most light aircraft---Cessna, Piper and Mooney, for example, add the two lower instruments to the basic T; the turn coordinator and vertical speed indicator. This equips the airplane for basic instrument flying. Some technicians call this the "six pack."

Large Aircraft
Turbine-powered airplanes are often outfitted with a "suite" of avionics from one manufacturer, as in this EFIS system.

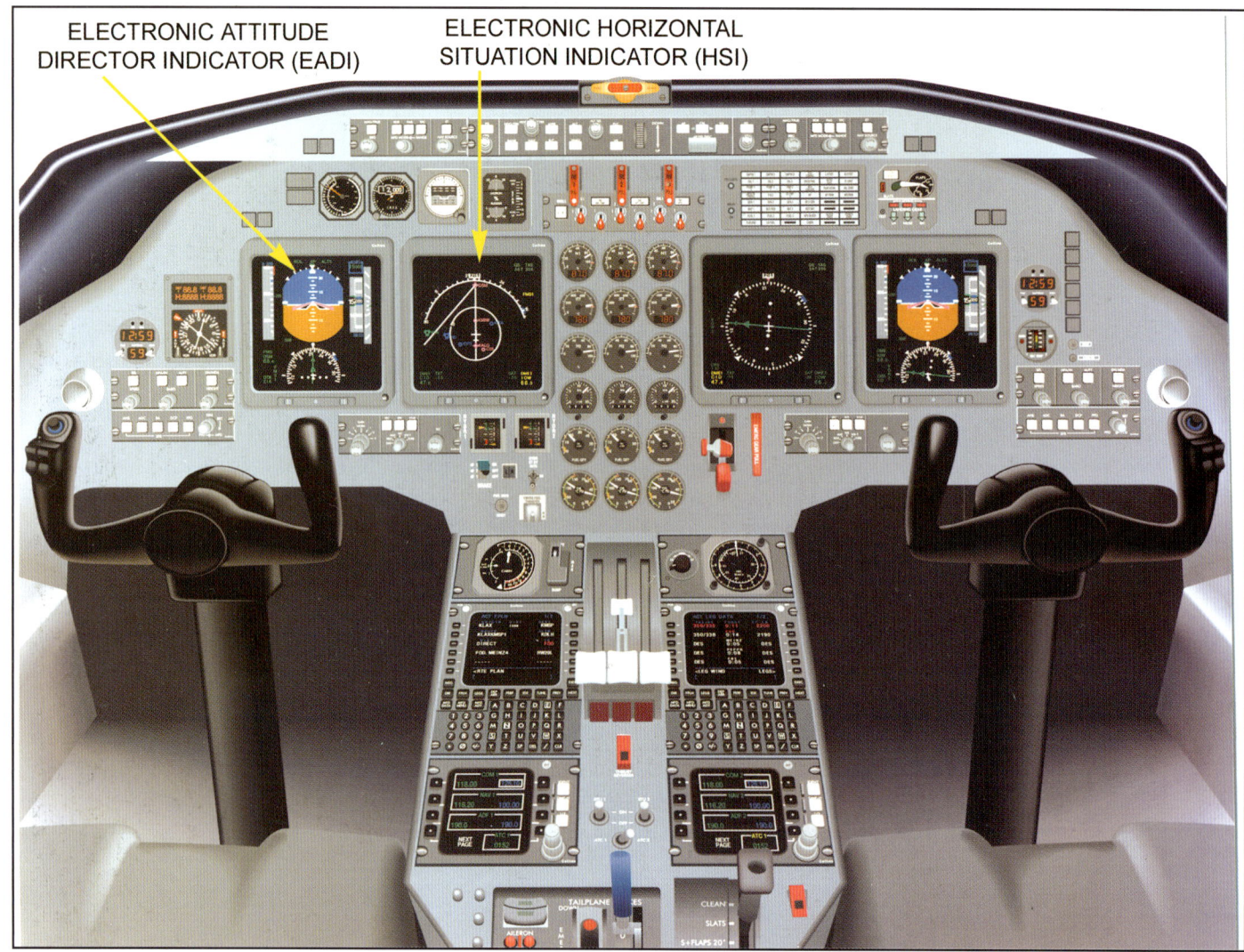

ELECTRONIC ATTITUDE DIRECTOR INDICATOR (EADI)

ELECTRONIC HORIZONTAL SITUATION INDICATOR (HSI)

Collins

An early EFIS system, introduced in the mid-1990's, is still flying aboard many business aircraft and regional airlines. It has four cathode ray tubes in the instrument panel, with two more tubes down in the pedestal. Pilot and co-pilot sides are nearly identical. The tube at the left, the Electronic Attitude Director Indicator, is mainly for flying the airplane manually or on autopilot. It also contains radionavigation information. Second tube from the left is the Electronic Horizontal Situation Indicator, which displays compass, waypoint, weather radar and other information The two tubes below, in the pedestal, are for flight management---mostly to store and fly routes, waypoints and airports---loaded on the ground before take-off. Shown here is a Collins Pro Line for the Falcon 50 .

Flat Panel, Integrated EFIS

(image label: PRIMARY FLIGHT DISPLAY)
(image label: MULTIFUNCTION DISPLAY)

Cirrus

By the year 2000, the future of instrument panels was clear. Flat panels (LCD's) would replace cathode ray tubes (CRT's). Instruments would become "integrated," that is, separate gauges merge into the electronic display.

The main part of the display shown here are two 10.4-inch (diagonal) flat panel LCD's The one at top left is the PFD, or "Primary Flight Display." Although it can depict almost any information, it is often used as shown; the top half for flying attitude, the lower half with compass and waypoint information.

The display on the right is the MFD, or "Multifunction Display," which shows moving map, traffic, weather and other data. The three small round instruments on the lower left are for backup---airspeed, attitude and altitude---but they are electronic, not electromechanical, displays. Over on the far right are engine instruments. The trend, however, is to merge these onto the main electronic displays.

This airplane, the Cirrus SR-22, eliminates the vacuum system usually required in production airplanes. Regulations require that flight-critical instruments have different power sources; usually accomplished with an electric-driven turn coordinator and a vacuum-driven artificial horizon. The all-electric Cirrus satisfies the rules by having two batteries and two alternators.

The EFIS system is the FlightMax Integra by Avidyne.

Typical Avionics Equippage

The instrument and radio chart shown earlier covers only equipment required by law. Aircraft owners often add systems to reduce workload or improve safety.

DAY VFR
- COM
- PORTABLE GPS (or GPS/COM)
- TRANSPONDER
- ALTITUDE ENCODER
- INTERCOM

BASIC IFR
- AUDIO PANEL, 3-LIGHT MB AND INTERCOM
- #1 GPS/COM (VFR)
- #2 NAV/COM
- VOR/LOC INDICATOR
- TRANSPONDER
- ALTITUDE ENCODER
- ADF

IFR
- AUDIO PANEL, 3 LIGHT MB AND INTERCOM
- MOVING MAP/GPS (VFR)
- NAV/COM
- VOR/LOC/GLIDESLOPE INDICATOR
- GLIDESLOPE RECEIVER
- TRANSPONDER
- ALTITUDE ENCODER
- AUTOPILOT, 1 AXIS (RADIO TRACK, HEADNG)
- ENGINE MONITOR
- ADF

FREQUENT IFR
- AUDIO PANEL 3-LIGHT MB, INTERCOM
- MULTIFUNCTION DISPLAY / MOVING MAP
- COM
- GPS (IFR)
- NAV/COM
- HSI WITH SLAVED COMPASS SYSTEM
- VOR/LOC/GLIDESLOPE INDICATOR
- GLIDESLOPE RECEIVER
- TRANSPONDER
- ALTITUDE ENCODER
- AUTOPILOT, 2 AXIS (TRK, HEADNG, ALT HOLD)
- ENGINE MONITOR
- STORMSCOPE AND/OR NEXRAD WX UPLINK
- FLIGHT TELEPHONE
- ENTERTAINMENT SYSTEM
- ADF

CORPORATE AIRCRAFT
- AUDIO PANEL, 3-LIGHT MB, INTERCOM
- MULTIFUNCTION DISPLAY / MOVINGMAP
- WEATHER RADAR
- GPS (IFR)
- NAV/COM #1
- NAV/COM #2
- HSI WITH SLAVED COMPASS SYSTEM
- VOR/LOC/GLIDESLOPE INDICATOR
- GLIDESLOPE RECEIVER
- TRANSPONDER #1
- TRANSPONDER #2
- ALT ENCODER #1
- ALT ENCODER #2
- AUTOPILOT, 2 AXIS (TRCK, HEADING, ALT HOLD)
- ENGINE MONITOR
- STORMSCOPE
- FLIGHT TELEPHONE
- COLLISION AVOIDANCE
- TERRAIN AVOIDANCE
- IN-FLIGHT ENTERTAINMENT
- ADF

TURBINE AIRCRAFT
- AUDIO PANEL, 3-LIGHT MB, INTERCOM
- MULTIFUNCTION DISPLAY / MOVING MAP
- RADAR INTERFACED TO MFD
- GPS (IFR)
- NAV/COM #1 NAV/COM #2
- VOR/LOC/GLIDESLOPE INDICATOR
- GLIDESLOPE RECEIVER
- TRANSPONDER #1, TRANSPONDER #2
- ALTITUDE ENCODER #1 and #2
- HSI WITH COUPLED COMPASS SYSTEM
- FLIGHT DIRECTOR
- AUTOPILOT, 3 AXIS WITH YAW DAMPER
- ENGINE MONITOR
- STORMSCOPE INTERFACED TO MFD
- SATPHONE OR FLIGHT TELEPHONE
- COLLISION AVOIDANCE
- TAWS: TERRAIN AVOIDANCE
- IN-FLIGHT ENTERTAINMENT SYSTEM
- EFIS OPTIONAL
- ADF
- HF FOR TRANS-OCEANIC AIRCRAFT

Manuals and Diagrams

The key to an installation is the manufacturer's manual on the specific model. Besides showing where each wire connects, pictorial drawings clarify difficult areas and give dimensions, power consumption and mounting hardware. There are schematic diagrams for troubleshooting.

A manufacturer's manual is also accepted by a government inspector (FAA, CAA) as "approved data." At some future time you may be questioned on what you used for installation guidance---and you can point to the manual.

There is no special format for manuals in General Aviation. In the airlines, however, manuals are written according to an ATA (Air Transport Association) "chapter." These documents include the "Component Maintenance Manual" and "Illustrated Parts Catalog."

The section in the manual used much of the time is the "pin-out diagram," which shows how wires run among various connectors and units during an installation. It's also used for troubleshooting later on.

Obtaining Manuals

There are several sources for installation manuals. If a maintenance shop is a dealer for an avionics manufacturer, it's usually required to have a library of

Typical manufacturer's manual for General Aviation, in this illustration a Bendix-King compass system with a horizontal situation indicator. Always check the model number on the unit and compare it with the manual. For example, the model name "KCS 55A" may not be the same as "KC 55," although the illustration may appear the same.

Installation Drawing

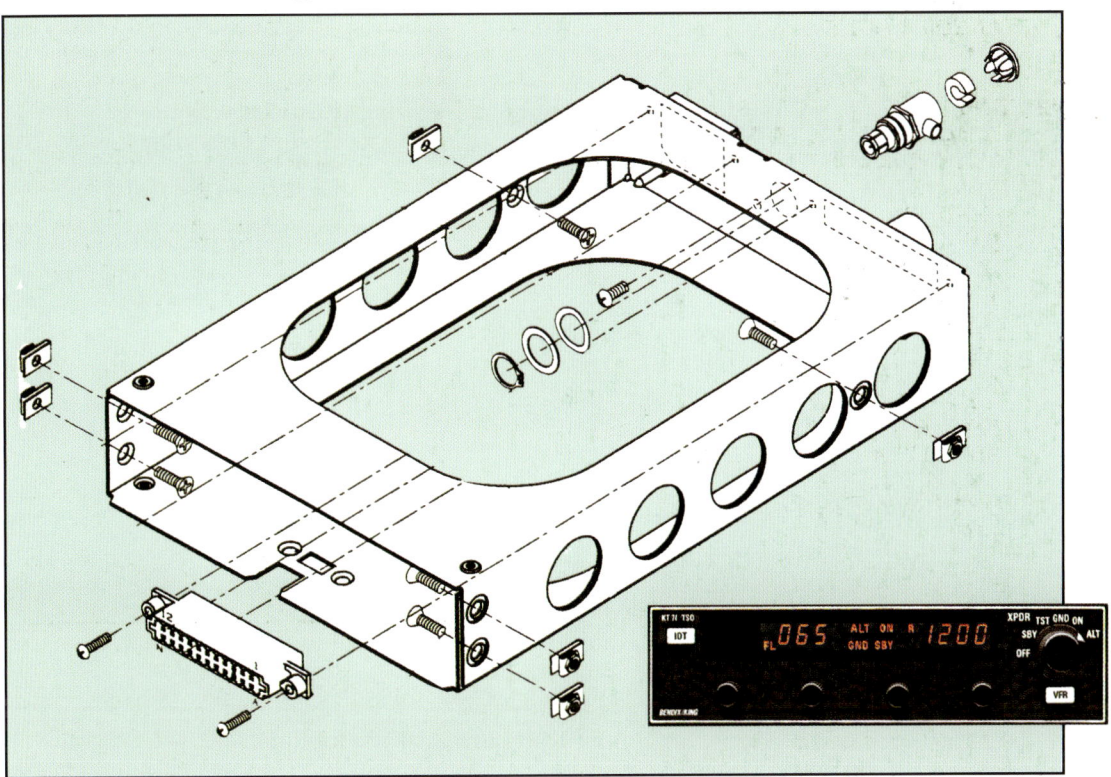

Pictorial illustrations in a manual, like this one for a Bendix-King transponder tray, are essential for mounting hardware. The drawing shows where to assemble connectors and gives details on fastening the tray to the instrument panel.

161

manuals for the equipment it installs. These books are purchased directly from the manufacturer.

Manuals are also available to members of "Resource One." This is an on-line service of the Aircraft Electronics Association (www.aea.net).

Manuals are sometimes available from resellers who list them in aviation publications.

In some instances, manufacturers make their manuals available on line at no charge.

Schematic (Circuit) Diagrams

Manuals contain schematics for troubleshooting down to the circuit board level and are not often required for installation work. An installer follows pin-out or interface diagrams like the examples shown on these pages. The schematic shows every resistor, capacitor, chip and other small component soldered to printed circuits inside the radio enclosure. The schematic is more useful for troubleshooting on the shop bench with specialized test equipment.

The Manual Locates Connectors and Pin Numbers

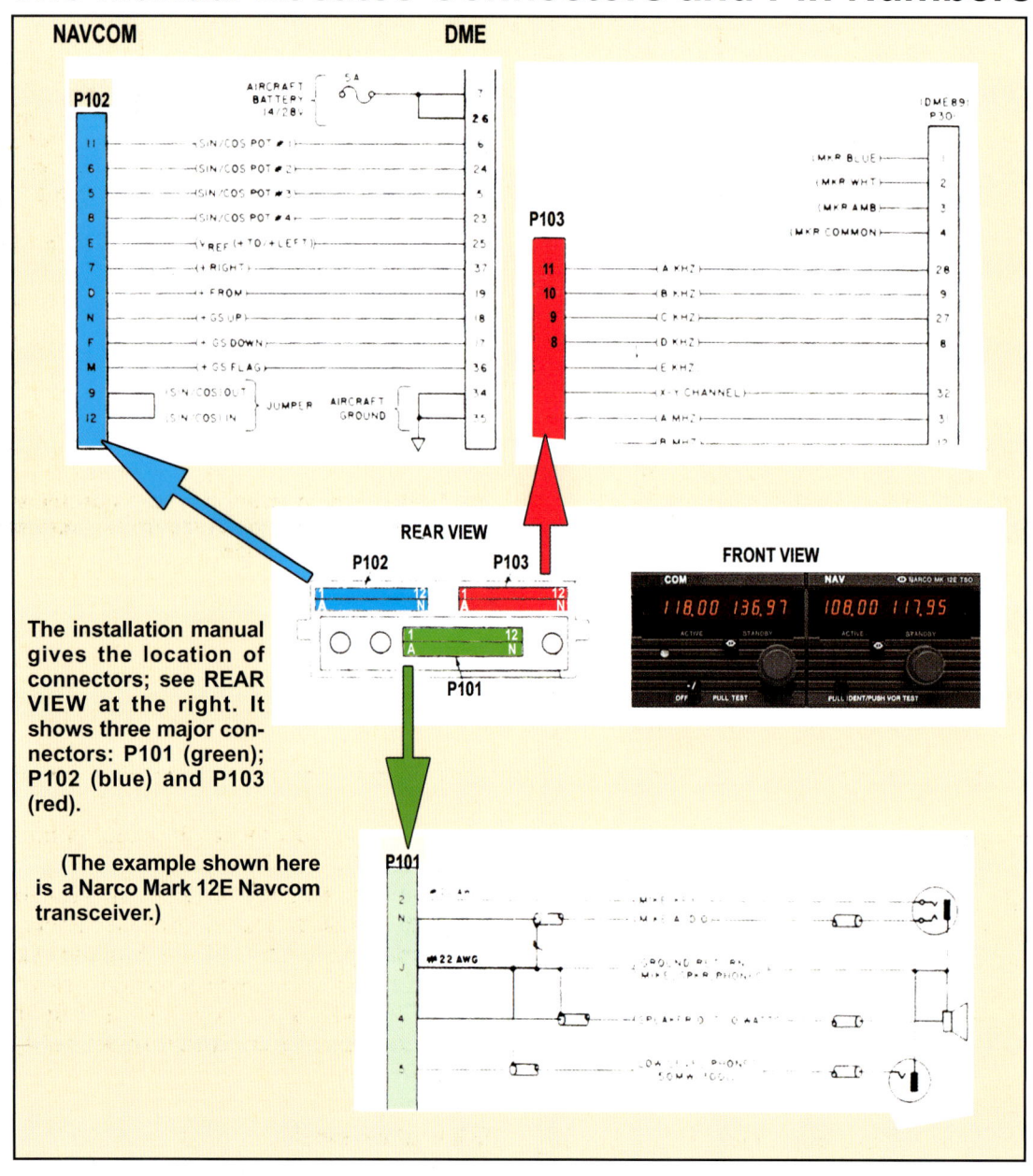

The installation manual gives the location of connectors; see REAR VIEW at the right. It shows three major connectors: P101 (green); P102 (blue) and P103 (red).

(The example shown here is a Narco Mark 12E Navcom transceiver.)

162

Pin Assignments

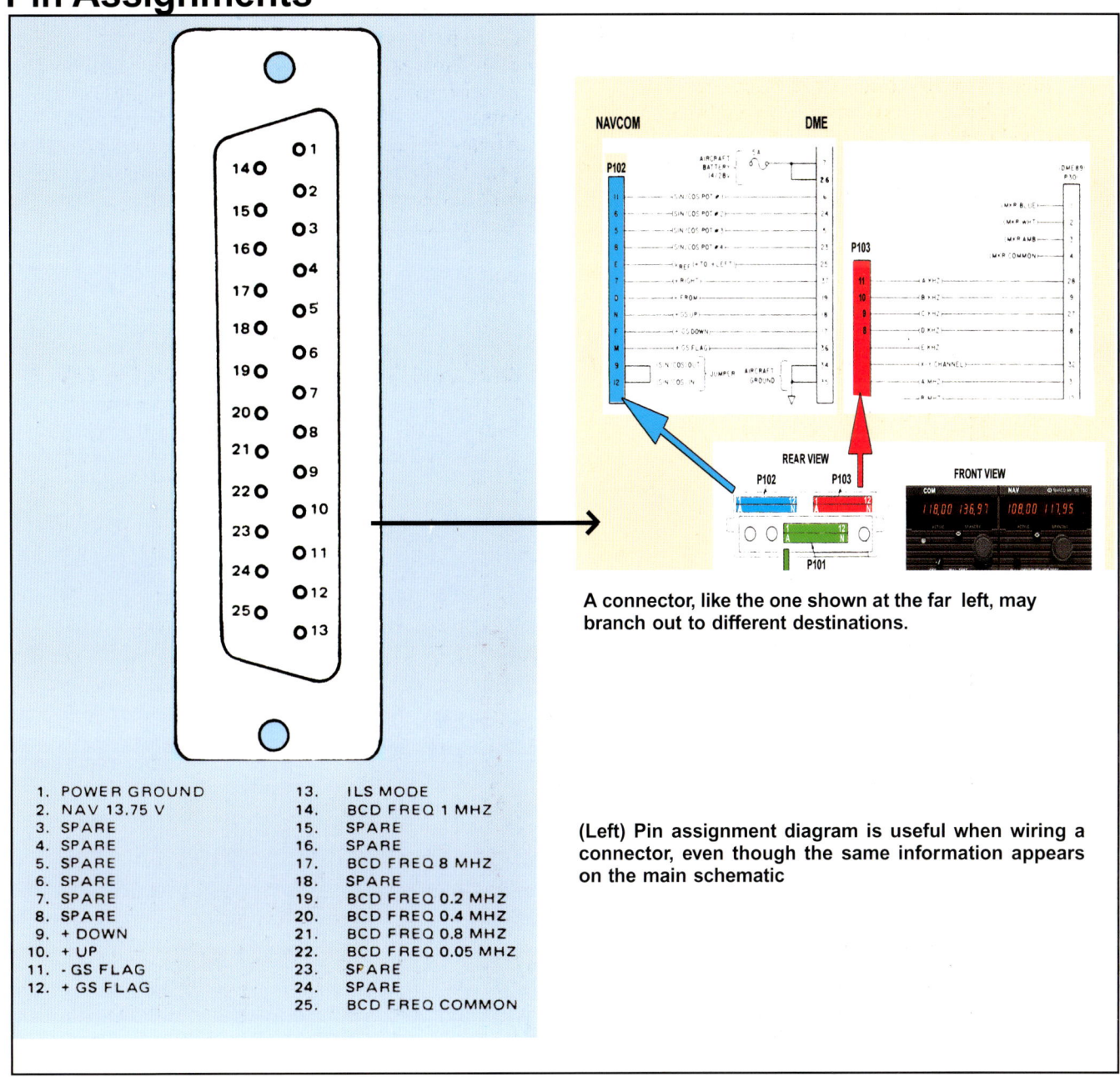

Pin	Assignment	Pin	Assignment
1.	POWER GROUND	13.	ILS MODE
2.	NAV 13.75 V	14.	BCD FREQ 1 MHZ
3.	SPARE	15.	SPARE
4.	SPARE	16.	SPARE
5.	SPARE	17.	BCD FREQ 8 MHZ
6.	SPARE	18.	SPARE
7.	SPARE	19.	BCD FREQ 0.2 MHZ
8.	SPARE	20.	BCD FREQ 0.4 MHZ
9.	+ DOWN	21.	BCD FREQ 0.8 MHZ
10.	+ UP	22.	BCD FREQ 0.05 MHZ
11.	− GS FLAG	23.	SPARE
12.	+ GS FLAG	24.	SPARE
		25.	BCD FREQ COMMON

A connector, like the one shown at the far left, may branch out to different destinations.

(Left) Pin assignment diagram is useful when wiring a connector, even though the same information appears on the main schematic

Reading the Wiring Diagram

The parts on a wiring diagram---connectors, cables, terminals, etc.---are not laid out like they appear in the actual radios. Components may be shown next to each other on a diagram, but lie at opposite ends of the radio enclosure. If the designer drew wires as they actually run, the diagram would be impossible follow. Wires would criss-cross everywhere and the diagram difficult to trace. In the diagram, wires are arranged to run in straight lines.

Some schematics look complex, but there are ways to make them simple. Don't begin the job by identifying the wires, but first look at where they originate and end. By far, most wires begin and end at connectors. Some connectors are part of the radio, while others are at the ends of wiring harnesses.

Sometimes it's difficult to match a connector with its symbol on the diagram. The connector may be identified only by, say, "P302." Look in the manual for other illustrations, such as photos or drawings, that show where P302 is found on the radio. It's helpful to identify the location of every connector before beginning a wiring job.

Schematic symbols

Symbols in schematics are not standard and vary from one manual to the next, but they're not difficult to learn. (Examples are shown in the illustration.)

Most important is to identity the type of wiring required; twisted pair, shielded pair, coaxial cable, for example. The manual gives wire size and type required by each connection. Be sure to read the fine print at the bottom of the schematic because that's often where the information appears. It may say, for example; "Use No. 22 wire except where noted."

Grounds

One item to be careful about is what the schematic says about grounding. Cables that have a shield need to be grounded (to provide a return path for one side of the circuit). But check the schematic carefully for where to make the ground. In some cases, it says in a tiny foot

Schematic Symbols for Wiring

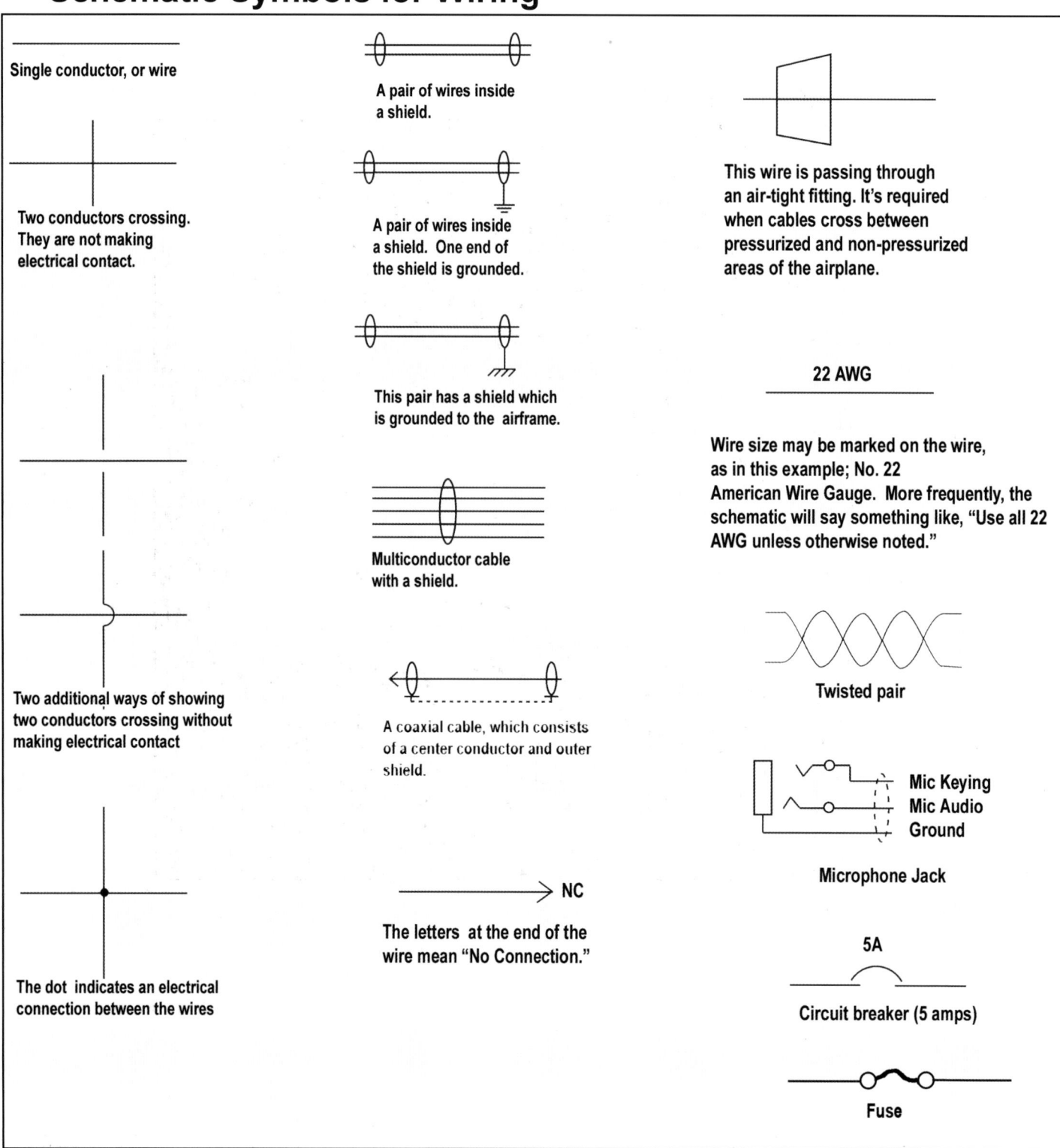

Single conductor, or wire

Two conductors crossing. They are not making electrical contact.

Two additional ways of showing two conductors crossing without making electrical contact

The dot indicates an electrical connection between the wires

A pair of wires inside a shield.

A pair of wires inside a shield. One end of the shield is grounded.

This pair has a shield which is grounded to the airframe.

Multiconductor cable with a shield.

A coaxial cable, which consists of a center conductor and outer shield.

→ NC
The letters at the end of the wire mean "No Connection."

This wire is passing through an air-tight fitting. It's required when cables cross between pressurized and non-pressurized areas of the airplane.

22 AWG

Wire size may be marked on the wire, as in this example; No. 22 American Wire Gauge. More frequently, the schematic will say something like, "Use all 22 AWG unless otherwise noted."

Twisted pair

Mic Keying
Mic Audio
Ground

Microphone Jack

5A

Circuit breaker (5 amps)

Fuse

164

note; "Connect to the nearest airframe ground."

Sometimes it is stated as: "Connect to A/C (aircraft) ground," which is also the nearest airframe ground.

Many circuits, however, require grounding at only at one end of the cable. Study the diagram and footnotes to be sure. Grounding incorrectly causes interference and poor performance.

Location Restriction

The manual warns about certain mounting limitations. FAA rules say that knobs, switches and controls operated by the pilot must be clearly labelled for function and lie within easy reach for operation. The installation manual may add a specific caution, like the one on viewing angle shown below.

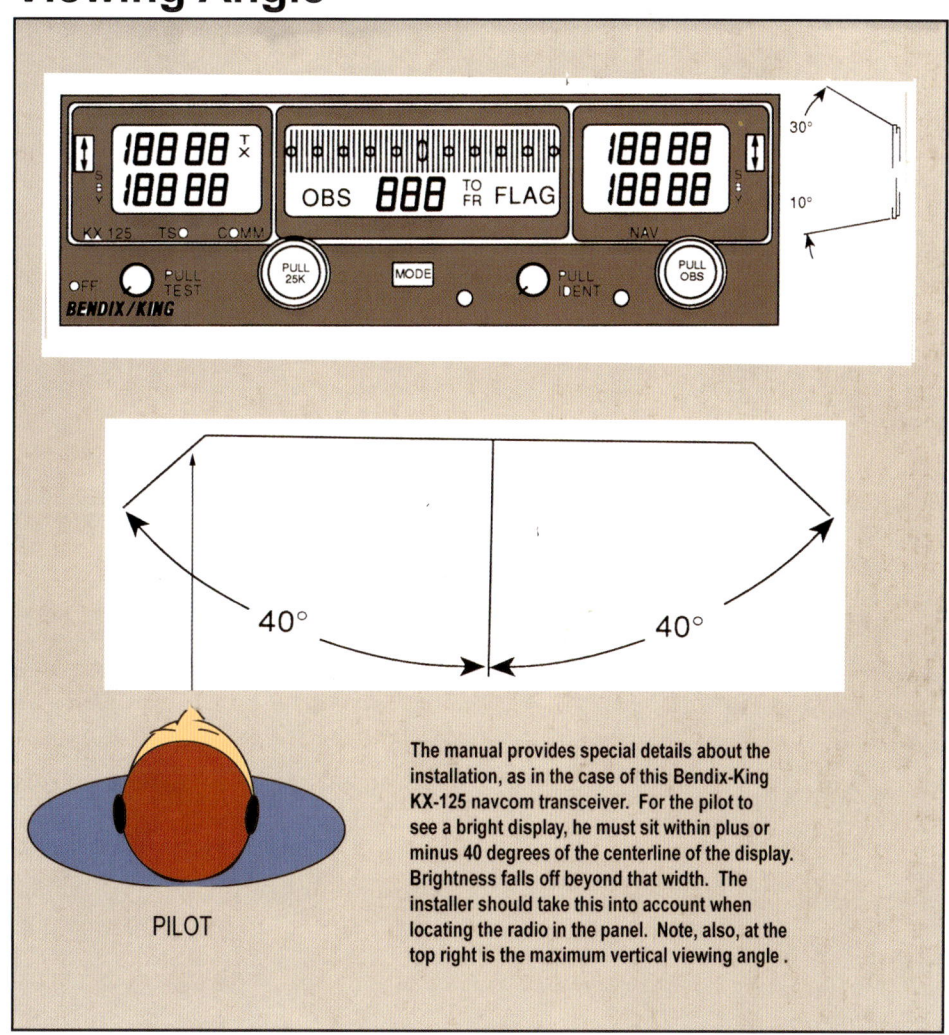

Viewing Angle

The manual provides special details about the installation, as in the case of this Bendix-King KX-125 navcom transceiver. For the pilot to see a bright display, he must sit within plus or minus 40 degrees of the centerline of the display. Brightness falls off beyond that width. The installer should take this into account when locating the radio in the panel. Note, also, at the top right is the maximum vertical viewing angle.

165

Typical Navcom Connections

> 1. POWER INPUT
> 2. GROUND
> 3. MICROPHONE KEY LINE
> 4. MICROPHONE AUDIO
> 5. COM AUDIO
> 6. NAV AUDIO
> 7. SPEAKER OUTPUT
> 8. AUX AUDIO
> 9. INSTRUMENT LIGHTING
> 10. SWITCHED POWER
> 11. KEEP ALIVE LINE
> 12. DME
> 13. VOR/LOC

1. Power Supply

The positive (+) side of the 14- or 28-volt DC source from the airplane electrical system. Also called "A+" it comes from a fuse or circuit breaker designated "navcom."

2. Ground

There are several types of grounds in an airplane. Here, the ground is a path for DC power to return to the negative side of the aircraft electrical system. In aluminum airplanes, it is sometimes the metal structure (which is connected to the negative side of the battery). Some metal parts are insulated from ground by shock mounts made with rubber. To reach ground, mounts can be bypassed with short lengths of metal braid.

There will be many grounds required behind the instrument panel and technicians often prefer to run a separate ground wire from each radio, instrument, etc., to a terminal block behind the panel. A heavy common ground is then run from the block to the negative side of the aircraft battery.

The structure of a composite airplane prevents the airframe's use as a common ground because it will not conduct electricity. Grounds are provided by running a "bus bar," a heavy copper strip or wire from the negative side of the battery to the device being grounded.

When you are required to ground the shield of a wire carrying audio, consult the manual. In nearly all cases, audio leads (for microphones, for example) must be grounded only at one end.

3. Microphone key line.

This lead runs to the press-to-talk button on the microphone. It connects to the microphone jack, specifically to the terminal that connects to the tip of the mike plug.

4. Microphone audio

This carries audio (voice) from the microphone jack to the radio. On the mike jack, this is the center terminal.

5. Com Audio

The voice signal from the receiver. Fed to a headphone jack for listening, it is called "headphone" or "low level" audio. In most aircraft, however, the voice is fed to an audio panel so it can be amplified for a cabin speaker or used with an intercom.

6. Nav Audio

This is audio from the VOR receiver, heard by the pilot to identify the station by a Morse Code or voice identifier. VOR audio may also carry the voice of a Flight Service Station. Most aircraft feed nav audio to an audio panel for listening on headphones or cabin speaker.

7. Speaker Output

Some navcoms have a built-in amplifier for driving a cabin loudspeaker. Otherwise, an audio panel must be added.

8. Aux (Auxiliary) Audio

If the radio has a built-in speaker amplifier, it can take low level audio from other radios and boost it to speaker level. It's usually done in aircraft without audio panels.

9. Instrument Lighting

Also known as the "dimmer" line, it runs to an instrument lighting controller. It enables the pilot, with one knob, to dim radio lights along with other lights on the instrument panel.

10. Switched Power

Use this line when you need to turn on accessories from the power switch on the radio. The VOR indicator, a separate instrument, is one example.

11. Keep Alive

This line bypasses the power switch and goes directly to aircraft battery power. When the radio is turned off, the keep alive line continues powering receiver memory that stores frequencies and other data.

12. DME (Distance Measuring Equipment)

The DME is a separate radio, but is tuned by the VOR receiver. When the pilot selects a VOR, the DME is automatically "channeled" to the correct frequency.

13. VOR/LOC Composite

These are the navigation signals (VOR and localizer) processed by the receiver. They are sent through this line to an indicator for display to the pilot.

Review Questions
Chapter 21 Planning the Installation

21.1 Any major rebuilding of an instrument panel must conform to the airplane's _____.

21.2 When installing new equipment critical to flight, the work must conform to a _____.

21.3 In planning a major avionics installation, it is important know under what conditions the airplane will be flown. What are three general categories?

21.4 What instruments are in the "Basic T" layout?

21.5 Name two additional flight instruments for instrument flying?

21.6 Before beginning a wiring job, it's helpful to locate and identify every _____.

21.7 Before wiring, determine the size and type of each wire by referring to the _____ _____.

21.8 Wire sizes are often described as "AWG". What does it mean?

21.9 Is a ground wire always connected to the metal airframe?

21.10 When selecting a location on the instrument panel, what is the consideration for viewing angle?

21.11 What is the purpose of "nav audio?"

21.12 What is a "keep alive" line?

Chapter 22

Electrical Systems

AC and DC Power

Avionics and instruments require a variety of voltages and frequencies but they all begin with "primary" power. Most aircraft require low-voltage DC (direct current), which starts at the battery. A major difference is how primary power is distributed throughout the airplane. In light twins and smaller aircraft, power is distributed as 12- or 28-volt DC. Most radios work directly from that source. But in large turbine aircraft, DC is for starting engines and powering some devices. Most electrical power in these airplanes is taken directly from engine-driven generators which produce 115 volts AC. That high voltage is not only distributed throughout the airplane, but is stepped down for recharging batteries. As we'll see, 115 VAC is an efficient method to power a large airplane with hundreds of feet of wire.

A more recent system generates primary power at 270 volts DC. Designed for military aircraft, it looks ahead to the "all-electric" airplane, where electric motors replace today's heavy hydraulic and pneumatic actuators for gear, flaps, flight controls and other mechanical devices. The high voltage---270---carries electrical power with less loss from heating in the wiring.

12 VDC. Adapted from the automobile industry, this system consists mainly of an alternator and storage battery. It's called a 12-volt system, but has other names, as well. On some diagrams it's a 14-volt system---on others it's 13.75 volts. But the system is usually called "12 volts." When the alternator is recharging the battery voltage rises to over 13. On schematic diagrams, you may see "13.75 volts" because the circuit designer wants you to use that voltage while operating the radio on a test bench. By adjusting to

12-volt Battery: Percent Charge
- 12.70 volts   100%
- 12.50 volts   90%
- 12.42 volts   80%
- 12.32 volts   70%
- 12.20 volts   60%
- 12.06 volts   50%
- 11.90 volts   40%
- 11.75 volts   30%
- 11.58 volts   20%
- 11.31 volts   10%
- 10.50 volts   0%

Voltages measured at the terminals of a storage battery drop with the state of charge. To measure the capacity of the battery, however, use a tester that puts a load on the battery.

DC System

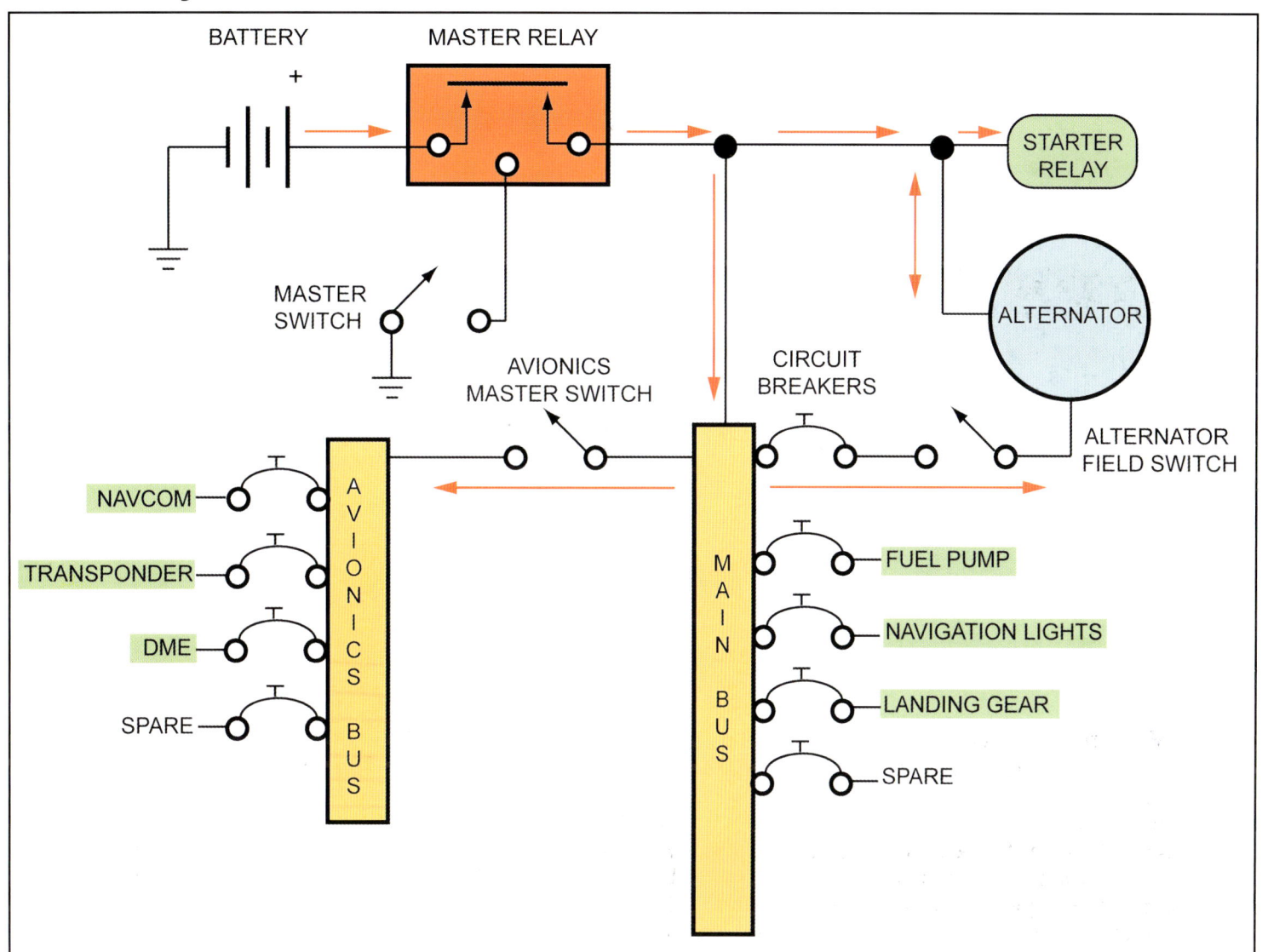

The electrical system for a single-engine airplane. Starting at the upper left, there is a storage battery with its negative terminal grounded to the metal airframe and engine block. To energize the electrical system, the pilot turns on the master switch which operates the master relay. The relay keeps heavy starting currents from moving through the master switch---and delivers those currents to the starter relay. The red arrows show the distribution of current through the system.

When the engine is running, the alternator generates voltage for recharging the battery and to power the two buses; main and avionics.

The main bus---a heavy copper bar--- powers various electrical devices such as lighting and pumps. Each has a circuit breaker. The "T" at the top of each breaker symbol shows it can be reset by the pilot. In older aircraft there are fuses.

Note at the bottom of each bus a "spare" position. A breaker is installed here when adding future equipment.

The avionics bus originally protected electronic equipment from sudden "spikes" (short bursts of high voltage) while starting the engine and turning on other electrical devices. Modern radios, however, are hardened against such voltages. The main reason for the avionics bus today is a convenience for the pilot: he may turn on all avionics with one switch. Because all avionics are lost if that one switch fails, a second switch is often installed as a backup to restore power.

169

13.75 volts, the test points measured on the radio should agree with those on the schematic.

28 VDC. As aircraft grew larger, 28 VDC systems were developed. The reason is longer wiring runs and more electrical systems. Because wire has resistance, it wastes part of the current as heat. By raising primary voltage to 24, less current flow is required (for the same power). It's the same reason cross-country transmission lines operate at nearly 1 million volts to carry electricity hundreds of miles with little heating loss. In an airplane, higher primary voltage means less weight and less copper. It also allows more wires to bundle together without causing excessive heat.

By the 1960's light aircraft also switched over to 24-volt systems for the same reasons.

Don't Shock the Airplane. When powering an airplane on the ground during maintenance, be sure the ground power unit will deliver the correct voltage, frequency and amperage. The plug and socket shapes make it difficult to make a mistake, but there are enough instances of "smoking the electronics" on an airplane to observe this precaution.

115 volt Systems. With the arrival of large aircraft, airframe manufacturers began installing 115-volt AC electrical systems. This introduced two power-saving techniques. First, voltage went higher---from 24 to 115---raising the efficiency of power distribution throughout the aircraft. Note, too, that power is now "AC"---alternating current---instead of 12- or 24-volt DC, direct current. The advantage of AC is an ability to easily step it up or down to any voltage and convert (or rectify) it to DC.

115 VAC @ 400 Hz. You may recognize "115 VAC" because it's the voltage in many countries for ordinary house current. This voltage, however, is delivered at 50 or 60 Hz (cycles per second). In an aircraft the voltage is 115 VAC, but frequency is 400 Hz. The higher frequency reduces the size and weight of transformers which change the voltage for various aircraft electrical and electronic equipment.

Any power-generating system aboard an airplane, large or small, is held to a constant voltage by a regulating system. This is important since generators are driven by an aircraft engine that is changing RPM during climb, cruise and descent.

Because a 115 VAC system operates at 400 Hz, it requires *frequency* regulation to hold the 400 Hz steady as the engine changes speed.

Constant Speed Drive. A system to solve the problem is the CSD, or constant speed drive. It contains an oil-driven hydraulic unit and a (mechanical) differential. A governor senses when generator speed is too high or low, and adjusts hydraulic pressure accordingly to keep RPM constant to the generator. When the constant speed drive is constructed in one case with the generator, the system is known as an IDG, for "integrated drive generator."

Check Power Supply Voltage. When selecting equipment for installation, determine the required supply voltage. Old equipment usually works on only one voltage and the manufacturer offered two different models, one for 12 (or 14) VDC and another for 24 (or 28) VDC. The trend today is to offer models with selectable 14 and 28 supplies built in. In some equipment, the manufacturer simply states the radio works on any voltage between 10 and 30 VDC.

Low voltage caution. When doing avionics work on an airplane in a hanger, it is convenient to turn on the master switch to test the installation. Do it for only brief periods (if at all) to avoid discharging the battery. As shown by the chart, a fully charged battery puts out 12.7 volts; a battery with only 10 percent charge produces 11.31 volts (and there may be other losses in the system, such as corroded connections to bring down the voltage further).

Another problem is that some radios automatically switch off to protect themselves during low voltage. It may lead you to believe the radio is bad, when the fault is low primary power. This can waste a lot time during troubleshooting. The cure is to plug a ground power unit into the airplane and be sure voltage is adequate.

Auxiliary Power Unit (APU). Located in a wheel well or near the tail, the APU can power the airplane while on the ground. It is driven by a small turbine engine which turns a generator similar to those mounted on the engines. In many aircraft, the APU may be started and operated in flight to supply emergency backup power.

Tranformer-Rectifier Unit

AMPS	VOLTS DC	APPLICATION
20	28	USAF F-15E
50	12	747-400, 777
65	28	727, 737
75	28	747, L1011
75	28	MD-10, MD-11
120	28	757, 767, 777
125	28	S-92, KC-135
150	28	USAF F-15 Fighter
150	28	USN F/A-18 Fighter
150	28	Global Express®, KC-10
200	28	UH-60/SH-60 Heli, KC-135
250	28	Gulfstream G-V, VC-10

ELDEC

Transformer-Rectifier Units are in a variety of aircraft to reduce the output of engine-driven generators (115 volts AC) to low voltage DC (12 or 28).

Airline Electrical System

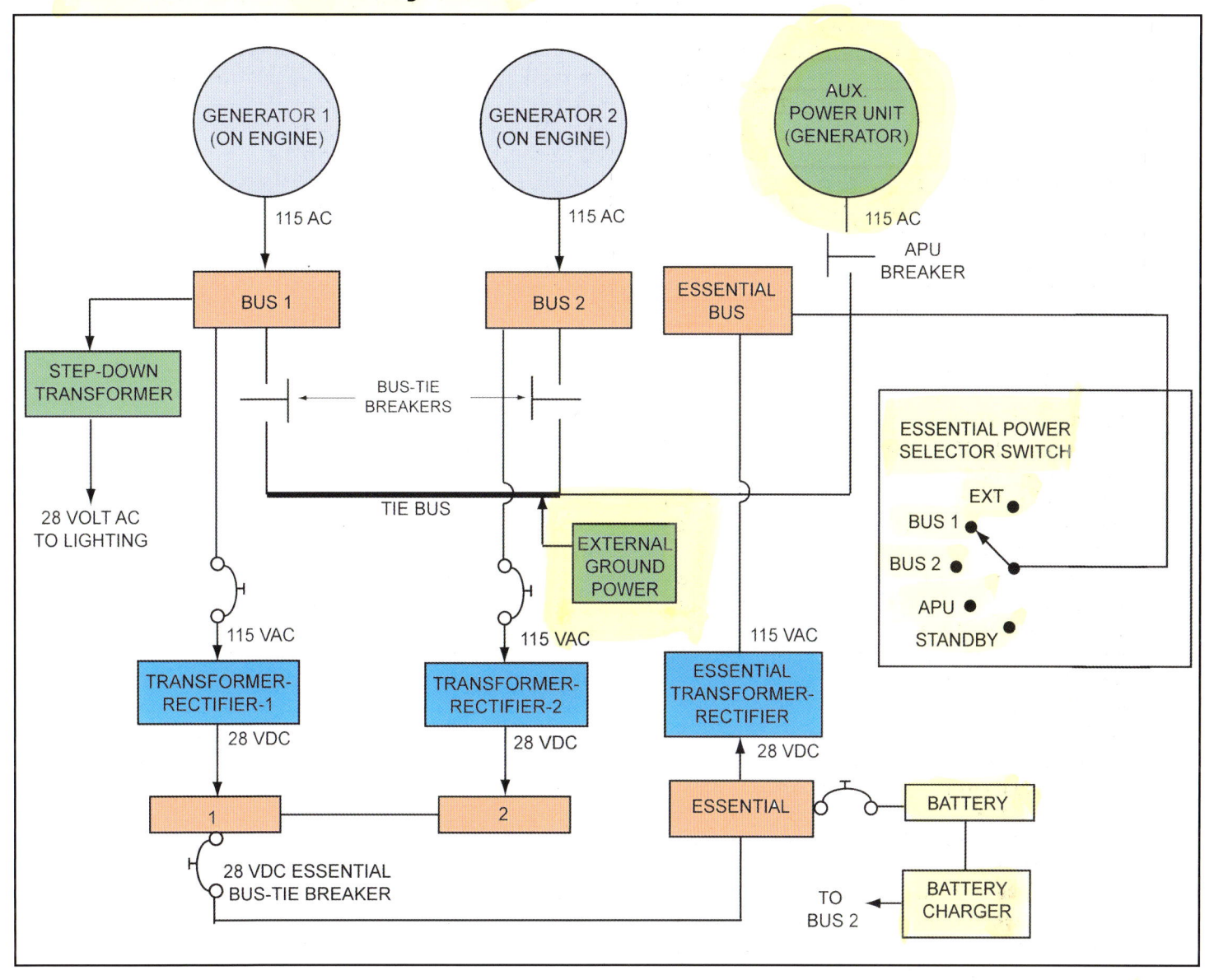

Simplified diagram of the electrical system for a large air transport twinjet. It is designed to give the pilot many options for restoring power in the event of engine failure or other interruption to electrical power. The main features of the system:

Power is produced by two engine-driven generators at 115 VAC, 400 Hz. This is applied to Bus 1 and Bus 2. Normally, the buses are tied together.

A third generator is the Auxiliary Power Unit (APU), operated by a small gas turbine engine located in a wheel well or tail area. The APU provides power to the instrument panel, lighting and other devices while the airplane is on the ground and main engines are off.

The "essential" bus can power equipment essential to flight (that is, enable the pilot to make a safe landing after power failures). The pilot can directly connect to the aircraft batteries which supply 28 VDC to a standby inverter (not shown). The inverter changes 28-volt battery power to 115 VAC.

Note at the right side of diagram an "Essential Power Selector Switch." This gives the pilot a choice of sources:

EXT: External refers to power obtained on the ramp from a Ground Power Unit.
BUS 1 (Engine generator)
BUS 2 (Engine generator)
APU: Auxiliary Power Unit. (Some aircraft cannot operate the APU in flight.)
STANDBY: Power is drawn from a standby inverter which is driven by battery voltage. The inverter produces 115 VAC which powers essential equipment. This is selected if both engines fail.

Total engine failure in multiengine aircraft is not common, but can happen. Recent examples in airliners include fuel exhaustion due to damaged fuel lines and running out of fuel because of long holding patterns. One Boeing 747 had all four engines fail when the airplane penetrated the ash cloud of an erupting volcano.

171

RATS. Meaning "Ram Air Turbine System," this is a small propeller that drops out of the belly (in flight) when all other power sources fail. It generates just enough power to keep the pilot from losing control of the aircraft, plus a few more amps. How can a modern air transport with three or more engine-driven generators and storage batteries ever run out of power? By losing all engines. This happened when a fuel leak on a large twinjet sprayed fuel overboard while the airplane was over the mid-Atlantic. The airplane had enough altitude to glide over a half-hour and make a dead-stick landing at an airport, with no injury to crew or passengers. The "RATS" supplied just enough power to control and communicate.

Switches

Switches give long and dependable service, but they contain mechanical contacts and springs which wear during each operation. Electrical arc'ing erodes the contacts and airborne grease enters the housing. Mechanics report that switches on the pedestal between captain and co-pilot are especially vulnerable. The pedestal is used as a convenient tray for coffee and soft drinks. One popular cola is said to be the most corrosive liquid a pilot can spill into the switches.

New avionics equipment often have buttons of the "membrane" type, with no cracks for liquid to enter. But tens of thousands of airplanes will have old-fashioned unsealed switches for a long time.

The loss of a switch is serious, even in aircraft with much redundancy. When pilots operate at a high workload (such as approaching a busy airport during low visibility) it's no time to deal with a switch malfunction. Much of the problem is eliminated by using aircraft-rated (Mil-spec) switches, which are more rugged and reliable than switches for other industries.

Caution on mounting position. If you look at a switch it may be difficult to tell which is the "on" position. The terminals on the back are symmetrical, and may look the same either way. Some switches have a small nameplate with "on" that slips over the handle when the switch is bolted to the panel. Failing to observe the correct "on-off" position while mounting a switch can have serious consequences. In one actual incident, a technician installed a new magneto switch and reversed its position. With the handle "up," the magneto was off. In the down position, it was "on." This is opposite to the standard "up is on, down is off." When another mechanic fueled the airplane he pulled the prop through by hand with the magneto switch "off." The engine fired---the switch was actually "on"--- and spun the prop around with great force. Fortunately, the blade did not strike the mechanic but disaster was only inches away.

Check a switch before installation with an ohmmeter. Select the "R x 1" scale, place the probes across the terminals and you'll read zero resistance for contacts that are closed, infinite resistance when open. And when working around airplanes treat every prop as if it is alive.

When checking a switch that's been in service look for any sidewise movement of the handle. Even though the switch can turn the power on and off, replace it. Wobbling is a sure sign of early failure.

Select the Switch Rating. Different loads have different effects on switches. Turning on a lamp sends high current into the filament because it has low resistance when cold. This also sends a large inrush of current through the switch, which may be 15 times greater than when the lamp is operating. Switch contacts must withstand that by "derating"---selecting a higher current rating. Otherwise, contacts may weld together or corrode as heavy current flashes over.

When a switch controls a device with a coil of wire, such as a relay, it also needs derating because of "inductive kickback." As the coil is energized, it stores energy as a magnetic field. When the switch is opened the field collapses and "cuts across" the coil, inducing high voltage across the switch. The contacts burn and pit, which is avoided by derating the switch.

Nominal System Voltage	Type of Load	Derating Factor
28 VDC	Lamp	8
28 VDC	Inductive (relay, solenoid)	4
28 VDC	Resistive (heater)	2
28 VDC	Motor	3
12 VDC	Lamp	5
12 VDC	Inductive (relay, solenoid)	2
12 VDC	Resistive (heater)	1
12 VDC	Motor	2

1. How to find required "Nominal" switch rating
     A. Obtain from the equipment manual the current rating of the lamp, motor or other load the switch will control
     B. Select (above) "Nominal System Voltage" (28 or 12).
     C. Select "Type of Load"
     D. Multiply switch rating by the "Derating Factor." The answer is the switch "Nominal" rating in amperes

2. Next find the "Continuous" current that switch can handle.
     A. *Divide* the "Nominal" rating (obtained above), *by* the "Derating Factor" (using the same voltage and type of load.

Avionics Master. This switch originated as a method for protecting early models of solid-state radios. It enabled the pilot to keep all avionics off while cranking the engine to prevent damaging voltage spikes from reaching the radios. It also protected radios against low voltage as the starter drew heavy current from the battery. These problems were common to the first generation of transistorized equipment; avionics today have built-in devices to protect against surge and low voltage protection.

But the avionics master is still widely used because it is a convenience for the pilot. Instead of flipping a half-dozen switches or more, he throws one switch to turn on all avionics.

This convenience, however, comes at a price. It's the *single-point failure*. If the avionics master switch fails, it disables all radios in the airplane. A solution in light aircraft is to wire a second switch across (in parallel with) the avionics master. Using the back-up switch restores the lost power.

There are more advanced methods for keeping radios working in the event of failure of the master switch, the master relay or from other interruptions to the primary power. Some technicians install an "essential" bus, which is a wire directly from the battery. Multiengine aircraft have dual electrical systems to prevent the single-point failure.

CIRCUIT BREAKERS

The purpose of a circuit breaker is to protect wiring, not the equipment. Most recent avionics already have overvoltage protection. The hazard is that a wire carrying excessive current may heat and cause smoke or fire. By placing the breaker close to the power source, more of the wiring is protected.

The size of a circuit breaker---its rating in amperes (amps)---is selected so it opens before current exceeds the capacity of the wire. The chart shows the size breaker or fuse for different levels of current in DC circuits.

Switch Types

1. **Toggle Switch**
This switch has a "bat" handle (resembles a baseball bat). Sometimes several are ganged together with their handles linked. It assures that all switches are thrown at once in same direction.

2. **Pushbutton switch.**
Used when a circuit is operated for a short time, such as the push-to-talk switch on a microphone. It has a spring-loaded button. Depending on the circuit, the switch may be push-to-make (contacts close) or push-to-break (contacts open).

3 **Rocker switch**
The pilot pushes the top half of the rocker switch to energize the circuit, the bottom half to break the circuit.

5. **Rotary switch**
Enables several circuits to be selected with one knob. It also is more resistant to being knocked off its position (which can happen in rocker and toggle switches).

6. **Microswitches**
The name is from the few thousandths of an inch between contacts on the make and break. They have snap action. These switches are operated by the pilot, or located at different points on the airplane to sense a mechanical position. An example is a retractable landing gear; when the wheels are down and locked they operate microswitches that illuminate lights on the instrument panel.

7. **Pressure switch.**
These are often used to warn when pressure in a hydraulic or pneumatic system is too high or low. It's usually done through a flexible disk (diaphragm) which moves with pressure. It operates switch contacts and indicator lamps.

8. **Thermal Switch**
Used to warn of overheating in a component (a generator, for example) and to sense and indicate an engine fire. It works on a bi-metal thermostat that curves with heat and makes an electrical contact.

9. **Proximity Switch**
Are doors and hatches on the airplane closed and locked before take-off? This is done with proximity switches. One half is a housing containing two strips of metal mounted parallel with each other. This may be fixed to the door frame. On the door is a permanent magnet. When the door is secured, the magnet pulls the metal strips together to operate a warning light in the cockpit.

Description	Off	Illuminated
Hidden Legend	⬛	AAI (green text on black)
Hidden Legend Lighted Background	⬛	AAI (on green)
Lighted Background	AAI (on white)	AAI (on orange)
Lighted Letters	AAI (white on black)	AAI (red on black)

Variety of choices for illuminated pushbutton switches. Note in the lower two examples, the label ("legend") is visible even when the switch is turned off.

Recessed button type. When an overload occurs it will, as commonly said, "Pop the breaker." This is because heat in the breaker opens a pair of contacts. The button on the breaker pops out and the circuit remains open until the pilot pushes it in. However, if this is done quickly, the breaker may not remain engaged because it is still cooling down. Wait a moment and try again.

A pilot or technician will always attempt to reset the breaker to see if the problem is gone. After three attempts to reset the breaker with no success, you can assume the short-circuit didn't go away.

Resettable breaker ("Push-Pull") This breaker style has a knob so the pilot can pull it and break the circuit, as if it's a power switch. This becomes a diagnostic tool for the pilot who experiences an electrical problem in flight. He can "pull breakers"---one by one--- in an effort to make the problem go away. If it does, he can leave the breaker open to disable the defective equipment.

If the breaker is "trip-free," it means it cannot be reset if the overload still exists when the button is pushed in.

Automatic reset breaker. Some circuit breakers are designed to break the circuit, then automatically reset later, when the internal element has cooled. These breakers *are not recommended* for aviation. Although the problem may have cleared, there may be enough heat remaining in surrounding metal objects to keep the breaker open for a long period time. This happened to one pilot with a problem in a landing gear motor. An automatic circuit breaker sensed excessive heat in the electrical motor and opened the circuit. This prevented the wheels from dropping and locking. Although the overload had cleared, the breaker was mounted on the case of the motor, which is a large mass. It acted as a heat sink and continued to hold the breaker open. As a result, the pilot made a wheels-up landing that damaged the prop, engine and belly. If he had known the landing gear motor would cool and reset the breaker---probably within 20 minutes---the accident wouldn't have happened. A better design is to run the gear motor electrical power through a breaker under control of the pilot.

Pulling breakers and nuisance alarms. Experts in human factors know about "nuisance" alarms. If a warning keeps sounding when there is no problem, pilots will turn off the system. They do this, even though the system has no off switch. They simply "pull the breaker." This happened in the early days of "ground prox" (ground proximity warning system), which had many false alarms. Instead of blaming the pilots, the manufacturers went back to their drawing boards and redesigned the system.

Breaker locks. There is a device that can be put under the button of a circuit breaker to prevent it from being pushed in. It's a temporary measure used by technicians during maintenance to be sure a defective piece of equipment will not have power applied accidently.

Maintaining breakers. A good practice is to pull the knob of a circuit breaker in an out several times if it appears unreliable. Do this while the equipment is off and no current is flowing. Operating the breaker cleans the contacts and reduces electrical resistance.

Lighted Pushbutton Switch

Improved lighted pushbutton switches, like this Korry model, eliminate a major source of maintenance; replacing lamps with burned-out filaments. Lighting is supplied by bright LED's (light emitting diodes) that can last the life of the airplane. They also run cooler than conventional lamps.

A pull-off cap (lower left) provides access to the switch, which can be removed by a mounting screw and cam. The switch itself may be four independent microswitches, an arrangement which eliminates switch "chatter" during shock and vibration.

The external connector module (upper right) carries wiring to the switch through crimped connector terminals. A "poke home" push engages the switch into the connector module.

These switches are made for 5 or 28 volts commonly found in instrument panel lighting. Electronics for dimming the light are built into the switch.

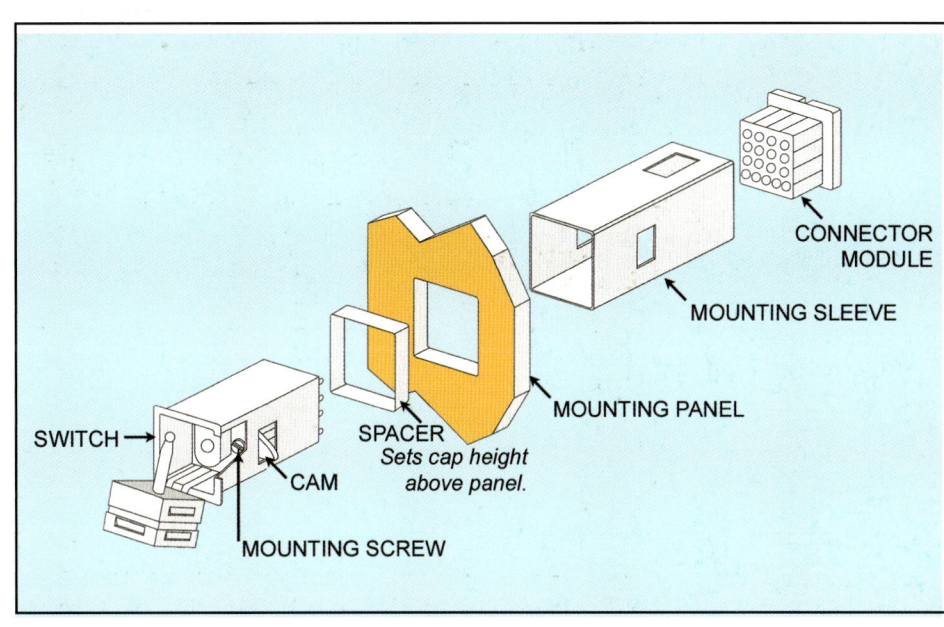

Switch Guards

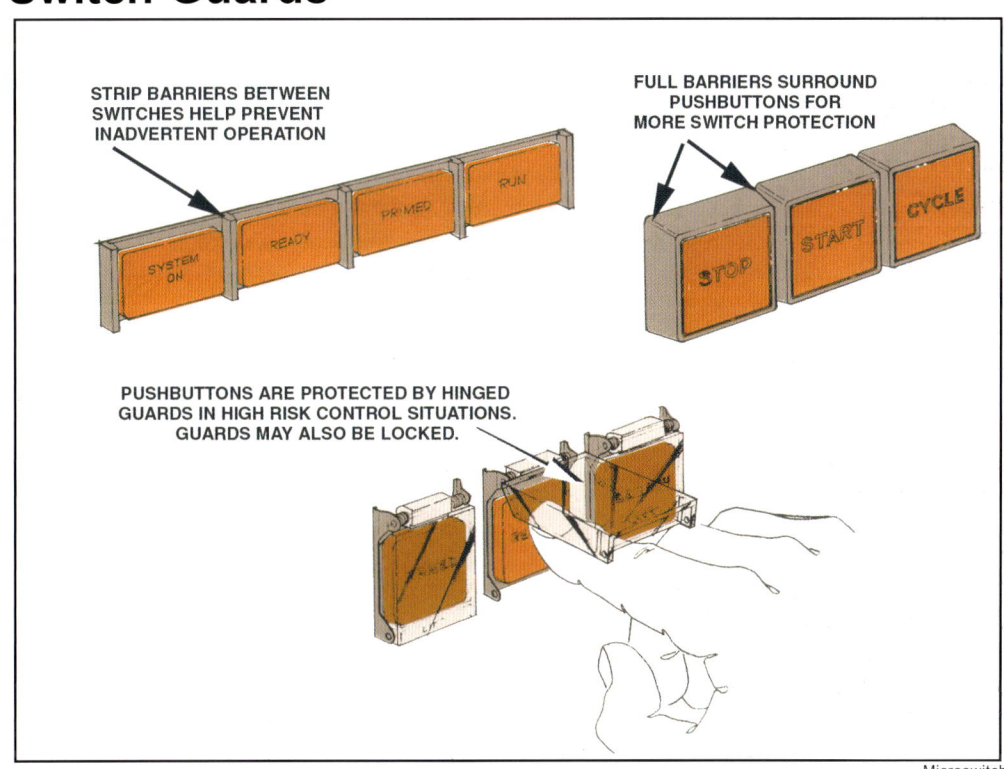

A pilot may push the wrong switch while flying in turbulence. Barriers between rows of switches reduce that problem. In the bottom drawing, switches are covered by a guard which must be lifted.

175

Fuses

The convenience of circuit breakers has reduced the number of fuses---which burn out after one overload. For this reason, fuses should be accessible to the pilot for changing in flight.

Heavy duty fuses are found where large amounts of electrical power concentrates at distribution centers on the aircraft. In some installations, a built-in lamp illuminates when the fuse "blows."

Another form of fuse is the current limiter, also used in high wattage locations. It can pass a large overload without breaking the circuit, aided by a ceramic housing which can take the heat.

Circuit breakers during construction of a panel Note that each breaker is temporarily labelled with a felt-tip pen. This helps identify the wiring and final testing. After the panel is painted, machine-made labels are applied.

Recessed Button Breaker

Circuit breaker panel using "recessed button" type breakers. Having no buttons protruding from the panel, they cannot be pulled out by the pilot to disable a circuit.

When a breaker "pops," the button comes out and remains out so long as there is an overload. It cannot be reset (pushed in) until the problem is cleared.

Occasionally a breaker cannot be reset, even though the overload is no longer present. If this happens, wait a minute or two for the thermal element in the breaker to cool and reset.

If a breaker pops and you reset it, the circuit may continue to operate for a while, then pop the breaker again. When that happens, it is good practice to reset the breaker only three times. After that, the source of the overload needs investigation.

Review Questions
Chapter 22   Electrical Systems

22.1 The primary electrical source in most airplanes is low voltage DC (direct current). What are the two most common voltages?

22.2 In a light aircraft, connection to the battery is completed by the _____.

22.3 A heavy copper bar that distributes power to most electrical systems aboard the airplane is called the _____ _____.

22.4 A heavy copper bar that distributes power only to radios and related equipment is called the _____ _____.

22.5 Large aircraft distribute power in the form of alternating current (AC) at _____ volts. The frequency is _____.

22.6 What is the main purpose of an APU (auxiliary power unit) in a large aircraft?

22.7 What is the meaning of RATS, and how does it work?

22.8 Why must switches have the ability to handle more current than required during normal operation?

22.9 What is the primary purpose of a fuse or circuit breaker?

Chapter 23

Mounting Avionics

Most vehicles operate in two dimensions (left and right, forward and back). Airplanes move *six* ways; around their own axes; pitch, roll and yaw (3)---and they move forward, climb and descend. Combine these motions with gravity and acceleration, and it's easy to see why an airplane is not airworthy unless its components are secured in a strong structure.

Airplanes are also short on space. Equipment is squeezed into crowded areas behind instrument panels, inside a nose or in a small equipment bay. Mounting equipment in compact spaces can generate sufficient heat to cause early failure.

Another hazard is that avionics share space with cables, control columns, chains, gears, levers, motors, pedals, ducts and pilots' feet. One pilot rolled down a runway for take-off and, at flying speed, pulled the control yoke to raise the nose. The yoke would not move. There was enough runway left, fortunately, to re-land the airplane. Looking behind the panel we found an antenna cable wound around the control column, securely locking it in place. Frequent incidents like these prove the critical nature of mounting and wiring avionics and instruments.

Installation problems are not limited to light aircraft. In one close call, three hundred airline passengers nearly met disaster because of a minor mounting bracket. While in cruise, the right engine of a big twin jet suddenly flamed out. Any modern airliner can continue to fly with one engine out, but a half-hour later, the left engine died. The pilots were unaware that a large quantity of fuel was venting overboard. They had enough altitude to land at an airport, suffering no more damage than eight blown tires.

Investigators soon found the problem. Five days earlier, a bracket was installed to support a hydraulic line. Although the bracket looked OK, it was slightly

In this construction of a new instrument panel, radio trays are being riveted to brackets. The brackets were fastened earlier to the panel.

different in size from the correct one. This reduced clearance between hydraulic and fuel lines, causing contact between them. After five days of flying, vibration and rubbing cut the fuel line, which spilled a large quantity out the trailing edge of the wing.

These cautionary tales show how minor discrepancies lead to major problems. In an avionics installation or retrofit, it is not unusual to mount a few hundred brackets, clamps, cable ties, supporting structures and other hardware.

New or Old Installation?

There are various levels of installation. A pilot may want to replace one old radio---or need an upgrade where the panel is extensively refurbished. There is also the "tear it all out and start over again" job.

If the airplane is a light aircraft, the radio shop often designs the installation. When the work affects structures and control systems, large shops depend on the skills of their A&P (airframe and powerplant) mechanics to fabricate brackets, shelves and other supports for the new radios.

Before tackling an installation, the technician runs a weight-and-balance calculation to see if removing and/or adding new equipment exceeds the aircraft's CG (center of gravity) or adds excessive weight.

Airplanes in flight also develop *dynamic* loads while maneuvering, especially in turbulence. If there is any question whether additional support is required, the job should be reviewed by a DER (Designated Engineering Representative) with a specialty in structures. He can design a system that meets acceptable installation practices. A DER is required when cutting through a structural member (such a rib or bulkhead). Large shops often have a DER on staff, while smaller facilities may hire an independent DER on a per-job basis.

STC's. Avionics equipment that is flight-critical is STC'd, meaning the manufacturer applied for a supplemental type certificate. The installation has been worked out in great detail and a kit of components may be supplied. Wherever an STC applies, the installer follows the document carefully for structural and wiring details.

Hostile Areas. Beyond the cockpit and passenger cabin, there is an unfriendly environment for electronic equipment. Engines generate heat and vibration, the airplane flies through large temperature and humidity changes and low areas on the airframe accumulate oil, hydraulic fluid and water. Many radios and instruments are fortified against these hazards with a TSO (technical standard order), which certifies they are built to meet environmental conditions encountered in aircraft.

The manufacturer's maintenance manual is the most valuable source of installation information. Not only does the company have the engineering resources to design the installation, its manual is recognized as "approved data."

Selecting Metal

Offering a good combination of lightness, strength, electrical conductivity and workability, sheet aluminum is the common choice for panels. Aluminum angle and flat bar are selected for fabricating structures such as brackets and supports.

Look for the label "Alclad," which refers to a thin coating (on both sides) of 99 percent pure aluminum. The coating has high resistance to corrosion. Oxidation on aluminum does not have a distinctive color (like rust on iron) and may be difficult to see.

Consider various aluminum stock:

2024. This alloy is often selected for sub-panels that hold instruments, for example. It is not heavy enough for the overall instrument panel. 2024 has good resistance to metal fatigue and bends without cracking.

2024 aluminum can be polished to an almost chrome-like finish.

3003. This grade of aluminum is alloyed with manganese for strength. The material is still easily workable and resists cracks while bending. It is also corrosion-resistant (without Alclad)

3003 is excellent for forming brackets, housings and structures that require a lot of bending and forming, especially where the higher strength of 2024 is not required.

6061. This popular alloy is strong, corrosion-resistant, workable and relatively low in cost. It is available in Alclad for added protection.

6061 is useful in building structures which support the weight of a stack of radios or instruments.

Thickness. Aluminum comes in many thicknesses, measured in thousandths of an inch, for different applications:

.040. This size is good for forming lightweight brackets for supporting cooling fans, switches and other lightweight equipment such as cover plates and glove box doors.

.063 makes sturdier brackets for supporting heavy equipment (a radio stack, for example), panels for headphone and mike jacks and equipment shelves.

.080 This heavy material is suitable for the overall, or main, instrument panel. In this thickness, however, you can make bends only over a large radius.

Cutting Holes

Aluminum is not difficult to punch, cut and file. There are several options, from working it in the shop to sending it out to a specialty house.

Manual cutting. It is possible to cut instrument and other holes with a jig saw and metal-cutting blade. A set of round, half-round and flat files are for finishing touches and smoothing burred edges.

So long as holes are round or rectangular, cutting them may be time-consuming, but not impossible. The problem arises in odd shapes, such an ARINC-type instrument that has bevels on all four sides. This can take much effort to shape by hand-filing.

Router. Some shops use a router to cut panel holes. The sheet aluminum is placed on a routing table and, with the aid of a template, holes cut with a router bit designed for aluminum.

Punches. Metal punches are made to the exact size and shape for cutting panels. A pilot hole is drilled in the panel and a bolt inserted to hold a cutting die. Tightening with a wrench drives the halves of the die together for a neat hole. In large shops, a hydraulic driver operates the punch, greatly shortening the time to make the hole.

If an avionics manufacturer produces a radio with an odd shape, he may also offer for sale the punch to make that hole.

Laser Cutting. In the newest technique, holes are made by laser beam. Because the machines are costly, such work is often sent out to a specialty shop. The holes are extremely accurate and clean.

To work with an outside house, you design the panel on a PC, using software such as Autocad. The file may be sent to the laser company by e-mail.

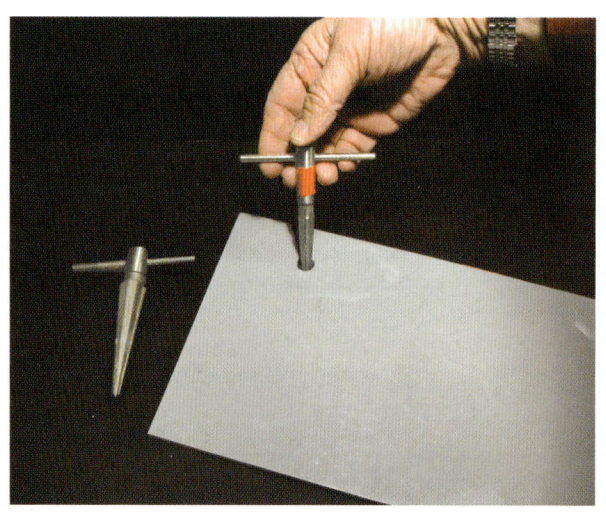

A hand reamer cuts holes in panel for switches, controls and other small items. First, a pilot hole is made with an electric drill.

A hole punch cuts neat round or square holes in panel. A square one is at the left, the other is a round one. Punches are also made for cutting odd-shaped holes required by some instruments.

The hole punch cuts into the panel as its nut and bolt are tightened with a wrench.

Structures

"Structures" refers to avionics enclosures and mountings, including cabinets, cases, racks and supports. There are few standard sizes for General Aviation equipment, but for airline and military service, manufacturers are held to strict dimensions. The installation of equipment in any airplane must comply with airworthiness standards for weight, power, secure mounting, labelling and others.

General Aviation

Although GA has no standard sizes, certain customs are followed by most avionics manufacturers. In light aircraft, where radios are usually mounted in a center "stack," the width is typically 6.25 inches. To avoid the mismatch of radios of different widths in the same stack, most radios observe that dimension. Thus, in a simple stack there might be an audio panel at the top, one or two navcoms next and a transponder at the bottom. Their heights and depths are different, but they all fit in the stack.

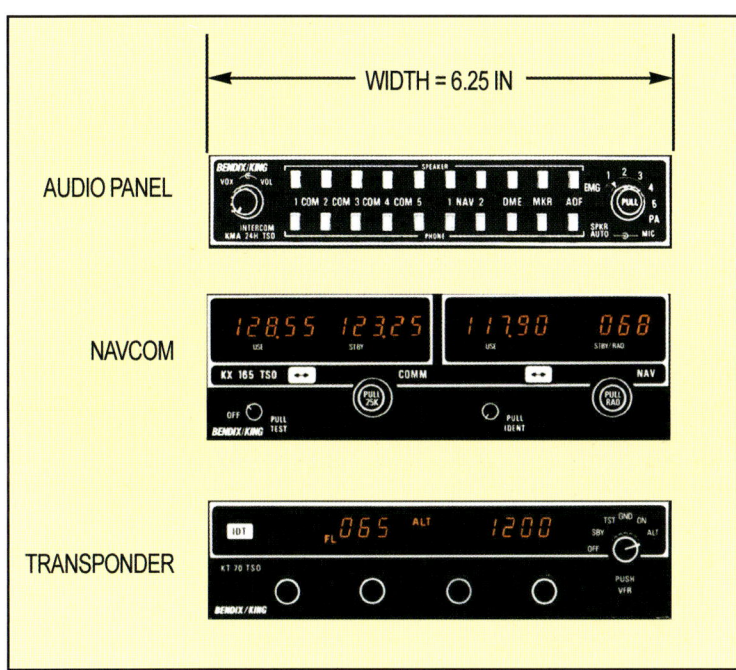

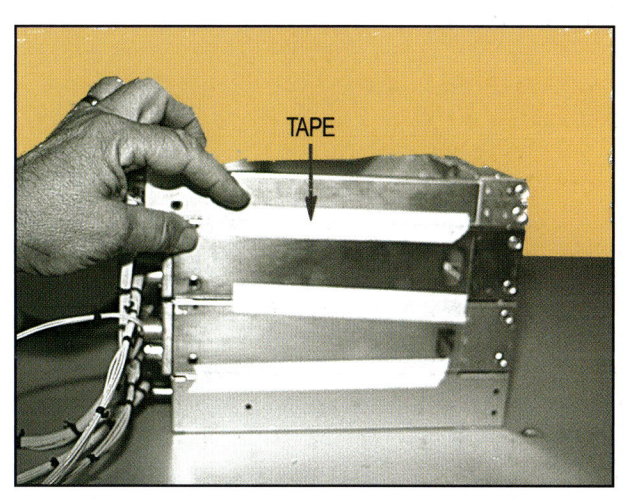

While wiring a radio stack on the workbench, tape the sides of the trays together. This keeps the trays stable while you wire the harnesses at the rear. Remove the tape just before installing the stack in the airplane.

After the stack is installed in the airplane, the forward section (at the right) bolts to brackets on the instrument panel. Brackets should also fasten to the back of the stack (left) to keep trays locked together.

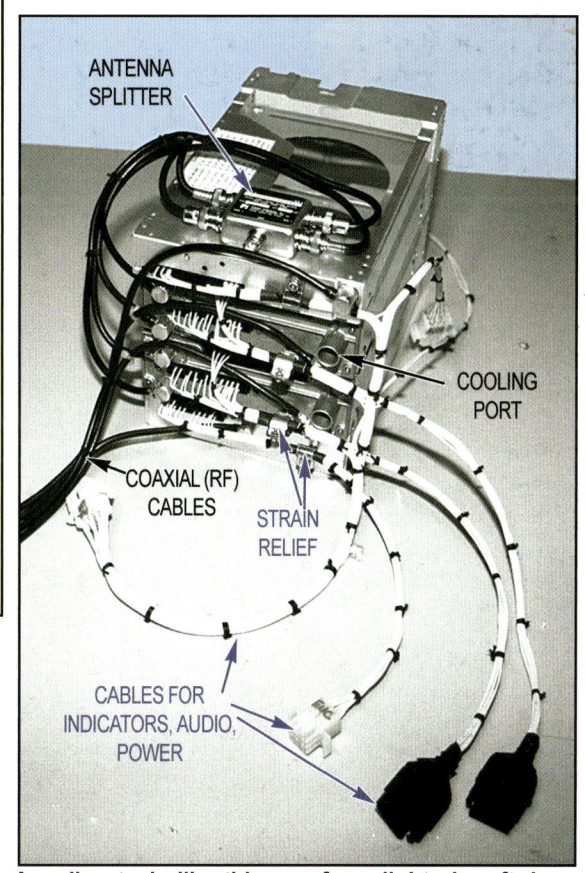

Radios in the center stack of a light aircraft are usually 6.25 inches wide.

EFIS

The future for General Aviation is the electronic instrumentation system (EFIS). The airlines have transitioned to the "glass cockpit" and the trend is well-established in General Aviation. Conventional 6.25-inch wide radios will be here for several more generations, but nearly all new production and experimental aircraft began outfitting with EFIS by 2004.

A radio stack, like this one for a light aircraft, is pre-wired on the workbench before installing in the airplane. Three avionics trays are shown; two navcoms on top, with a transponder on the bottom. Wiring is done to connectors mounted on the back of the trays. The radios are slid into place later, and make contact with the rear tray connectors.

EFIS systems are often supplied by the manufacturer with pre-cut and pre-wired cables, with all connectors attached at the factory. Because so many systems appear on one screen, fewer holes are cut in the instrument panel.

Will this mean less installation work for the technician? Probably not. From the end of World War II, there's been an ever-increasing stream of new avionics systems, government requirements and airborne telecommunications services (fax, telephone, Internet, etc).

Corporate and Business Aircraft

Larger commercial aircraft---the turboprops and jets flown by corporations--- also do not follow common avionics standards. These systems are usually remote-mounted, with control-display units in the instrument panel, and remote radios mounted in the nose, belly or near the tail.

The old "radio stack" will disappear as new and upgraded airplanes, of even the smallest size, are outfitted with EFIS (Electronic Flight Instrument System). This Blue Mountain system combines flight instruments, moving map and terrain warning on one display.

Avionics Bay of a Corporate Jet

Gulfstream

Most LRU's (line replaceable units) are in the lower fuselage. Examples shown include: transponder (XPDR). DME, NAV (VOR receiver) and Com (VHF). Each radio is duplicated for safety. The large dark area in the center ("INS") is the rack for an inertial navigation system (which has been removed for repair).

The "Warning" placard near the center (in red) cautions against excessive weight. The text says: "Maximum load of radio area not to exceed 750 pounds." This avoids exceeding the airplane's weight and balance limits.

The handset at the upper left enables the technician to talk over the aircraft intercom system.

Airline (ARINC) Structures.

The airlines solved their structures problems back in 1929 when they formed ARINC (Aeronautical Radio Inc.). The organization developed standards (called "Characteristics") for mounting and interconnecting each piece of avionics. The specifications, however, apply only to the radio's "form, fit and function." This means an airline can buy a DME from one manufacturer, then 10 years later buy an improved DME from another manufacturer and plug it into the same tray or rack. There's no rewiring or modification. Both old and new radios have the same form, fit and function--- even though the inside of the new radio may have a different design and internal components.

Two important Characteristics for airline radio structures are ARINC 404 and ARINC 600. The first, 404, contains sizes known as ATR. Although some people interpret ATR as "Air Transport Radio," ARINC says it means "Austin Trumbull Radio," after the developer.

ARINC 600 came into existence with digital avionics. Thus, ARINC 404 represents an earlier, analog era, while 600 is the digital successor. However, it is common for airliners to have a mixture of both 404 and 600 structures and avionics.

MCU Case Sizes (ARINC 600)

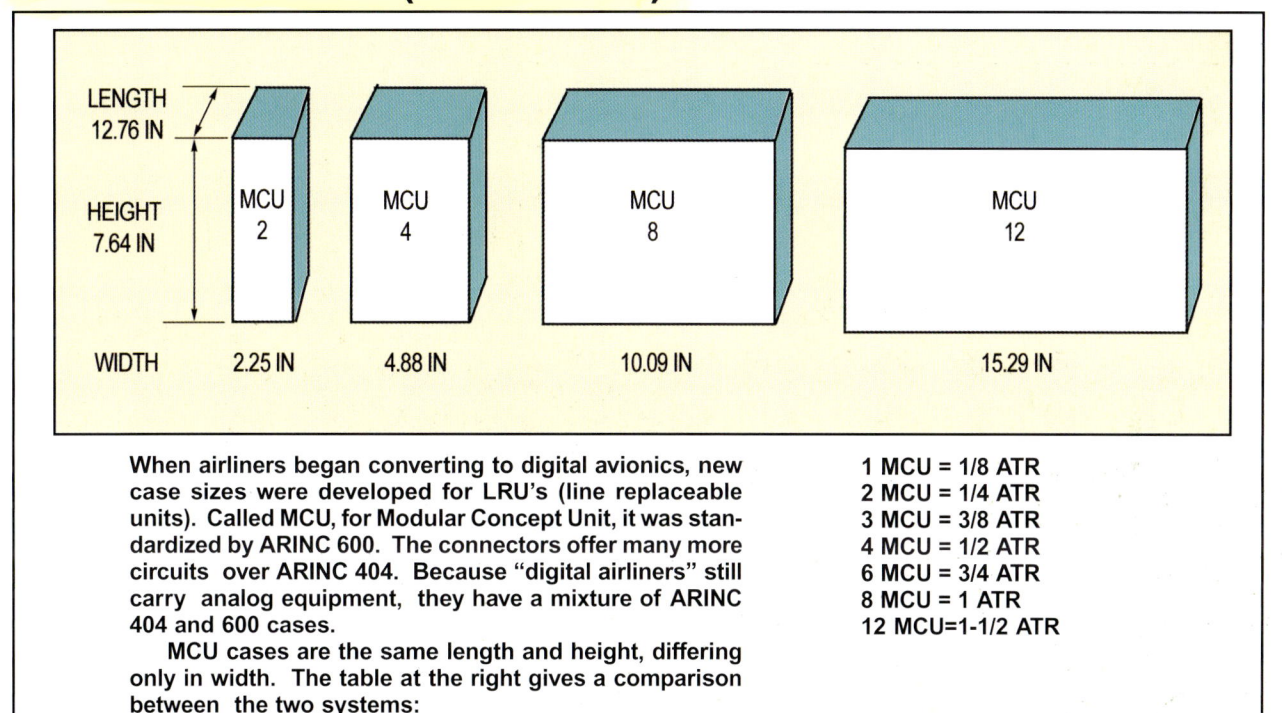

LENGTH 12.76 IN
HEIGHT 7.64 IN

| MCU 2 | MCU 4 | MCU 8 | MCU 12 |

WIDTH: 2.25 IN, 4.88 IN, 10.09 IN, 15.29 IN

When airliners began converting to digital avionics, new case sizes were developed for LRU's (line replaceable units). Called MCU, for Modular Concept Unit, it was standardized by ARINC 600. The connectors offer many more circuits over ARINC 404. Because "digital airliners" still carry analog equipment, they have a mixture of ARINC 404 and 600 cases.

MCU cases are the same length and height, differing only in width. The table at the right gives a comparison between the two systems:

1 MCU = 1/8 ATR
2 MCU = 1/4 ATR
3 MCU = 3/8 ATR
4 MCU = 1/2 ATR
6 MCU = 3/4 ATR
8 MCU = 1 ATR
12 MCU = 1-1/2 ATR

ATR Case Sizes (ARINC 404)

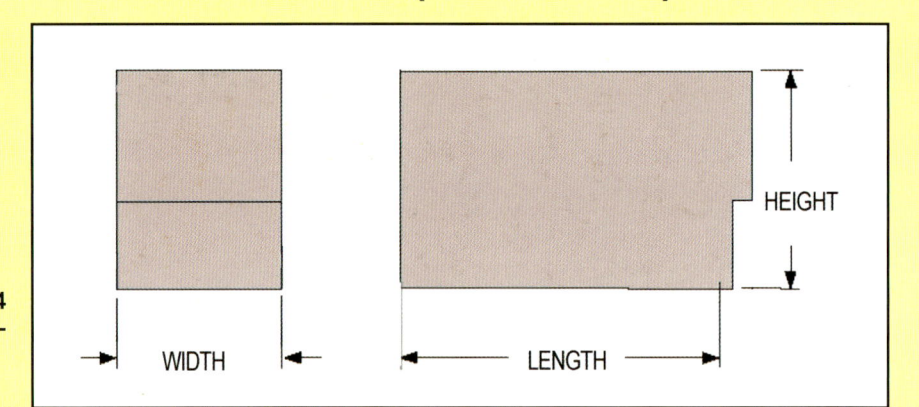

Earlier ARINC 404 case used for analog avionics.

ATR SIZE	WIDTH INCHES	WIDTH mm	LENGTH INCHES	LENGTH mm	HEIGHT INCHES	HEIGHT mm
Dwarf	2.25	57.15	12.52	318.0	3.38	85.8
1/4 Short	2.25	57.15	12.52	318.0	7.62	193.5
1/4 Long	2.25	57.15	19.52	495.8	7.62	193.5
3/8 Short	3.56	90.41	12.52	318.0	7.62	193.5
3/8 Long	3.56	90.41	19.52	495.8	7.62	193.5
1/2 Short	4.88	123.95	12.52	318.0	7.62	193.5
1/2 Long	4.88	123.95	19.52	495.8	7.62	193.5
3/4 Short	7.50	190.50	12.52	318.0	7.62	193.5
3/4 Long	7.50	190.50	19.52	495.8	7.62	193.5
1 Short	10.12	257.05	12.52	318.0	7.62	193.5
1 Long	10.12	257.05	19.52	495.8	7.62	193.5
1 1/2	15.38	390.65	19.52	318.0	7.62	193.5

Electrostatic Discharge (ESD)

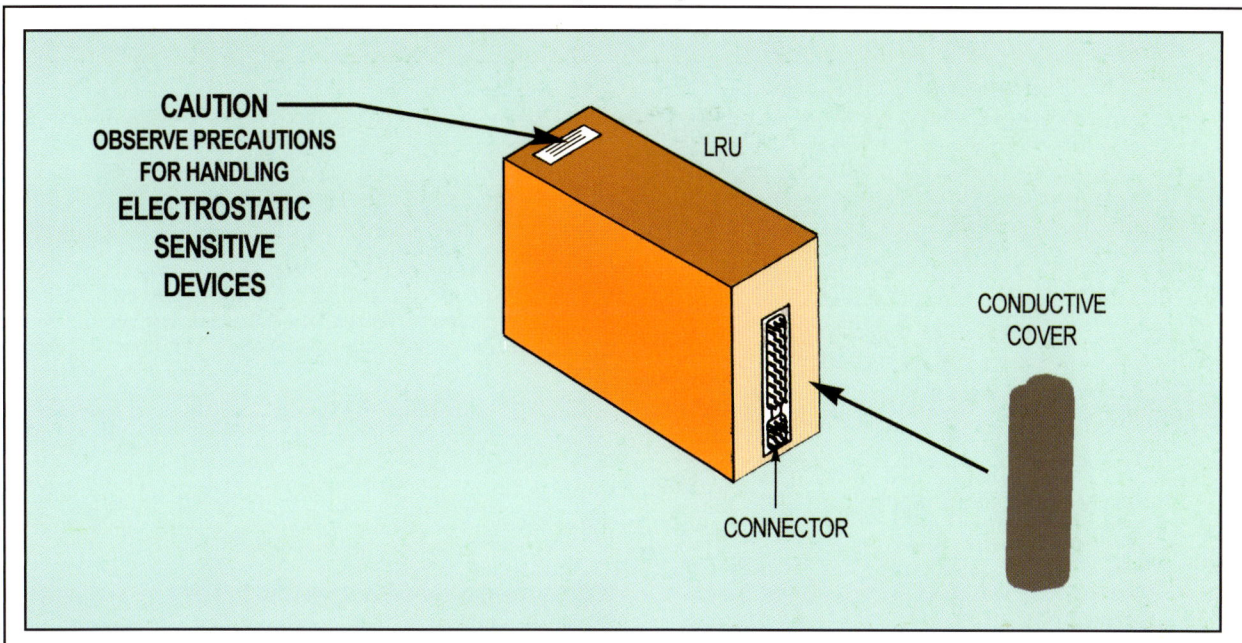

[Illustration: An LRU (Line Replaceable Unit) with a CAUTION label reading "OBSERVE PRECAUTIONS FOR HANDLING ELECTROSTATIC SENSITIVE DEVICES," showing the CONNECTOR and a separate CONDUCTIVE COVER.]

Microcircuits bring great benefits to avionics, but they create a new problem; "ESD," for electrostatic discharge. Components are so tiny, they are susceptible to static electricity built up on a technician's body, especially in dry parts of the country or during the low humidity of winter. The electrical charge builds to several thousand volts but the technician is unaware because the current is so low. The charges, however, can puncture thin layers of semiconductor material on circuit cards inside.

When installing or removing an LRU (line replaceable unit) check if it has an ESD warning label, as shown in the illustration. Here are some precautions:

If you're handling an LRU with its connector removed, don't touch the bare pins. Also, first touch the metal case (ground)---to drain off charges that accumulate on your body.

Before transporting an ESD-sensitive radio back to the shop (or manufacturer) obtain a conductive cover and place it over the connector (as shown in the illustration).

Before you remove a circuit card from an LRU, use a wrist strap that connects you to ground (the airframe). Place the card in a conductive bag made for the purpose.

The ESD problem could worsen as more components are squeezed into smaller spaces. There is also a trend to build larger circuit cards to accommodate integrated modular avionics on new aircraft. (A single module can cost the equivalent of four years of a technician's annual salary!) These simple grounding techniques, however, prevent damage.

Cooling

The greatest threat to the life of avionics is overheating.

The heating problem grows worse as more systems are installed in limited space and the number of circuit components per square inch rises. Few people realized the full impact of temperature until military investigators in the 1960's proved that overheating was the number one cause of avionics failure.

There are several solutions to cooling. If the aircraft is air transport or military, cooling systems are carefully designed at the airframe manufacturer. As shown by the illustration, cooling for B-737 avionics is built in as part of the airplane.

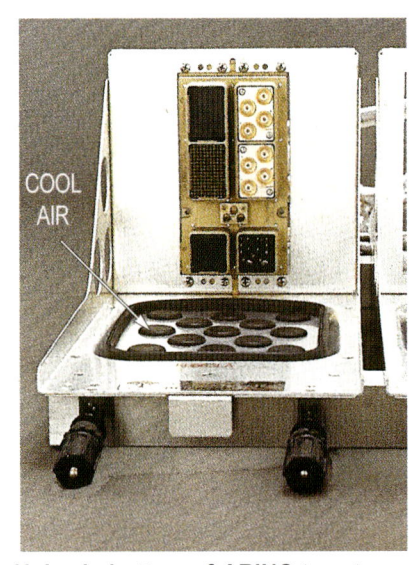

Cooling fan is built into ARINC tray for some airline avionics. A filter removes particles to extend equipment life and raise MTBF (mean time between failure).

Holes in bottom of ARINC-type tray admit air to cool avionics.

In General Aviation, there is no standardization because light aircraft vary widely in how they're outfitted. There is little official guidance so the solution is left to the installer.

Some single-engine airplanes come from the factory with small air scoops on the fuselage where they catch the air blast from the propellor. It's the technician's responsibility to hook the air ducts from the scoops to the avionics. Some technicians believe the scoops also deliver water to the radios when flying in rain. This has never been proven and, after flying an airplane with scoops for many years, personal experience shows no bad effects on the radios. It is more prudent, however, to cool every avionics installation with forced air from one or more fans. Check to see if any piece of avionics has already been fitted with an internal fan and what the manufacturer says about ducting the air flow. When the short life of overheated avionics is explained to a pilot or owner, he invariably will want to pay for adequate cooling.

Warranty Warnings. A turning point in cooling happened when radios started using digital electronics. Some manufacturers will not honor the warranty if the radio shows signs of overheating (meaning it was in-

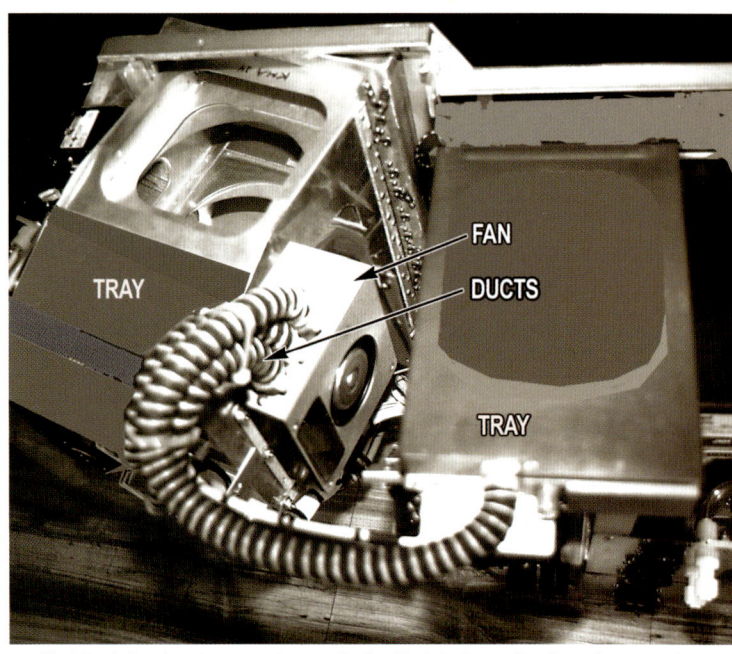

Behind the instrument panel of a light aircraft, showing cooling fan and ducts. The fan is attached to the side of the radio stack at the left. Several air ducts are seen emerging from the fan. They connect to ports on the radio trays, where they deliver cool air.

stalled without cooling). In earlier radios, warranty repair might be replacing a ten-cent resistor. The new radios, however, are loaded with integrated chips that are expensive to replace. In some instances, a whole circuit board is required, an expense the manufacturer wants to avoid.

Newer radios make the job of cooling much easier. They are often designed with nozzles for fastening air ducts leading to fans located some distance away. As shown, one fan may be rated to cool several radios.

Cooling for Airline Avionics

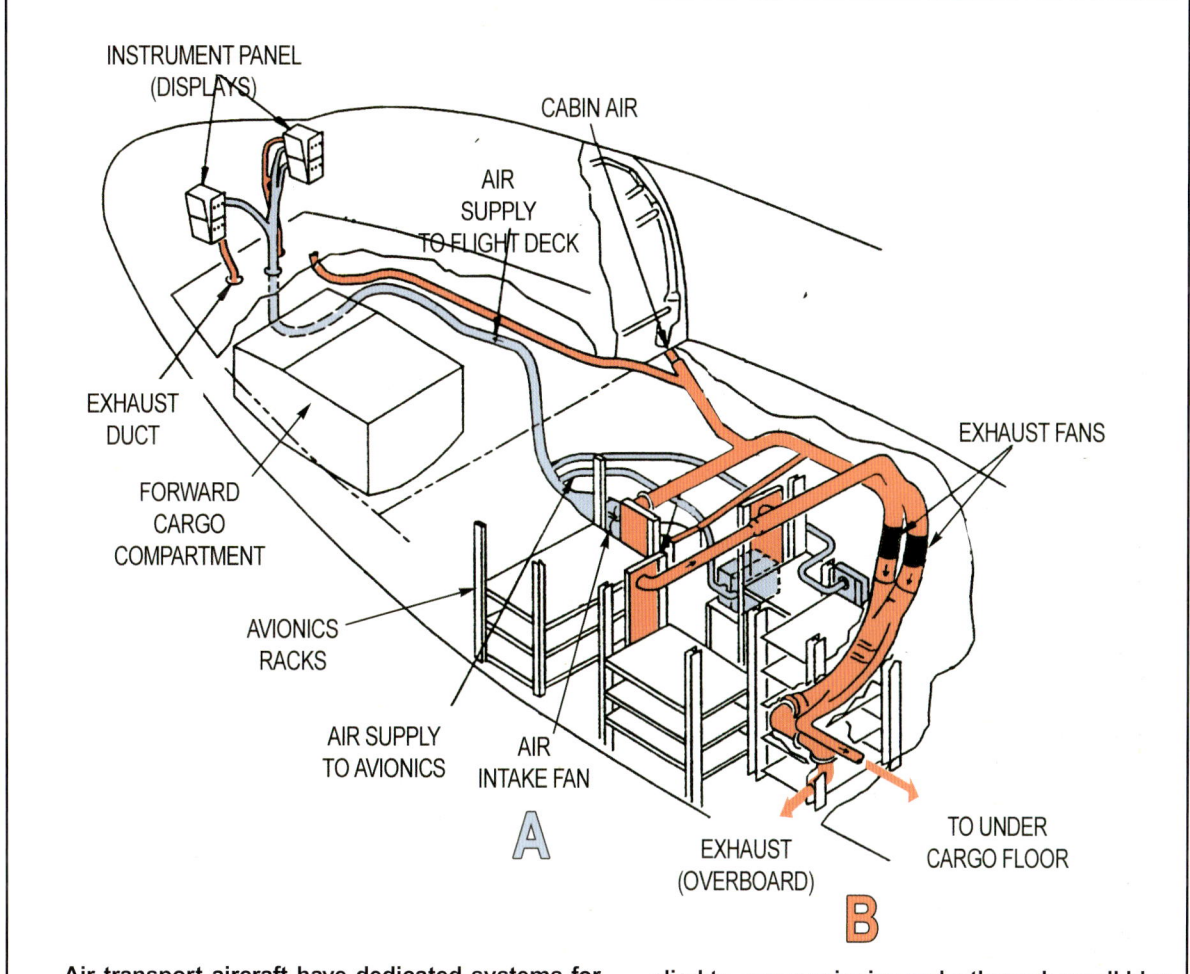

Air transport aircraft have dedicated systems for cooling avionics, as shown in this simplified diagram (based on a B-737). To assure reliability, both sides of the system---intake and exhaust---have two fans each. If one fan fails, an alarm warns the pilot to switch to the alternate.

In the illustration, blue ducts supply cool air, red ducts carry hot exhaust air from the equipment. The operation begins at "A - Air Intake Fan." The arrow points to a fan that draws air from the avionics equipment compartment. Cool air is supplied to some avionics racks through small blue ducts.

A large duct runs up to the flight deck and cools the instrument panel (mainly EFIS displays). Other equipment is cooled by air drawn through the avionics rack by an exhaust fan, and vented overboard or into the cargo compartment (see "B").

Hot air is vented overboard only when the airplane is on the ground or at low altitude. At higher (colder) altitudes, that air is sent to warm the forward cargo compartment.

Fans. Choose a cooling fan designed for avionics. To be sure, look at the catalog description; it should say PMA and TSO (Parts Manufacturers Authorization and Technical Standard Order---both FAA certifications.) They have brushless DC motors (which eliminate sparking and interference) and are 14- or 28-volt DC.

Fans are typically made with 1 to 5 outlets (or ports) which connect to radios through 5/8-inch hoses. If a fan has more ports than you need, unused ones are capped (which increases air flow to the other ports). Ample air is delivered, regardless of how many ports are connected. One port can typically put out about 26 CFM (cubic feet per minute) to cool one radio, such as a navcom or GPS.

Some fans come in kits, including mounting brackets and hoses. The manufacturer may also have a fan designed for a specific-model radio.

The fan is usually mounted near the rear of the radio stack and, in the simplest arrangement, hoses are brought near and aimed at the rear of the radio. Avionics of more recent design have a fitting for directly attaching the hose. In some installations, the hose attaches to a plenum, which is a metal chamber that runs alongside the radio stack, with holes that direct air to the radios.

An avionics fan may be expected to have long life, with ratings of nearly 80,000 hours of continuous operation.

Locking Radios in Racks

It not unusual for a pilot to taxi up to the radio shop and say; "My navcom doesn't work---no transmit, no receive." Before the technician reaches for any tools, he takes the palm of his hand and presses it against the face of the radio. The radio starts playing! The pilot is amazed. A large number of failures are simply due to vibration causing a radio to slide out of its connectors. Just a fraction of an inch does it. The remedy is to check the security of all radios when the airplane is brought to the shop.

Large aircraft have more effective locking devices, but even here there are problems if a radio is forced into its mounting tray. Designers in recent years have produced sturdier trays which resist bending and deforming. The latest approach is known as "zero insertion force," where the technician doesn't push the radio home. He operates a lever that causes the connectors to mate. This development followed a great increase in the number of terminals within a single connector, which increases chances of mis-mating pins and sockets by forcing the radio into the tray.

In the Instrument Panel

Avionics that mount in a panel are usually retained with a latch operated by a tool inserted into a hole in the front of the radio. Commonly used tools are an Allen wrench, size 5/64 or 3/32.

When the radio is installed, the latch is first positioned correctly by turning the screw all the way counterclockwise. Look at the underside of the radio and there should be a notch (or cut-out) which receives the latch. After the radio is slid into the rack, the latch should be aligned with the cut-out. Turn the wrench just enough to feel the latch engaging

Next, be sure the connectors at the back of the

Panel-Mounted Radios

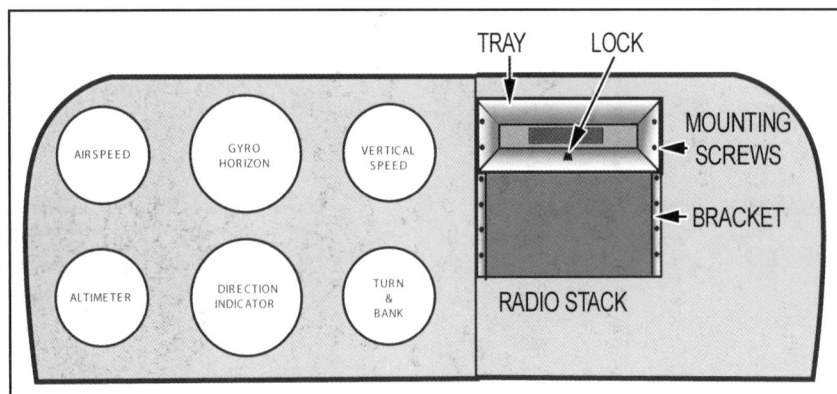

Although panel-mount radios are associated with small aircraft, they're also found in commercial aircraft (commuter and regional airlines). The panel mount uses the instrument panel as the support structure. A rectangular hole is cut in the panel and vertical brackets riveted to the sides of the opening. The tray (which receives the radio) is bolted to the brackets. After the tray is mounted, the radio is slid in and locked by turning a front panel screw, as shown.

After radios are mounted, they may be too heavy to be supported by the instrument panel. In this event, the installer adds brackets from the *back* of the radio trays to the airframe.

Releasing the Radio

Typical radio stack in a light aircraft. In this example, there are two navcoms at the top, with a transponder on the bottom. They slide into trays which are fastened behind the instrument panel. The usual method for inserting or removing these radios is inserting a tool (Allen wrench, Torx, screwdriver or other) at the lock release points.

188

radio and on the tray are ready to engage.

While you turn the wrench clockwise, place a hand on the radio front panel and gently push to help the radio into the connectors. In other words, don't just depend on the latch to draw the radio in.

Be gentle with the final tightening. With the radio completely in the rack, tighten the screw only to snug up the radio to the back of the tray. Tighten too much and the latching mechanism can be damaged.

If you're installing a stack of radios vertically, sometimes one opening is too narrow and blocks a last radio. You can usually solve this by loosening all the radios in the stack and sliding them back in a different sequence.

Instead of Allen wrenches, some radios use an ordinary slotted screw. Often they require only a quarter-turn to engage or release. If the radio will not release, hold the front panel at the sides and try to work it out with a slight sidewise motion.

Some radios require a very long screwdriver with a 1/8-inch flat tip to operate the locking mechanism. Others require a special tool inserted through the front to activate a release device.

Regardless of the system for locking the radio to its tray, the rule is: don't force it. If something is stuck (a frequent problem with old radios) try to coax, rock, wiggle or gently pry until you find a path of least resistance out of the rack.

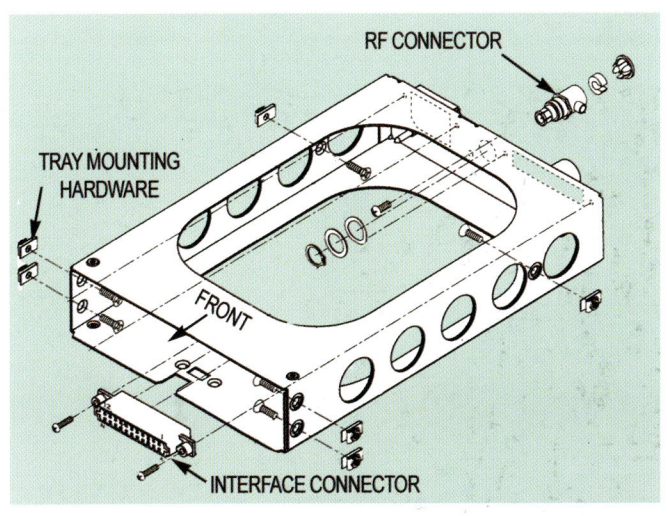

Panel-Mount Details

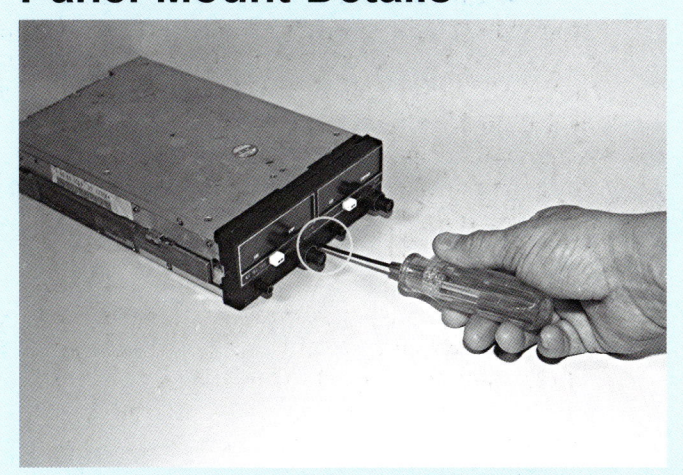

A panel-mounted radio is removed from the instrument panel by inserting a tool into an opening on the front. It may be a screwdriver, hex wrench or other tool.

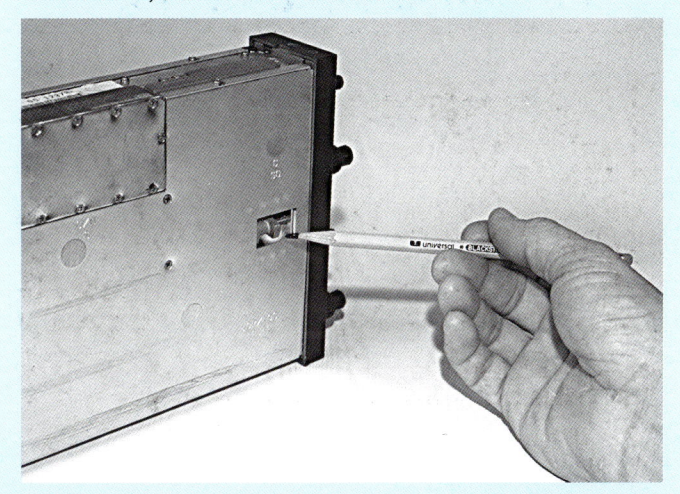

Pencil points to locking mechanism at underside of radio. As it turns, it engages a slot in the tray and pulls the radio in. To avoid damage, never force the radio into the tray.

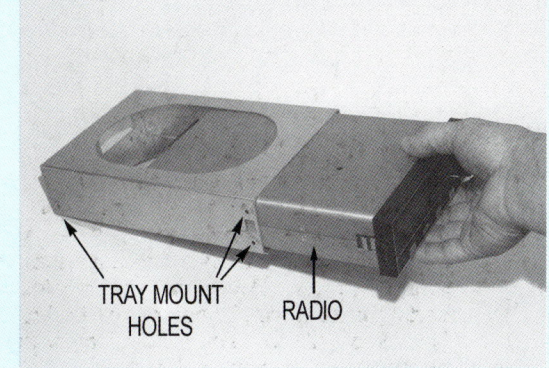

Radio is shown sliding into mounting tray. Holes in the tray are for fastening the tray to the instrument panel.

Preparing the tray for mounting. Hardware on the sides of the tray holds it to brackets in the instrument panel. The RF connector, which goes to an antenna, is fastened at the back. The interface connector will also mount on the rear of the tray, next to the RF connector. Holes along the sides of the tray lighten the structure.

189

Remote-Mounted Radios (Corporate)

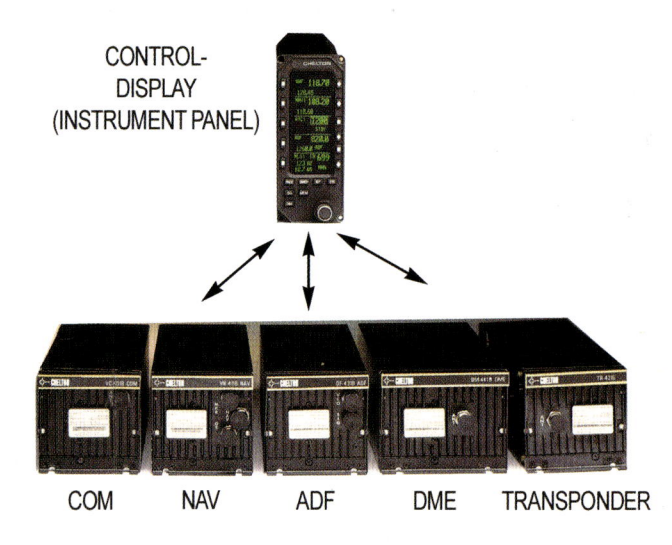

CONTROL-DISPLAY (INSTRUMENT PANEL)

COM NAV ADF DME TRANSPONDER

REMOTE-MOUNTED AVIONICS

In remote-mounted avionics, only a control-display unit (CDU) is in the instrument panel; the rest is in a remote location. For small business aircraft, the location is often in the nose; larger jets have a compartment in the fuselage.

The reason for remote mounting originally was that radios were too large to fit behind the panel. With microminiaturization, however, avionics are now often less than half their original size.

The example in the illustration is a Chelton radio management system. The small control-display manages the large remote boxes (LRU's, or line replaceable units).

Mounting Tray

The remote radio is supported in a tray located away from the flight deck. (Shown in this example is a Collins glideslope receiver.) At the bottom, the tray is fastened to the airframe by screws. That structure can be a shelf fabricated by the technician or one that already exists in the airplane.

After the tray is in place, the radio slides in and engages the rear lock. The front lock is tightened to complete the installation.

Connectors on the front of the radio go to the control head in the instrument panel, a power source and the glideslope antenna.

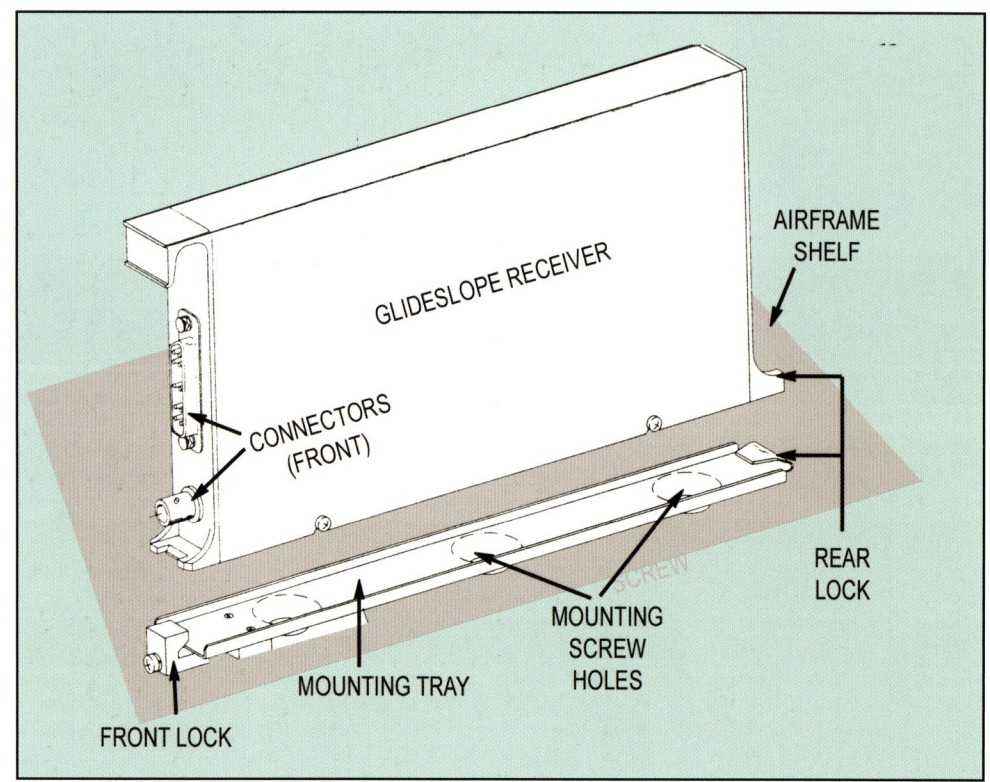

GLIDESLOPE RECEIVER

AIRFRAME SHELF

CONNECTORS (FRONT)

REAR LOCK

MOUNTING SCREW HOLES

MOUNTING TRAY

FRONT LOCK

Airline Mounting

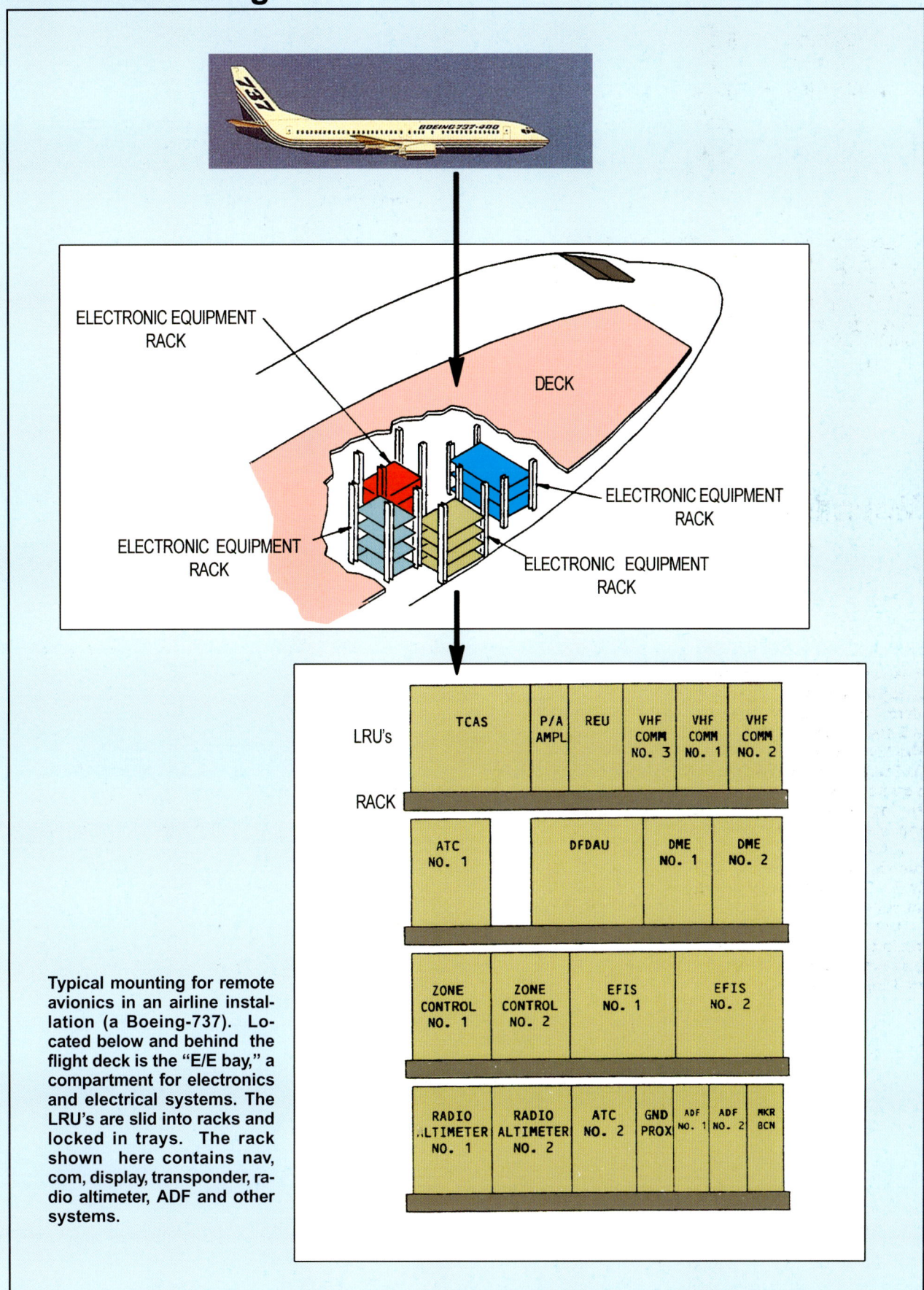

Typical mounting for remote avionics in an airline installation (a Boeing-737). Located below and behind the flight deck is the "E/E bay," a compartment for electronics and electrical systems. The LRU's are slid into racks and locked in trays. The rack shown here contains nav, com, display, transponder, radio altimeter, ADF and other systems.

Locking Systems (Airline)

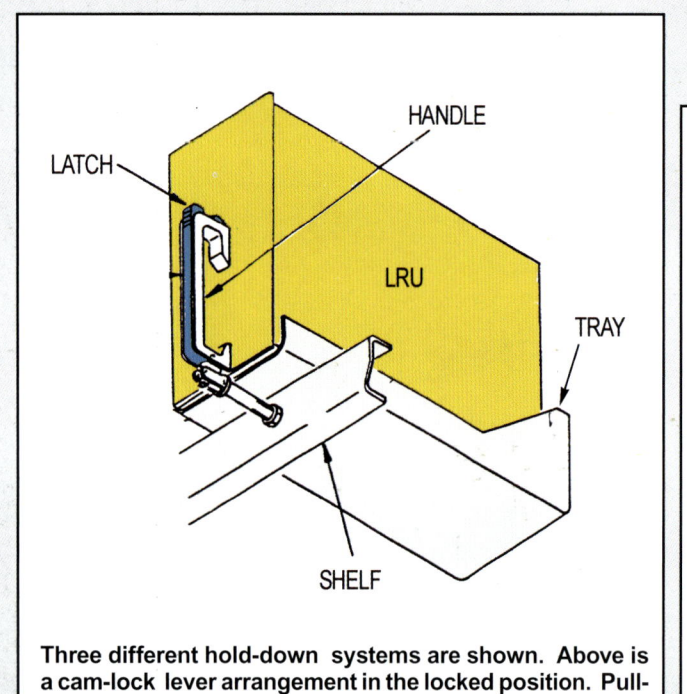

Three different hold-down systems are shown. Above is a cam-lock lever arrangement in the locked position. Pulling down the handle releases the LRU from the tray.

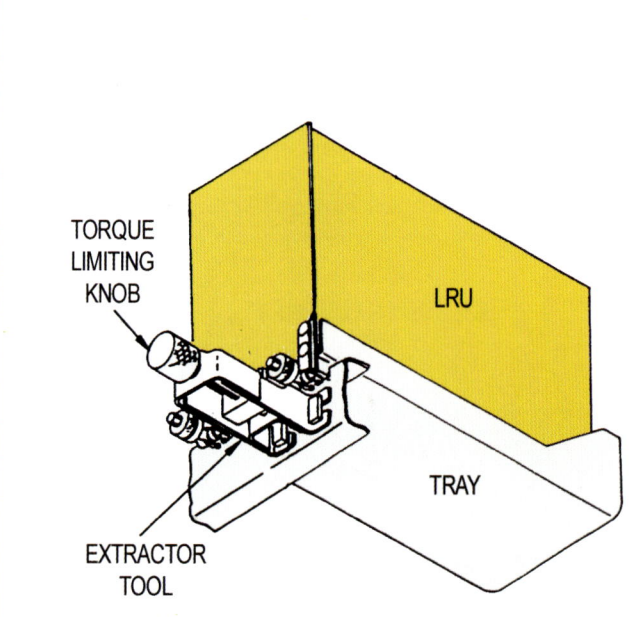

A knob and extractor tool release the LRU. To prevent damage from overtightening by the technician, the thumbscrew slips after proper torque is reached.

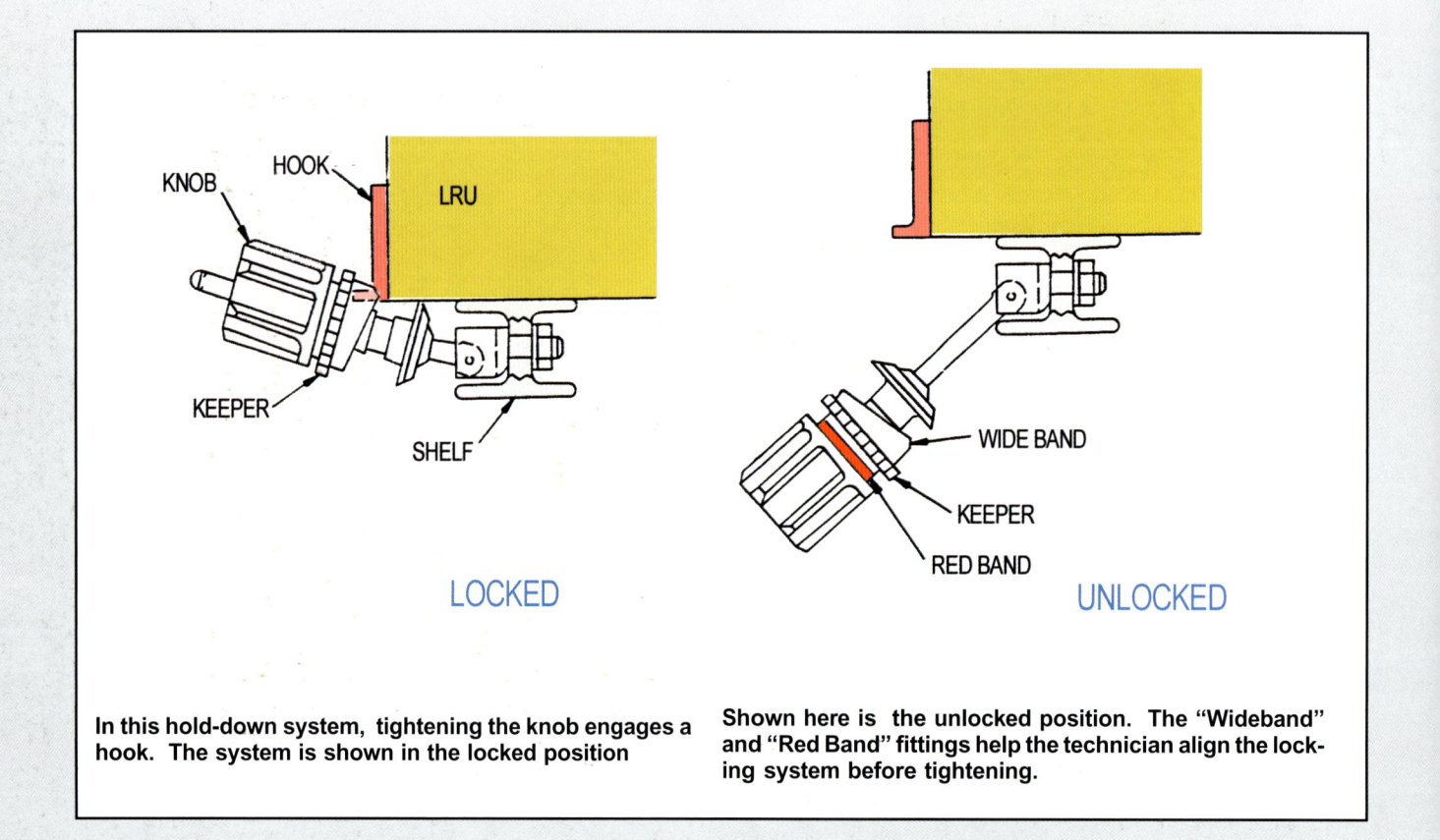

LOCKED

UNLOCKED

In this hold-down system, tightening the knob engages a hook. The system is shown in the locked position

Shown here is the unlocked position. The "Wideband" and "Red Band" fittings help the technician align the locking system before tightening.

Indexing Pins Prevent Error

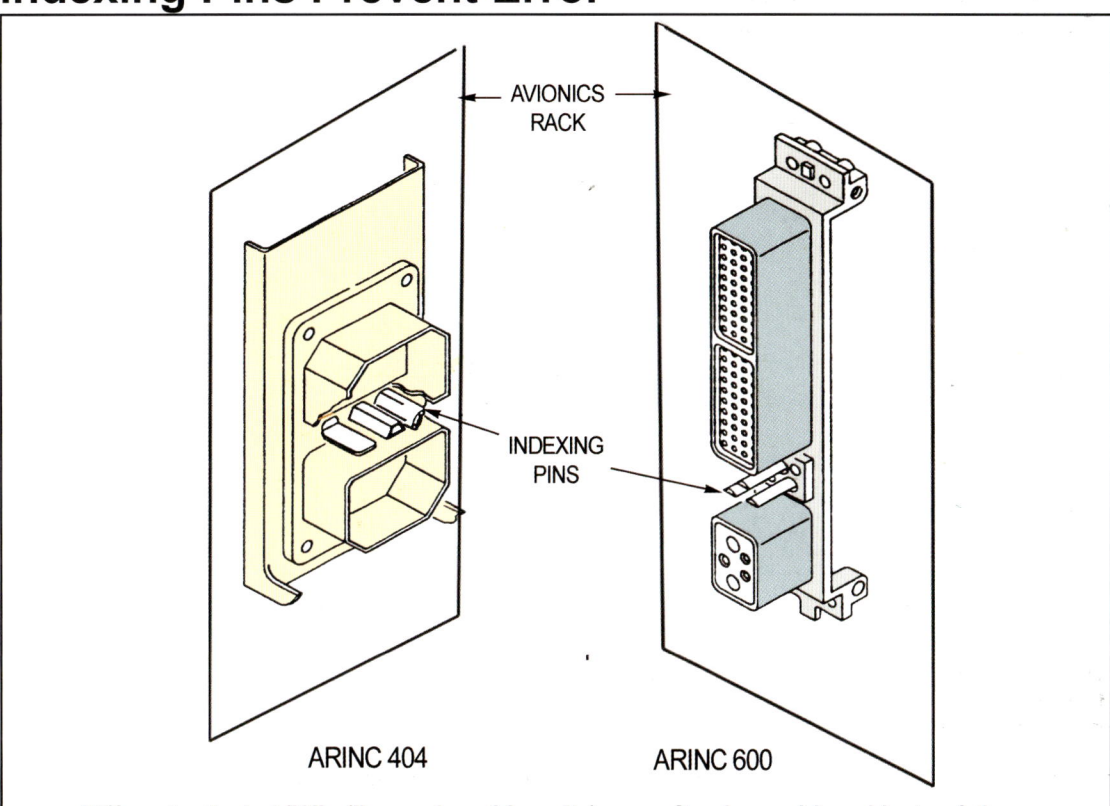

ARINC 404

ARINC 600

Different avionic LRU's (line replaceable units) are often housed in cabinets of the same size---which could cause installation error. To prevent it, a connector has an indexing pin at an angle that matches only one LRU. Unless all pins line up with connector holes, the radio cannot be pushed in. To prevent damage, however, avoid forcing a radio into the rack.

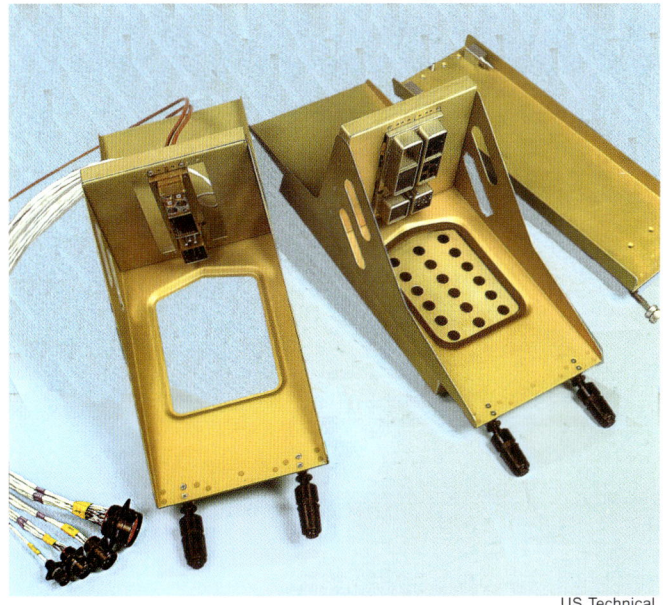

ARINC trays are designed with variations to accommodate different cooling, connector and radio sizes. The black knobs, which lock in the equipment, have a mechanism which cannot be overtightened and damage the connectors.

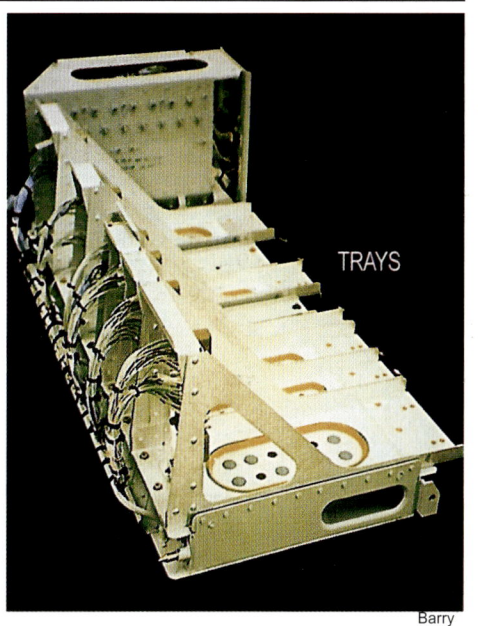

Several ARINC trays are often mounted together to form a "rack" (sometimes called an "equipment cabinet").

Integrated Modular Avionics (IMA)

Boeing 777

EQUIPMENT CABINET

HIGH SPEED BACKPLANE BUS

LINE REPLACEABLE MODULES (LRM)

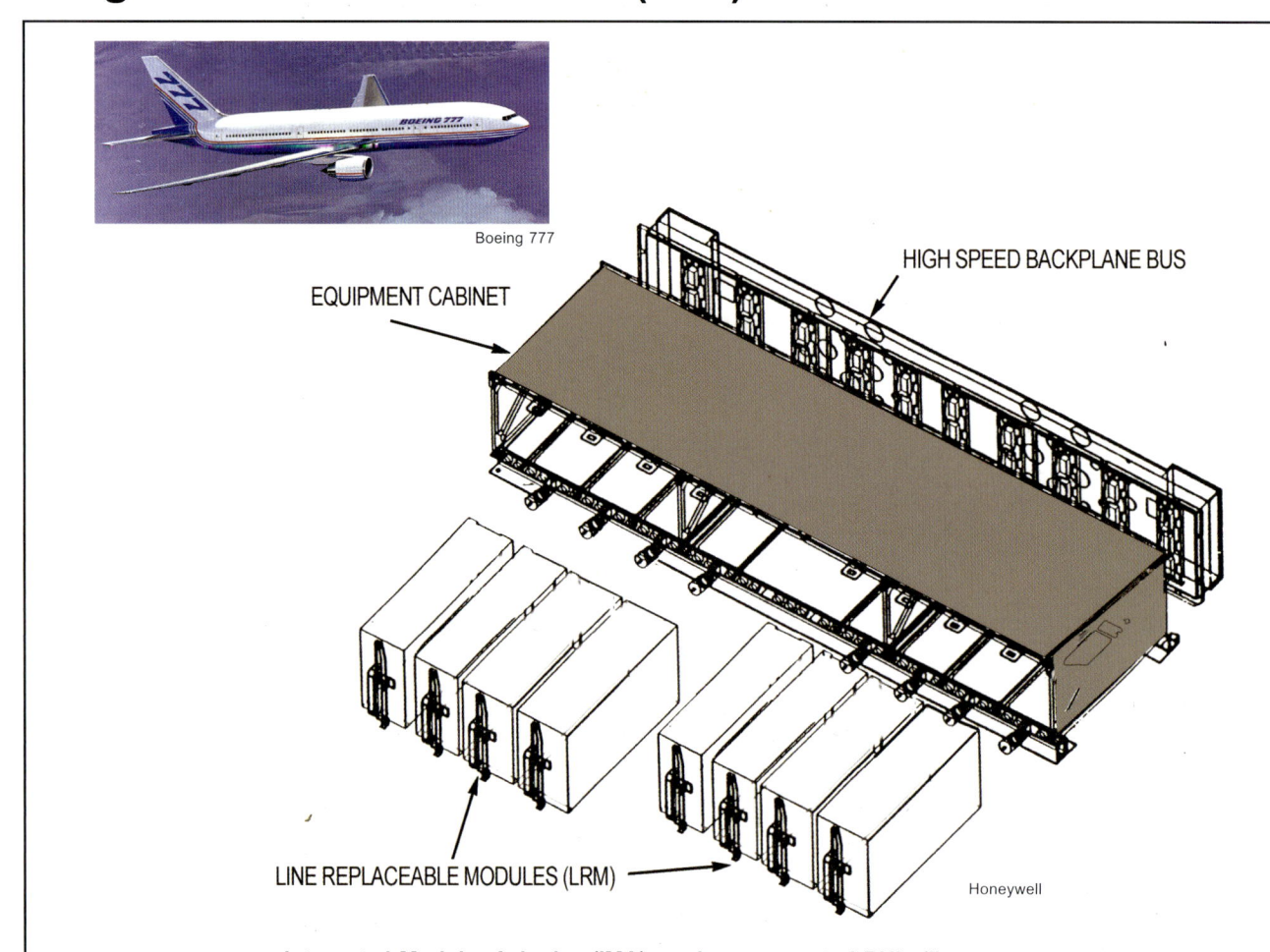

Honeywell

Integrated Modular Avionics (IMA) replace separate LRU's (line replaceable units), with LRM'S, "line replaceable modules." Unlike earlier systems, LRM's do not have one function per unit, such as receiver, transmitter, etc. They are more like computer resources that share and process information over a high speed databus. This provides smaller size and weight, and greater reliability.

For the technician, troubleshooting is simplified by a built-in central maintenance computer that identifies problems and indicates which module to replace.

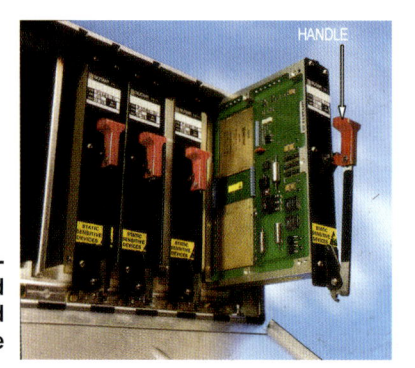

Example of a cabinet for Integrated Modular Avionics. Red handle is used to unlock and remove the Line Replaceable Module (LRM)

Instrument Mounting

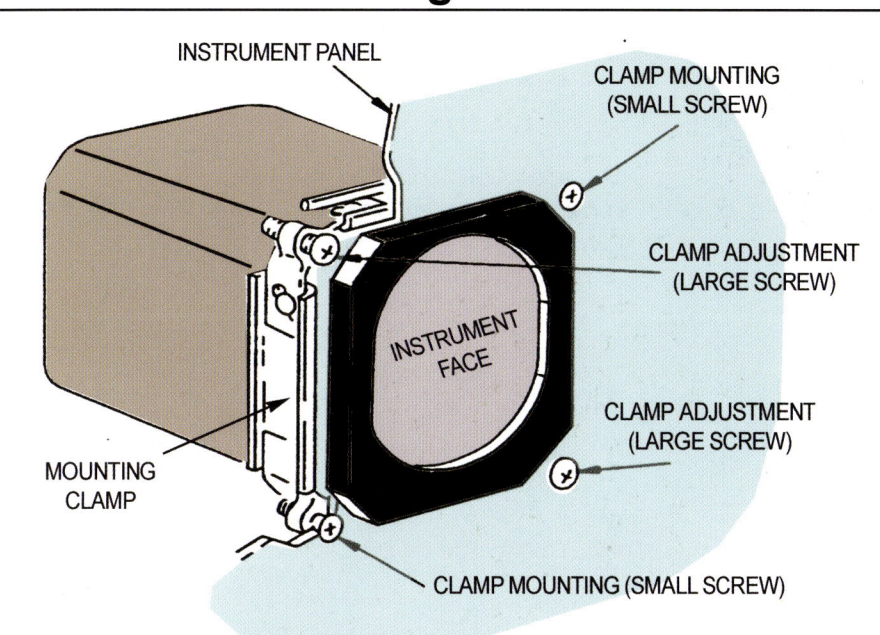

Instruments like this 3-inch rectangular are often held by a mounting clamp behind the panel. The clamp is slid over the case and two sets of screws are adjusted. Two screws have large heads labelled "Clamp Adjustment". They tighten the clamp around the case. The other pair, labelled "Clamp Mounting," hold the clamp to the back of the instrument panel.

Round instruments are installed in similar fashion with a round clamp. However, there are only two screws; one to tighten the clamp on the instrument, the other to hold the clamp to the panel.

Some instruments have tapped holes on their cases and need no clamps. Check the manufacturer's literature on using the correct screws. If too long, they can penetrate the case and damage the instrument,

Instrument screws are often made of brass, especially when mounting a magnetic compass. As a non-magnetic metal, brass will not cause deviation in the compass.

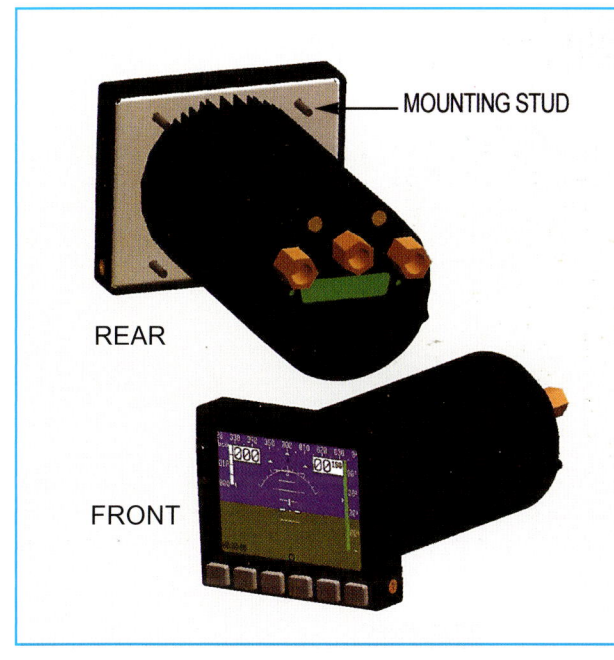

Some instrument cases are fitted with mounting studs, as in this Dynon EFIS display. Four holes are drilled in the instrument panel according to the template (below) supplied by the manufacturer. The large hole receives the instrument case.

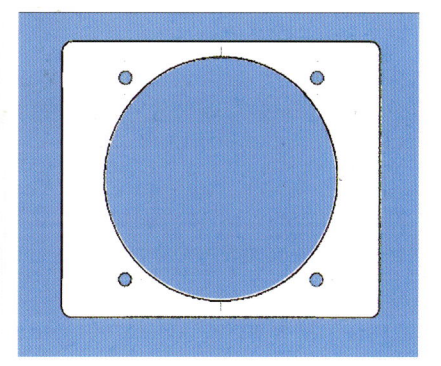

Round Instruments: 2- and 3-inch

Many flight instruments in General Aviation mount in round holes. The two main sizes are 2- and 3-inch diameters (actually 2-1/4 and 3-1/8-inches).

An example of each is shown in this Mooney panel; a 2-inch chronometer and a 3-inch airspeed indicator.

The instruments are held to the panel by four screws, as seen around the instrument face. The screws are held behind the panel by threaded fasteners ("grasshopper nuts"). Because the fasteners are easily lost during installation, there are mounting kits like the one shown below to simplify the job.

2-INCH ROUND CHRONOMETER

3-INCH ROUND AIRSPEED

Instrument Mounting Kit

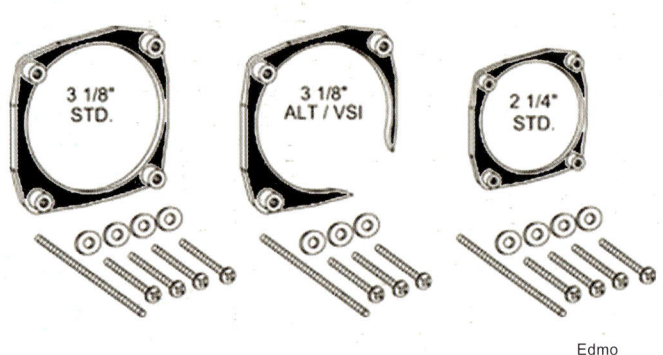

3 1/8" STD.

3 1/8" ALT / VSI

2 1/4" STD.

Edmo

"Nut rings" make the installation job easier. They come in standard 2- and 3-inch sizes. There are two versions of the 3-inch; note the one in the center, "ALT/VSI" which has a cut-out at the lower right. This allows space for the altimeter knob after the instrument is installed. ALT is for altimeter, which has a knob adjusted by the pilot (for barometer setting). VSI (vertical speed indicator) has a small screw adjustment for zero'ing the needle.

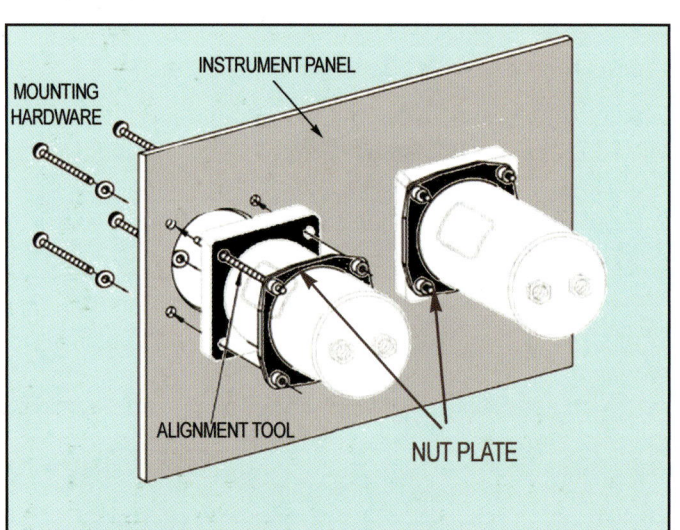

MOUNTING HARDWARE

INSTRUMENT PANEL

ALIGNMENT TOOL

NUT PLATE

A nut plate is installed by sliding it onto the back of the instrument. To make it match holes in the panel with holes in the nut plate, the installer inserts an alignment tool through one hole, as shown. It's removed when all holes line up, and mounting screws can be inserted.

Airline Instrument Mounting

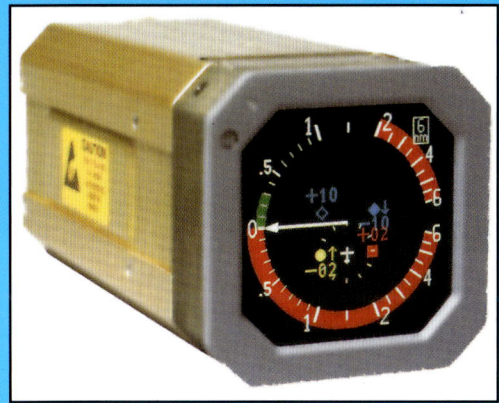

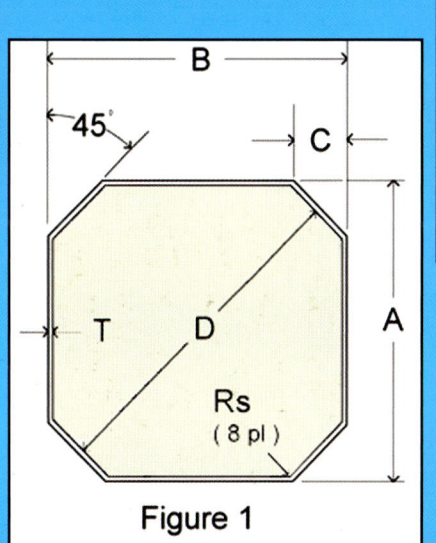

Instruments designed for the panel of airliners have case sizes that conform to "ATI" dimensions. Developed by ARINC, they assure the instruments will match holes in the panel.

Generally, a "3 ATI" instrument is approximately 3 inches wide, a "4 ATI" is 4 inches wide (but check the manufacturer's information for exact dimensions).

An example is the instrument pictured above; a VSI-TCAS (vertical speed and anti-collision display). It is offered by the manufacturer (Sextant) in either a 3 ATI or 4 ATI size. Because ATI instruments have bevelled (slanted) edges, it is difficult to make the cut-out with ordinary hand tools. Special cutters are available.

Although ATI sizes are an airline specification, they are also found in corporate and light aircraft, as well.

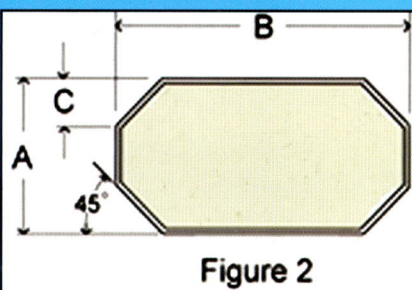

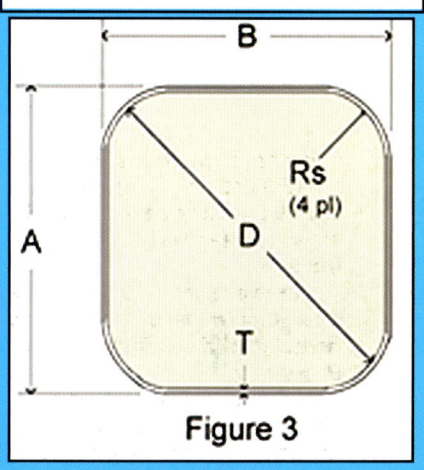

Figure 1

Figure 2

Figure 3

ARINC 408 Instrument Housings

ATI SIZE	A	B	C	D	E	Rs	FIG.
	±.010	±.010	REF	±.010	±.010	REF.	
1.5 X 3 ATI	1.457	3.175	0.428	2.67	2.775	0.125	2
MM	37	80.84	10.87	67.81	69.97	3.17	2
1.5 X 4 ATI	1.457	3.975	0.428	3.236	3.321	0.125	2
MM	37	100.96	10.87	82.19	84.35	3.17	2
2 ATI	2.175	2.175	0.407	2.5	2.585	.063, .695	1 & 3
MM	55.24	55.24	10.33	63.5	65.65	1.6, 17.653	1 & 3
2 X 4 ATI	2.175	3.975	0.407	3.773	3.858	0.063	2
MM	55.24	100.96	10.33	95.83	97.99	1.6	2
3 ATI	3.175	3.175	0.428	3.885	3.97	.063, .725	1 & 3
MM	80.64	80.64	10.87	98.67	100.83	1.6, 18.415	1 & 3
3 X 4 ATI	3.175	3.975	0.428	4.451	4.536	0.063	2
MM	80.64	100.96	10.87	113.05	115.21	1.6	2
4 ATI	3.975	3.975	0.429	5.015	5.1	0.063	1 & 3
MM	100.96	100.96	10.89	127.38	129.54	1.6	1 & 3
4 X 5 ATI	3.975	4.975	0.429	5.722	5.737	0.068	2
MM	100.96	126.36	10.89	145.33	145.71	1.6	2
5 ATI	4.975	4.975	0.478	6.36	6.445	0.063	2
MM	126.36	126.36	12.14	161.54	163.7	1.6	1
5 X 6 ATI	4.975	5.975	0.478	7.067	7.152	0.063	2
MM	126.36	151.76	12.14	179.5	181.66	1.60	2
6 ATI	5.975	5.975	0.513	7.725	7.81	0.063	1
MM	151.76	151.76	13.03	196.21	198.37	1.6	1

MSP Aviation

Review Questions
Chapter 23 Mounting Avionics

23.1 Who should be consulted if an avionics installation will affect structures in the airplane?

23.2 What designation assures that a piece of equipment has high resistance to heat, humidity and other environmental conditions of flight?

23.3 "Approved data" for an installation may be found in the _____

23.4 When selecting aluminum for making structures, what label indicates resistance to corrosion?

23.5 What are efficient methods for cutting odd-shaped instrument holes in an aluminum panel?

23.6 When mounting new equipment in an instrument panel or in the avionics bay of a large aircraft, do not exceed the _____ and _____ limitations of the airplane.

23.7 What is a major advantage of ARINC cases in large aircraft?

23.8 What are the two basic types of ARINC cases?

23.9 What are two precautions when handling avionics that are sensitive to electrostatic discharge (static electricity)?

23.10 What is the greatest threat to the life of avionics?

23.11 What techniques are used in large and small aircraft to prevent overheating?

23.12 When mounting a magnetic compass, always use _____ screws to avoid _____ in the compass.

Chapter 24

Connectors

Working with connectors often takes up more of an installation technician's time than any other task. A light aircraft has connectors in the dozens, while larger airplanes count them in the hundreds. Without connectors, avionics can't be removed for maintenance or modifications.

There is a trend in avionics to reduce the amount of connectors and wiring. They add weight, take up space and cause trouble when improperly installed. It is now possible to blend signals of many systems on a single pair of wires or fiber optic cable and send them around the airplane. Applications increase with each generation of new aircraft but we will have to live with connectors for another 30 or 40 years.

Connectors look like simple mechanical devices-- --an outer shell, terminals (pins or sockets) and insulating material. Nevertheless, connectors are a major cause of equipment failure. Pins are wired incorrectly, bare wires touch and short-circuit or connector pins are accidently bent. All can be avoided by careful wiring technique.

Technicians have different approaches to avoiding errors in wiring. Some follow the old carpenter's warning; "Measure twice, cut once." In wiring, it means double-checking for the correct pin, marking each pin on the diagram as it's done and making a final check after all wiring to the connector is complete. Finding trouble after the installation is done takes far more time than checking for error as you build up the wiring harness.

Reading pin connections. Some errors are due to the way pins are identified. Because of their high number and small size, markings on connectors are not only tiny, but often the same color as the background. You may have to hold the connector up to a bright light to make the number legible.

(continued p. 202)

Typical Connectors

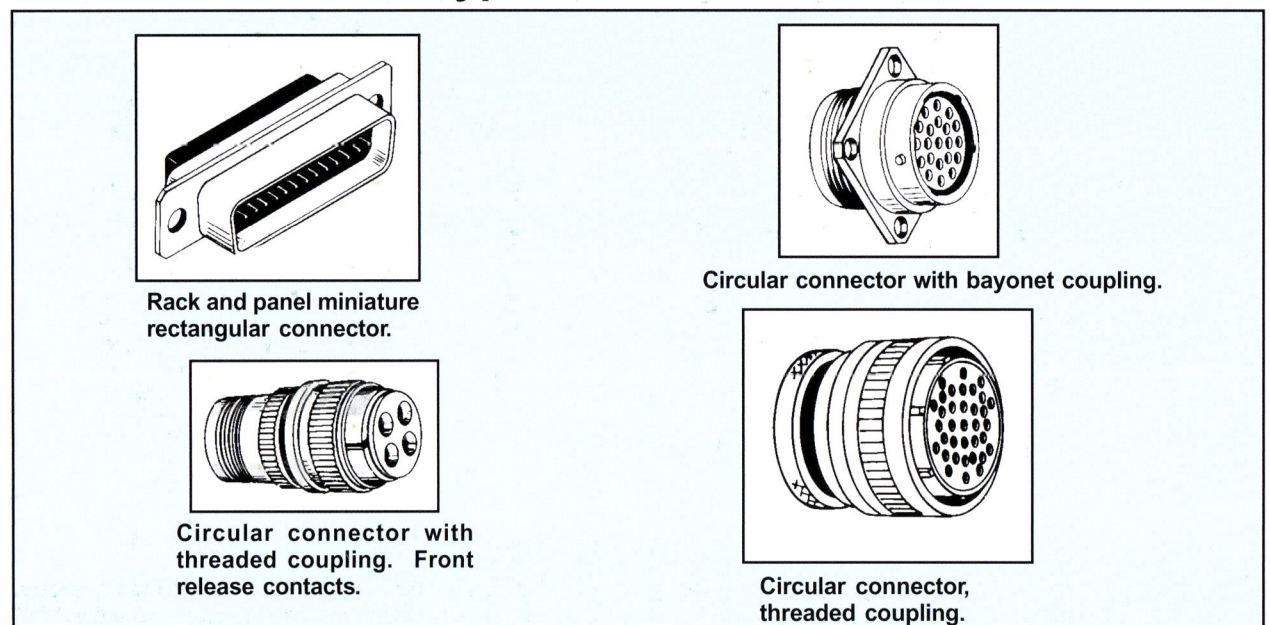

Rack and panel miniature rectangular connector.

Circular connector with threaded coupling. Front release contacts.

Circular connector with bayonet coupling.

Circular connector, threaded coupling.

RF (Radio Frequency) Connectors

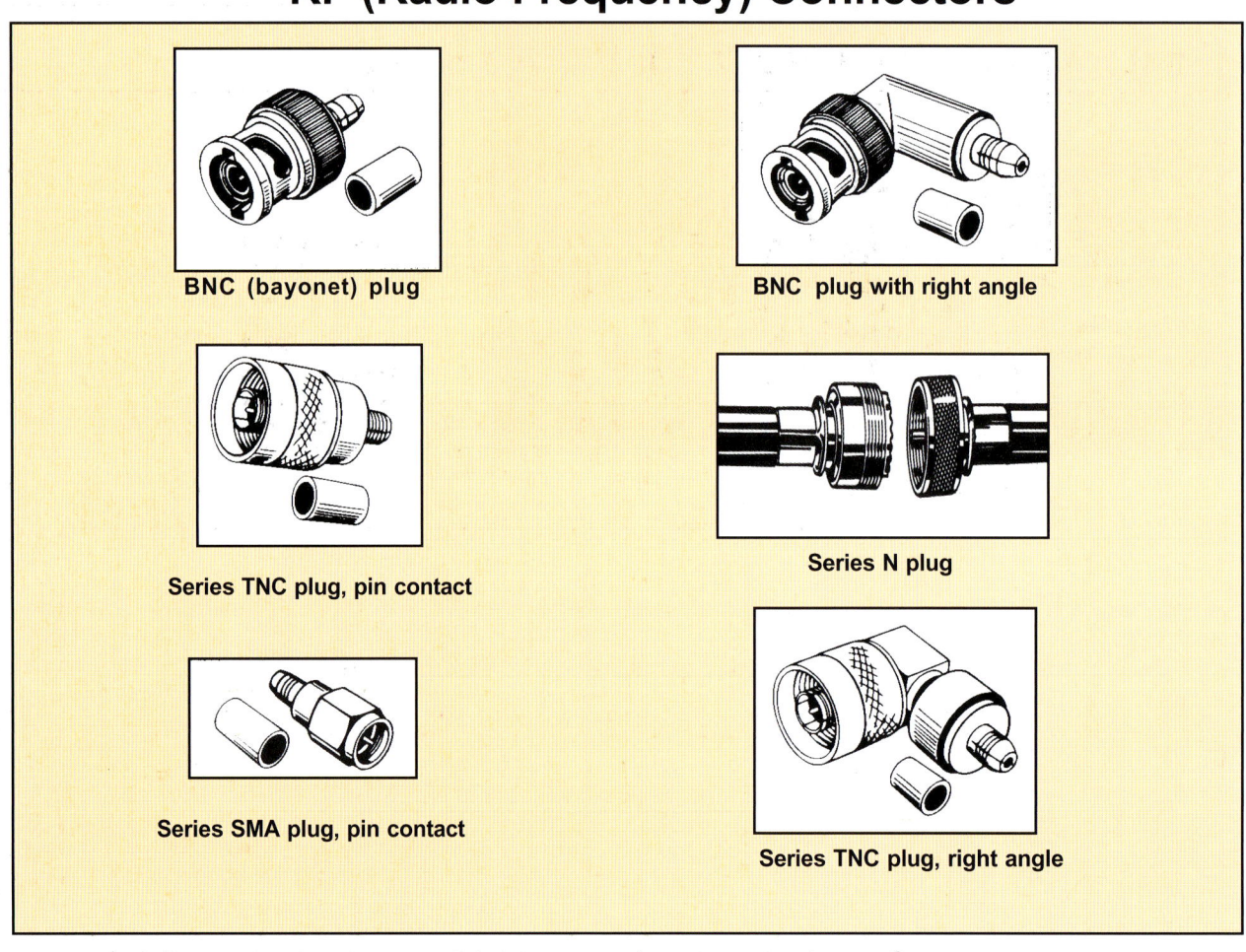

BNC (bayonet) plug

BNC plug with right angle

Series TNC plug, pin contact

Series N plug

Series SMA plug, pin contact

Series TNC plug, right angle

How to Identify Connector Contacts

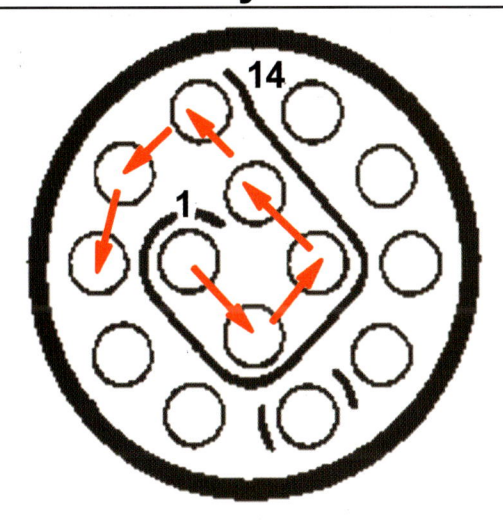

1. The end of a circular connector, where contacts will be inserted. Some connectors identify every contact, but limited space may allow only starting and ending numbers like "1" and "14," shown above. Start at "1" and follow the guide line around to "14," as shown by the arrows (which are not on the connector).

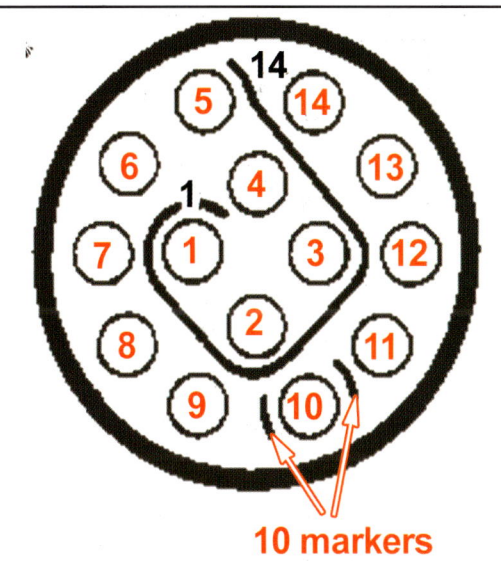

2. The same connector is shown; with red numbers added to clarify the numbering system. Note that "10" is surrounded by two markers. They speed up your counting. You can jump to the first marker and know it's 10. When looking at the back of a connector (where the terminals are inserted or wired) counting is usually done in a counterclockwise direction.

10 markers

Contacts are Selected to Fit the Application

A circular connector, like this RMS bayonet receptacle, can be obtained with a variety of contacts (pins and sockets) to fit the application.

Shown below is an excerpt from an RMS specification sheet. By knowing maximum current rating and wire size, you can order the contact size for insertion into the connector. The contacts in this model are rear insertion and front release.

Contacts
For use with Series R0715, R0716, R0717, R0718 and R0719

Contact Size	Wire Size AWG	Max. Current Rating Amps.	MS Part No. Pin	MS Part No. Socket
No. 20	24 22 20	3.0 5.0 7.5	M39029/31-241	M39029/32-260
No. 16	18 16	16 22	M39029/31-229	M39029/32-248
No. 12	14 12	32 41	M39029/31-235	M39029/32-254

Several examples of how letters and numbers identify pins are shown in the illustrations. Some use numbers, such as 1 through 15 or 1 through 34. Others use letters A through Z. If the total number of connections goes beyond Z, the next pin may be "a" or "aa" (lower case letters).

Caution: Some connectors omit letters such as G, I, O and Q. Therefore, don't simply count the pins to get to a desired one unless you are sure the numbers or letters are consecutive.

Soldering Connectors. The classic method for attaching a wire to a connector terminal is with a soldering iron. It is a more difficult skill than it appears. A soldering iron in tight spaces with small objects easily causes heat damage. Also, many beginners believe that solder is "pasted" onto a wire by dabbing it with the iron. This results in a "cold solder joint" which soon crumbles. Good soldering technique requires that the iron heats both wire and terminal so solder turns liquid and flows freely between them. The technique has proven troublesome enough for airlines and military organizations to run "soldering schools," taught by a skilled operator (often from a manufacturer of soldering tools).

Now the good news; soldering wires to connectors has been largely replaced by a faster, more convenient and effective joining method.

Crimping. This is the process of squeezing a metal contact around a wire with a special tool. As shown in the illustrations, the wired contact is then inserted into the connector until it snaps into place. You *must* have the crimping tool designed for that size and type of terminal. A good crimping tool has a mechanism which applies the correct force to crimp the terminal no matter how hard you squeeze the handles.

Back shell. Some connectors have a back cover, or shell, which protects the wire where it enters the connector. The back shell may also have a clamp that goes around the wire bundle to relieve strain on the pins. Strain relief for all wires entering a connector is important. If there is stress from a wire pulling on its terminal, the connection may not last long.

Connector Trends

Aviation borrows heavily from connectors for the computer industry (similar to those on the back of a PC). They accommodate large numbers of wires, and provide reliable, fast methods of attaching connections.

(Continued p. 206)

Identifying Mil-Spec Connector Part Numbers

Many connectors in avionics comply with a Military Specification. Using the "MS" number, you can decode the connector's specifications. Consider the example (by Glenair):

MS3402DS28-21 PY

MS =	Military Standard
3402 =	Box Mount Receptacle (Designation)
D =	High Shock (Environmental)
S =	Shell Material
28 =	Shell Size
21 =	Contact Arrangement
P =	Pin type (Male)
	or "S" (Socket, or female)
Y =	Polarization keying

The first four numbers after "MS" (3402 in the example above) indicate physical type. Other types include:

- 3400 Wall mounted receptacle
- 3401 In line receptacle
- 3402 Box mount receptacle
- 3404 Jam nut receptacle
- 3406 Straight plug
- 3408 909 plug
- 3409 45Q plug
- 3412 Box mount receptacle with rear threads

The single character which follows indicates the connector service class:

- D High Shock
- K Firewall
- L High Temperature
- W General Purpose

The next character, S in our example, indicates the shell material; in this case, stainless steel.

The next two characters, 28, identify the shell size.

The following pair of numbers, 21, identifies the contact arrangement. If this pair of characters is followed by an "S", it indicates female style (socket) contacts. If they are followed by a "P", it indicates male contacts (Pin).

The final character, Y, indicates the choice of polarization keying.

Typical Coaxial (RF) Connectors

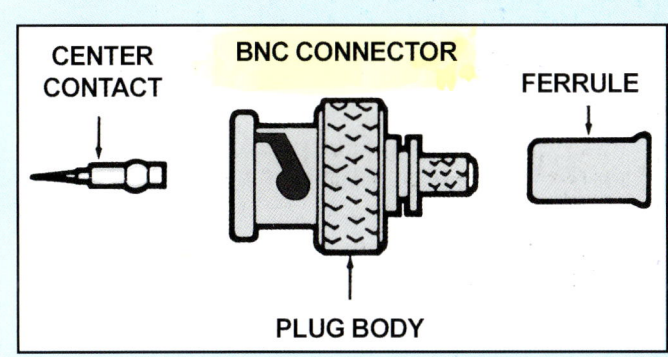

Exploded view of plug assembly

BNC
This is among the most common RF connectors for avionics. A bayonet coupling makes it easy to make or break the connection with a push and half-twist. BNC's are typically rated for 50-ohm coaxial cable.

TNC
This is similar to BNC but replaces the bayonet with a threaded coupling. The TNC is a higher-performance connector, especially under high vibration.

Both BNC and TNC connectors come in a variety of mounting styles, including bulkhead, straight and right angle.

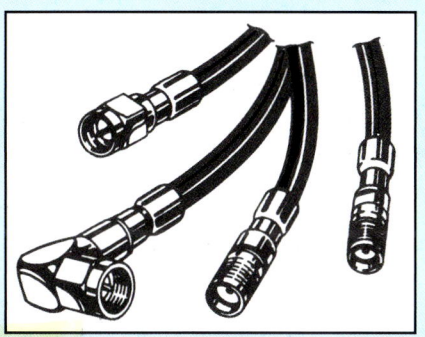

SMA SERIES
Widely used in avionics, especially for military applications. It's a high performance connector for subminiature coaxial cable.

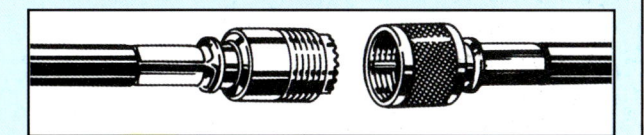

N SERIES
A screw-on connector, the N type is available for crimp connection to coaxial cable.

Connector Illustrations: Anixter

ARINC Connectors (Airline)
ARINC, the airline avionics organization, sets the standards for connectors aboard nearly every airline in the world. The two most common series are ARINC 404 and ARINC 600.

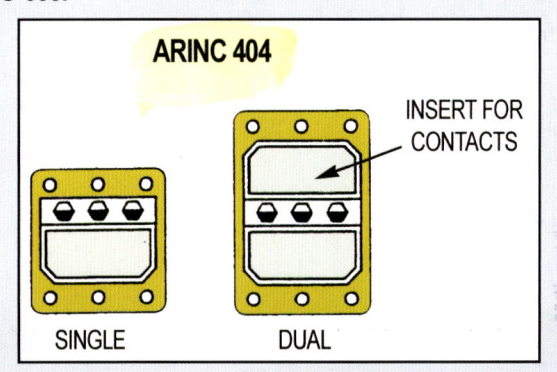

ARINC 404 is aboard airliners that began production during the 1960's. This includes first models of the Boeing-727, -737, -747 and Airbus-300. Instruments and radios operate on conventional (analog) principles, which require fewer pins than today's avionics. These connectors are set up for various equipment by different inserts (which hold the contacts). The ARINC 404 connector is still needed on most recent airliners.

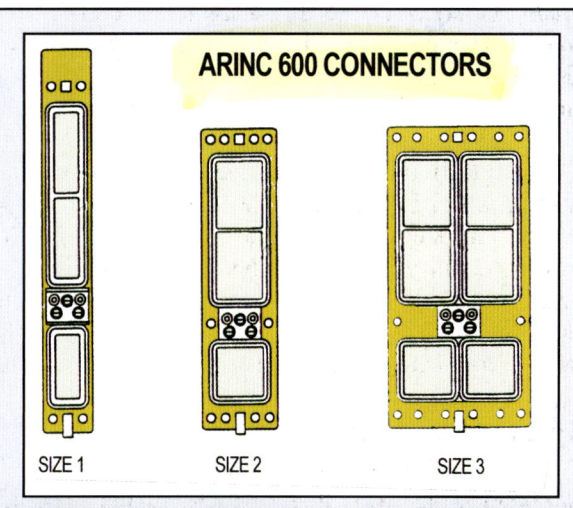

ARINC 600 appeared with the new generation of "digital" airliners during the 1980's, including the Boeing-757, -767 and Airbus A-320. Because of the advanced systems, ARINC 600 provides more contacts in a small area

When old airline aircraft are upgraded with digital avionics (such as the Boeing 747-400) the flight deck becomes a "glass cockpit," meaning EFIS, the electronic flight instrument system. Digital equipment requires ARINC 600 connectors.

Coaxial Connector: Attaching to Cable

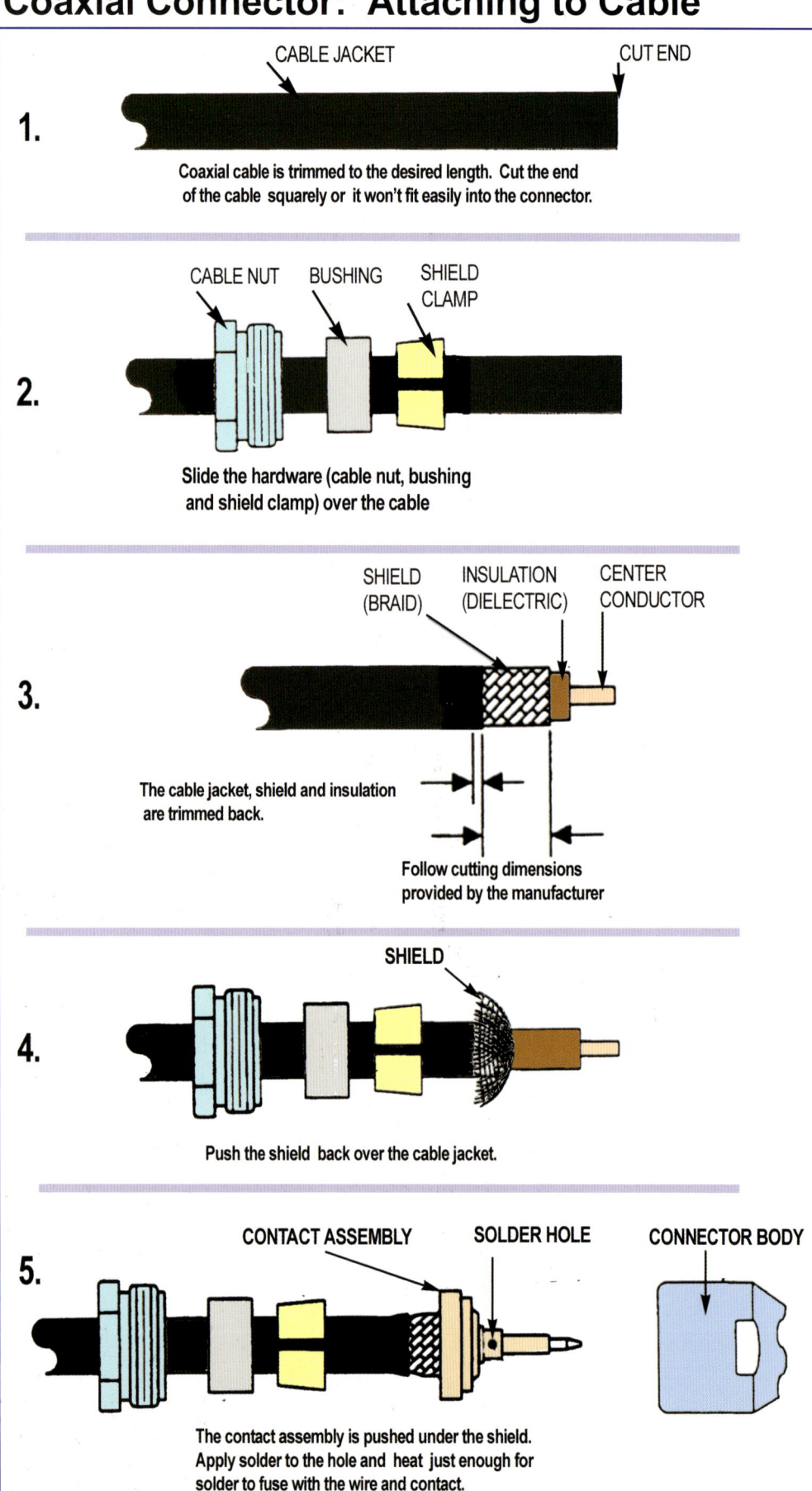

There are several methods for wiring a coaxial connector to a cable, including crimping and soldering. The method shown here uses a cable nut to squeeze together the connector parts. Regardless of method, follow the manufacturer's cutting dimensions carefully when trimming back the cable.

During the final step, 5, the cable nut is threaded into the connector body. This tightens the shield clamp over the shield for good electrical contact, and connector parts are tightly held together.

The solder hole is heated and solder added to connect the center conductor to the pin. Avoid overheating to avoid damage to the wire insulation.

1. CABLE JACKET — CUT END
Coaxial cable is trimmed to the desired length. Cut the end of the cable squarely or it won't fit easily into the connector.

2. CABLE NUT — BUSHING — SHIELD CLAMP
Slide the hardware (cable nut, bushing and shield clamp) over the cable

3. SHIELD (BRAID) — INSULATION (DIELECTRIC) — CENTER CONDUCTOR
The cable jacket, shield and insulation are trimmed back.
Follow cutting dimensions provided by the manufacturer

4. SHIELD
Push the shield back over the cable jacket.

5. CONTACT ASSEMBLY — SOLDER HOLE — CONNECTOR BODY
The contact assembly is pushed under the shield. Apply solder to the hole and heat just enough for solder to fuse with the wire and contact.

Crimping: Attaching Wires to Connector Contacts

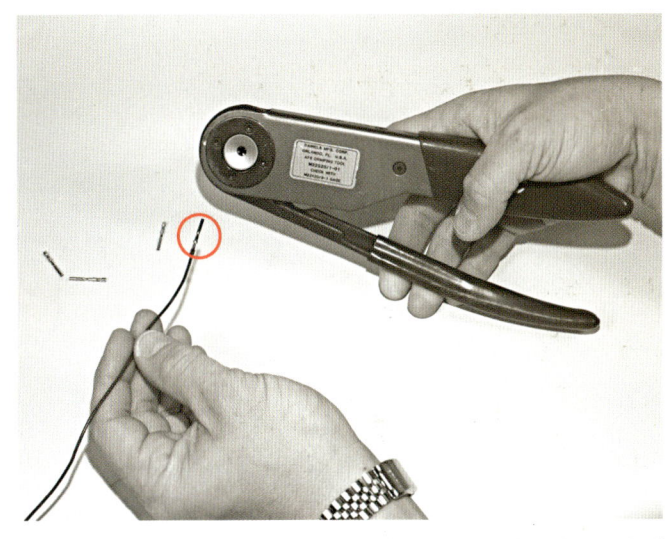

Connector contact (in red circle) is crimpled to the end of a wire with a crimping tool (type DMC AF8).

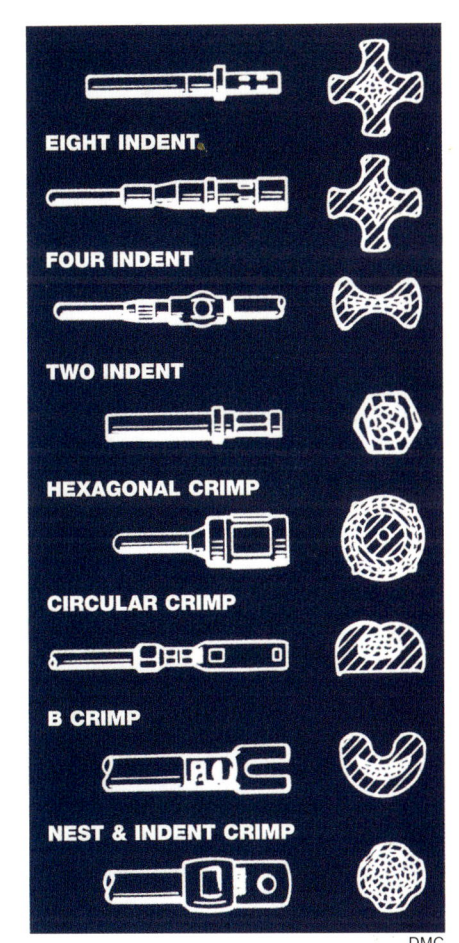

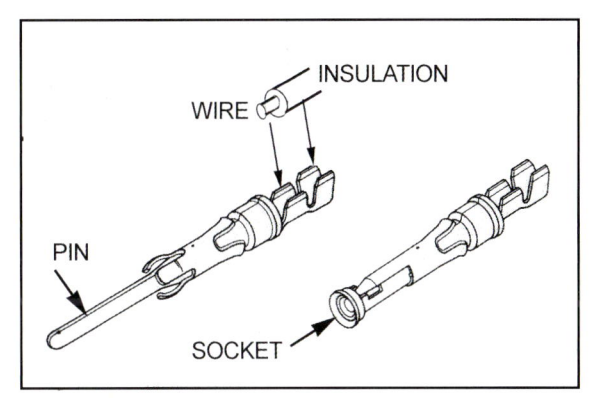

Pin and socket contacts that will be inserted into the connector. The wire is crimped into tabs on the contact. Another set of tabs clamps the insulation to relieve strain. These terminals are used in a "Molex" connector.

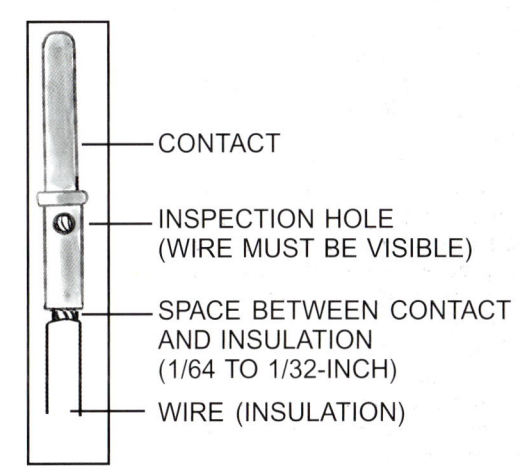

- EIGHT INDENT
- FOUR INDENT
- TWO INDENT
- HEXAGONAL CRIMP
- CIRCULAR CRIMP
- B CRIMP
- NEST & INDENT CRIMP

Crimp tools are fitted with dies for making different crimp patterns. The most common for avionics work is the "Eight Indent" shown at the top.

- CONTACT
- INSPECTION HOLE (WIRE MUST BE VISIBLE)
- SPACE BETWEEN CONTACT AND INSULATION (1/64 TO 1/32-INCH)
- WIRE (INSULATION)

In this type of contact, there is an inspection hole. Bare wire must be visible through the hole. There must also be a small space between the contact and the insulation.

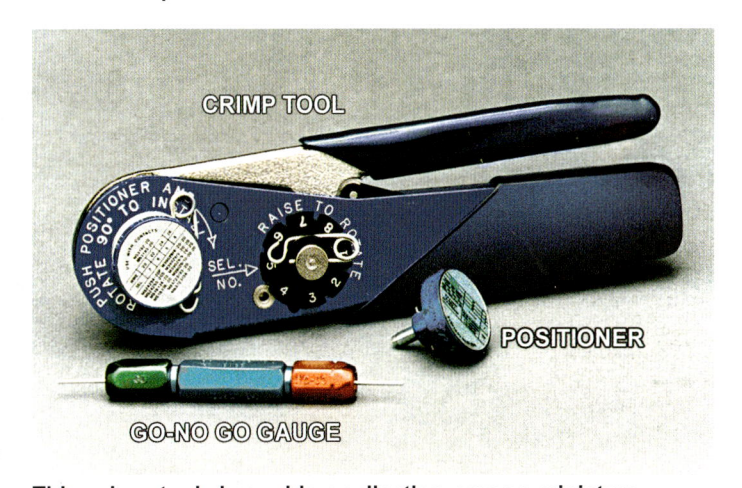

This crimp tool has wide application among miniature and sub-miniature connector types. It delivers a standard 8-impression crimp. When the handles are squeezed, a ratchet controls maximum pressure. A selector knob sets the correct wire depth. The "go-no go" gauge checks the tool's accuracy. The "positioner" holds the contact in the correct position. The model shown is the DMC AFM, also known as "Little Blue."

D subminiature. The "D" refers to the shape of the connector, which is wider on one side to prevent the plug from being inserted incorrectly. The connector is often used in circuits under about 5 amps, such as power, audio, digital signals and ground.

The D subminiature is made in several sizes, with 15- and 25-pin models common in avionics.

Avoid using D subminiature connectors sold in local stores, the ones intended for a PC. They may work well in the quiet environment of a home, but prove unreliable in aviation service. A good connector will also have a sturdy system for removing strain on the cable.

Molex Connector A plastic block that accepts crimp-type pins, the Molex connector is often found on the rear of mounting trays for avionics equipment. When the radio is slid into the tray, the pins mate with the tray-side connector.

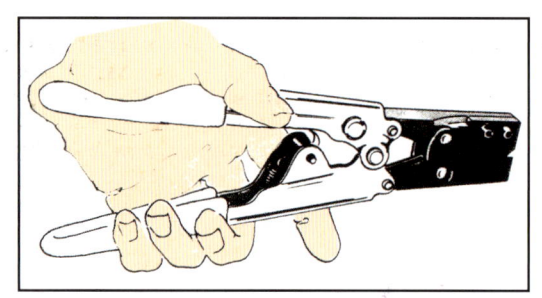

Crimping tool for Molex connectors.

Wire is laid into the Molex pin and the pin crimped in the tool shown above.

Amphenol 57. This series of connectors found in aircraft resembles the D subminiature type. It's used for radio mounting trays, remote-mounted avionics (outside the instrument panel) and for in-line cable-to-cable connections. The pins are available in 14- 24- 36- and 50-pin connector sizes.

To complete the connector, a metal hood is slid over the wire bundle and screwed to the back..

Amphenol 126. Another common number in avionics, this connector is hexagonal. It is often used for autopilots and other applications. Small size, easy assembly and reliability make it a good choice for aircraft. The connector comes in varying numbers of pins; 4, 5, 7, or 9 gold-flashed pins. (A thin layer of gold on a pin resists corrosion.)

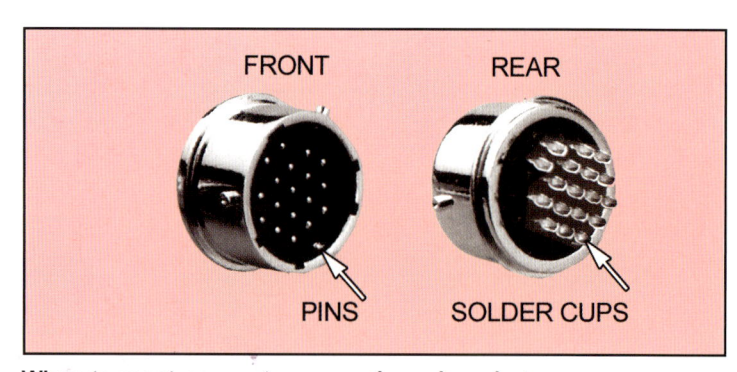

FRONT — PINS
REAR — SOLDER CUPS

Wires to most connectors are crimped on, but some require soldering. The back of this connector has "solder cups" for holding solder and wire. After the connection is made, shrink-tubing is slid over bare wires to prevent short-circuiting.

Releasing Connector Pins

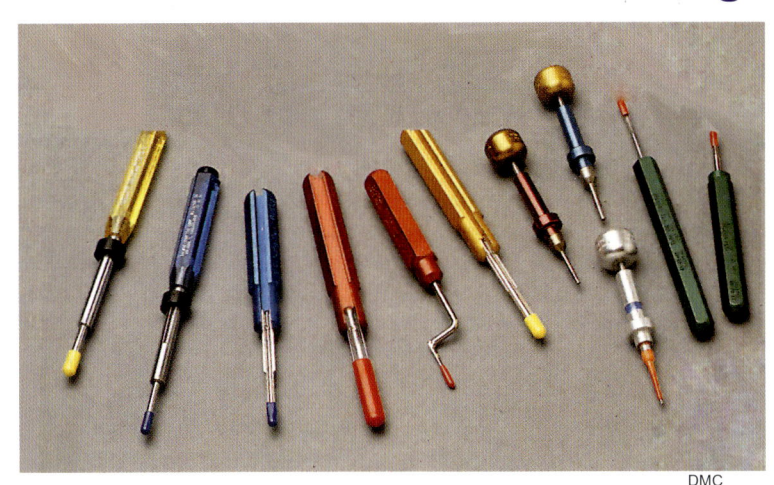

A variety of tools is available for installing and removing connector pins, either front or rear release. Although these tools may be offered in plastic, metal is preferred for durability.

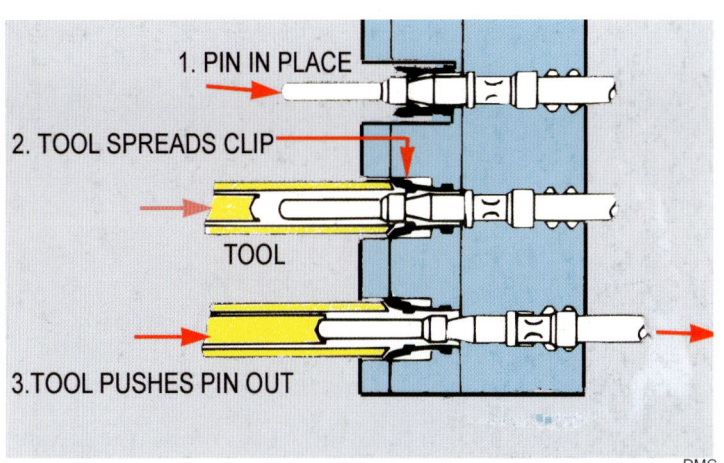

To release pins (1) from a connector, the removal tool is inserted over the pin (2). When the tool seats, it spreads a spring which retains the pin. Now when the tool is pressed further, the pin is pushed (3) out. Shown here is a "front release" connector. Other connectors may be "rear release," but the principle is similar

Heat Gun for Shrink Tubing

A heat gun (also called a "hot air" gun) is essential for heat-shrink tubing and solder-sleeving.

Effectiveness is greatly increased by adding nozzles, shown at right. They curve and concentrate hot air on the work, which creates equal and faster heating. Nozzles come in different sizes for work of various diameter. The nozzle at the top, however, should handle most avionics jobs. It has a 1-1/2-inch diameter.

A heat gun like this Steinel 1802 has selectable temperature from 120 F to 1100 F (50 C to 650 C). It consumes 1500 watts at 120 VAC.

207

Safety Wiring Connectors

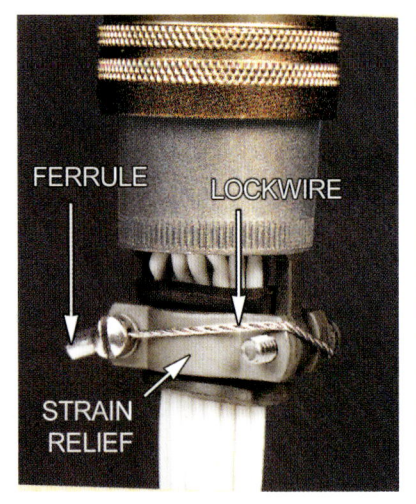

FERRULE **LOCKWIRE**
STRAIN RELIEF

Strain relief for cable is provided by clamp at back of connector. Screws are secured by safety wire applied by tool.
(Illustrations: DMC)

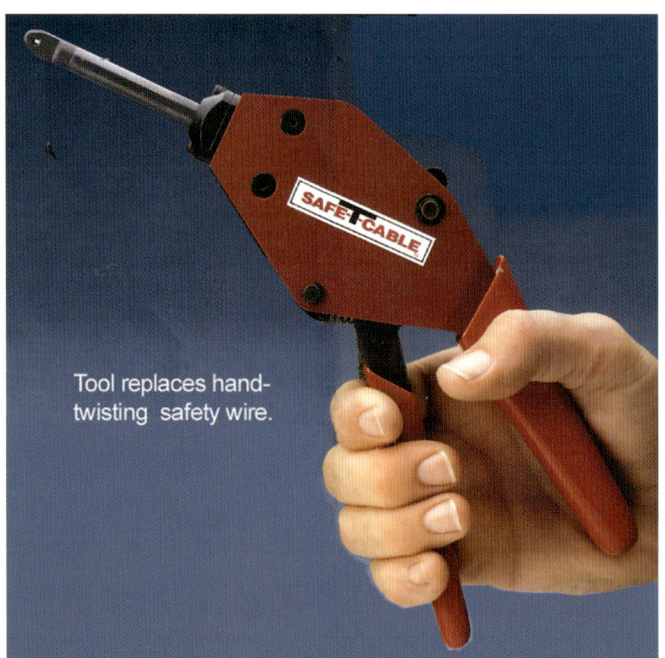

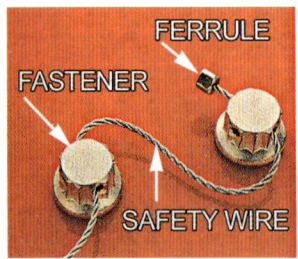

Tool replaces hand-twisting safety wire.

FERRULE
FASTENER
SAFETY WIRE

1. Pre-twisted safety wire is inserted into fasteners. A ferrule has been crimped to one end (upper right).

2. The other end of the wire is inserted into the tool nose, which stores ferrules.

3. As the tool is squeezed it crimps a ferrule on the wire and applies correct tension. The wire is trimmed flush with ferrule.

4. Completed job. It takes a fraction of the time required by manual safety wiring and eliminates sharp ends.

Besides safety wiring connectors, as shown above, it is recommended in other areas. If the covers of junction boxes, panels, shields or switch housings cannot be accessed in flight, and they are not fastened by self-locking hardware, they should be safety wired.

Review Questions
Chapter 24 Connectors

24.1 What is a major cause of failure when newly-installed equipment is first powered up?

24.2 What is one of the most common RF (radio frequency) connector types?

24.3 Soldering wires to connectors has mostly been replaced by _____.

24.4 Pins are released from a connector with a _____ tool.

24.5 Shrink-tubing is installed with a _____ _____.

24.6 What is the purpose of a safety wire (also known as a "lock wire")?

Chapter 25

Wiring the Airplane

During the 1990's, following major accidents, investigators raised questions about wiring in aging aircraft. The result was a government-industry task force that examined 120 jet transports flying in regular service. The results were surprising. Thousands of cracks were found in wiring insulation in just *one* airplane. Metal shavings were seen in wire bundles, wires were tied to fuel lines or attached to hot air ducts. They found contamination by fluids and chemicals and improper use of clamps.

These faults are time bombs---ticking away until they might explode into a disaster. In one B-747 accident, investigators determined that sparks from a high-voltage cable arc'ed over to a low-voltage wire and travelled to a fuel tank. So widespread were such problems that any aircraft over 10 years old was said to have an aging wire problem.

In the SWAMP

Routing wires is so important that certain places on an aircraft are known as SWAMP areas, meaning "Severe Weather and Moisture Prone". These include engine compartments, leading and trailing edges of the wing, landing gear and wheel wells.

Researchers frequently observed poor installation techniques. Cables were bent too sharply, wire bundles not properly supported, high and low power cables run in the same bundle and improperly-installed connectors. They discovered that certain wire types were prone to cracking and carbonizing, which spreads the danger to other cables.

In some cases, when mechanics performed maintenance on an airplane, they unknowingly damaged wire by stepping on it. They also grabbed wire bundles to use as hand-holds---which cracks the insulation.

The investigation learned a lot about aircraft wiring, improved insulation and a greater awareness of installation techniques. Many of their recommendations appear throughout this chapter.

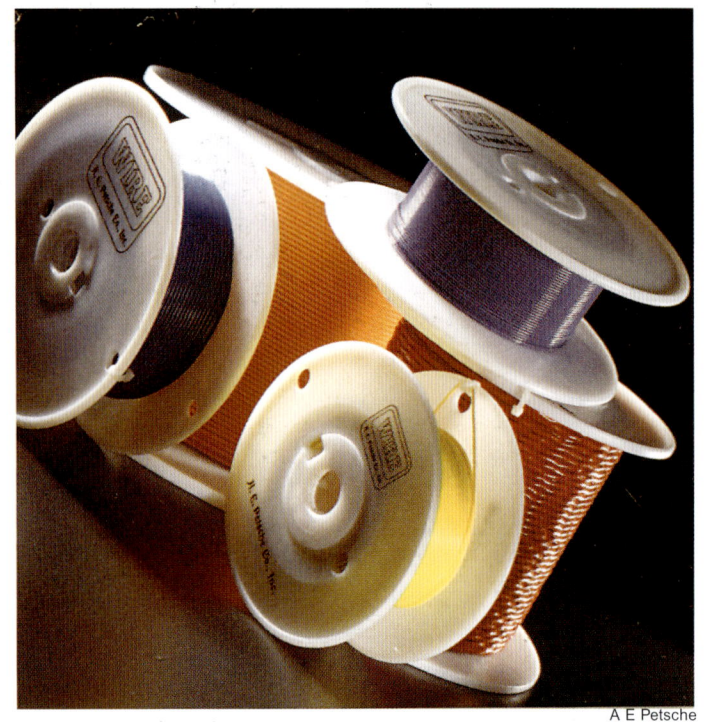

A E Petsche

Wiring made for aircraft is tough and heat-resistant. To avoid long-term problems, avoid anything less than aviation-grade.

High Risk Areas for Wiring

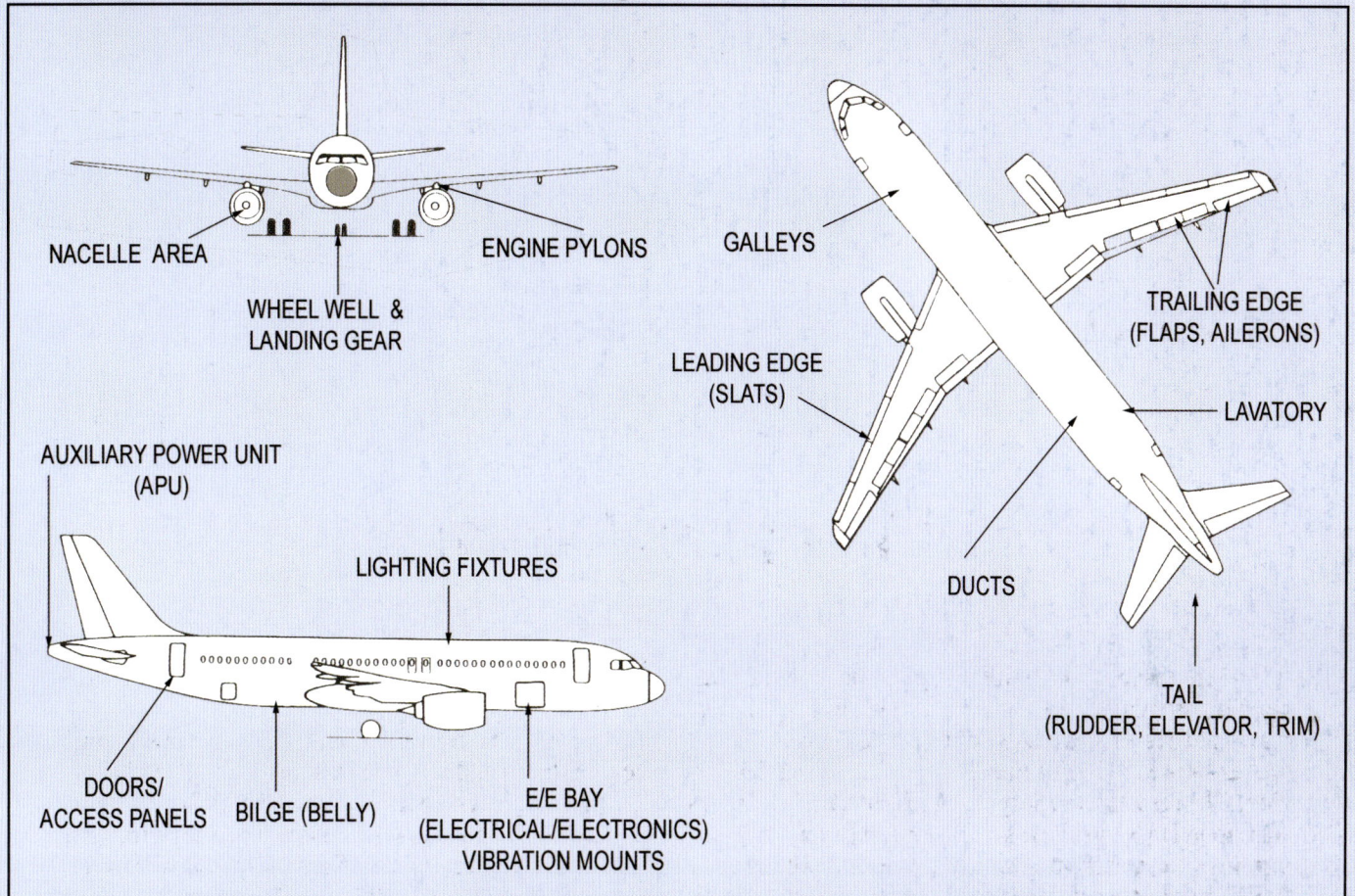

NACELLE AREA
ENGINE PYLONS
WHEEL WELL & LANDING GEAR
GALLEYS
TRAILING EDGE (FLAPS, AILERONS)
LEADING EDGE (SLATS)
LAVATORY
DUCTS
TAIL (RUDDER, ELEVATOR, TRIM)
AUXILIARY POWER UNIT (APU)
LIGHTING FIXTURES
DOORS/ ACCESS PANELS
BILGE (BELLY)
E/E BAY (ELECTRICAL/ELECTRONICS) VIBRATION MOUNTS

Aircraft wiring must operate in a hostile environment. Passengers on the ground in Phoenix, Arizona, in summer are comfortable in the cabin, but a few feet away, wiring may be heated to over 100 degrees F. Minutes after takeoff, temperatures drop below 0 degrees F. Vibration is continuous and humidity swings over a wide range, often causing moisture to condense in hidden places. Certain areas, pictured above, have proven particularly damaging to wiring which has not been carefully installed and inspected.

Wings: leading and trailing edges.
The problem is flaps, slats and ailerons. Because they extend during takeoff and landing, they expose the inside of the wing to the environment.

Engines
Heat, vibration and chemicals are hazards in areas which house the engine; such as nacelles and pylons. This also applies to the engine in the tail---the Auxiliary Power Unit (APU).

Landing Gear
Rocks, mud, water and ice are thrown against wheel well and landing gear, where numerous harnesses run.

Galleys and Lavatories
The drains below these areas must be kept clear and flowing. Otherwise, wiring is damaged by water, coffee, food, soft drinks and lavatory fluid.

Doors and Windows
Look for signs of water damage on wiring in these areas: below a cockpit side window that slides open, under doors used for passengers, cargo and service entry.

Ducts
If hot air escapes from a broken duct, it may not burn the wire but weakens the insulation until cracking causes problems.

Bilges
Liquids---water, fuel, oil, hydraulic fluids--- flow to the lowest point, which is the bilge, or low point in the belly.

SWAMP
Many of these areas are known in the aviation trade as SWAMP, "Severe Wind and Moisture Problems."

Failures in wiring may be sudden and catastrophic. When a radio is turned on for the first time, a wiring error may cause a short-circuit. Trouble appears in an instant and, hopefully, a circuit breaker prevents further damage. But most wiring problems don't happen that way. More often, a slowly building condition reaches a critical stage years later and causes a failure. Unfortunately, they create the most difficult symptom to deal with---the intermittent connection. A pilot squawks the problem to the maintenance department, but when the technician checks the airplane, he finds nothing wrong.

It's important to note that nearly all problems that appear in new wiring can be avoided without spending much extra installation time or material.

Poor wiring is often called a "rat's nest" but that's not the case in this example. The technician is carefully labelling every wire. Note that all wiring is formed into neat harnesses.

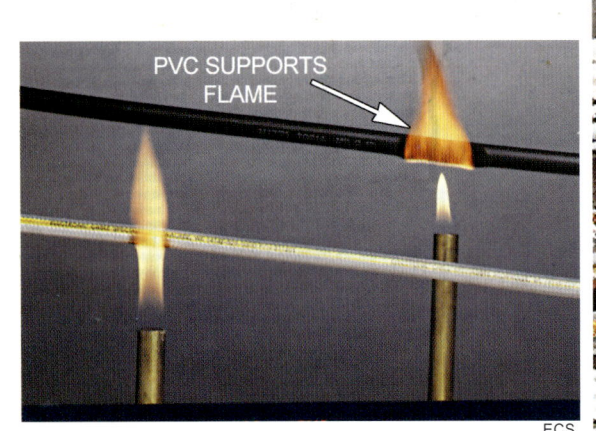

PVC wire was banned by the military, then in commercial aircraft. Besides supporting flame, PVC spreads toxic fumes. Aviation wire is now related to Teflon (left). In photo at right, wire bundles are carefully supported by clamps and cable ties.

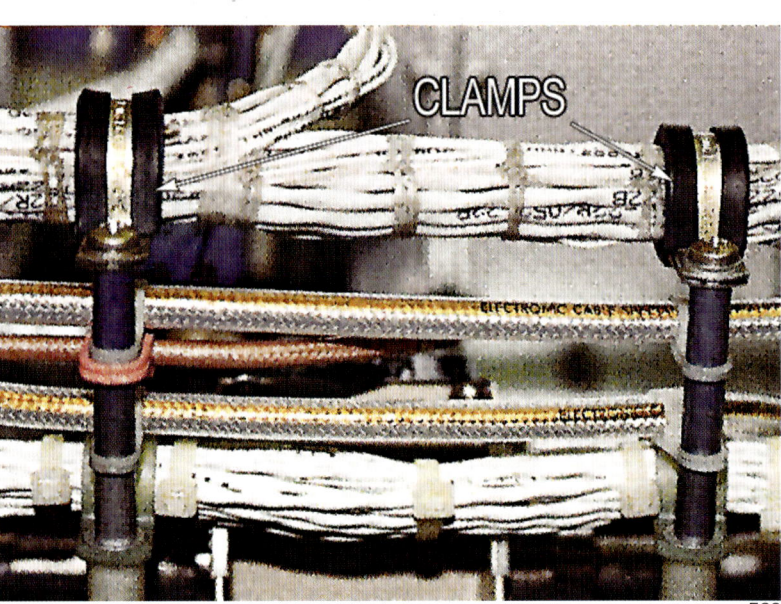

212

Selecting Wire

Although wire is a fraction of the cost of an avionics installation, it is critical to safety of flight. Skimping on wire quality makes little sense considering the amount of damage it can cause.

When asked to quote on an extensive avionics upgrade, some shops will not re-use existing wiring in the airplane. They've learned from experience that old wiring harbors many potential defects and makes it difficult for a shop to guarantee its work. When this is explained to the airplane owner, he often accepts the decision to re-wire. In fact, it may cost more, in the long run, to maintain old wiring, especially if older wire types were installed.

Wire quality has made much progress. Copper is the conductor of choice because it combines good conductivity at a reasonable price. In aviation, conductors are usually stranded because multiple wires absorb vibration better than solid wire. Copper is typically plated (or "tinned") with silver or tin for good solderability and resistance to corrosion.

Aircraft makers have used aluminum wire to save weight (and cost). That effort failed when they discovered that aluminum corrodes at the connecting terminals. Increasing electrical resistance here generates heat and a fire hazard. (For the same reason, aluminum wire was banned in house wiring many years ago.) Today, the technician may find aluminum still used between the starter and battery in some light aircraft, but many have been converted to copper.

A jumbo, like the Boeing 747, has nearly 150 miles of wiring which weighs a ton and a half. With rising fuel prices, wire is a target for reducing weight. Military aircraft are even more sensitive because weight reduces performance and payload.

Over 50 years, wire producers responded by reducing the weight of wire by 25 percent. It's been done with wires of higher temperature rating, which allows copper to be reduced in diameter. There are now better materials for insulation that can be applied in smaller thickness.

High-Grade Aircraft Wire

Wire for aviation often has "Tefzel" insulation (in the Teflon family), with copper conductors plated with tin. A typical rating is operation up to 150-degrees C. Fire-resistant wire may have nickel-plated copper to withstand higher temperature---up to 260 degrees C and multi-wall insulation. Made to Mil Specs, aviation wire is typically rated to 600 volts. These wires are the choice of major airframe builders for installation in new aircraft and used throughout General Aviation for upgrading avionics.

Below is an excerpt from a *Wiremasters* spec sheet describing characteristics of a 2-conductor shielded cable.

Wiremasters

Tin Plated Copper Tefzel Shielded Cable

Conductors: 2
Gauge: AWG 22
Shielding: Round Tin coated Copper 85% Min Coverage
Jacket: Tefzel
Conductor Color Code: White, White/Blue
Voltage Rating: 600 Volts
Temperature Rating: -55 to +150 Degrees C
Weight: 12.40 lbs/Mft
Conductor OD: 0.030" Nominal.
Outside Diameter Over Finished Cable: 0.124 inch
Insulation: ETFE (Ethylene Terafluoroethylene)
Mil Spec: MIL-DTL-27500-22TG2T14

Recommended Wire

Before looking at wire types, consider what *not* to use.

Avoid PVC. This is the common plastic-covered hookup wire sold in local radio and auto stores. FAA tests show that PVC insulation burns nearly twice as fast as the legal limit of 3 inches per second. It burns with large amounts of smoke and produces hydrochloric acid when exposed to moisture.

Mechanics have reported that simply moving wire bundles with PVC in old aircraft caused wires to break and short.

Avoid Poly-X. Both civil and military users have had problems in cracking and abrasion.

Do not use Kapton wire. It's caused problems in civil and military aircraft.

Tefzel: Aircraft Wire

At the time of this writing, Tefzel is a recommended wire for aircraft installation. It is extremely resistant to abrasion and does not support flame or fire. It won't generate large amounts of smoke if overheated. It resists the attack of moisture, chemicals and cleaning compounds. Tefzel is in the Teflon family and is also known as ETFE (Ethylene Terafluoroethylene). It's available from aviation distributors and wire manufacturers.

Wire Size

Most wire sizes are shown in the chart at the right, and run from AWG OO (over one-third inch thick) to AWG 38, which is like a strand of hair. For avionics work, sizes mainly fall within the range of AWG 14 to 22. For example, No. 22 gauge wire is often used in audio, mike keying, headphone and instrument lighting. Higher current devices such as landing and navigation lights and pitot heat (in light aircraft) may require No. 14 gauge wire. An alternator, which generates large currents (60 amps or more) may require 8 gauge, while a starter motor, which draws the most current, may call for a No. 2 conductor.

The most important rule is to follow the equipment manufacturer's guidance. The maintenance manual states the correct wire size and type (shielded, twisted, etc.) for each connection. That information is on the wiring diagram, but often in tiny letters that may be hard to read, as in this example:

ALL WIRES ARE 24 AWG MINIMUM UNLESS OTHERWISE NOTED

Note the word "minimum," which implies you can use a larger size. That may not harm electrical performance, but large wire presents other problems. First, a bigger conductor may not fit into the connector or terminal. It also takes up more room in a clamp. In large aircraft, it adds weight and size and most airframe builders are actively against this.

Wire Sizes
American Wire Gauge (AWG)

AWG 00   AWG 20

As the AWG number goes *up*, the wire becomes narrower and resistance (in Ohms) increases.

AWG Wire Size	Ohms per 1000 ft	Diameter, Inches
00	.078	.3648
0	.0983	.3249
1	.1239	.2893
2	.1563	.2576
3	.1970	.2294
4	.2485	.2043
5	.3133	.1819
6	.3951	.1620
7	.4982	.1443
8	.6281	.1285
9	.7925	.1144
10	.9987	.1019
11	1.261	.0907
12	1.588	.0808
13	2.001	.0720
14	2.524	.0641
15	3.181	.0571
16	4.018	.0508
17	5.054	.0453
18	6.386	.0403
19	8.046	.0359
20	10.13	.0320
21	12.77	.0285
22	16.20	.0253
23	20.30	.0226
24	25.67	.0201
25	32.37	.0179
26	41.02	.0159
27	51.44	.0142
28	65.31	.0126
29	81.21	.0113
30	103.7	.0100
31	130.9	.0089
32	162.0	.0080
33	205.7	.0071
34	261.3	.0063
35	330.7	.0056
36	414.8	.0050
37	512.1	.0045
38	648.2	.0040

Wire diameter is measured without insulation.

Wire and Cable Types

Single conductor has center wire and insulating jacket.

Twisted pair is less susceptible to picking up, or radiating, interference. Twisted pair cables are often shielded for further protection.

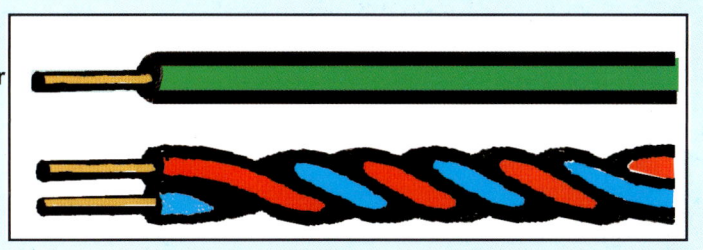

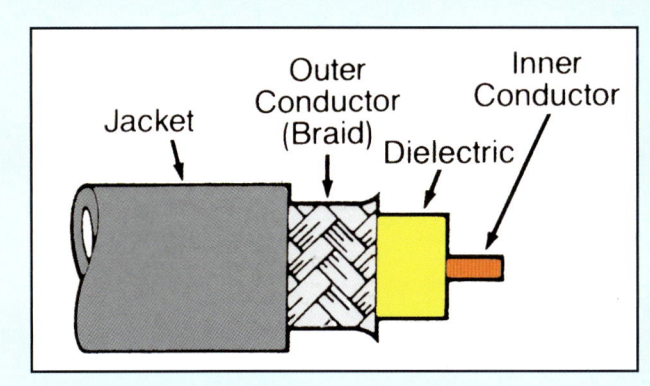

The most common coaxial cable has one center conductor surrounded by a dielectric (insulation) and an outer jacket. Until not long ago, the jacket was made of PVC, a material now banned for new installations but still found in airplanes. The old cable, "RG-58" is replaced by higher-temperature cables (such as RG-400) which are also more resistant to abrasion. Coaxial cables are mostly used to connect transmitters and receivers to antennas. The most common rating in aircraft is "50 ohms impedance."

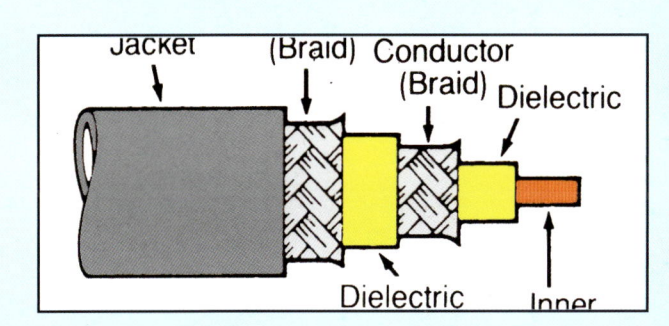

"Twinax" cable has a pair of insulated conductors inside a common shield. The inner conductors may or may not be twisted, depending on the application. Twinax is used where the cable must have excellent immunity to electrical noise---high-speed data transmission, for example, a type of signal that's growing rapidly in aircraft.

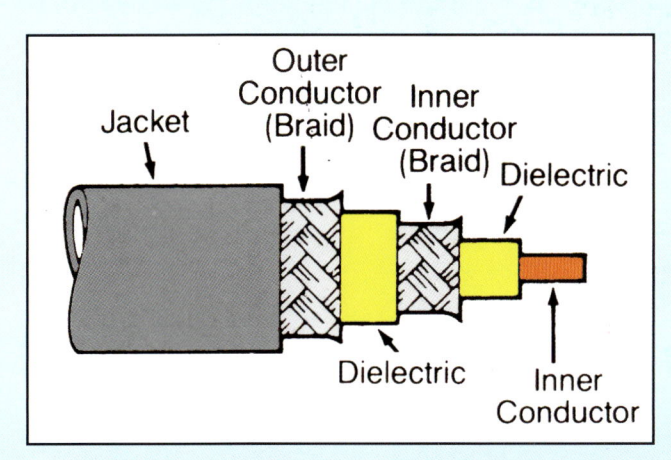

Triaxial cable, with two outer conductors separated by insulation. The outer conductor (braid) serves as a signal ground, while the other is an earth ground. This arrangement provides very high immunity to electrical noise.

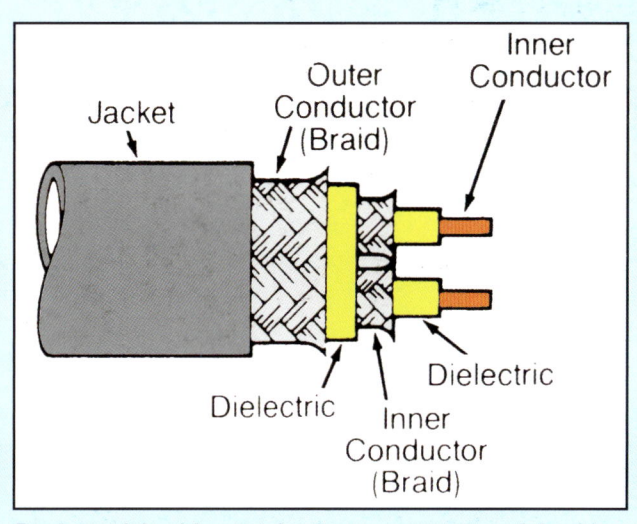

Dual coaxial cable contains two separate coaxial cables covered by a common outer jacket.

Coax Illustrations: Anixter

In large aircraft, wire sizes are part of the airplane's Type Certificate and must be observed for legal reasons. When a manufacturer builds a new piece of avionics for an old aircraft, he may obtain an STC (supplemental type certificate). In that document, wire sizes are described in detail.

In the absence of manufacturer information, the FAA provides guidance on choosing an exact wire size. It is described in detail in Advisory Circular 43.13-1B-2B. If you need to look further into the design of an aircraft wiring system, this is the primary reference.

Consider these factors for installing cables, harnesses, bundles and other wiring methods:

Stranded vs Solid

Solid wire is usually to be avoided in aircraft. As mentioned earlier, stranded wire is flexible and less affected by vibration. If stranded wire is called for, don't attach it directly under a screwhead, or the strands might break. First connect a ring terminal to the end of the stranded wire.

Single and Bundled wires.

If wires are strapped together in a harness (or bundle) they are unable to dissipate heat as readily as in free air. This affects the amount of current allowed to flow in the wire; a bundled wire is rated to carry less current. It's most important when wires carry high currents of several amperes or more.

Length

The length of a wiring run affects current-carrying capacity. If the run is long, wire size might have to be larger (smaller AWG number) to prevent excessive heating and voltage drops. (See table.)

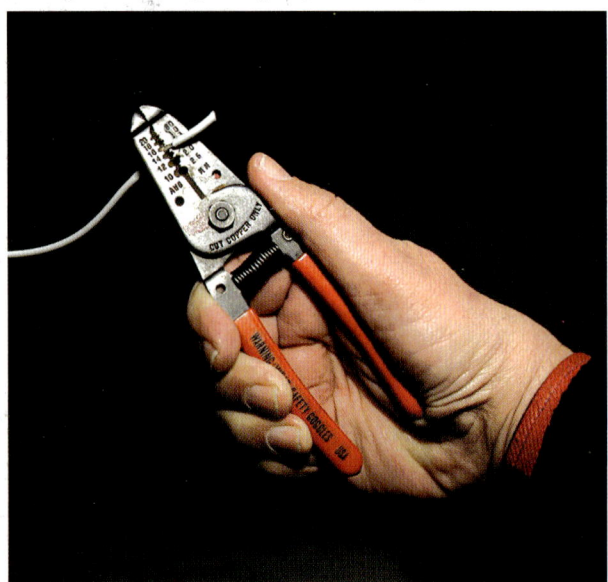

A wire stripper, like this manual type, should have notches to match different wire sizes. A high-quality stripper remains sharp and has a return spring. At the right is a semi-automatic stripper. It grips the wire as blades cut and remove the insulation.

Wire Stripping

Removing insulation from wire takes skill---as shown by the fact that FAA permits a wire to be installed with damaged strands. (The illustration gives the details.) That damage usually happens during wire stripping, when the wire is nicked (or scratched).

Strippers. A cheap wire stripper causes trouble and wastes time. It may have adjustable jaws but no method for setting to the wire size. Low-cost strippers do not retain sharp cutting edges, which increases the risk of damage.

A well-designed stripper has V-notches for different wire sizes. This prevents cutting past the insulation and into the wire.

The semi-automatic wire stripper is effective in both holding the wire, then stripping it at a squeeze of the handle. Be sure the wire is inserted into the correct "V" notch.

Some experienced technicians don't like any kind of automatic stripper and prefer the simplest type. Over the years they've developed a sensitive feel by hand and know just how deeply to cut into the insulation, before pulling it off the wire. But today's wire has tougher insulation and is more difficult to cut. A well-made precision stripper takes away the guesswork.

Coaxial cable is delicate because just below the insulating jacket is a braid of fine wire (the shield) that is easily damaged. The shield must be in tact because an incomplete shield changes electrical properties of the cable.

Cutting coaxial cable requires a special technique. You can buy an automatic stripper or use a razor blade to carefully score and remove the jacket.

The two hazards in any wire stripping are strands that are cut completely cut through or nicked (cut part way through). Nicked wires usually break after bending several times or are subject to vibration. When a connection is made, loose strands of wire may touch and short out nearby circuits.

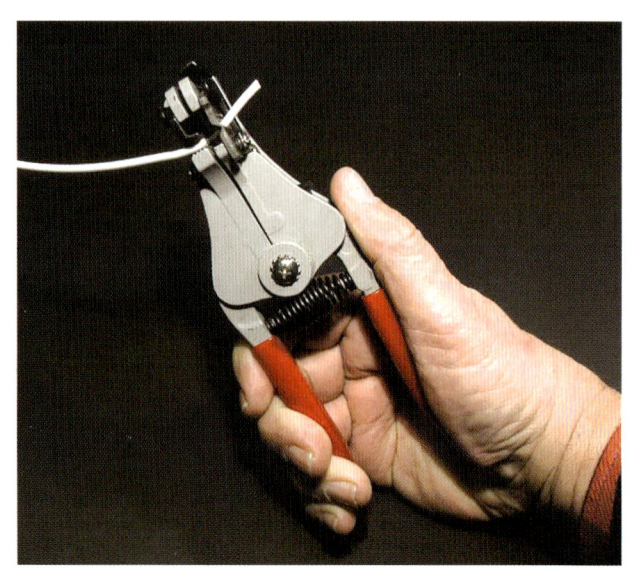

What To Do About Nicked or Broken Wires

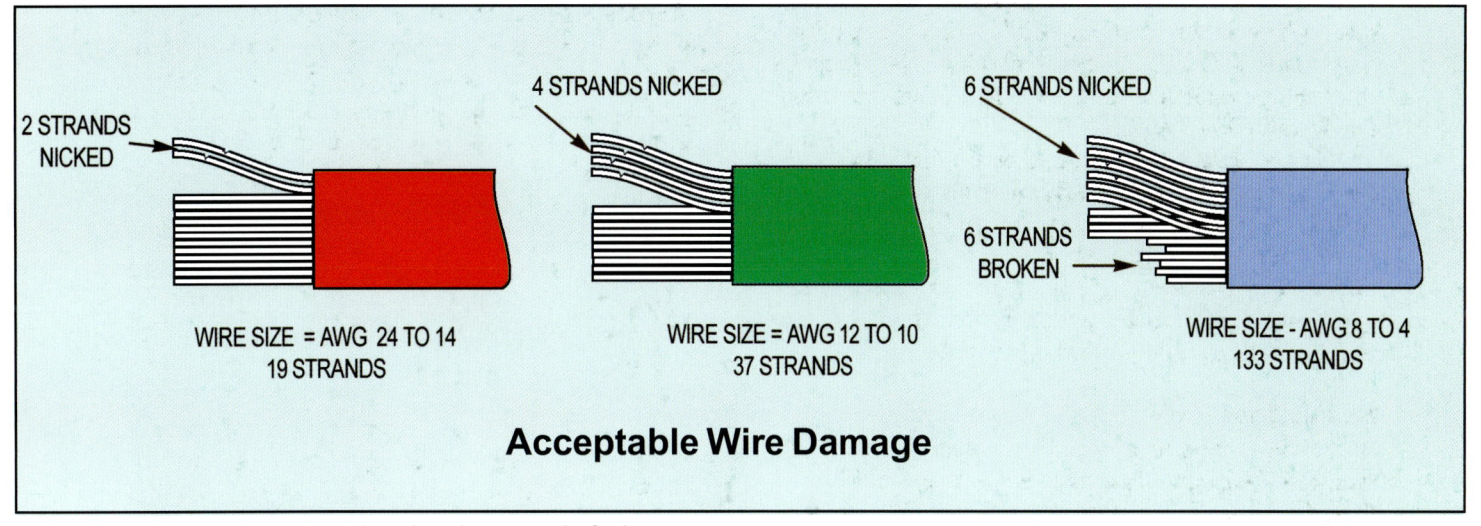

Acceptable Wire Damage

- 2 STRANDS NICKED — WIRE SIZE = AWG 24 TO 14, 19 STRANDS
- 4 STRANDS NICKED — WIRE SIZE = AWG 12 TO 10, 37 STRANDS
- 6 STRANDS NICKED, 6 STRANDS BROKEN — WIRE SIZE - AWG 8 TO 4, 133 STRANDS

If you accidently nick or break a strand of wire, you may be able to install it anyway by following FAA guidelines shown above. Determine the number of strands in the wire (a figure usually available from the supplier, or simply count them). As shown in the first example, at the left, there can be two nicked strands in wires from AWG 24 to 14, so long as all other strands are in tact. Broken wires are allowed in larger wires at the right, which can contain 133 strands.

Precut Cables

For some avionics systems, prewired cables may be supplied by the manufacturer or obtained from a company which specializes in fabricating cable harnesses. Some come with connectors, others require the technician to install the connectors.

Factory precut or prewired cables should never be shortened or lengthened unless the manufacturer indicates otherwise. Some cables are supplied in several lengths- --10 feet, 20 feet, etc. If the cable is too long, the excess is coiled up and secured. (Avoid coiling too tightly, as shown in the illustration.)

Cable sets are often made for installation on a fleet of identical aircraft, In this case, cables are already cut to proper length.

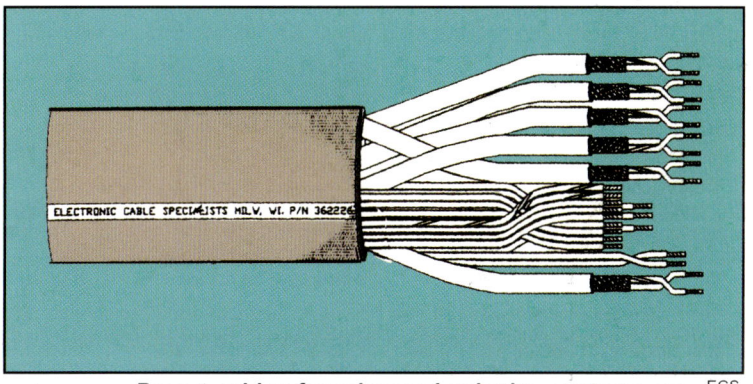

Precut cables for advanced avionics systems are often available from suppliers. ECS

Some cables are extremely sensitive to length. Coaxial cables that go to antennas for TCAS and radar altimeters, for example, also act as timing devices. If they're altered, the pilot will see targets in the wrong place or incorrect altitude above ground.

Splicing Wires

A large part of a technician's job is joining wires to connectors and other terminal devices. Once the connector is wired, there are requirements about splicing the cable to other wires.

Coaxial cable. The efficiency of this cable depends on precise spacing between its outer shield and inner conductor. It is difficult to splice without affecting those dimensions.

Power wires. Heavy copper cables from the battery, alternator or starter cannot tolerate even a small resistance from an imperfect splice. It lowers voltage of the whole electrical system or heats up and causes a fire hazard.

Databus (Multiplex) cables. Increasingly, avionics systems communicate with each other with digital signals sent through a twisted, shielded pair. A poor splice changes electrical properties and distorts the shape of the signals.

Location of Splices

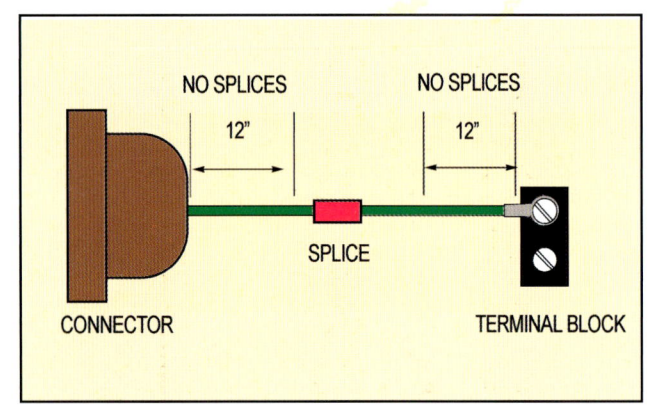

Avoid splicing a wire more than once in a segment, which runs between any two terminal points, such as a connector, terminal block or disconnect point. (There are certain exceptions.) Note that a splice should not be less than 12 inches from the terminal points at either end.

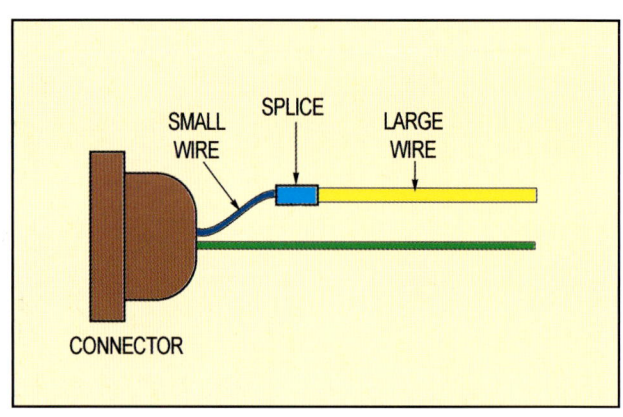

An exception to "one-splice-per-segment" is shown above. If a wire is too large for the connector, splice it to a smaller wire. The small wire from the connector, known as a "pigtail," can be crimped to a connector contact.

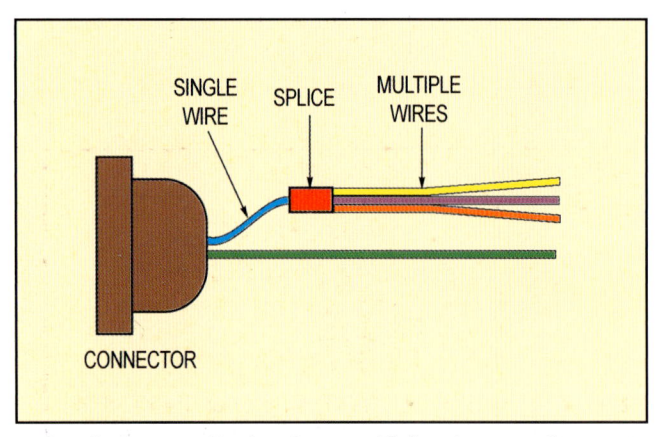

Another exception is when multiple wires need to go to one pin on a connector. They can be spliced to a single wire at the connector.

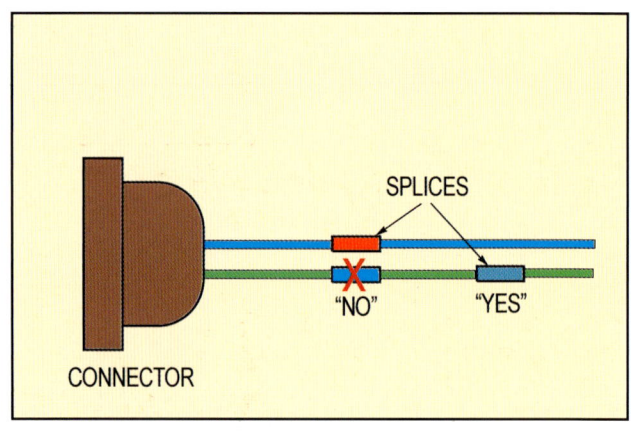

Each of the wires from this connector has a splice. Do not locate splices adjacent to each other. Stagger them to prevent overlapping, which causes bulges in the wire bundle that might not fit into tight spaces. Bulges also make future maintenance more difficult.

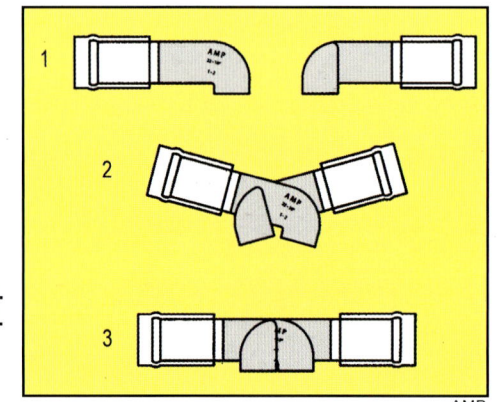

Knife splice is used for quick disconnect.
1. Two ends of the splice are shown apart.
2. The ends are angled together.
3. Push down and the splice is locked.

Ring Terminals

Ring terminals are color-coded according to the range of wire sizes they accept. For example, red takes any wire from AWG 22 to 16. Also choose the stud size the ring must fit over. For example; the first stud size on the list takes a #4 screw. All ring terminals are crimp-on and self-insulated, except the bare one at top right.

WIRE SIZE	STUD SIZE	COLOR
22-16	#4	Red
22-16	#6	Red
22-16	#8	Red
22-16	#10	Red
22-16	1/4	Red
22-16	5/16	Red
22-16	3/8	Red
16-14	#4	Blue
16-14	#5	Blue
16-14	#8	Blue
16-14	#10	Blue
16-14	1/4	Blue
16-14	5/16	Blue
16-14	3/8	Blue
12-10	#6	Yellow
12-10	#8	Yellow
12-10	#10	Yellow
12-10	1/4	Yellow
12-10	5/16	Yellow
12-10	3/8	Yellow
12-10	1/2	Yellow

Terminal Strip (or "Block")

STUD

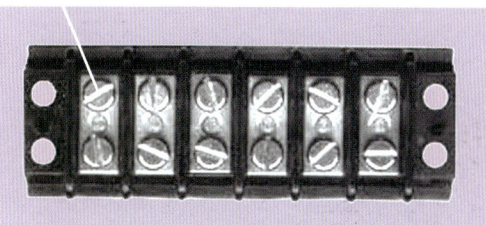

A terminal strip is a junction for aircraft wiring, providing easy access for wiring changes and troubleshooting. This is a 6-terminal strip with 12 studs (each vertical pair is connected together). Be sure to obtain a terminal strip with a raised barrier between studs to prevent short circuits. If the strip is open, use a protective cover made for the purpose. It's a good idea to obtain a strip with extra contacts for future additions, and if a stud is damaged (by stripped threads, for example).

Terminal strips develop corrosion and loose screws over time and need to be checked. Always select a strip that has the size and current rating to fit your terminals.

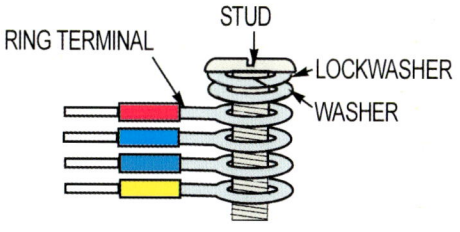

Do not put more than 4 ring terminals under the head of one stud. If more wires must connect, put *three* ring terminals under a stud, plus a *bus bar* (see below). The bar is a short heavy wire or jumper that joins two adjacent studs. This allows three more wires on the second stud, as shown.

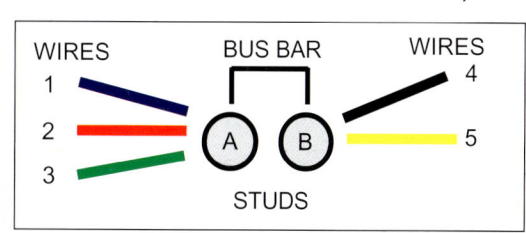

In this example, five wires must connect to one point. Three are connected "A". The bus bar, or jumper, connects "A" to "B". Up to three wires can connect to B without exceeding the limit of four connections per stud.

Marking Wires

No technician should attempt an installation without marking the beginning and end of cables, harnesses and wires. After each is connected at one end, it's simple to identify the other end and join it to its destination.

Besides marking the *ends* of a wire, it's important to label it all along its run. If this isn't done, when it comes time to troubleshoot the airplane, tracing wires becomes infinitely more difficult. Wires snake through the airframe in inaccessible places and are nearly impossible to trace with your eye. Thus, it is strongly recommended that all new wiring be clearly identified all along its route.

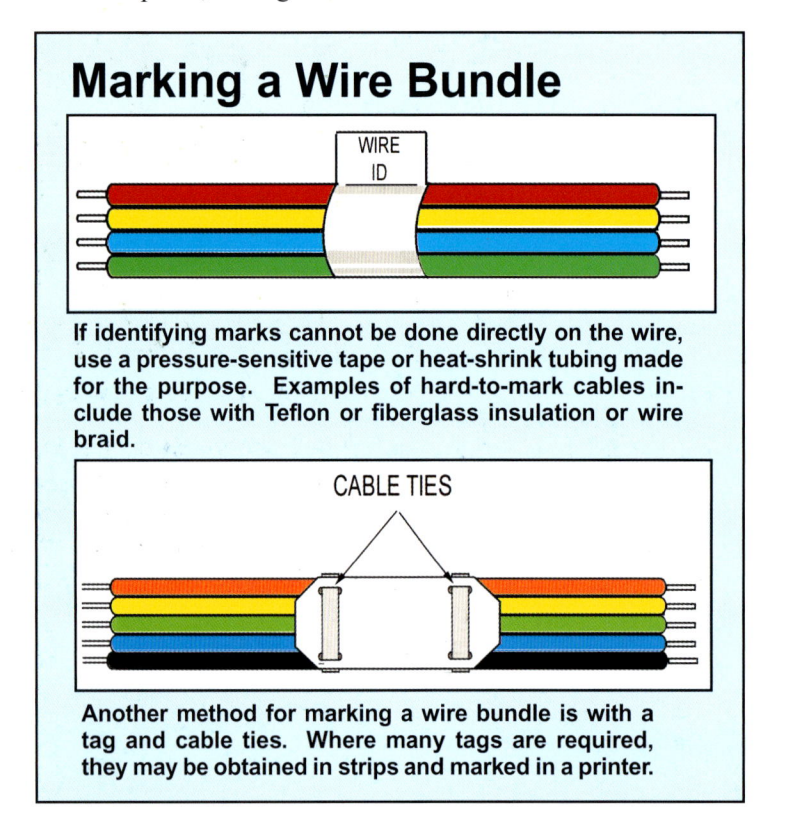

Marking a Wire Bundle

If identifying marks cannot be done directly on the wire, use a pressure-sensitive tape or heat-shrink tubing made for the purpose. Examples of hard-to-mark cables include those with Teflon or fiberglass insulation or wire braid.

Another method for marking a wire bundle is with a tag and cable ties. Where many tags are required, they may be obtained in strips and marked in a printer.

Wire Marker Intervals

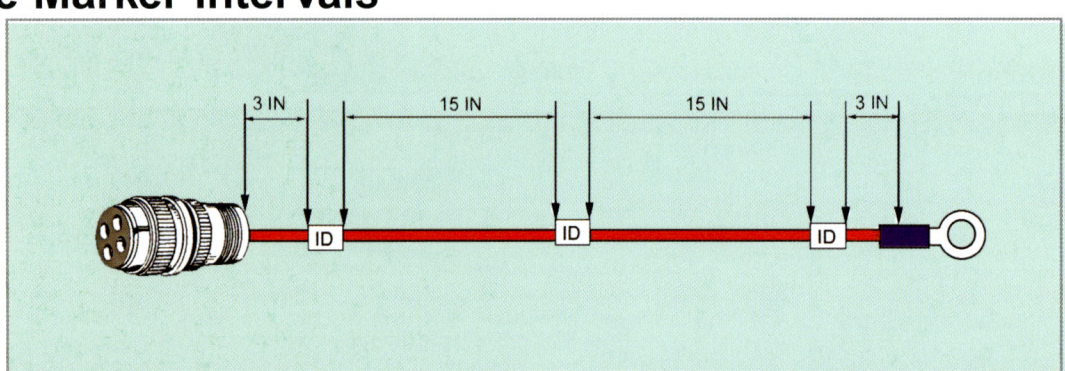

Individual wires should be marked. The recommended spacing is an identification ("ID") marker within three inches of where the wire originates (in this example, a connector, and within three inches of where it ends at the ring terminal at the right). Along the wiring run a marker should be installed at a maximum of every six feet. Wires less than three inches long require no ID. Wires between three and seven inches should be marked at about the center.

Marking Methods

In marking a wire, the information should identify the wire, the circuit it's part of and the AWG, or gauge, size.

Because some wires are sensitive to the surrounding area, don't use metal bands for the ID. Any marking method must not deform a coaxial or databus cable to prevent electrical losses.

The preferred method of marking is *directly* on the wire insulation or jacket. Many jackets, though, especially those made of Teflon, are difficult to mark without expensive laser equipment. Wire with a bare shield is also difficult to mark, as are multiconductor and thermocouple (heat sensor) wires. Where the surface is difficult, you can use the *indirect* method, where you apply a label to a sleeve, then slip it over the wire. The sleeve might be a tube that you heat-shrink onto the wire.

Stamped Marking. In this method, wire insulation is stamped (indented) with a tool and hot ink applied in the depressions. This works well but the technician must follow the manufacturer's instructions and adjust the machine carefully to avoid harming the wire. To avoid damage, the indentation in the wire must be no deeper than 10 percent of the insulation thickness. Very small gauge wire, therefore, cannot be hot-stamped.

Stamped Sleeving. In this indirect method, a sleeve is imprinted with a laser, inkjet or other printer and slid onto the wire. The sleeve is held by various fasteners.

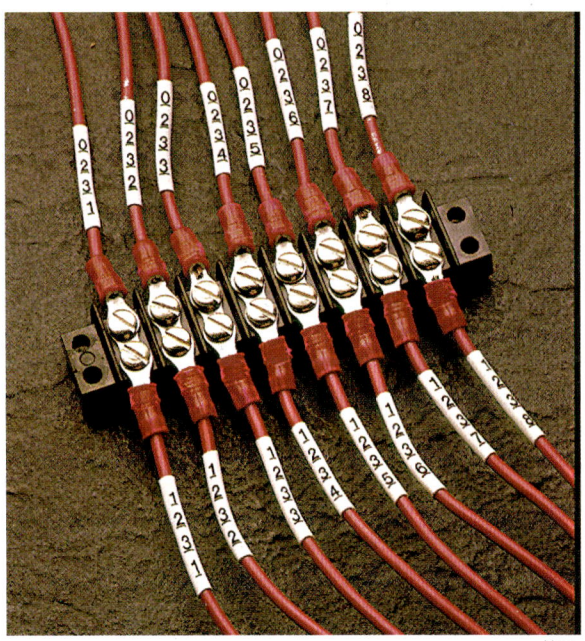

Labels marked on white shrink tubing

Wire Bundles. The marking systems just mentioned are for single wires. When they are grouped into a bundle, the ID may not be visible. Sleeving that can be marked and fastened around the whole bundle are readily available.

Harnessing the Wire Bundle

From each radio or instrument, wire branches run to main trunks, then fan out to their separate destinations. A neat, squared-off harness is easier to install and troubleshoot than running each wire directly from source to destination. Direct wiring which criss-crosses in every direction creates the well-known "rat's nest," and is a sign of poor workmanship. A neatly-bundled, squared-off harness also takes up less room behind the panel and is much easier to service later on. (Check for any restrictions in the manual on changing cable lengths, especially for coaxial type).

When an airplane is wired at the factory, creating the harness is easy because it's done on a wiring jig. This is a large board with pegs that guide the wires along neat paths. The engineering department figures out the pathways for one airplane and the pattern is used for all production aircraft.

But many installations are not done to fleets, but custom-built---each one is different. One approach to designing a harness is to make a rough drawing of the panel to determine where each radio or instrument is located, then pencil in the most efficient, obstruction-free route for the harnesses.

EMC/EMI. All wires may not run in the same bundle because they could interfere with each other. It's the problem of EMC, or electromagnetic compatibility. (Sometimes it's known as EMI, for electromagnetic interference.) Wires carrying signals of widely varying power levels transfer energy among themselves. It was once simply known as "crosstalk," but as the number of wires aboard aircraft multiplied, EMC became a major subject of new regulations.

The EMC problem grew worse as avionics became more digital. These signals are very low in level and susceptible to interference.

Many cables (and connectors) are now designed to prevent EMI. Some are protected by braided shields, others are twisted and some have both forms of protection. But shielding may not completely contain the signal, and interference can occur when a transmitter cable (which usually carries high power) is bundled with a low-level cable carrying receiver signals.

Another source of interference is from cables carrying power from an alternating or pulse-type source. In large aircraft, this includes the inverter (part of the power generating system). Strobe lights are frequent interference generators because they operate with short, repeating bursts of power. Pulse-type current is troublesome because it also produces harmonics, signals of much higher frequency that interfere with receivers and lightning detection systems

DME's and transponders also send pulses of high power through antenna cables and raise the possibility of interference.

Avoiding Electromagnetic Interference

To reduce chances of interference, keep power and transmitter cables outside the bundles that carry low-level radio, audio, digital and control signals. Several inches of spacing may be sufficient. If those cables run at right angles to the wiring harness, much less energy is transferred.

Dealing with noise is most effective when done at the source. Wires causing interference can be treated by shielding, bonding, grounding and filtering, as described in a later chapter.

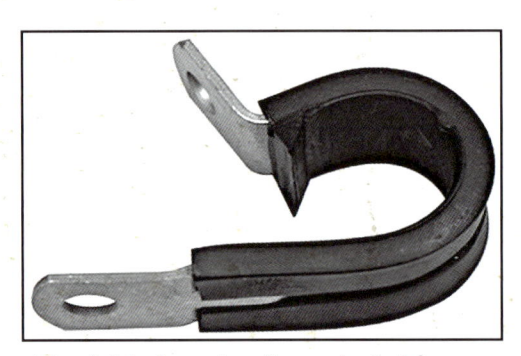

The Adel clamp is often selected for supporting wire harnesses in aircraft. It has a cushioned liner to reduce chafing on wire insulation. The clamp is made in several sizes to accommodate wire bundles of varying thickness.

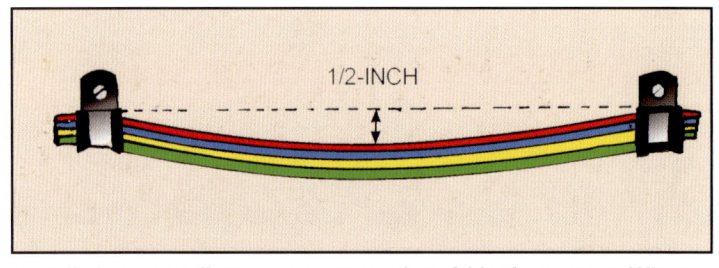

Install clamps at distances no more than 24 inches apart. When you grab the wire bundle and give it a *slight* pull, it should not move axially (to the left or right in this illustration). The bundle may droop up to 1/2-inch under normal conditions. That slack may be exceeded if you are sure the bundle cannot touch a nearby surface and suffer damage from abrasion (rubbing).

Tie Wraps (Cable Ties)

Call them "cable ties," "tie wraps" or "plastic ties"--they're all the same---but these little helpers reduce installation time and make the finished job look neat and professional. Thread the tie wrap, pull, and wires are instantly bundled. Tie wraps can even be attached after the harness is installed.

Installing Tie Wraps. There are precautions. Many a technician has put his arm behind an instrument panel, only to have it scratched or cut by the sharp end of a tie wrap. It happens when the end of tie wrap wasn't properly trimmed (see illustration).

Another precaution: when installing a tie wrap avoid the temptation to pull it very tight. (It's easy to do with little effort.) This squeezes the wires, changes their diameter or cuts into the insulation, making it more susceptible to vibration. Simply pull the tie wrap until it is snugly around the wires---and retains the wire bundle in place---without crushing.

How many tie wraps are required? Use enough to hold the bundle together, as well as support the harness where it changes direction. Install one where small bundles break out from larger ones. In general, if the bundle is not supported for more than 12 inches (by a clamp, for example) install a tie wrap.

Don't overload the harness with tie wraps before laying it into the airplane. If ties are too close, you may find it difficult to curve the harness around tight corners.

Cable lacing. Before the introduction of tie wraps, wires were laced into bundles by a special cord with a wax coating. Lacing cord is still available, but it has been mostly replaced by tie wraps. Besides requiring more time and labor to install, cable lacing cannot be added after the harness is in place behind the instrument panel. Another problem with lacing happens during maintenance and upgrading. Making changes to a harness is quick and easy with tie wraps, in or out of the airplane. Nevertheless, some technicians cling to lacing cord as a sign of craftsmanship Lacing looks good, but takes considerable effort and time.

Clamping the Harness

Cable ties keep wires together in a bundle. The bundle, however, must also be supported along its run to prevent damage or interference to moving parts of the airplane.

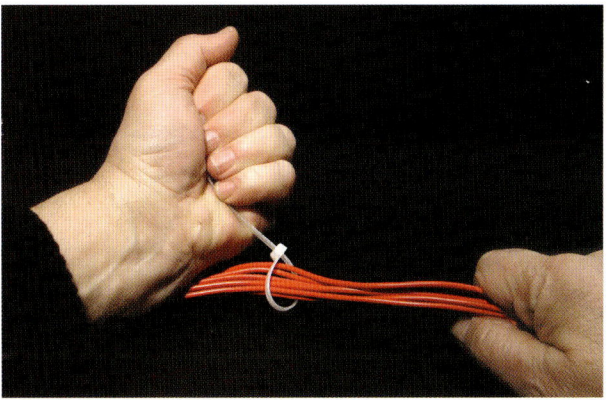

When installing a tie wrap, don't overtighten it. Pull the tab until the tie wrap is snug around the harness.

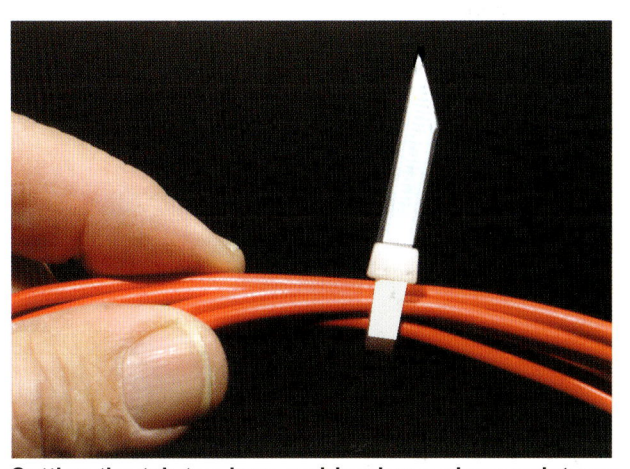

Cutting the tab too long and leaving a sharp point can injure the next person reaching for the harness under the instrument panel.

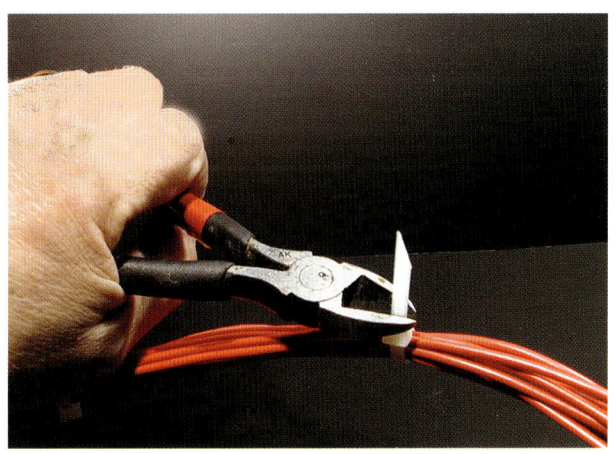

Cut the tab flush with the locking part of the tie wrap.

Where Wiring Problems Begin

Chafing and Abrasion

The steady vibration of an aircraft in flight is transferred to wire bundles, rubbing away insulation. This is avoided by supporting wire bundles away from sharp objects and other surfaces. It's done by cable clamps installed along the wiring run, especially where the wire runs through holes in the airplane structure.

1. If wiring harness clears the edges of the bulkhead hole by at least 3/8-inch, only the cable clamp is required at the hole.

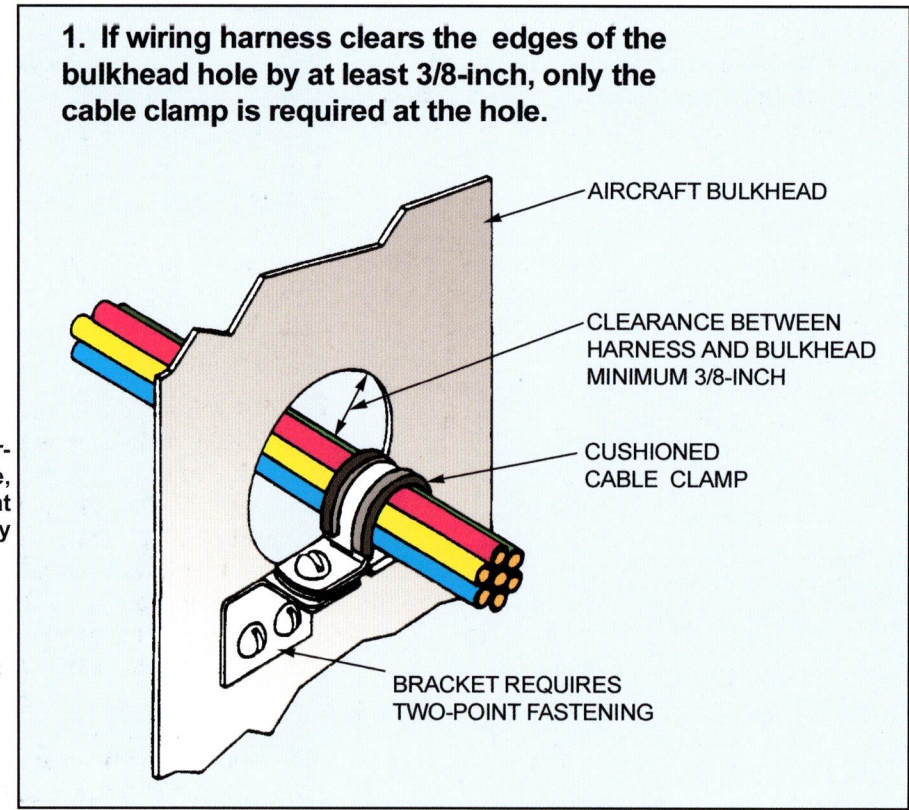

- AIRCRAFT BULKHEAD
- CLEARANCE BETWEEN HARNESS AND BULKHEAD MINIMUM 3/8-INCH
- CUSHIONED CABLE CLAMP
- BRACKET REQUIRES TWO-POINT FASTENING

When running a harness through a hole, wires must remain at least 3/8-inch away from the edge.

2. If wiring harness does not clear edges of bulkhead hole by 3/8-inch, a grommet is required for further protection.

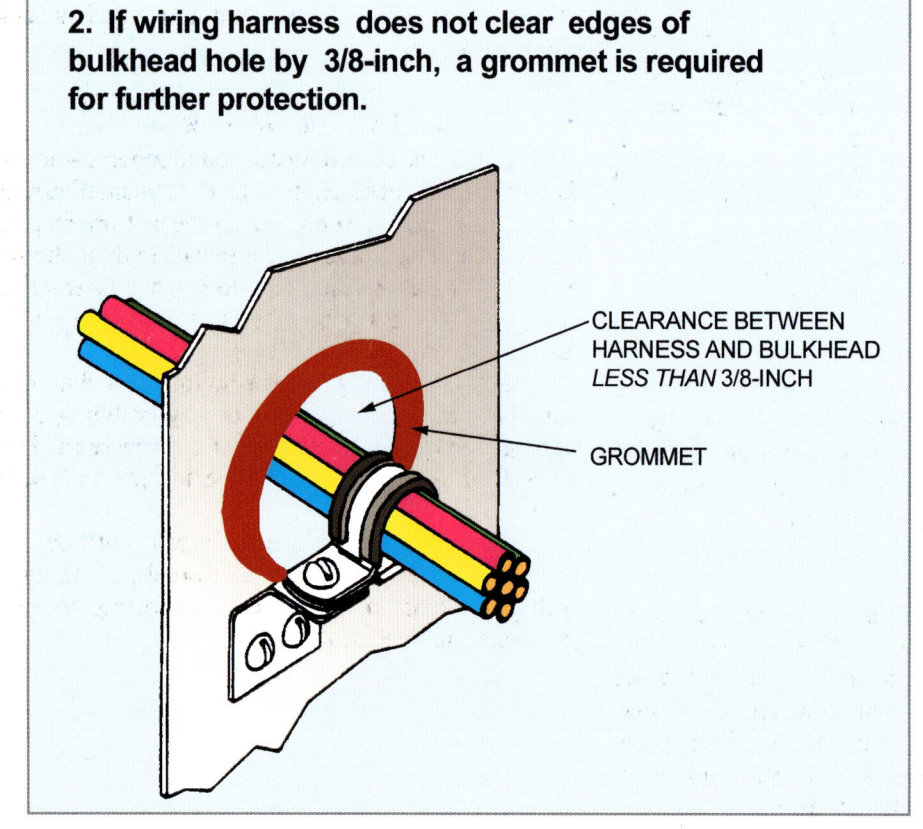

- CLEARANCE BETWEEN HARNESS AND BULKHEAD *LESS THAN* 3/8-INCH
- GROMMET

If there is less than 3/8-inch clearance, add a grommet to the hole. There are many openings in an airframe for running wire harnesses. They're known as "lightening holes" because they lighten the airframe. They are not "lightning" holes.

Moisture

Aircraft move through great variations in temperature and humidity, often in the same day. It encourages moisture to form, which then flows to the lowest point. The bottom of the fuselage (the belly of the airplane) is a moisture-prone area. Another troublespot is near air conditioning ducts.

Areas of moisture and high humidity corrode connector pins, terminals, sockets and hardware attached to wiring. Avoid it by supporting harnesses above and away from these areas.

Conduit

In large aircraft, wiring is protected by running inside conduit, but moisture forms inside the tube. After conduit is laid out and fastened, find the lowest point in the system. Make a 1/8-inch diameter hole at the lowest point for liquid to flow out. When wires are installed inside conduit, do not fasten them together with cable ties.

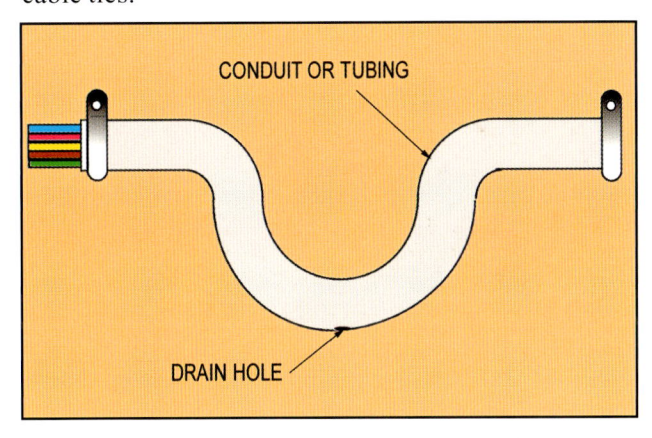

Conduit comes in metal and non-metal versions, rigid and flexible. When selecting a diameter make it about 25% larger than the wire bundle that goes inside. After cutting to length, remove any burrs which might cut the wire. When flexible conduit is cut with a hacksaw, it can have a ragged end. This is avoided by wrapping the end of the conduit with tape before sawing. Conduit also needs to be supported with clamps.

Broken Ducts

A torn or broken duct can direct hot air onto a wire bundle and, over a sufficient period, cause cracks. Even small breaks in insulation might enable a spark to jump across two wires. Look for such breaks in the duct during installation and maintenance.

High Temperature

The way to avoid heat damage is to space the harness away from such high heat zones as heating ducts and engine exhaust. When that's not possible, high-temperature sleeving might reduce the problem. If it's an old airplane, it may contain coaxial cable with a polyethelene jacket, which melts at elevated temperature. Other heat-damaging areas are galleys and lighting fixtures.

Corrosive Chemicals

An airplane is nearly a flying chemical factory and wiring is always under attack. Here are the leading offenders:

Battery acid	Jet fuel
De-icing fluid	Cleaning materials
Hydraulic fluid	Lavatory waste systems
Paint	Soft drinks

Besides injuring wires, dirt, grease and grime make it difficult to read labels on wires and prolong troubleshooting time. If you don't have the manufacturer's recommendation on cleaning wires, use the industry practice; a soft cloth and general-purpose detergent. Check to be sure that cleaning doesn't remove the labels.

Connectors

A vulnerable point is where wiring enters a connector. First, be sure the strain relief is working to prevent pulling on the end of the wire. Look for missing hardware and replace it.

Look at how the wire enters the connector. If moisture forms on the wire can it run down and flow into the connector? To avoid corrosion, form a drip loop in the wire so water cannot run downhill to the connector.

There are times when a connector is removed from a radio and not immediately reconnected. This leaves connector openings exposed to contamination. They need to be covered with a plastic cap. Never force connectors to mate. Be sure the plug is seated in a socket before tightening.

Harnesses, Not Handholds

An airplane is not a friendly environment---not only for wiring, but technicians, as well. In small aircraft, technicians lie upside down under the instrument panel, legs on the seats--- and rudder pedals in their shoulder blades. In large aircraft, the avionics bay is small and cramped. Radios are mounted in tight corners of the fuselage.

There is a strong temptation for the technician to hoist himself out of tight places by grabbing a wire harness like a subway strap. Using harnesses as handholds is discouraged by FAA inspectors, as well as stepping on wires.

Yet another hazard is leaving metal cuttings, tools and waste material among wire bundles. As an aircraft grows older and wire becomes brittle, these objects cut into insulation.

Clamping Near Fuel and Other Lines

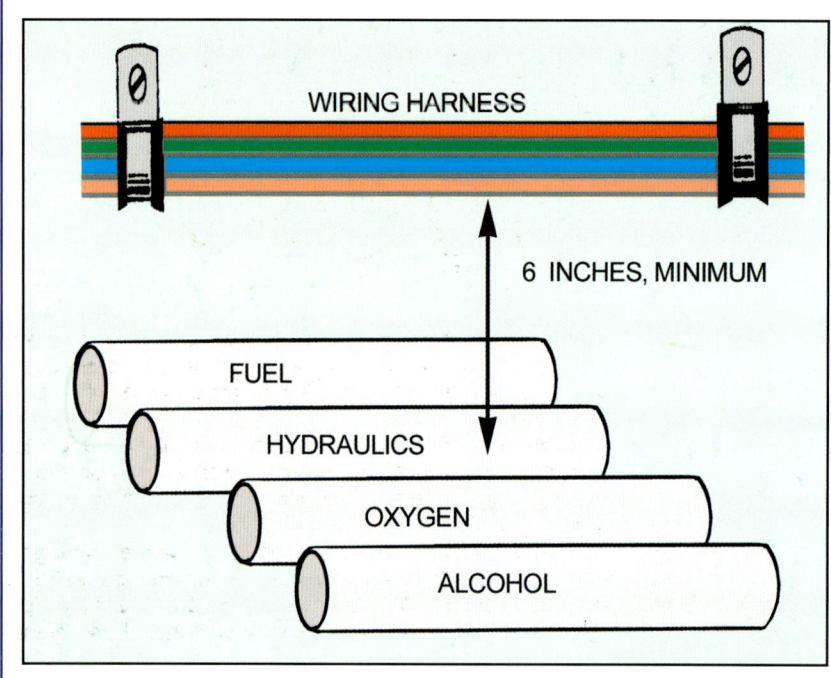

When runnning a harness near plumbing lines that carry flammable liquids or gases, mount the harness *above* them. Try to maintain a clearance of least six inches. If that's not possible, avoid running the harness parallel to those lines and maintain a minimum clearance of two inches

If you can install a clamp near a crossover point, clearance may be as small as one-half inch. When the harness must be connected directly to a plumbing line, use a clamp, as shown here. Don't use that clamp as a regular support for the harness, but use additional clamps. To avoid movement between harness and plumbing lines, install the additional clamps on the same part of the aircraft structure.

Mounting Clamps

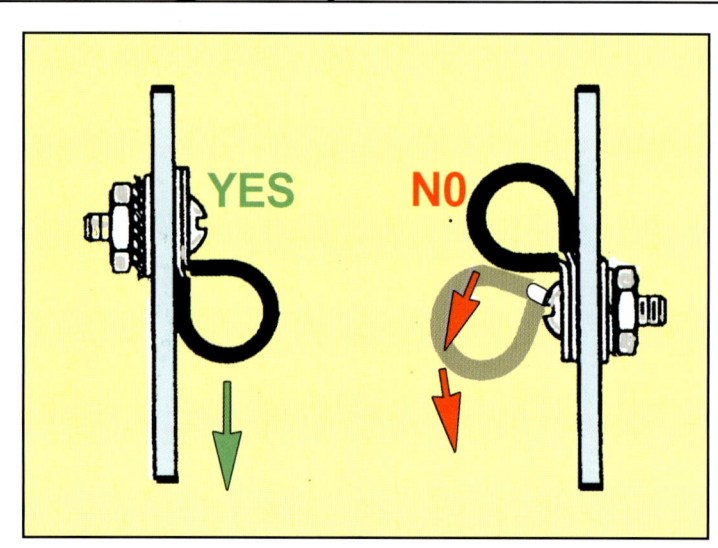

When mounting cable clamps on a vertical surface, locate the nut and bolt *above* the loop that holds the wire (see "Yes"). Placing them below, as shown at "No," may cause wiring to sag if the hardware loosens. This could cause trouble.

To be sure hardware is secure, use lockwashers (external teeth) or self-locking nuts.

"Adel" clamps come in many sizes and mounting types to support wire harness. They are lined with cushion material to hold the bundle, without chafing or rubbing.

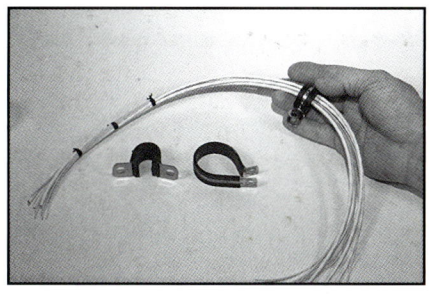

Chafing and Abrasion

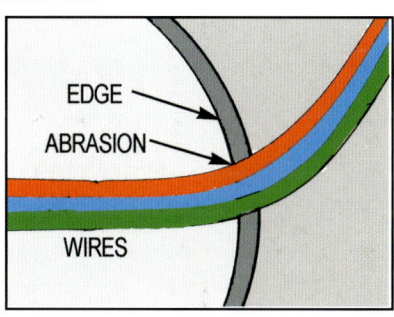

The large wire bundle at the top is supported by a clamp. Wires contained in the bundle carry low-level signals (receiver, data, control, audio, etc.) that don't interfere with each other.

Note that an RF (radio frequency) cable is supported with a tie wrap outside the main bundle. It is good practice to keep transmitter signals (high power) away from low-level bundles.

Large cables that carry power from the electrical generating source (alternator, battery, inverter, etc.) should also be isolated. Wires from any part of a strobe light system are best kept out of the main wiring harnesses.

Grounding to Airframe

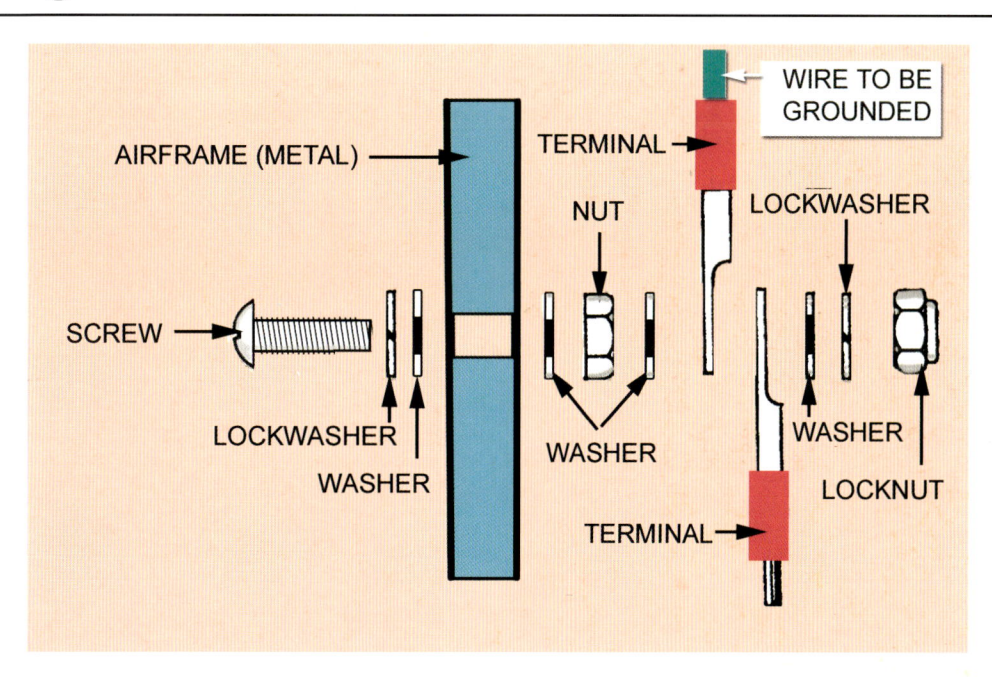

The procedure for attaching wires to the metal structure of the airplane. It serves one of two purposes. One is "grounding," which provides a return connection to the power source for a radio or other electrical device. The whole airframe (if made of metal) is the return connection. The other is "bonding", which connects two surfaces (flaps and wing, for example) with the lowest possible electrical resistance. This reduces the chance of generating interference.

To obtain a good ground, the contact area must be clean. Remove paint, primer, grease and corrosion. If aluminum is protected by a coating of Alodine, remove it.

Two terminals (red) are shown being grounded in the illustration. Do not connect more than four terminals to a single grounding point.

Bending a Coaxial Cable

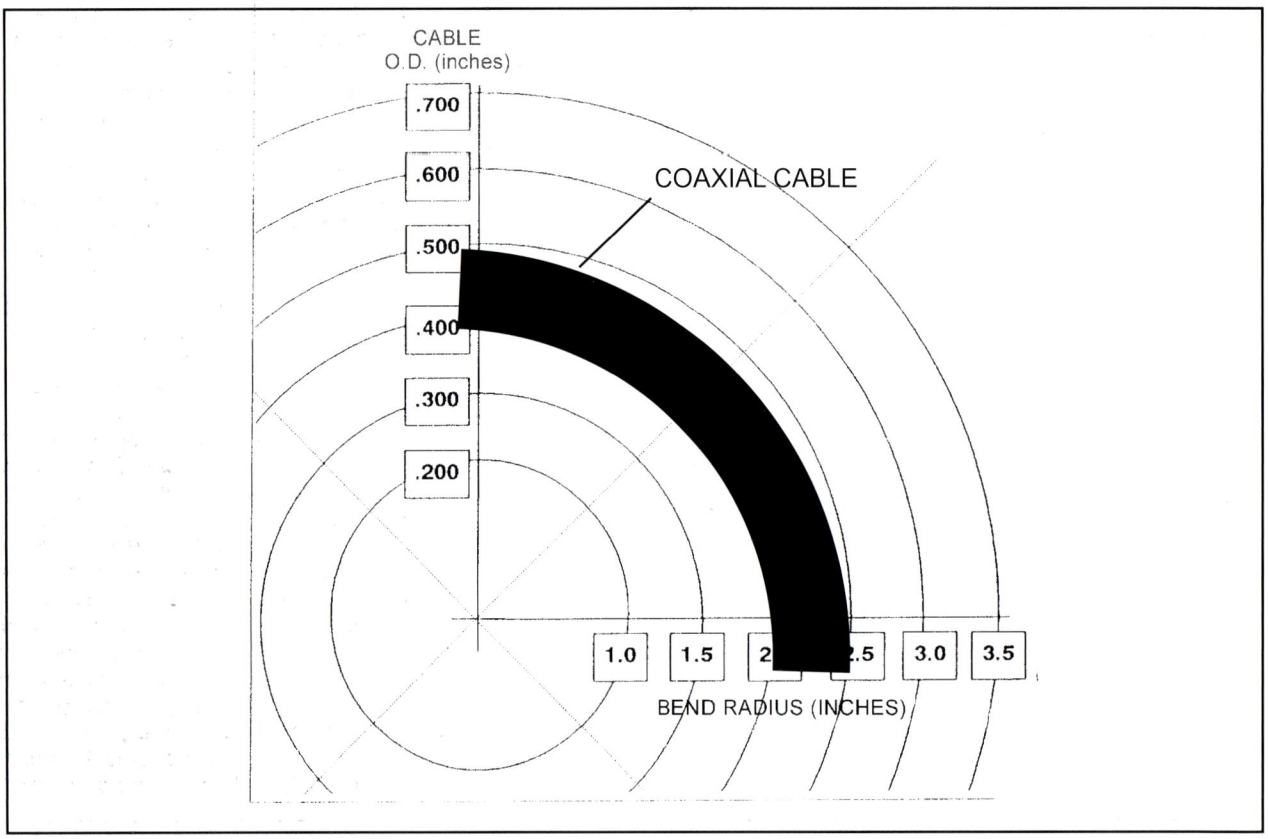

Bend Radius Template

Pic Wire and Cable

Coaxial cable has two conductors which share the same axis; a center wire surrounded by a shield. Unless they remain perfectly spaced ("concentric") they lose electrical performance and power is reduced.

A common problem occurs when coaxial cable is bent too sharply, which can happen during an installation in the limited space of an airplane. Tight bends shift the position of the center conductor and the result is an electrical (or impedance) "bump" that steals energy.

Another problem occurs when coaxial cable runs near an edge. If the cable pushes against the edge, a kink can form.

To prevent losses, cable designers recommend bending coaxial cable over a radius no less than five times its diameter.

In the illustration is a template which shows the minumum bend for cables of various sizes, with an example using a .4-inch diameter. It is laid over the curve that leads to a 2-inch bend radius.

TYPICAL CABLES	PIC CABLES	O.D.	BEND RADIUS
RG58*	----	0.195 (inches)	1.0
RG142	----	0.195	1.0
RG400	----	0.195	1.0
---	S44191	0.195	1.0
---	S44193	0.195	1.0
SF142B		0.195	1.0
---	S66162	0.230	1.2
---	S33141	0.270	1.5
---	S55122	0.300	1.6
(Example)	T556124	0.385	2.0
RG393	----	0.390	2.0
RG214*	----	0.425	2.5
---	S22089	0.435	2.5
---	R11062	0.640	5.0

*PVC insulation, not recommended for aircraft

(PIC)

Coaxial cables are shown with their outside diameters (O.D.) and recommended bend radius. By knowing the O.D. the cable may be laid over the template to see the minimum bend radius. The second column, "PIC Cables", shows model numbers from that company's catalog.

The highlighted "Example" is used in the template.

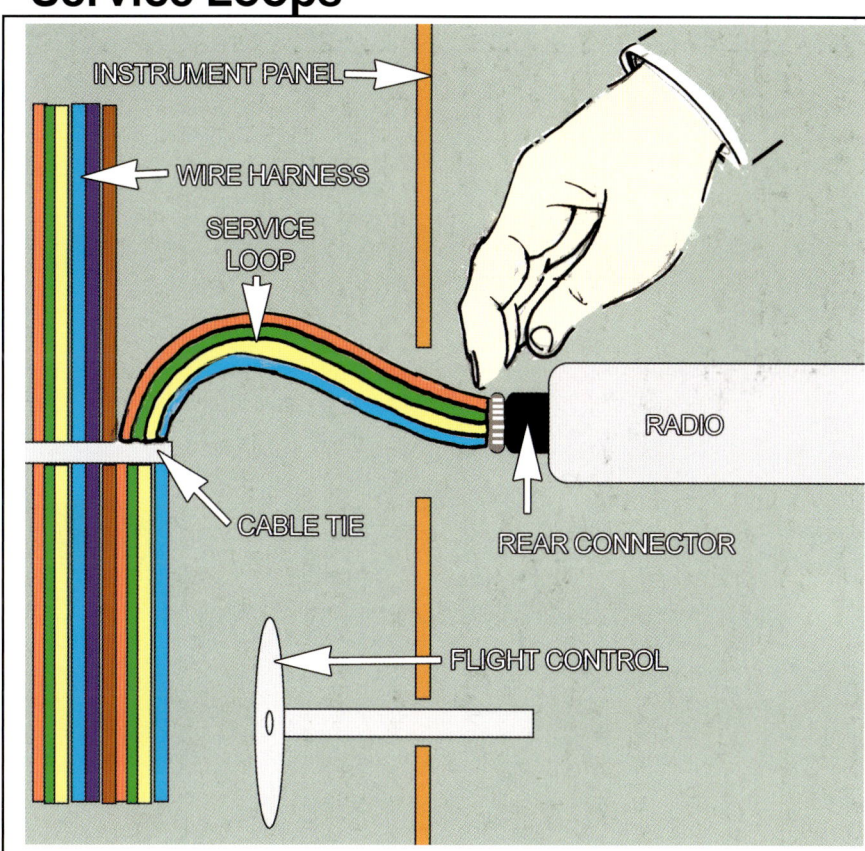

Service Loops

INSTRUMENT PANEL
WIRE HARNESS
SERVICE LOOP
CABLE TIE
RADIO
REAR CONNECTOR
FLIGHT CONTROL

Every radio or instrument should be installed with a "service loop." It is extra slack that permits the radio to slide out of the panel, allowing you to unfasten the connectors. Otherwise, you could spend hours groping behind the panel.

But service loops must be carefully installed to avoid introducing their own problems. Install a cable tie where the service loop breaks out from the harness (there are 90-degree or Y-types for this). Don't bend the wires sharply where they come out of the main harness. Tie the service loop every 4 to 6 inches. If there's any chance of one service loop touching another, cover them with expandable sleeving.

Strain relief is required where the service loop enters the connector at the back of the radio. Frequently, this is provided by the backshell of the connector, but a cushion (Adel) clamp can also mount on the back of the radio.

How long is a service loop? Make it so the radio can be pulled out of the panel by about 3 to 6 inches. This provides clearance to put your hand inside the panel and remove connectors.

Because you are lengthening wires when making a service loop, extra care is needed to prevent the loop from touching moving parts behind the panel (cables, pulleys, gears, etc.). Most movement behind the panel is from the yoke, so move it back and forth and side to side, full travel, to check for rubbing or tangling.

Review Questions: Chapter 25 Wiring the Airplane

25.1 Hazardous areas for wiring on large aircraft are known as "SWAMP." What does it mean?

25.2 What type of wire should never be used for new work on an airplane?

25.3 Why are stranded wires preferred over solid wire for aircraft?

25.4 What type of insulation is found on wire used in many aircraft today?

25.5 As the AWG number for a wire size goes up, the wire diameter _____.

25.6 Which wire has the larger diameter; OO or 36?

25.7 (A) Is it permissible to use wire of greater diameter than required? (B) Are there disadvantages?

25.8 What is the most important rule for selecting wire size and type?

25.9 Unshielded wires are more susceptible to picking up or radiating _____.

25.10 What should you avoid when stripping insulation from wire?

25.11 How many splices may you insert in a length of wire running between two terminals?

25.12 Why should a wire be labelled every 15 inches along its length?

25.13 What two types of cables should be kept apart to avoid interference?

25.14 Why is it important to keep wire bundles from touching aircraft structures?

25.15 Why should wiring be supported away from the bottom of the fuselage?

25.16 Every wire entering a connector must have some form of _____ to prevent it from breaking out of the connector.

25.17 A wiring harness should run 6 inches or more *above or below* lines that carry fuel, oxygen, alcohol or hydraulic fluid?

25.18 Before grounding a wire to the airframe, what steps will insure a good ground?

25.19 What is the purpose of a service loop?

Chapter 26

Aviation Bands and Frequencies

Many problems affecting avionics occur in antenna systems that communicate between the airplane, ground stations, satellites and other aircraft. Antennas operate in the most hostile environment; 500-knot winds, wide temperature swings, ice, hail and other foul weather. Inside the aircraft, antennas connect to cables that run through extremes of temperature, humidity and corrosive chemicals. They are sensitive to location, are easily contaminated and require special test equipment to check their operation.

A single-engine aircraft may have ten antennas; airliners have about 20. It was once believed that antennas would drop in number as designers developed new systems that didn't need radio signals, namely the laser gyro. In spite of advances in gyroscopic and laser instruments, the success of satellite navigation and communications increased the number antennas to meet the demand for new passenger, airline and air traffic services.

Radio Frequencies (RF)

Antennas operate in the world of RF. Radio frequencies form when electrical currents are driven back and forth at about 10,000 times per second and higher. The source is a transmitter, which applies the energy to an antenna. Each time current rushes into or out of the antenna, a field of energy, an *electromagnetic wave*, travels outward.

All radio waves move the same speed; 186,000 miles per second, and consist of electromagnetic en-

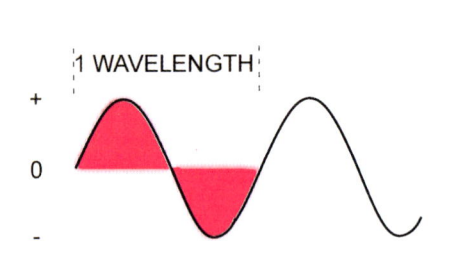

A radio signal can be pictured as an alternating flow (the shape of a sine wave). Electrical energy for the first cycle begins from zero (at left) and rises in strength in the positive direction. Next, it drops to zero, reverses and strengthens in the negative direction. This completes one cycle, measured in Hertz (Hz). If the cycle repeats 122 million times per second, the frequency is 122 MHz (megahertz), which falls in the aviation com band.

Distance from one cycle to the next is also the wavelength of the signal. There is a direct relation between frequency and wavelength. By doubling the frequency the wavelength is reduced by one half.

Radio Frequency Bands

Band		Frequency	Aviation Services
Very Low Frequency	VLF	3-30 kHz	1. Omega (now terminated). 2. VLF (Active in submarine operations, no longer used in aviation).
Low Frequency	LF	30-300 kHz	1. Non-Directional Beacons (NDB) used by airborne ADF 2. Loran 3. Stormscope, lightning detection,
Medium Frequency	MF	300-3000 kHz	1. Non-Directional Beacons (NDB) 2. Standard AM broadcast band (which can be tuned by aircraft ADF).
High Frequency	HF	3 -30 MHz	Long-range voice communications for oceanic and flight in remote areas. Some data communications.
Very High Frequency	VHF	30-300 MHz	1. VHF Communications (air-to-ground, air-to-air) 2. VOR ground navigation stations 3. Instrument Landing System (ILS) 4. Marker Beacons (for ILS) 5. Emergency Locator Transmitters (ELT). Will be relocated to the UHF band.
Ultra High Frequency	UHF	300-3000 MHz	1. Distance Measuring Equipment (DME) 2. Tacan (Military navigation) 3. Glideslope (ILS) 4. Global Positioning System (GPS), U.S. 5. Glonass (Russia, similar to GPS) 6. Galileo (European Union, similar to GPS, under construction) 7. Transponder 8. Traffic Alert and Collision Warning System (TCAS) 9. Emergency Locator Transmitter (2nd generation to be implemented)
Super High Frequency	SHF	3-30 GHz	Air Traffic Control Radar, Airborne Weather Radar, Radar Altimeter,
Extremely High Frequency	EHF	30-300 GHz	Millimeter wave radar (for experimental enhanced vision systems)

Aviation radio began at the lower end of the radio-frequency spectrum. Nearly all communications and navigation before 1940 occurred on Low and Medium Frequencies because devices for higher frequencies hadn't been invented.

By the end of World War II (1945), advances in High Frequencies expanded avionics further up the spectrum.

Note that each band begins and ends with the digit "3". This was determined by international agreement to provide a global structure.

Frequencies which fall within any band behave similarly. Low frequencies hug the earth, following the curve over the horizon. Higher up, frequencies act like light---travelling in straight lines.

ergy. But their behavior varies depending on frequency (number of Hz per second) and wavelength.

Bands. Aviation services are inserted into segments called "bands," as determined by international agreement. Each frequency within a band may also be called a "channel."

Some aircraft bands border on other services. One navigation band (for VOR and ILS) begins at 108 MHz, just above the FM broadcast band (88-108 MHz). At the lower end of the radio frequency spectrum are aircraft beacon stations below 530 kHz, the beginning of the AM broadcast band. Such close spacing is important to know because interference to avionics often originates just outside the aircraft band. As recently as 2008, pilots reported interference from high-intensity signals emitted by FM broadcasters. These events forced the avionics industry to tighten specifications on radios to resist such interference but it is occasionally troublesome.

Low Frequencies

Disruption to aircraft radio from other sources is tied to where the signal occurs in the spectrum. For lower frequencies, VLF (Very Low Frequency), Low Frequency (LF) and MF (Medium Frequency) bands the aircraft receiver is more susceptible to electrical noise from generators, alternators, spark plugs and lightning from thunderstorms. This was especially difficult for pilots during the early days of instrument flying because all navigational aids (navaids) were low in frequency. These services, in fact, performed worst when needed most; at night and in areas of lightning and thunderstorms. Fortunately, avionics moved to higher frequencies which are far more resistant to such interference.

Radios working on low frequencies are also susceptible to "P-static"—(P for Precipitation). Electrical charges build on the skin of an airplane flying through snow, ice particles and other visible signs of moisture. Besides noise, it can also cause complete loss of signal in a Loran receiver, which operates low in the spectrum.

Low frequencies also suffer from "shore effect." When they move between land and water, the difference in conductivity speeds or slows the radio wave. This bends the wave, causing the aircraft receiver to see the signal arriving from a different angle. The result is navigational error.

There is also "night effect", troublesome to the low frequencies of ADF (automatic direction finder.) It's caused by "skipping," a phenomenon that carries signals from hundreds or thousands of miles away, causing interference to the desired station. It's the same problem you hear at night on an AM car radio; stations from across the country, silent during the day, arrive at great strength and compete with local stations. In an

Higher Bands: *Named by Letters*

Microwave Bands

Name	Frequency
L Band	1 to 2 GHz
S-Band	2 to 4 GHz
C Band	4 to 8 GHz
X Band	8 to 12 GHz
Ku Band	12 to 18 GHz
K Band	18 to 26 GHz
Ka Band	26 to 40 GHz

Millimeter Wave Bands

Q Band	30 to 50 GHz
U Band	40 to 60 GHz
V Band	46 to 56 GHz
W Band	56 to 100 GHz

When frequencies rise above 1 GHz, they are further divided into bands identified by letters. Because wavelengths are so short, they are termed "microwaves," and are measured in a few inches or centimeters.

Microwaves are valuable because the radio signal penetrates haze, snow, clouds and smoke. For detecting thunderstorms, weather radar operates on microwave frequencies which reflect from rainfall. Several services depend on the fact that microwaves travel in straight lines; GPS, radar, DME and transponder, for example.

Another property of microwaves; they can be focussed into a beam with a small antenna. This provides sharp images for radar weather, and broadband for TV, Internet, and data services between airplane and satellites.

Above the microwave region are shorter signals of Millimeter Wave Bands. One of the first applications in avionics is the "Enhanced Vision System," which creates runway images for low-visibility landings.

Skipping Through the Ionosphere

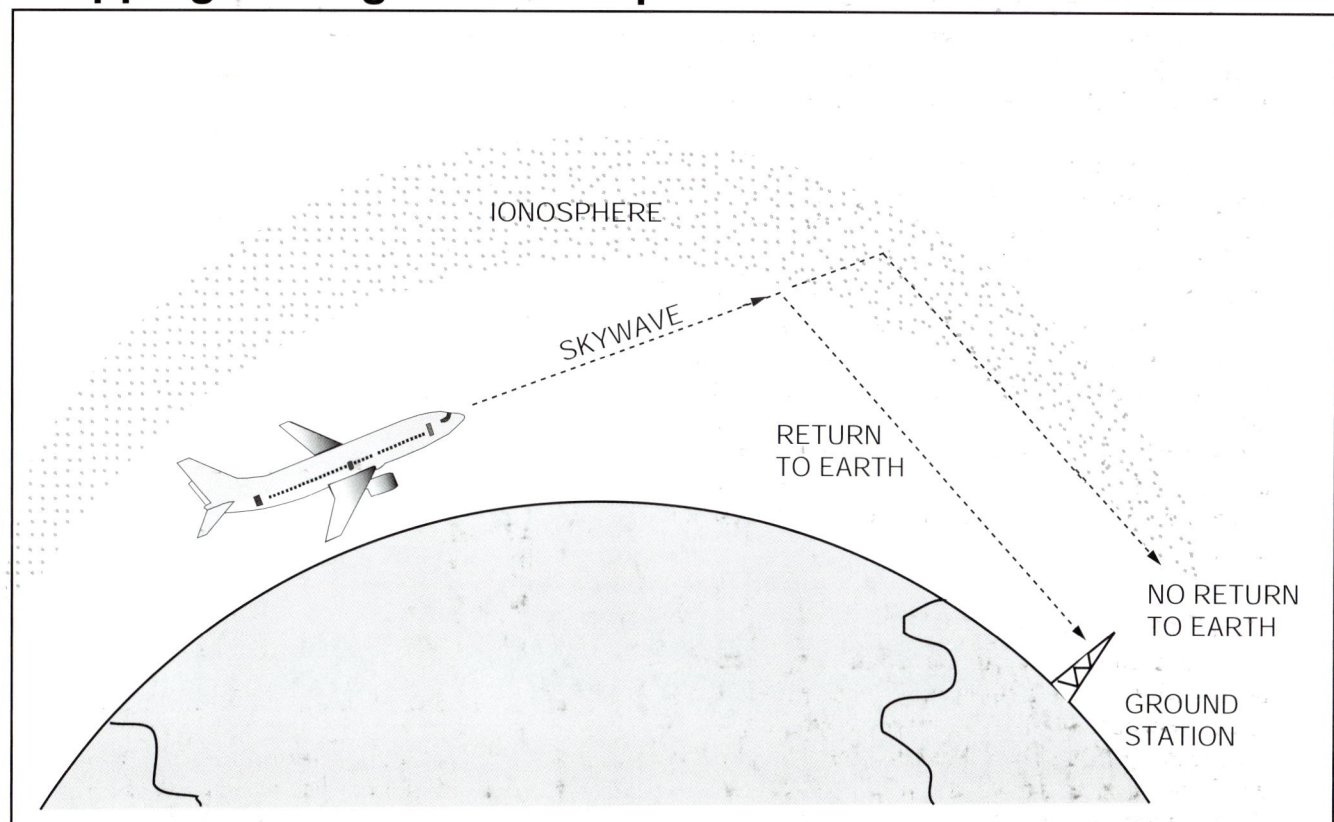

The High Frequency (HF) band communicates across oceans and remote areas by "skipping" through the ionosphere. Radio signals leaving the aircraft reflect after striking ionized layers of air from about 60 to 200 miles high. Because HF signals hardly touch the earth, they lose little strength. However, angles must be correct for the signal to reach the ground station. Created by the sun's ultraviolet radiation, the ionosphere rises and lowers each day. Different frequencies reflect at different angles so the pilot selects the most favorable HF channel. Some frequencies will not bend enough and never return to earth (at certain times of the day), as shown by the skywave at the upper right; see "No Return to Earth."

HF radio is congested and sometimes difficult to use. It will be replaced by satellite communications, which move through the ionosphere with little effect.

aircraft ADF (Automatic Direction Finder), it causes the pointer on the instrument to wander, flutter and have difficulty locking on to the NDB (Non-Directional Beacon) station.

High Frequencies

High Frequencies (HF) played a major role in long-range navigation because they once were the only way an aircraft over an ocean or remote area could communicate to a land station. The long-distance capability of HF arises from "skip", as shown in the illustration.

Although HF sends signals thousands of miles, it has shortcomings. For HF to "hop" from the aircraft antenna and strike the desired ground station, the angles must be correct. And the ionosphere is hardly cooperative. Created by ultraviolet rays striking the top of the atmosphere it is in constant change. As the sun rises the ionosphere thickens (grows deeper)—then, after sunset, it thins out. This has the effect of lowering and raising the bottom of the ionosphere from about 60 to 200 miles above the earth. Not only does this cause a daily variation in skip angles, but there is also an 11-year sunspot cycle that interferes with signals.

HF radios try to solve this several ways. A pilot can select among a half-dozen or so HF frequencies to find a path where ionospheric conditions produce the correct angle to reach the distant station. HF radios also have automatic tuners which quickly adjust the antenna to a selected frequency. The HF band, however, has always irritated pilots because it doesn't provide the instant connection of other radio services. Adding to the problem is a shortage of HF frequencies, causing crowding on the channels.

There was an attempt to improve HF by replacing (analog) voice with digital signals. After years of development a system became practical but HF will eventually be replaced by satcom, which now provides instant contact anywhere on the globe.

VHF Band

Very High Frequency offers many improvements over lower frequencies. It has excellent immunity to electrical interference, thunderstorms, P-static, strobe lights and sparks from rotating machinery aboard the aircraft. Within its band, VHF provides hundreds more channels than lower frequencies. It is not affected by "night effect" and meets the navigational requirement of traveling in straight lines.

VHF, on the other hand, has limitations. Most important, it is line-of-sight; it does not follow the curve of the earth. The range of a communications (com) or navigation (nav) signal mostly depends on the altitude of the aircraft. A Cessna at 3000 feet will "see" a VOR ground station at about 50 miles; an airliner at 39,000 feet picks it up at about 200 miles.

L-Band

What is becoming the most important part of the radio spectrum for aviation is the L-band (1-2 GHz). At these ultra high frequencies, bands are named by letters.

L-band has long been the region for DME and transponder, now joined by GPS as the major L-band service. Signals easily penetrate rain, cloud and fog, which is essential for all-weather navigation.

Future Bands

Major changes will affect both radionavigation and radiocommunication. Frequencies are in short supply and there is competition among countries of the world to use them for purposes other than aviation. This is causing a worldwide shift to satellites for both navigation and communication. Because of the large bandwidth at the higher end of the radio spectrum, many more channels are available, as well as techniques to squeeze more information into digital messages, rather than voice.

The only challenge to radionavigation has been by inertial reference systems (IRS). In their first generation, they used "spinning iron" gyroscopes and other devices to provide navigational guidance. The equipment, however, requires much maintenance. A major improvement is the laser gyro, which uses light beams and no moving parts; the gyro's are aboard most airliners of the current generation equipped with EFIS.

Laser gyros's, however, are limited to en-route navigation because their error (about one nautical mile per hour) cannot allow precision landings. Also, an inertial system must be loaded with a known position at

From Hertz (Hz) to Gigahertz (GHz)

The unit of frequency is the hertz (Hz), or cycles per second. It ranges from about 20 Hz, the low range of human hearing, up through trillions of Hz, where waves behave like light.

1000 Hz = 1 kilohertz (1 khz)
1 million Hz = 1 megahertz (1 MHz)
1 billion Hz = 1 gigahertz (1 GHz)
1 trillion Hz = 1 terahertz (THz)

To convert hertz to kilohertz, move the decimal three places to the left:

1000 Hz = 1 khz

To convert kilohertz to megahertz, move the decimal three places to the left:

3000 khz = 3 MHz
200 khz = .2 MHz

To convert megahertz to gigahertz; move the decimal three places to the left.

100 MHz = .1 GHz

How the term hertz is applied for radio frequencies is shown in bands and frequency charts in this chapter. Other common frequencies in avionics include:

300 - 3000 Hz: The range of audio frequencies for voice communications. It's a narrow range compared to high fidelity used for music recording because that would broaden the signal and cause interference on adjacent channels. Researchers found that most of the ability to hear the human voice is in the 300-3000 Hz range.

400 Hz: In large aircraft, 400 Hz is the frequency for distributing power, much like 60 and 50 Hz are for power distribution in homes. Using 400 Hz in aircraft reduces the size and weight of power components.

90 Hz and 150 Hz are found in the ILS system for carrying navigational signals of the localizer and glideslope.

Line of Sight Communications

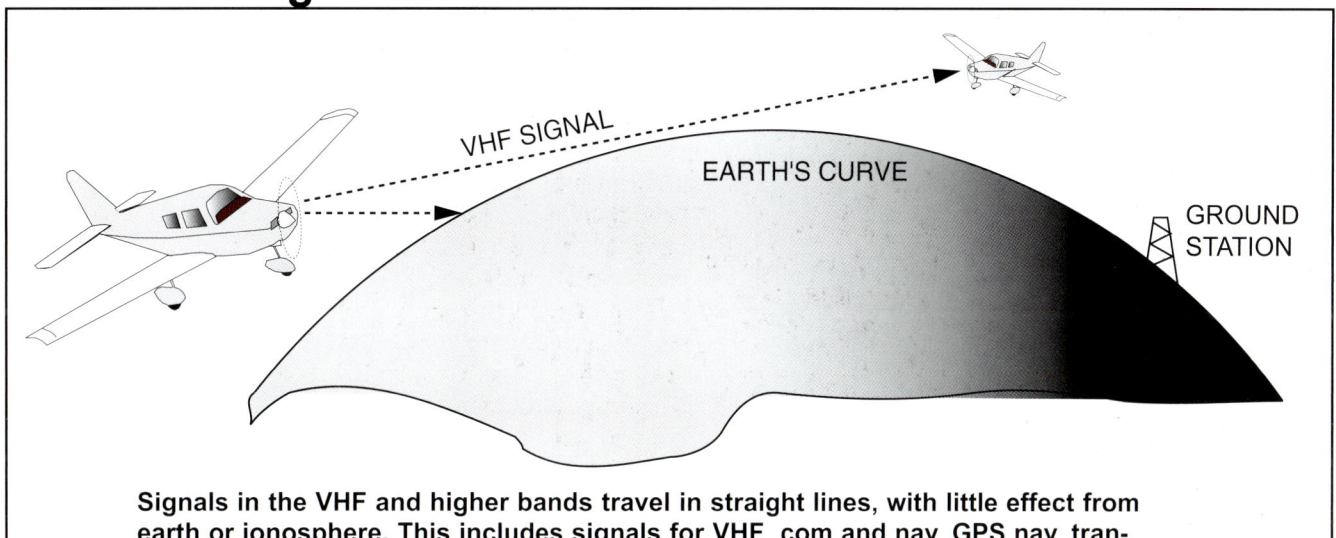

Signals in the VHF and higher bands travel in straight lines, with little effect from earth or ionosphere. This includes signals for VHF com and nav, GPS nav, transponder, DME, radar and other services above 30 MHz. Thus, the airplane at the left cannot communicate with the ground station on the far right because the signal is shaded by the earth. The two airplanes, however, can communicate because they "see" each other above the earth's horizon.

the beginning of a trip and updated against other known positions during long trips. GPS, on the other hand, not only finds itself in minutes almost anywhere on earth, but is far more accurate.

New navigation systems based on extremely high frequencies in the "millimeter bands," where radio waves take on the properties of light, are in development. They will create runway images (synthetic vision) for landing, solve the long-term problem of clear air turbulence and provide other services not possible on lower frequencies.

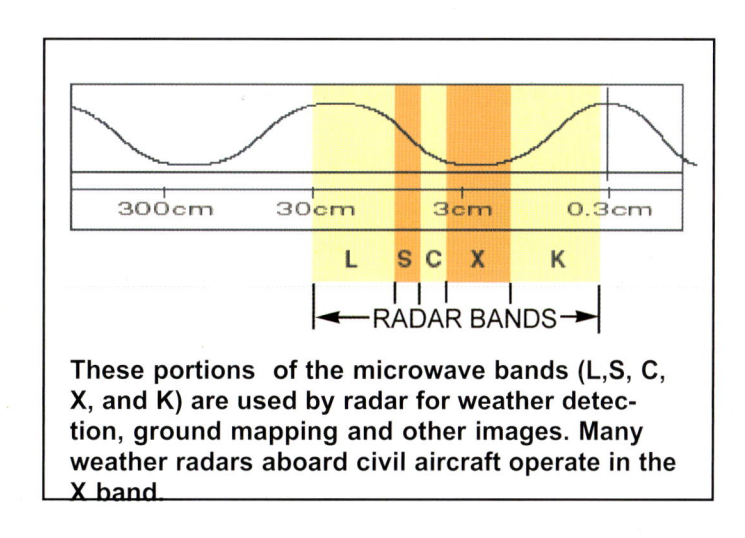

These portions of the microwave bands (L, S, C, X, and K) are used by radar for weather detection, ground mapping and other images. Many weather radars aboard civil aircraft operate in the X band.

235

Control and Display of Bands and Frequencies

This stack of control-display units are found on airline instrument panels and represent five different bands of operation. Knobs marked "TFR" mean "transfer"---which takes a frequency stored in one window and transfers it to the active position ("ACT"). All the controls shown here conform to the ARINC 500 characteristic.

VHF - NAV
1. This is a combined control head. The left half is "VHF," and selects frequencies in the VHF communications band. In some radios, "Com" is used instead of "VHF."
   Navigation on VOR or localizer frequencies is selected by the knob at the right, under "NAV." (The localizer is part of the ILS, Instrument Landing System.)

DME
2. This control head selects stations in the DME band, but frequencies shown on the dial (108 and 117.95) are *not* DME frequencies. They are VOR (nav) frequencies. Because DME stations are paired with VOR stations, selecting a VOR automatically "channels" the DME.

HF
3. A control head for a High Frequency (HF) radio, used for long range communication. Each frequency, or channel, is selected to operate in one of three modes; USB (Upper Sideband), LSB (Lower Sideband) or AM, (Amplitude Modulation), an earlier form of transmission.

ADF
4. The Automatic Direction FInder (ADF) has a control known as a "BFO," for "beat frequency oscillator." It is switched on to identify the station in countries which do not transmit an audio ID tone. The BFO makes the ID audible (in Morse Code).

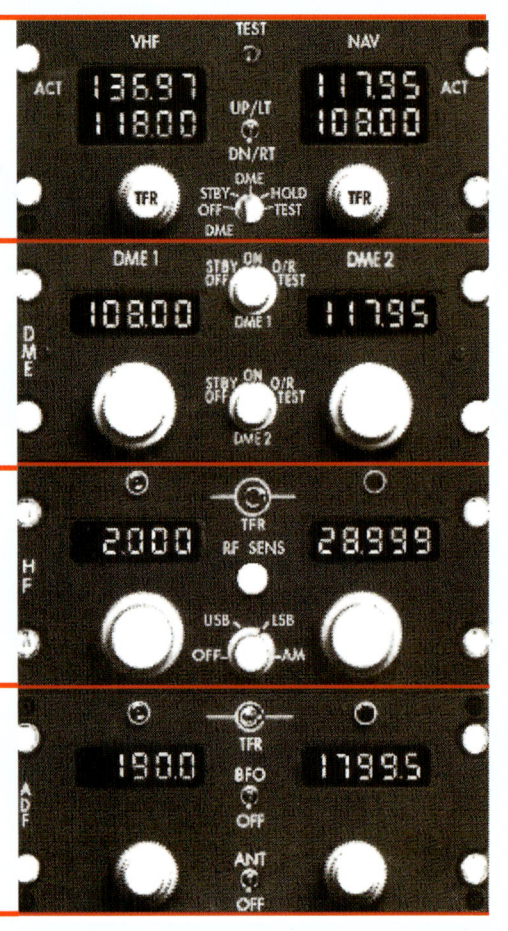

Sigma-Tek

Aviation Frequency Assignments

Navigation (Markers, VORs, ILS Localizers, VOR/ILS Test)

75 MHz	Transmitting frequencies of fan markers, Z markers, and ILS markers.
108.0 to 108.05 MHz	Commonly used for VOR ramp testers.
108.1 to 108.15 MHz	Commonly used for ILS localizer ramp testers.
108.2 to 111.85 MHz	Transmitting frequencies of VORs. Operated on even tenths.
108.3 to 111.95 MHz	ILS localizers with or without voice. Operated on odd tenths.
112.0 to 117.95	Transmitting frequencies of VORs.

Communications (Air Traffic Control, emergency, advisory, support, ARINC, air-to-air, flight schools flight inspection, military advisory, manufacturer test, special use)

118.0 to 121.4 MHz	ATC
121.5 MHz	Emergency frequency (search and rescue-SAR), emergency locator transmitter (ELT) signals (five-second operational check)
121.6 to 121.925 MHz	Airport ground control, ELT test
121.95 MHz	Aviation support
121.975 MHz	Private aircraft advisory (FSS)
122.0 to 122.05 MHz	FSS EFAS (Flight Watch)
122.075 to 122.675 MHz	Private aircraft advisory (FSS)
122.7 to 122.725 MHz	Unicorn, non-tower controlled airports
122.75 MHz	Air-to-air communications (fixed wing aircraft)
122.775 MHz	Aviation support
122.8 MHz	Unicorn, non-tower controlled airports
122.825 MHz	Aeronautical en route (ARINC)
122.85 MHz	Multicom
122.875 MHz	ARINC
122.9 MHz	Multicom, SAR training, airports with no tower, FSS, or unicorn
122.925 MHz	Multicom, special use (forestry management/fire suppression, fish and game management/protection, etc.)
122.95 MHz	Unicorn, tower-controlled airports, airports with full-time FSSs
122.975 to 123.0 MHz	Unicorn, non-tower controlled airports
123.025 MHz	Air-to-air communications (helicopter)
123.05 to 123.075 MHz	Unicorn, non-tower controlled airports
123.1	SAR, temporary control towers
123.125 to 123.275 MHz	Flight test stations of aircraft manufacturers
123.3 MHz	Flight schools
123.325 to 123.475 MHz	Flight test stations of aircraft manufacturers
123.5 MHz	Flight schools
123.525 to 123.575 MHz	Flight test stations of aircraft manufacturers
123.6 to 12.65 MHz	Air carrier advisory (FSS)
123.675 to 26.175 MHz	ATC
126.2 MHz	Military common advisory
126.225 to 128.8 MHz	ATC
128.825 to 132.0 MHz	ARINC
132.025 to 134.075 MHz	ATC
134.1 MHz	Military common advisory
134.125 to 135.825 MHz	ATC
135.85 MHz	FAA flight inspection
135.875 to 135.925 MHz	ATC
135.95 MHz	FAA flight inspection
135.975 to 136.075 MHz	ATC
136.1 MHz	Future unicorn or AWOS
136.125 to 136.175 MHz	ATC
136.2 MHz	Future unicorn or AWOS
136.225 to 136.25 MHz	ATC
136.275 MHz	Future unicorn or AWOS
136.3 to 136.35 MHz	ATC
136.375 MHz	Future unicorn or AWOS
136.4 to 136.45 MHz	ATC
136.475 MHz	Future unicorn or AWOS
136.5 to 136.975 MHz	ARINC

Ground Wave Transmission

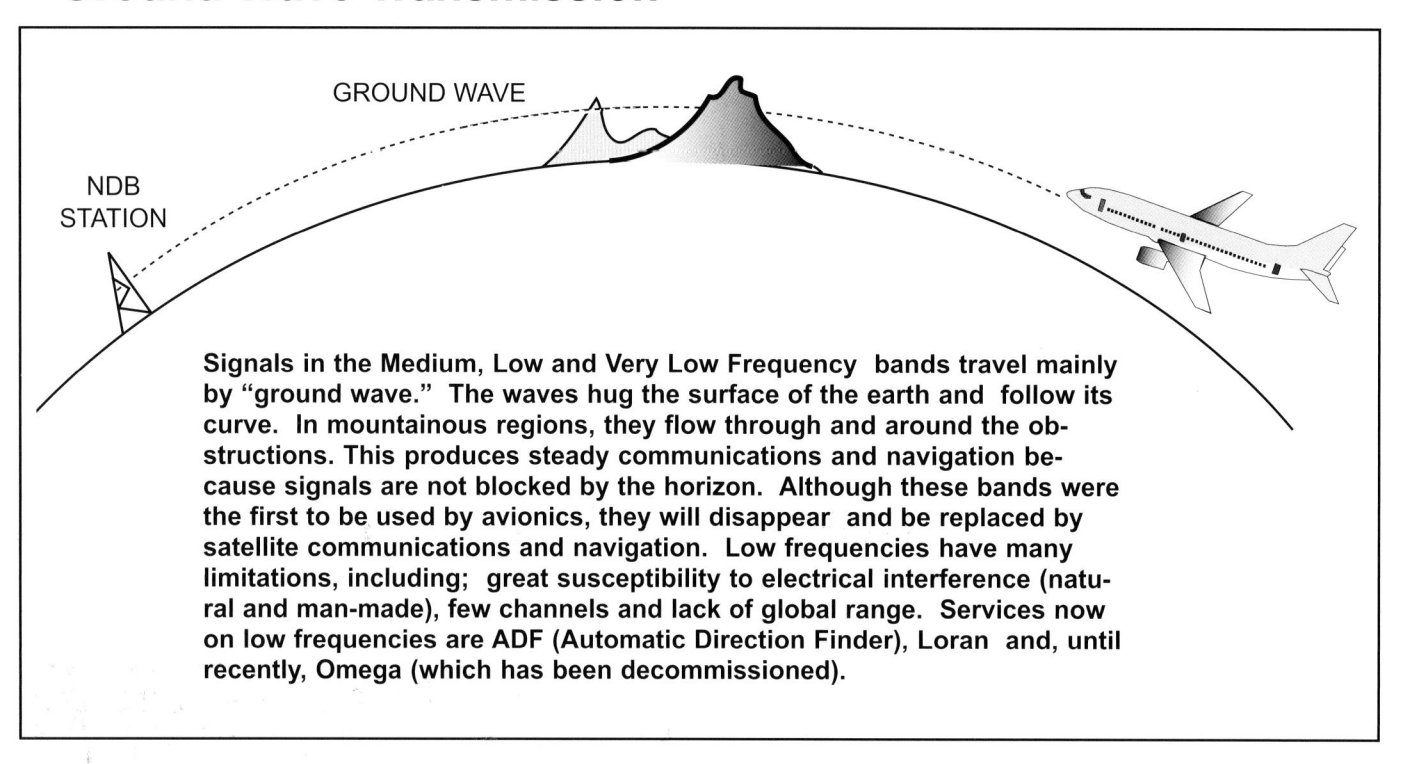

Signals in the Medium, Low and Very Low Frequency bands travel mainly by "ground wave." The waves hug the surface of the earth and follow its curve. In mountainous regions, they flow through and around the obstructions. This produces steady communications and navigation because signals are not blocked by the horizon. Although these bands were the first to be used by avionics, they will disappear and be replaced by satellite communications and navigation. Low frequencies have many limitations, including; great susceptibility to electrical interference (natural and man-made), few channels and lack of global range. Services now on low frequencies are ADF (Automatic Direction Finder), Loran and, until recently, Omega (which has been decommissioned).

Review Questions
Chapter 26 Bands and Frequencies

26.1 A radio wave travels at the rate of _____.

26.2 Aircraft communication frequencies between 118-137 MHz are in the _____ band.

26.3 Satellite navigation, on approximately 1.5 GHz (1500 MHz) are in the _____ band.

26.4 What problems made lower-frequency bands---VLF, LF and MF----difficult to use in aviation?

26.5 Frequencies in the HF band are able to "skip" long distances. What is the name of the HF signal that leaves the antenna, reflects off the ionosphere and returns to earth?

26.6 How many kilohertz (kHz) equal 3 MHz (megahertz)?

26.7 Frequencies below the Medium Frequency Band mainly travel via _____ waves.

26.8 The localizer frequency in an ILS system operates within the band 108.3 to 111.95 MHz. How are localizer frequencies assigned within the band?

Chapter 27

Antenna Installation

A light aircraft that occasionally flies on instruments ("light" IFR) typically has these antennas:

Com 1	DME
Com 2	GPS
VOR	ADF Loop
Localizer	ADF sense
Glideslope	Marker Beacon
Transponder	Emergency Locator

For low IFR (low ceilings, extended flight in clouds) a well-equipped General Aviation airplane may add antennas for:
Weather detection (Stormscope)
Traffic detection (TCAS I)
Datalink for weather (XM radio, WSI)
Antenna for emergency handie-talkie

Airliners and corporate aircraft carry most of the above, plus antennas for:
Weather radar
TCAS II (collision avoidance)
Satcom (satellite communication)
Radar altimeter
Passenger telephone and entertainment

Not mentioned above is Loran, a popular naviga-

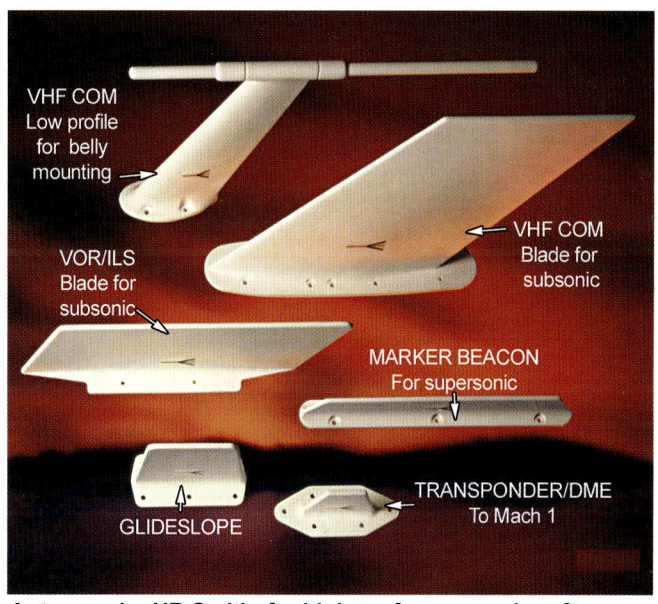

Antennas by HR Smith for high-performance aircraft are encased in thermoplastic composite material to resist rain erosion, impact, chemicals and temperature changes.

tion system that rapidly declined with the arrival of GPS. Loran does not cover the world and suffers interference from precipitation. Loran, however, is still aboard many light aircraft---which means there is a need for replacement antennas. The installation of a Loran antenna is the same as that described for a VHF com.

In 1971, a system known as Omega provided world-wide long-range navigation. But with the rise of GPS, Omega lasted only until 1997, when its eight global stations were taken off the air. Omega antennas were difficult to install because the airplane had to be "skin-mapped---tested all over to find an electrically quiet place that would not interfere with the Omega signal.

Antennas for Airline, Corporate and Military Aircraft

VHF Com, will operate a power rated at at 100 watts. Has internal duct for hot air de-icing. Used on Boeing, Lockheed and Douglas.

VHF Com with low profile. Has filter to prevent VHF interference to GPS receiving antenna.

Glideslope antenna rated for Cat III (instrument) landing. Used on Gulfstream, Regional Jet, others.

UHF com antenna for military aircraft covers 180-400 MHz. Rated to 70,000 ft and 35 G's

TCAS (collision avoidance) antenna determines direction of target. Arrow shows antenna mountng direction.

Directional antennas for TCAD (collision avoidance) system. Mounts on top and bottom of fuselage.

GPS models may look the same but some have built in preamplifiers to overcome losses in long cable runs

This amplified GPS antenna is aboard Boeing aircraft for GPS reception in multimode receivers.

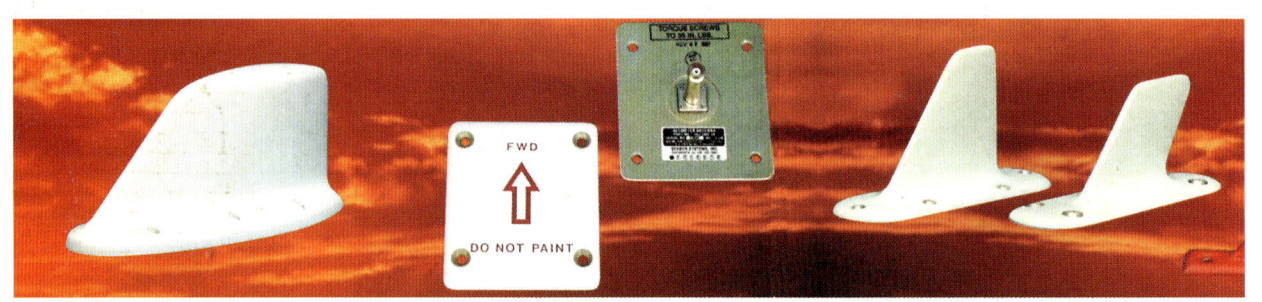

Satcom antenna for operating in Aero-L and Aero-C bands of INMARSAT satellite. It is a low-gain antenna for moving data on 1.5-1.6 GHz.

Outside and underside views of radio altimeter antenna. Operating on 4200-4400 MHz, it is flush-mounted on belly of aircraft.

Antennas for passenger radiotelephone service operating on 830-900 MHz.

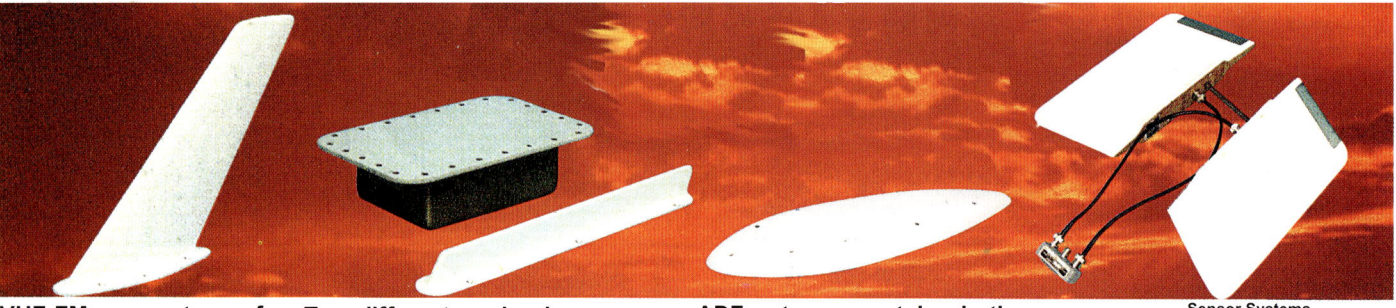

Sensor Systems

VHF-FM com antenna for 140-180 MHz (outside the aviation bands). Provides communications with non-aviation services.

Two different marker beacon antennas (75 MHz). Top one is a flush-mount used on the Boeing 747. Bottom one is low-profile, used on the Boeing 737 and Pilatus.

ADF antenna contains both loop and sense antennas for automatic direction finder. Applications include several Boeing, Douglas and Airbus airliners.

Two blades, known as a "balanced loop," mount on rudder fin to provide VOR, localizer and glideslope reception.

240

Hostile environment. Aircraft antennas operate under the worst conditions. Sitting on the ramp in Arizona on a summer day, they bake in over 100 degrees F. After the airplane takes off and reaches altitude 15 minutes later, antennas are chilled to 50 degrees F below zero and buffeted by winds over 500 kts. If the airplane flies in clouds at or near the freezing level, ice may form on the antenna. And it's not the *weight* of the ice, but the change in antenna shape that causes "flutter," a vibration that can break the structure. As if that weren't enough, the antenna is mounted on an aluminum fuselage which flexes as the airplane pressurizes. Add such hazards as fluid sprayed on the airplane for de-icing or hydraulic oil in the belly and you can see why antennas require rugged construction and follow-up maintenance.

What goes wrong? Despite the hostile environment, most problems (antenna-makers say) result from poor installation. Materials used by every reputable antenna maker are tested and well-proven. It's the installer's responsibility to provide a good location, a secure mounting, a seal against weather and a low-

How to Read an Antenna Spec Sheet

First, consider the type of aircraft. Catalogs often divide them into such categories as General Aviation, Commercial (business jets), air transport (airline) and military. These categories are divided into type of service; communications, navigation, transponder, DME and others.

In the example shown here, we are seeking a communication antenna for a business jet. Looking down the list of specifications (see numbers in red);

1. Frequency. The aviation "com" band extends from 118 to 137 MHz. The band is also called "VHF" or "VHF com."

2. VSWR. Meaning "Voltage Standing Wave Ratio," VSWR indicates antenna efficiency. The lower the first number (2), the higher the efficiency. Manufacturers produce antennas with a VSWR usually less than 3.0:1.

3. Polarization. VHF com antennas operate with "vertical" polarization, which helps concentrate signals toward the horizon, rather than angling up to space.

4. Radiation Pattern. "Omnidirectional" sends signals in every direction, a requirement because the ground station may be anywhere.

5. Impedance. The AC electrical load of the antenna, 50 ohms, is standard in aviation. The cable feeding the antenna is also 50 ohms, which produces a correct match to the antenna.

6. Power. The amount of radio-frequency power that can be handled by the antenna. Light aircraft transmitters generate about 5 to 10 watts. A transmitter for an airliner or business jet may run 25 watts.

7. Connector. Many antenna connectors are the "BNC" (bayonet) type, but be sure the coaxial cable is also fitted with a BNC connector. TNC is also used.

8. Altitude. Rated to 50,000 feet, this antenna can operate at altitudes flown by a business jet. (The highest altitude for air traffic control is 60,000 feet.)

9. Air speed. The antenna can operate at 600 knots at 25,000 feet. This is accomplished by the "blade" design, which is stronger than narrow whips or rods for slower aircraft. Any requirement above 600 knots would be for military, or aircraft flying at supersonic speeds

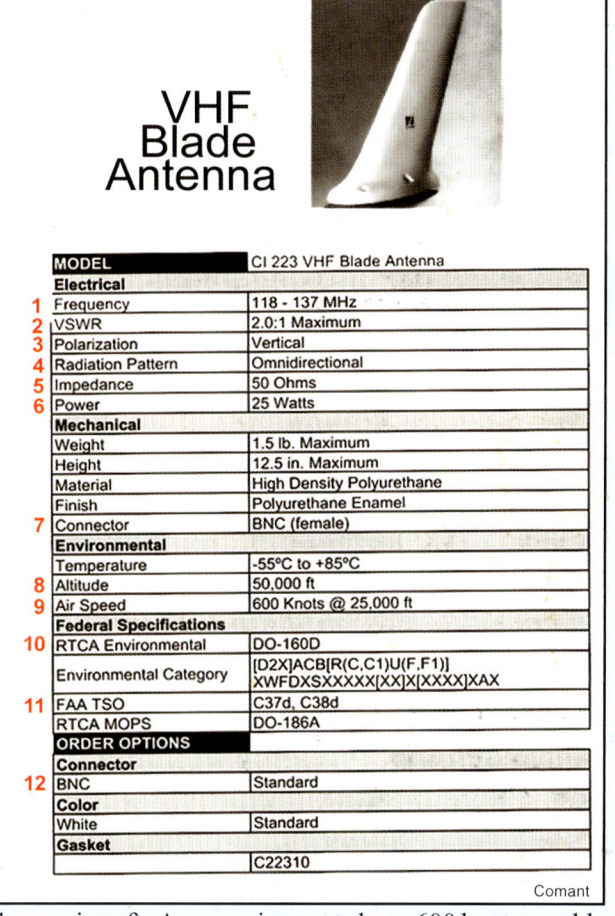

10,11 Certifications. These ratings mean the antenna will meet environmental and performance standards for this application.

12. BNC The antenna accepts the common BNC connector used on most coaxial antenna cables.

VHF Blade Antenna

	MODEL	CI 223 VHF Blade Antenna
	Electrical	
1	Frequency	118 - 137 MHz
2	VSWR	2.0:1 Maximum
3	Polarization	Vertical
4	Radiation Pattern	Omnidirectional
5	Impedance	50 Ohms
6	Power	25 Watts
	Mechanical	
	Weight	1.5 lb. Maximum
	Height	12.5 in. Maximum
	Material	High Density Polyurethane
	Finish	Polyurethane Enamel
7	Connector	BNC (female)
	Environmental	
	Temperature	-55°C to +85°C
8	Altitude	50,000 ft
9	Air Speed	600 Knots @ 25,000 ft
	Federal Specifications	
10	RTCA Environmental	DO-160D
	Environmental Category	[D2X]ACB[R(C,C1)U(F,F1)] XWFDXSXXXXX[XX]X[XXXX]XAX
11	FAA TSO	C37d, C38d
	RTCA MOPS	DO-186A
	ORDER OPTIONS	
	Connector	
12	BNC	Standard
	Color	
	White	Standard
	Gasket	
		C22310

Comant

241

Antennas for Light Aircraft

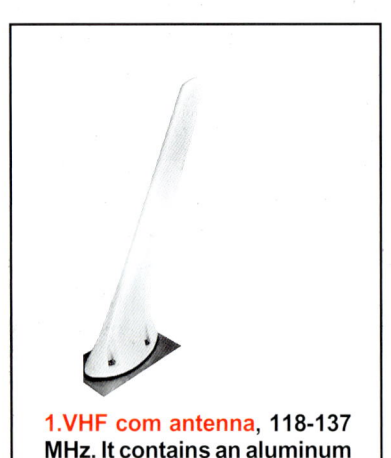

1. VHF com antenna, 118-137 MHz. It contains an aluminum radiator inside polyurethane foam.

3. "Cat's Whisker" VOR antenna found on most light aircraft. The round container at bottom is a "balun," which matches the coaxial cable to the V-shaped antenna rods. "Balun" means "balanced-unbalanced" (the antenna is balanced, the cable is unbalanced). The antenna also provides localizer and glideslope signals (see "couplers").

5- 6. Transponder and DME. The transponder antenna operates in the 1030-1090 MHz band to reply to interrogations from air traffic control. The tip of the antenna is a "corona" ball. If the antenna picks up static electricity, the ball discharges it with less interference than would a pointy tip. Height of the antenna is 3 inches.

A DME antenna is nearly identical to the transponder antenna in size and shape. It operates in the same UHF band, on frequencies 980-1220 MHz.

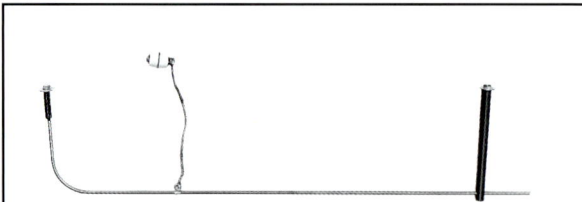

7. Marker Beacon antenna operates on 75 MHz and is 29 inches long. This is a "sled" type (it resembles a sled runner). If the antenna is enclosed, it's a "boat" type. Marker beacon antennas mount on the bottom of the aircraft to avoid shading the signal from the ground station.

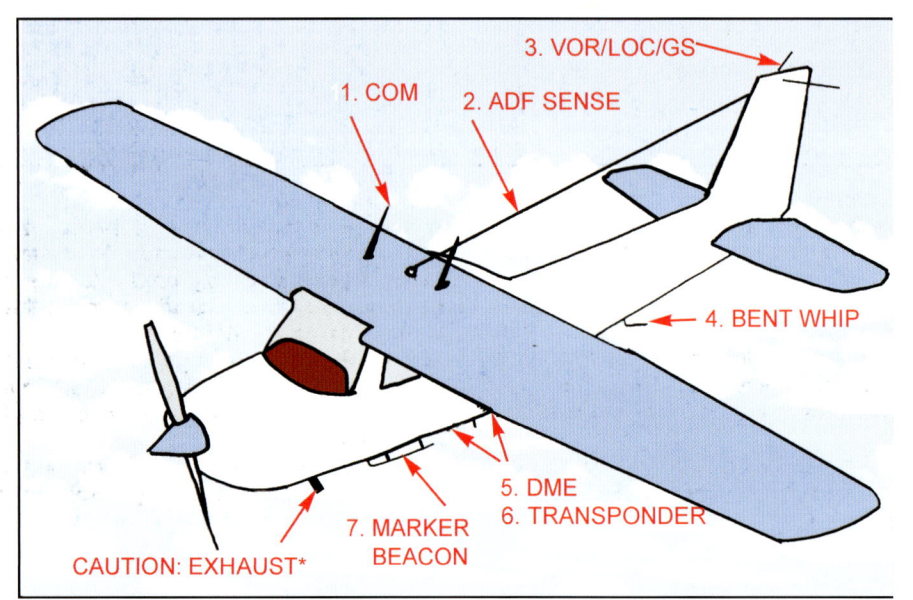

Don't locate an antenna on the belly in line with an exhaust pipe. It becomes coated and discolored. Locate antennas as close as possible to the centerline of the fuselage, but avoid cutting into structure like ribs and stringers.

242

resistance electrical ground to the airframe---all factors covered in this chapter.

Selecting an Antenna

Catalogs on antennas are filled with engineering terms; *isotropic radiator, dB gain, VSWR, impedance,* and more. It is not difficult, however, to choose the right antenna for the job, so long as you select a model from a reputable antenna manufacturer---for example, Comant, Dorne & Margolin, Sensor Systems, RAMI and HR Smith---the antenna should perform as described for its category (airline, military, general aviation, etc.). An explanation of important antenna terms is shown in the table.

An antenna with a "TSO" (Technical Standard Order) assures the antenna was tested to the limit of its specifications. The TSO is not a legal requirement for many installations, but antenna makers obtain this certification as a mark of quality. Business and airline operators usually require it.

Major specifications to know are shown in the table, "How to Read an Antenna Spec Sheet." Consider other details in selecting an antenna:

Speed. A dividing line between antennas depends on the speed of the aircraft. For General Aviation look for the rated speed, which may be 350 kts, 450 kts, etc. Slow aircraft (under 200 kts) often use antennas which

Airline Antenna Locations

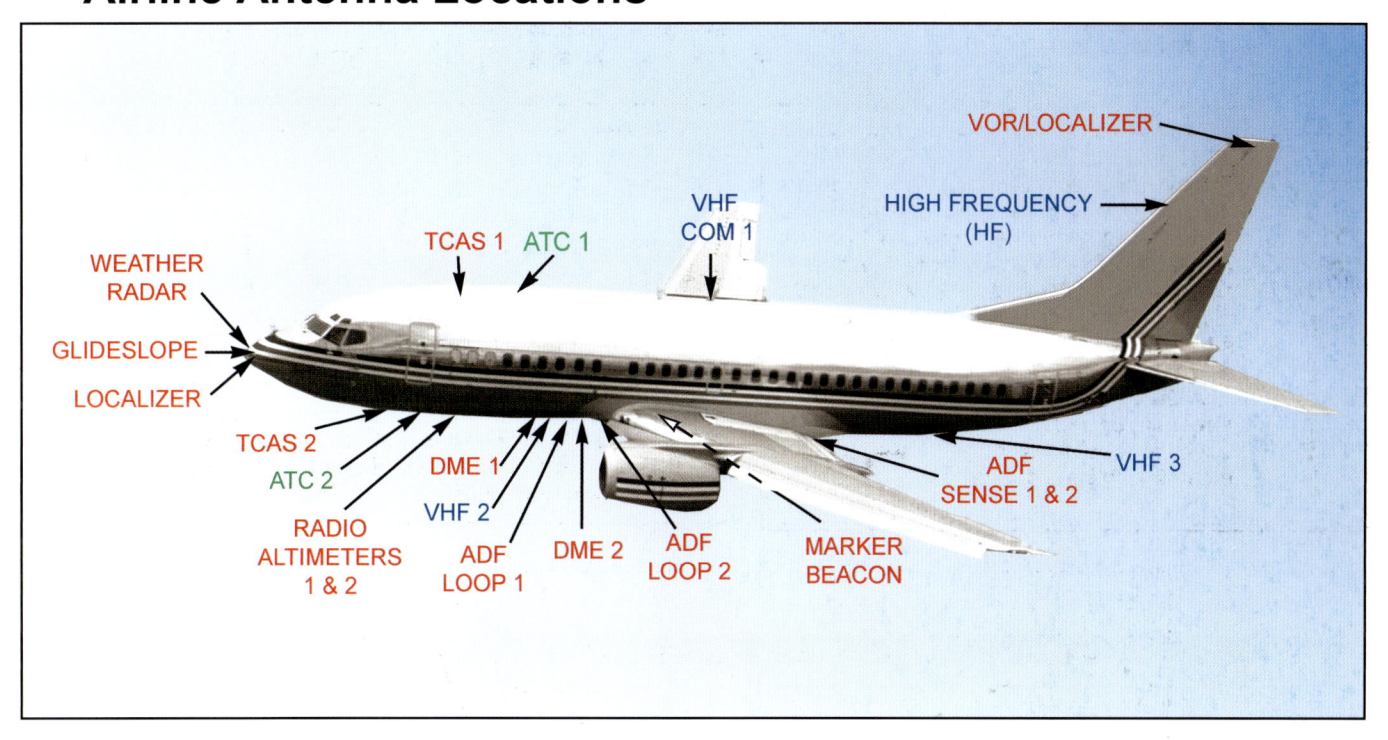

Shown above is a Boeing 737. Antennas divide into three categories, often known as CNI, for Communication, Navigation and Interrogation. Navigation antennas are shown in red, communications in blue. The two interrogation antennas (green) are for ATC, which is the airline term for "transponder."

Most antennas are dual installations because safety requires two or more radios for each function. Where possible, they are placed at top and bottom of the fuselage for greatest reliability.

The two radar altimeters must be at the bottom because they measure the few feet between the airplane belly and runway during a low-visibility landing (Category 2 ILS and higher).

In the nose are three antennas protected against weather by a plastic radome. The glideslope antenna in small aircraft is usually on the tail (as part of the VOR antenna). In large aircraft, however, the glideslope antenna is usually in the nose. During an approach, large aircraft pitch up at a high angle of attack, which could cause its wings and fuselage to block glideslope signals arriving from the ground. Putting the glideslope in the nose eliminates that problem.

New airline models add several more antennas; GPS for navigation and satellite (for voice and data).

are simple rods or whips. They are not sturdy enough for speeds of high-performance piston and turbine aircraft, so the antenna is encased in a fin-like shell filled with plastic or composite material. Some housings are sufficiently rugged to be rated for the Mach speeds of military aircraft.

Altitude. Antennas are rated for altitude. Non-turbocharged airplanes usually have a service ceiling around 20,000 feet. When turbocharged (with piston engines) they cruise in the low 20,000's and higher. Turboprop aircraft fly in the 30,000-ft range, while pure jets climb into the 40's.

For most single-engine aircraft, com 1 and 2 antennas are located atop the cabin. Try to space them at least 36 inches apart to reduce interaction. This could reduce reception and transmission in certain directions.

Connectors and Cable. Other items to check on the spec sheet are the cable and connector. If coaxial cable is not supplied with the antenna, check the type of connector at the base of the antenna---BNC, TNC or other. You'll need to match that connector when you make up the cable.

Antenna Types

VHF Com. Communications antennas in the VHF band (118-137 MHz) come in different models depending on airspeed. Slow aircraft can operate with a simple stainless steel rod. Faster aircraft need a rigid housing, often in the form of a blade. If you need to mount an antenna on the belly of a small aircraft, consider a "bent whip," which gives extra ground clearance.

VOR antennas. There is no power rating for the VOR antenna because it receives only low-level signals. Most light aircraft use the "cat's whisker," a V-shape antenna that mounts atop the rudder fin. For best operation, the open side of the V should point forward (which favors reception ahead of the airplane).

In the description of a VOR antenna, look for the word "balun." It means "balanced-unbalanced," referring to the fact that the antenna is "balanced," but the coaxial cable feeding it is "unbalanced." The balun is a small transformer which matches "balanced to unbalanced". In some VOR antennas, the balun is supplied inside the antenna.

Technicians once made their own baluns with lengths of coaxial cable, so these devices may re-appear if you open up an old airplane. The ready-made balun transformer is an easier solution.

Slow airplanes use the cat's whisker for VOR reception, but faster aircraft need added structural strength. As shown in the illustrations, this is accomplished with "towel bar" or "balanced loop" VOR antennas. These models also perform better during "RNAV" (area navigation); they are more efficient at receiving VOR stations to the right and left of the aircraft, rather just fore and aft. Airplanes once flew only toward or away from VOR stations, but RNAV receivers pick a VOR signal off one wing and electronically

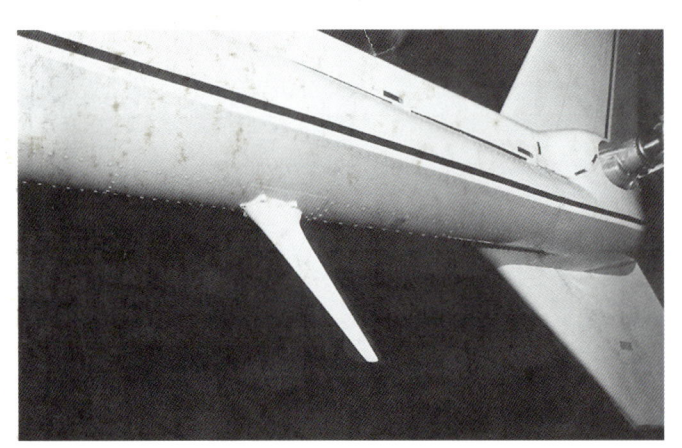

Com antenna on the bottom of a helicopter tail boom. Because of the main and tail rotor blades, antenna locations on helicopters are more difficult to find.

VHF com antennas are usually placed top and bottom, as in this King Air, when the aircraft is certified for flight in icing conditions. Ice on an antenna may cause vibration and breakage. It is unlikely that both top and bottom antennas will be affected the same way because of different slipstreams.

VOR "blade" is a strong, low-drag antenna for turbine-powered aircraft. With a blade on either side of the rudder fin, it is sensitive to VOR ground stations to the left and right of the aircraft course. This improves signals for area navigation (RNAV).

move it ahead of the airplane.

The VOR antenna in airliners is often hidden inside a fairing on the forward part of the rudder fin. Made of composite material, the fairing allows signals to pass through while protecting the antenna against weather.

Experimental and home-built aircraft often hide their antennas inside fuselage or wings because these airplanes are sometimes made of composites. However, some composites contain carbon fibers which block radio waves and rule out hidden antennas.

HF antennas. High Frequency antennas (for long range communications) began as a trailing wire reeled out by the pilot through a hole in the rear of the airplane. The length was adjusted to the frequency. With the development of antenna tuners, which electrically shorten or lengthen the wire, the HF antenna became a fixed wire strung from the tip of the rudder fin out to a wingtip or cabin roof. As big jets went into service, the HF antenna became a horizontal mast pointing forward on the rudder fin (as on the Boeing 707). HF antennas are now hidden behind plastic fairings and automatically tuned when the pilot selects a channel.

Location

The position of an antenna on an airframe greatly affects its life, performance and reliability. If the airplane is in corporate or air transport service, the airplane-maker has already determined the best antenna locations, so follow his guidance. Locations were carefully engineered and certified. If an antenna becomes defective, a replacement goes in the same location.

There are instances where new avionics are added to old airplanes and require new antennas. Again, if it

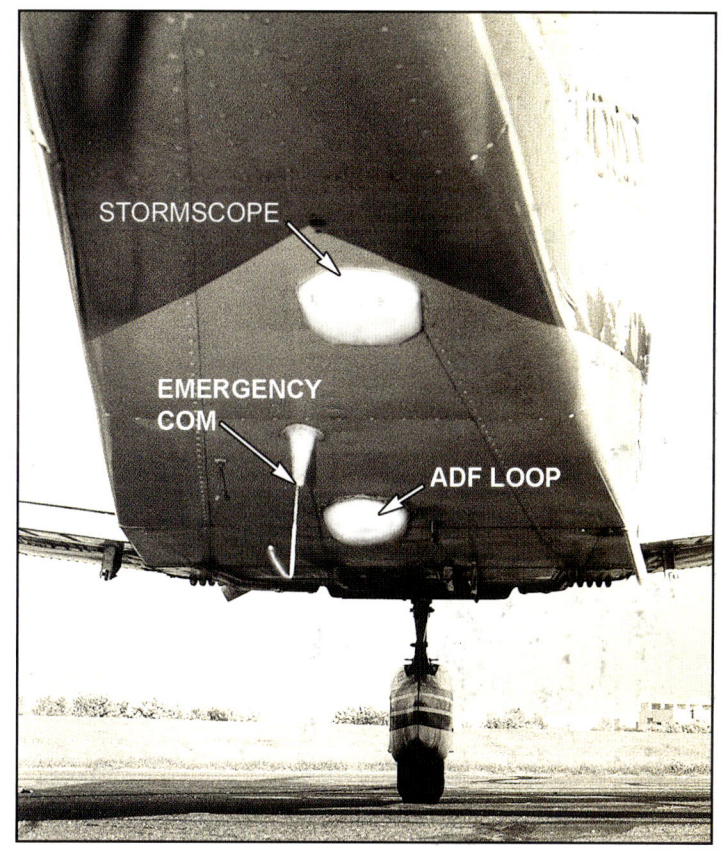

Belly of a single-engine aircraft. Stormscope antenna is for weather (lightning) detection. The ADF loop is forward. Between them is a "bent-whip" com antenna to connect a handie-talkie during radio or electrical failure. Some pilots also use it to listen for clearances while waiting on the ground with the engine off.

Notice that none of the three antennas straddles the center line of the airplane. Each is offset slightly to avoid cutting into a longeron, rib or other framing member.

245

is a large airplane, the manufacturer determines the location of the new antenna. He may have obtained an STC (Supplemental Type Certificate) which has engineering drawings of how and where the antenna is installed.

Never cut into the skin of a pressurized aircraft unless you have approved guidance material. Special techniques are required to keep the airframe air tight and weather proof.

For light aircraft in General Aviation, there is less guidance on antenna location. A starting point is using the same location already selected by the manufacturer for the airplane you're working on. If that's not practical, consider some recommended practices.

One precaution; walk around the tie-down area of almost any airport and you will see airplanes with poor antenna locations. The biggest mistake is putting antennas too close together---so don't use any aircraft as a model without considering some basic rules.

Flat base. Antennas work best when mounted on a flat surface. This keeps the rod or blade type pointing straight up. Little GPS antennas should lie horizontal for good 360-degree pickup. If an antenna is tilted, it will not operate equally in all directions. One exception

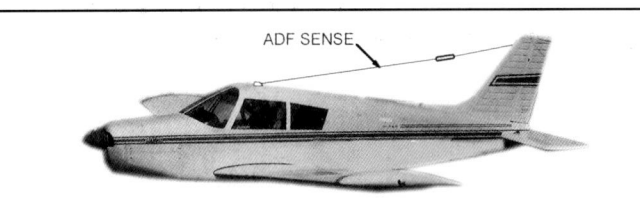

Nose cone is lifted on a corporate jet to reveal glideslope antenna. Weather radar antenna is inside the cone. Above glideslope is the weather radar transmitter-receiver.

is the "conformal" antenna, which is shaped to fit a curve on the airframe. Examples are shown in the chapter on satellite communication.

If an antenna is mounted on a slight curve on the fuselage, the manufacturer may provide a gasket to fill the crack between antenna base and aircraft skin. Don't overtighten mounting screws to eliminate the crack because deforms the aircraft skin. Some manufacturers offer a "mounting saddle" to match the base of an antenna to a deep curve. Even on flat surfaces there will be tiny cracks, but these are filled by a sealant.

Obstructions. Try to locate an antenna where it is not shaded by aircraft structure. This grows more important as frequency is higher; for DME, transponder and GPS, for example. A landing-gear door might block part of the signal on approach to landing, a critical phase of flight.

Spacing. Position antennas at least 36 inches apart. When com antennas are closer they interact and warp the signal pattern. The pilot notices this when he can't communicate the same distance in all directions.

If the antenna goes on the airplane belly, don't place it in a direct line behind an exhaust pipe. It will soon become covered with soot and scorched.

Hidden Structure. Even if you have an installation manual with the exact location of the antenna, always conduct your own survey. Look inside the skin where the antenna will mount to check for wiring harnesses, fuel lines, hydraulic pipes or other vital parts which could be struck by your drill.

Beware of the ribs, stringers and other primary structure of the airplane. Cutting them weakens the airframe. If there are no other antenna locations, you will need the services of a DER (Designated Engineering Representative) for the solution. Some shops employ a "structures" DER on staff, while others hire their services for each project.

When a mounting location is selected, see if you

ADF sense antenna on old aircraft is suspended between tail fin and top of cabin. The fitting on the wire contains a spring for taking up slack. By the 1990's, the outside sense antenna disappeared in new installations; it is now inside the ADF loop antenna housing, which is usually mounted on the belly.

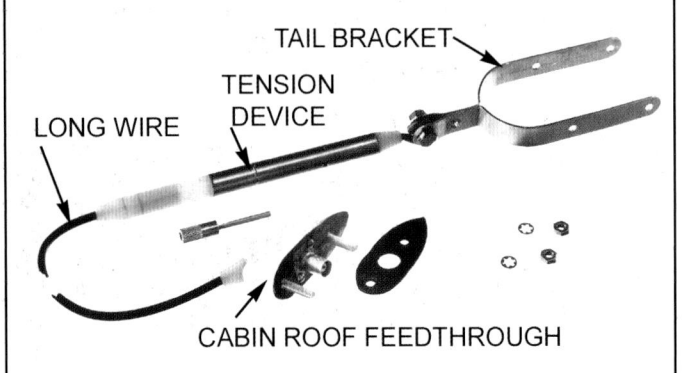

ADF sense antenna mounts above the cabin of a light aircraft. It consists of a long wire, tension device to keep the wire from sagging and a cabin-top feedthrough to bring the cable inside the aircraft.

Boeing 747 under construction. The nose radome is swung out for access to antennas; for localizer and weather radar.

Light aircraft usually place the localizer antenna on the tip of the rudder fin. But large aircraft approach at a high angle of attack, which would shadow a localizer antenna on the tail. The localizer antenna shown here is also on the Boeing 757/767/777 and Fokker 100.

The weather radar antenna also must have an unobstructed view ahead of the aircraft to "see" precipitation.

Although not shown, the glideslope antenna is usually placed in the nose for the same reason.

These antennas are protected from wind and weather by the radome, which is transparent to radio waves. Radomes operate in a hostile environment and need regular maintenance.

can reach (or crawl) inside the airplane to the area of the antenna base. You may need to install a doubler plate, attach an antenna connector or install strain relief for the cable from inside the airplane.

Doubler Plate

The thin skin of an airplane is not strong enough to support some antennas. To strengthen the mounting area, a metal plate, or "doubler," is installed under the skin, below the base of the antenna. Most VHF com antennas need such reinforcement, and smaller antennas for DME and transponder, as well.

Doublers (also called "backing plates") are available from three sources. Some antenna manufacturers provide them with the antenna; you can make one yourself from the drawing supplied with the antenna; or obtain a prefab doubler from an avionics parts distributor.

Examine the area under the skin, where the antenna will mount, and hold the doubler in place with your hand. Is it hitting any structure or components inside? It's alright to trim the doubler a small amount so it clears the obstruction---or move the antenna to a better location.

When a com antenna must be mounted on the belly of a light aircraft, use a "bent whip" to prevent striking the ground. Although not as efficient as a top-mounted antenna, the bent whip shown here was used for 10 years. It reliably communicated 50 miles in every direction at an altitude of 3000 feet,

247

Bonding the Antenna to the Airframe-1

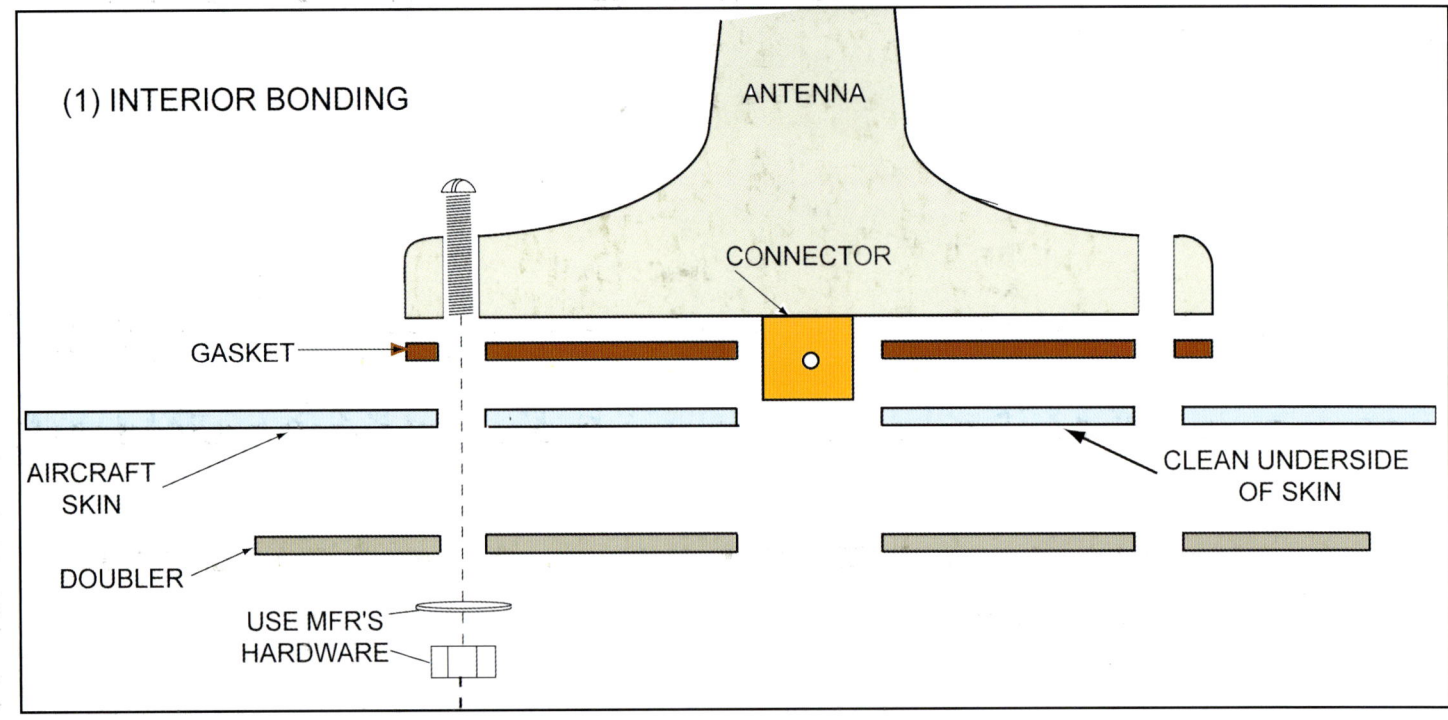

(1) INTERIOR BONDING

- ANTENNA
- CONNECTOR
- GASKET
- AIRCRAFT SKIN
- CLEAN UNDERSIDE OF SKIN
- DOUBLER
- USE MFR'S HARDWARE

In this approach, the antenna bonds through its mounting screws to the aircraft skin *inside* the fuselage. The inside area, therefore, must be clean and corrosion-free. After clean-up, apply Alodine to the bare skin to slow future corrosion.

The gasket keeps out moisture, vapors and contaminants. A bead of sealant is applied around the base of the antenna for further protection. In this mounting, the gasket does not have to be conductive.

Mark and Drill Antenna Location

Aluminum skin should *not* be marked with a pencil or other writing tool containing carbon. An investigation into one airline crash found that a row of rivets had been marked with an ordinary pencil. The pencil carbon interacted with aluminum, causing damage that led to the accident. A felt-tip marker is said to be harmless to aluminum.

Use the paper template supplied with the antenna to mark mounting and connector holes in the doubler (if it is not predrilled), or use the holes in the antenna base as a guide.

It's important to locate holes so the antenna is aligned with the fore and aft centerline of the airplane. This does not mean the antenna must be *on* the centerline and, in many instances, you cannot use the centerline because of primary structure below. That means the antenna is often offset slightly from the centerline---but be sure the antenna remains parallel to the centerline. You want the antenna to streamline with the wind and cause the least drag.

Hole sizes are shown on the installation drawing but common mounting screws are 8-32 or 10-32 stainless steel. Another consideration is the shape of the screw head; some antennas require a pan head screw, others call for a countersunk screw. Use the correct one to avoid damage to the antenna base.

To prevent loosening, a lock washer and flat washer are placed under the nut. If locknuts are used, only a flat washer is needed. Before tightening the antenna, however, there's another important step.

Bonding

Creating a tight contact between two metal surfaces with the least electrical resistance is "bonding." It's important to bond the base of the antenna to the skin of the airplane for several reasons. Most important, many antennas depend on the airframe as part of their electrical length. Without this action, there may be loss of signal or unequal signal in different directions. Bonding also affords some protection against lightning strikes by allowing the charge to flow through the metal skin.

Many antennas are "DC-grounded," which means that normal antenna signals move in and out of the antenna, but lightning is shunted directly to ground (the airframe), where it usually produces little damage. Bonding also improves radio reception because it uses the airframe as a shield against interference. It also helps

Bonding the Antenna to the Airframe-2

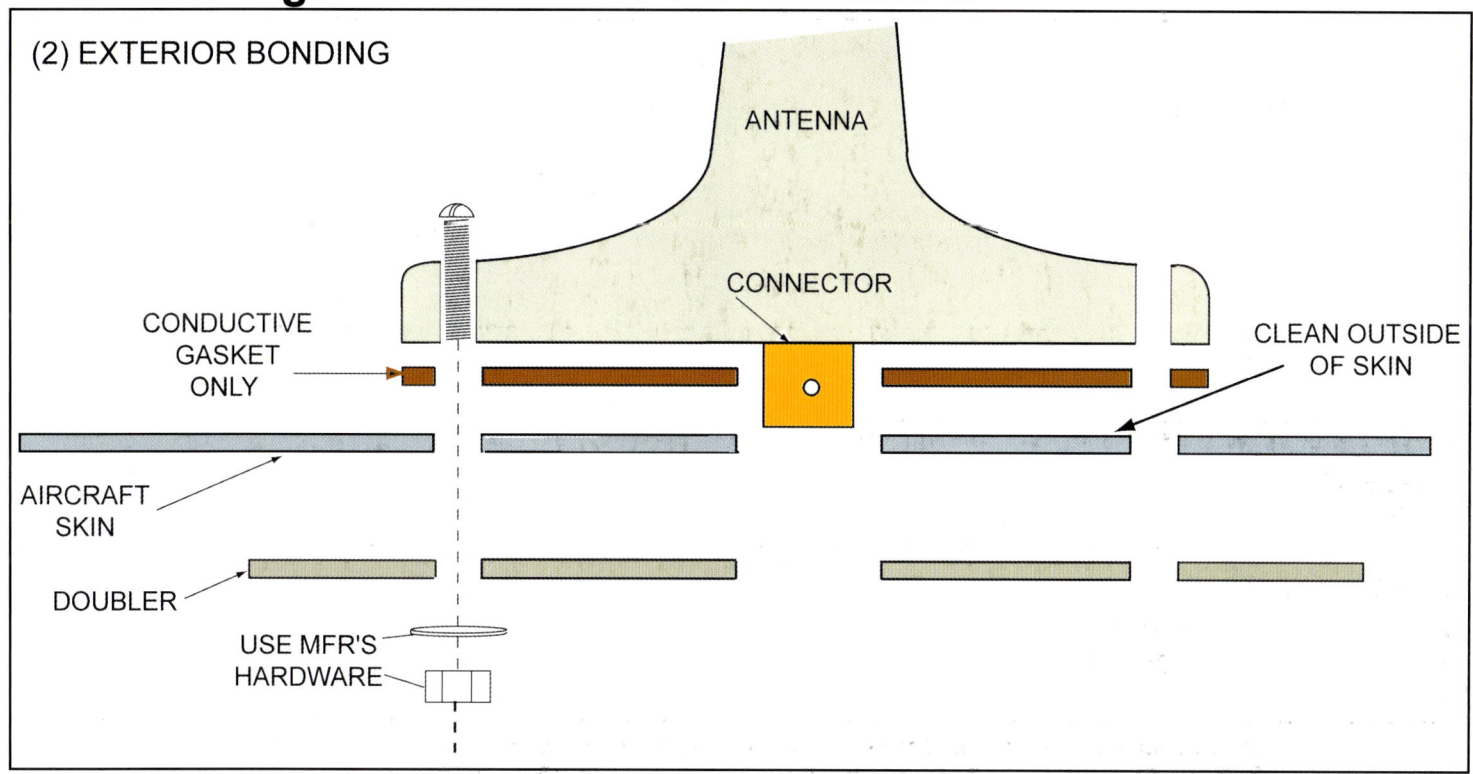

(2) EXTERIOR BONDING

- ANTENNA
- CONNECTOR
- CONDUCTIVE GASKET ONLY
- CLEAN OUTSIDE OF SKIN
- AIRCRAFT SKIN
- DOUBLER
- USE MFR'S HARDWARE

In the exterior method, bonding is made to the outside skin. Thus, the skin must be cleaned and treated with Alodyne to prevent corrosion. Only a *conductive* gasket can be used with exterior bonding; there must be good electrical contact with the outer skin.

CAUTION: **An antenna installation should be protected with sealant around the base and on the outside screwheads. Nuts or other mounting hardware inside the aircraft should also be sealed. This is important in pressurized aircraft. A well-sealed antenna prevents air and moisture from moving in and out of the fuselage during pressure changes in flight.**

drain away static electricity that builds during flight.

There are two methods for bonding an antenna and both are recommended by antenna manufacturers:

1. Interior bonding (see illustration). In this method, nothing is done to the outside skin of the aircraft except to drill mounting holes. However, the skin *inside* the airplane is carefully cleaned of paint, grease or other coating. To prevent corrosion, the interior area (where the doubler goes) is treated with Alodyne.

When the antenna is mounted, the doubler is squeezed against the inside skin for bonding. The electrical connection between antenna and doubler is made through the mounting screws. If a gasket is supplied, place it under the base of the antenna before mounting. The gasket, incidentally, does not have to be electrically conductive; the connection is made through the screws

2. Exterior bonding. In this approach, the area for the antenna base on the *outside* skin is cleaned and treated with Alodyne. If a gasket is placed at the base of the antenna, it must be the *conductive* type to maintain electrical bonding between antenna and airframe. The doubler is mounted inside, as described earlier.

Mount the Antenna

Before placing the antenna in its mounting position, the job is simplified if you pull the inside coaxial cable through the connector hole in the skin to the outside and fasten it to the antenna.

Another technique is to replace the antenna mounting nuts with "Rivnuts" (if not already supplied by the antenna maker). Rivnuts are riveted to the doubler plate and have threaded inserts to receive screws. This eliminates the need for another person to climb inside the airplane to hold the nuts while a person outside tightens or loosens them.

When using Rivnuts you must countersink them into the doubler plate so they lie flush with the surface. To prevent weakening the doubler, countersunk holes should not go deeper than one-half the thickness of the doubler.

Mount the antenna by inserting its connector into its hole in the skin, and line up the front mounting holes. Place the doubler plate into position inside the airplane and insert the two front mounting screws. Next, install the rear mounting screws and tighten the screws enough to draw the antenna base, gasket (if present), airplane skin and doubler plate together. Don't overtighten the hardware and, if possible, use a torque wrench:

    8-32 screws   20 in-lb
    10-32 screws   23 in-lb

Never paint an antenna. This detunes it and can cause loss of coverage and weak signals. The factory coating is all that's needed. Some antennas (radio altimeter, for example) are so sensitive, they carry the warning, "Do Not Paint."

Seal the Edges Any mounting system may produce a good electrical bond, but there will be tiny openings or cracks around the antenna base that could admit water. Seal the antenna by running a bead of RTV silicone sealant around the base.

For demanding applications, the installer can choose a high-performance sealant, such as "AC-236 Class B." It resists fuel and other contaminants found around airplanes and is applied in two parts.

Low Resistance. To check the quality of a bonding job, manufacturers suggest placing a meter between a bare screwhead on the antenna base and clean aluminum skin next to it. The reading on an ohmmeter should not exceed .003 ohms. Unfortunately, ordinary shop multimeters will not read that low. An accurate measurement requires a *milliohmmeter.*

On the subject of resistance, be aware that you cannot check an antenna by placing the probes of an ohmmeter from the center pin of the antenna connector to the outer connector shell. Many antennas read zero ohms (a dead short) because they contain a transformer that grounds the antenna for static electricity and lightning. The normal antenna signal ignores the transformer. Testing an antenna requires other techniques, described in the chapter on troubleshooting.

Coaxial Cable for Antennas

Nearly all antennas are connected to aircraft receivers and transmitters through coaxial cable. It carries radio frequencies with high efficiency through a center conductor surrounded by one or more shields. An important characteristic of coaxial cable is "impedance," a term which describes its resistance to alternating currents (in this case, radio frequencies). The standard impedance for most coaxial cable in aircraft is 50 ohms. Most aircraft antennas are also rated at 50 ohms, which provides an excellent match. Unless impedances are the same, there is little transfer of energy between the antenna, cable and radio. When something causes a mismatch (a broken wire, short circuit, etc.) radio energy during transmit fails to reach the antenna and reflects back to the transmitter.

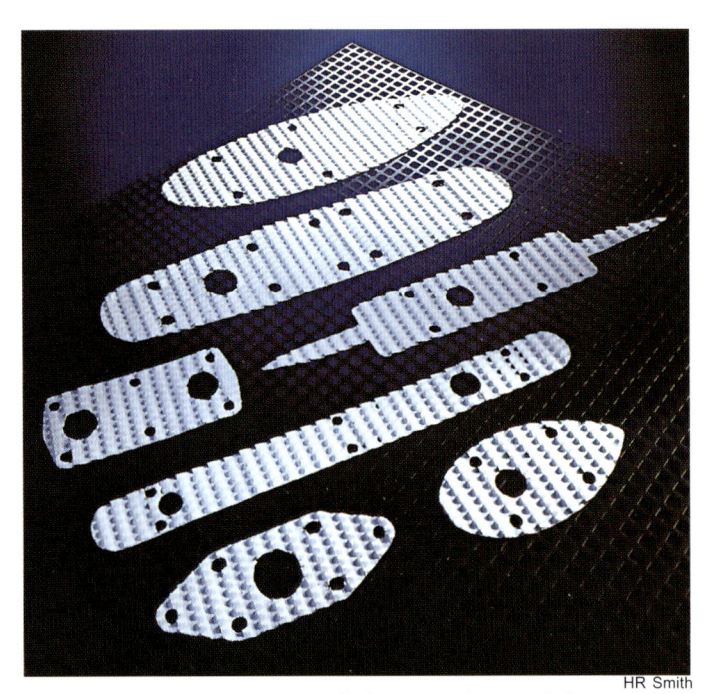

Conductive sealing gaskets fit between base of the antenna and airframe. Precut to size, they provide good electrical bonding.

The standard coaxial cable for avionics was RG-58 and there are still many miles of it in aircraft. But now it is ruled out for new installations because of a PVC jacket, which supports flame. Other coaxial cables, such as RG-400, have Teflon-based insulation that will not burn, as well as a double shield of copper braid for keeping desired signals in the cable and interference from getting in.

As frequencies rise higher in avionics equipment coaxial cable grows more important. When working with systems above the VHF band (satcom, satnav, TCAS and others) the manufacturer may call for a coaxial cable that performs more efficiently at higher frequencies. As shown in the illustration, never coil a coaxial cable tightly or its critical dimensions may change and cause losses.

As frequencies go higher, some manufacturers recommend RG-142 for coaxial cable. Typical applications include DME, transponder and UHF radiotelephones.

It's good practice to run RF coaxial cables separate from other wire bundles (which carry audio, DC power, data, etc.). RF cables usually contain heavy transmitter power that may induce interference into lower-level cables. Also, RF coaxial cables often carry weak receiver signals from the antenna, and are susceptible to picking up interference from other wiring.

Watch out for devices which require exact cable lengths. A "power combiner," which takes signals from a combined VOR/localizer/glideslope antenna, must have cables of exactly the same length. There are similar precautions for TCAS (collision avoidance), radar altimeters and other systems. These cables use their length as part of timing circuits so follow the manufacturer's precautions. Otherwise, run coaxial cables along the shortest practical route, but avoid sharp bends and tight coils.

Connectors. The most common connector for antennas is the bayonet-type BNC. If the application operates high in the radio spectrum, such as GPS, a lower-loss connector, such the TNC, is used. DME, transponders or UHF radiotelephone may also call for higher-performance connectors, such as TNC, C, N, or HN.

Duplexers. Although not commonly used, there is a device which enables two com transceivers to operate on one antenna. It has a switching arrangement that selects No. 1 or No. 2 transceiver. The circuit prevents the transmitted signal from one radio from overwhelming the receiver of the other, Most aircraft, however, have two separate VHF com antennas, which provide a backup if one antenna fails.

GPS Antennas

Antennas in the VHF band and below may install on the top or bottom of an airplane, but GPS antennas always mount on top. They need a clear view of the sky, as satellites rise from one horizon, climb to the zenith (overhead) then set on the opposite horizon.

Another reason for an unobstructed view is that GPS signals are extremely low in strength. Their frequencies are in the UHF band, where signals are easily shaded or reflected by aircraft structure.

A GPS antenna must lie flat. If mounted on a tilt, it will not "see" satellites low on the horizon.

No matter what location you choose there will be some obstructions at certain angles (a tail fin, a raised wing, etc.) but this is usually not critical. The GPS receiver typically gathers information from six or more satellites always in view.

Antenna Couplers

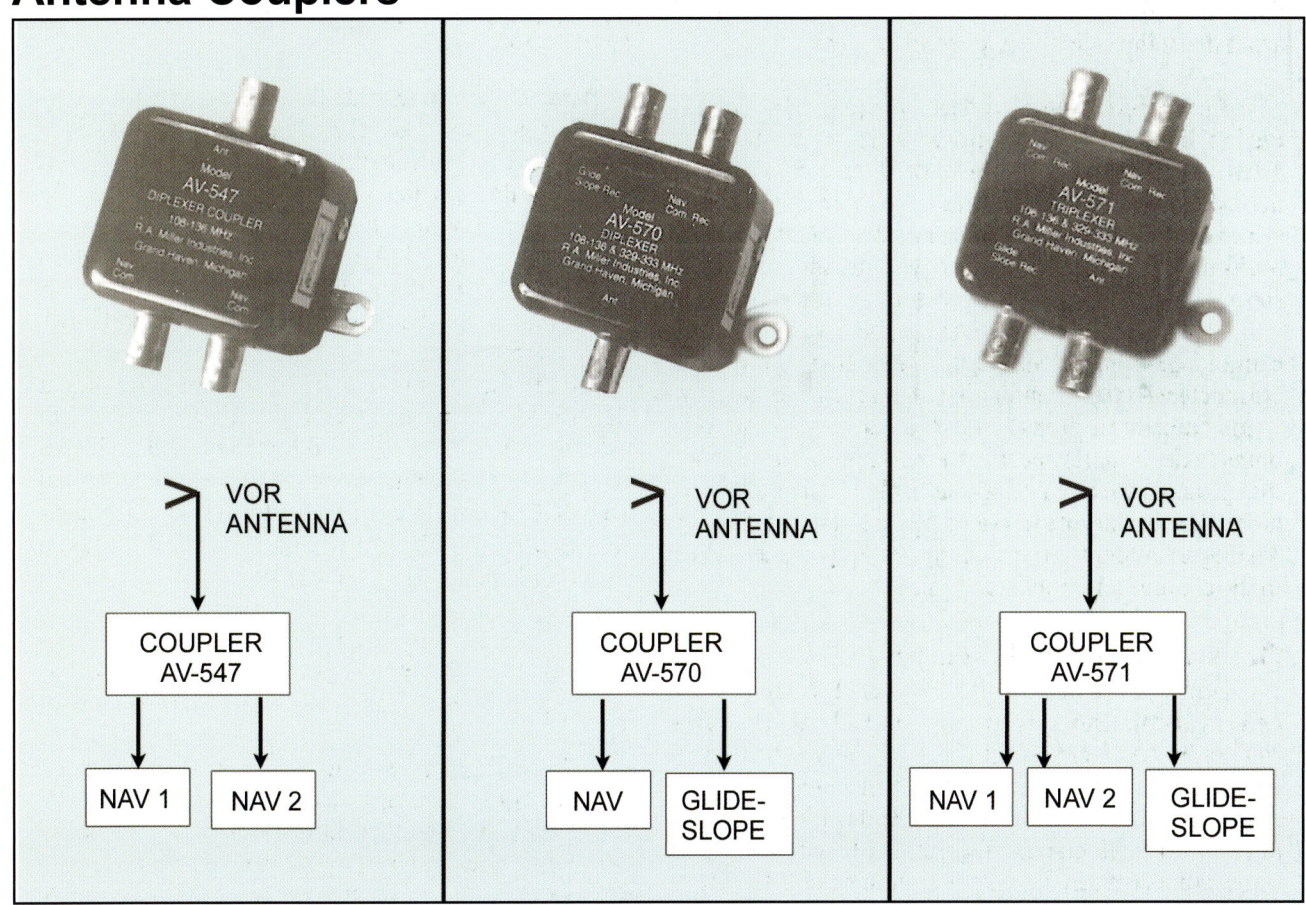

The coupler splits the signal from a single antenna to feed two VOR navigation receivers. As in most couplers, it is only for receivers, which operate on very low-level signals. All couplers reduce the antenna signal by splitting, but there is ample signal strength remaining to do the job.

This coupler splits the signal from a VOR antenna for one nav receiver (VOR) and one glideslope receiver. Although the VOR antenna is designed for a frequency of 108 to 118 MHz, the coupler retrieves glideslope signals on 329 to 333 MHz. All couplers shown here are "passive," they need no power source.

This coupler is found on more airplanes than any other type, especially light aircraft with basic IFR avionics. The coupler provides three outputs-- Nav 1, 2 and glideslope---all from a single VOR antenna.

251

Base Station and Mobile Antennas

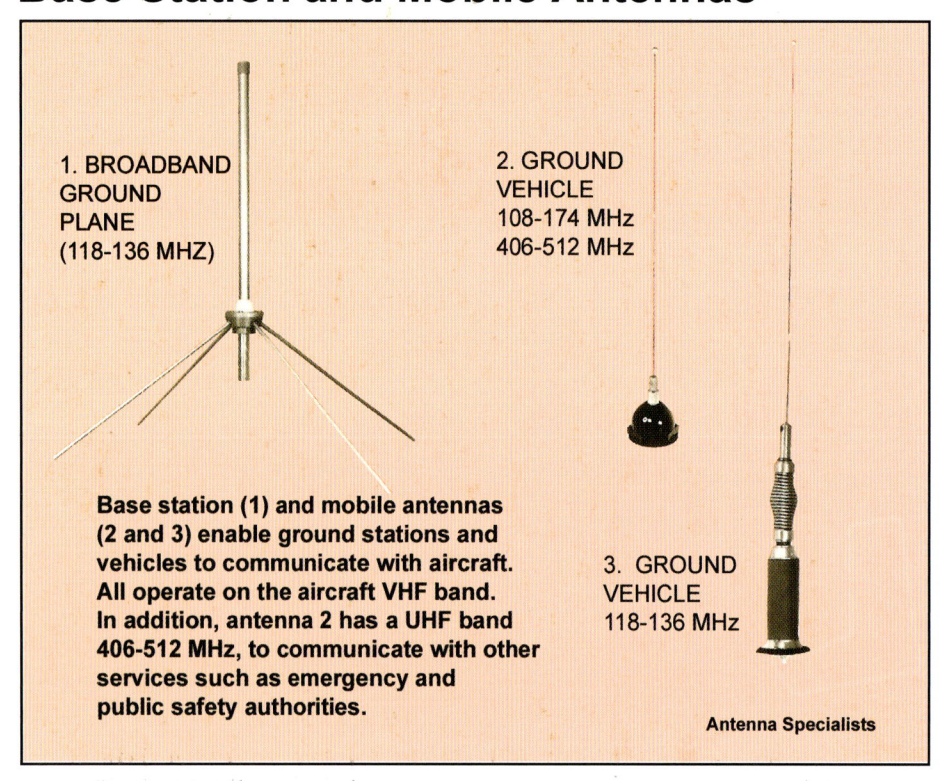

1. BROADBAND GROUND PLANE (118-136 MHZ)

2. GROUND VEHICLE 108-174 MHz 406-512 MHz

3. GROUND VEHICLE 118-136 MHz

Base station (1) and mobile antennas (2 and 3) enable ground stations and vehicles to communicate with aircraft. All operate on the aircraft VHF band. In addition, antenna 2 has a UHF band 406-512 MHz, to communicate with other services such as emergency and public safety authorities.

Antenna Specialists

The favorite location for a GPS antenna is atop the cabin, several feet behind the windscreen. Place the antenna too far forward, near the front edge of the windscreen, and you may induce interference. As wind moves over the windows it generates static electricity that affects GPS reception. Also, keep away from propellers and the shadow of larger antennas.

On high-wing aircraft, mount the GPS antenna on top, keeping it near the leading edge of the wing (but away from the windshield). This reduces shadowing of the GPS signal when the airplane banks and climbs.

On large aircraft with a satellite communications antenna keep the GPS antenna at least four feet away. These two services are close in frequency and may produce interference to the GPS.

It's good practice to avoid bundling the GPS coaxial cable with cables from transmitting equipment. VHF com transmitters operate lower in frequency than GPS, but they produce harmonics that can reach up to the GPS band and cause trouble.

Some manufacturers offer a combined GPS-VHF antenna, which saves the installation of one antenna. The problem of interference between them is solved by a "notch" filter, which reduces the power of VHF harmonics before they reach the GPS antenna.

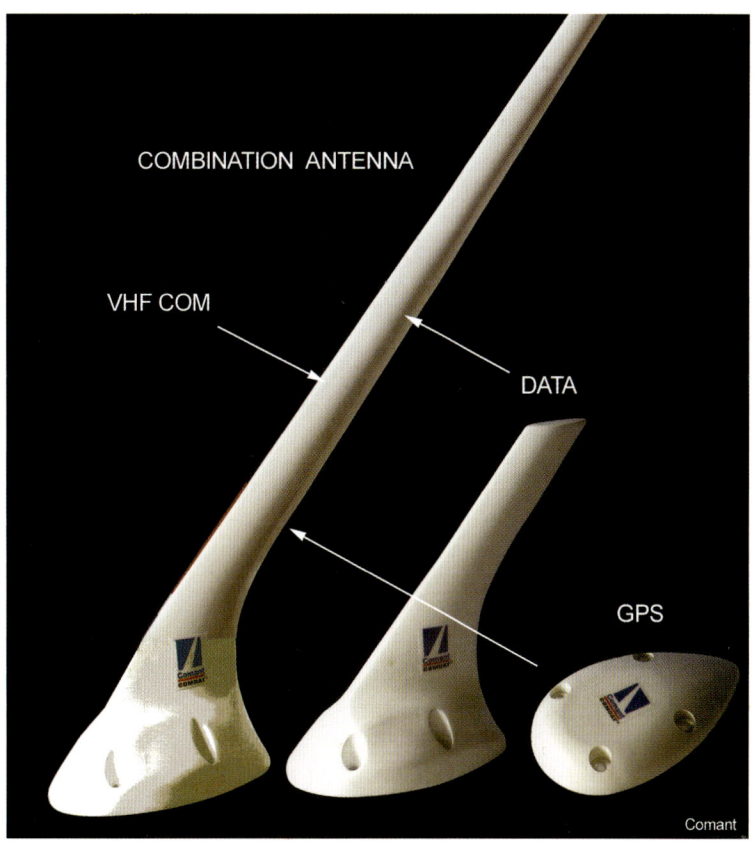

As antennas on aircraft multiply, manufacturers combine several into a single footprint. This example is ComDat; a VHF-com antenna at the left that also contains a GPS or data antenna. The combination not only saves space, it reduces installation time and effort. The antenna has two connectors in the base for attaching cables; a BNC-type for the com radio, and a TNC for GPS or data. The data radio handles such services as weather, messages and traffic information.

Review Questions
Chapter 27 Antenna Installation

27.1 What is the meaning of an arrow symbol on an antenna?

27.2 Why should antennas never be painted?

27.3 (A) What does VSWR mean? (B) What does it measure?

27.4 What is an "omnidirectional" antenna pattern?

27.5 What is antenna "impedance."

27.6 The matching device which connects a VOR antenna to its coaxial cable is called a _____.

27.7 What is a poor location for an antenna?

27.8 What is a major difference between antennas for aircraft that cruise below 200 kt and aircraft that fly above that speed?

27.9 To avoid interfering with each other, antennas should be located at least _____ inches apart.

27.10 Before mounting a new antenna, check the area inside the airplane for _____, and _____.

27.11 What is the purpose of a doubler plate?

27.12 How many ohms of resistance indicate a good electrical bond between antenna base and airframe?

27.13 What is the function of an antenna coupler?

27.14 Do not locate a GPS antenna on the cabin roof immediately behind the windshield. Why?

Chapter 28
Panel Labels and Abbreviations

According to regulation, all controls and indicators on an airplane must be clearly labelled and visible to the pilot. Unmarked items can be handled several ways:

Send it out
There are specialty shops which offer panel-finishing services that produce durable and high quality labels.

Silk Screen
After the panel is painted, you can apply lettering by the silk screen process. This applies painted letters directly on the surface of the panel, much like a factory-made panel. You first prepare a drawing showing the position and lettering and send it to a shop which produces the silk screen. The panel is placed against the silk screen and paint applied with a roller. The paint is forced through holes in the screen and applied to the panel.

Engraving Machine
Many shops use an engraving machine to make labels. It works like a router, using a spinning bit to cut into the top surface of the label. When the bit reaches the second layer, it exposes white material. The router is controlled by a guide to form each character as a white letter or numeral. The completed label has an adhesive back that adheres to the panel.

Tape Labels
You type the label on the keyboard of a small machine which creates lettering on a plastic tape. There are many different color tapes and type sizes. The tapes run from 1/4-in to 1-inch wide.

Preprinted labels
These are sheets of labels with labels printed on them. An adhesive backing attaches them. Such labels are the least durable method.

Terms on Labels
The terms or words on labels may read differently from one manufacturer to the next. If you don't know what an abbreviation should be, the following chart shows recommended standards.

Panel Abbreviations

Abbreviations for instrument panels on airline and military aircraft are standardized. In General Aviation, corporate and many commercial aircraft, the choice is often left to the installer.

Industry organizations recommend the following abbreviations, which may be used directly on the panel, the faces of electromechanical instruments, in symbols of an electronic display (EFIS), switches, buttons, placards, legends, controls and in diagrams.

When using abbreviations, there are several basic rules:

Do not use periods (.) in abbreviations; for example: the word **DECREASE** becomes **DECR**

An abbreviation might look confusing, as in **ISCTN** (for **INTERSECTION**). In these instances, a hyphen is inserted: **I-SCTN**

A forward slash (/) means the term has a dual function. Example: **OFF/RESET** is **OFF/R**

There are a few exceptions. Some terms, such as **GLIDESLOPE** have always been abbreviated as **G/S**. Although it is one function, the forward slash is still used.

Over 1,300 terms appear in the following table. Besides serving as a guide to marking instrument panels, they can also decipher abbreviations on existing equipment and in diagrams.

A

ABEAM	ABM
ABNORMAL	ABNORM
ABOVE	ABV
ABOVE GROUND LEVEL	AGL
ABOVE MEAN SEA LEVEL	AMSL
ABSOLUTE	ABS
ACCELERATION	ACCEL
ACCEPT	ACPT
ACCUMULATOR	ACCU
ACKNOWLEDGE	ACK
ACQUIRE	ACQ
ACTIVATE	ACTV
ACTIVE	ACT
ACTUAL GROSS WEIGHT	AGW
ACTUAL TIME OF ARRIVAL	ATA
ACTUAL TIME OF DEPARTURE	ATD
ACTUAL TIME OVER	ATO
ADDRESS	ADRS
ADJUST	ADJ
ADVANCE	ADV
ADVISORY	ADVY
ADVISORY ROUTE	ADR
AERONAUTICAL INFORMATION SERVICE, PUBLICATION	AIS, AIP
AERONAUTICAL MOBILE SERVICE, FIXED SERVICE	AMS, AFS
AILERON	AIL
AIR CONDITIONING	AIRCOND
AIRCRAFT	ACFT
AIRCRAFT INTEGRATED DATA	AID
AIRCRAFT NOSE DOWN	ND
AIRCRAFT NOSE UP	NU
AIRCRAFT OPERATING MANUAL	AOM
AIR CYCLE MACHINE	ACM
AIR DATA	AD
AIR DRIVEN GENERATOR	ADG
AIR DRIVEN PUMP	ADP
AIRPORT	ARPT
AIRPORT BEACON	ABN
AIRPORT REFERENCE POINT	ARP
AIRPORT TRAFFIC ZONE	ATZ
AIRBORNE REPORT	AIREP
AIRSPEED	AS
AIRSPEED INDICATOR	ASI
AIR TO GROUND	AG
AIR TURBINE MOTOR	ATM
AIR ROUTE TRAFFIC CONTROL (CENTER)	ARTC(C)
AIR TRAFFIC CONTROL	ATC
AIR TRAFFIC SERVICES	ATS
AIRWAY	AWY
ALARM	ALM
ALERT, ALERTING	ALRT
ALERTING SYSTEM DISPLAY, ALERTING DISPLAY	ASD
ALPHABET, IC	ALPHA
ALIGNMENT	ALN
ALTERNATE	ALTN
ALTERNATING CURRENT	AC
ALTIMETER	ALTM
ALTITUDE	ALT
ALL SPEED AILERON	ASA
ALL UP WEIGHT	AUW
AMBIENT	AMB
AMMETER	AMM
AMPERE	AMP
AMPERE HOUR	AH
AMPLIFIER	AMPL
AMPLITUDE MODULATION	AM
ANALOG TO DIGITAL	A/D
ANALYZER	ANAL
ANGLE OF ATTACK	AOA
ANGULAR	ANG
ANNUNCIATOR	ANN
ANTENNA	ANT
ANTI ICE	AICE ANTI
SKID	ASKID
APPROACH	APPR
APPROACH CONTROL	APP
AREA CONTROL CENTER	ACC
AREA POSITIVE CONTROL	APC
AREA NAVIGATION	RNAV
ARINC COMMUNICATIONS, ADDRESSING AND REPORTING SYSTEM	ACARS
ARRIVAL, ING	ARR
ARTIFICIAL	ARTF
AS SOON AS POSSIBLE	ASAP
ASSEMBLY	ASSY
ASSIGN	ASSN
ASYMMETRIC	ASYM
ATTENDANT	ATTND
ATTENTION	ATTN
ATTITUDE	ATT
ATTITUDE DIRECTOR INDICATOR	ADI
ATTITUDE- HEADING REFERENCE	AHR
AUDIO	AUD
AUDIO SELECTOR PANEL	ASP
AUGMENT, ER, ATION	AUG
AUTOMATIC	AUTO
AUTOMATIC BRAKE	AB
AUTOMATIC DIRECTION FINDER	ADF
AUTOMATIC FLIGHT CONTROL	AFC
AUTOMATIC FLIGHT GUIDANCE	AFG
AUTOMATIC FLIGHT	AF
AUTOMATIC FREQUENCY CONTROL	AFC
AUTOMATIC GAIN CONTROL	AGC
AUTOMATIC LANDING	ALAND
AUTOMATIC LANDING	AL
AUTOMATIC RUDDER CONTROL	A-RUD
AUTOMATIC TEST EQUIPMENT	ATE
AUTOMATIC THRUST	AT
AUTOMATIC POWER RESERVE	APR
AUTOPILOT	AP
AUTOPILOT AND FLIGHT DIRECTOR	AFD
AUTOPILOT/FLIGHT DIRECTOR	AP/FD
AUTOSTABILIZER	ASTAB
AUTOTHROTTLE	AT
AUXILIARY	AUX
AUXILIARY DATA ACQUISITION UNIT	ADAU
AUXILIARY POWER UNIT	APU
AVAILABLE	AVAIL
AVERAGE	AV
AVIONICS	AVNCS
AZIMUTH	AZ

B

BACK BEAM	BBM
BACK COURSE	BCRS
BACK LOCALIZER	BLOC
BAROMETRIC SETTING	BARO
BATTERY	BAT
BATTERY UNIT	BU
BEACON	BCN
BEARING	BRG
BEARING DEVIATION INDICATOR	BDI
BEAT FREQUENCY OSCILLATOR	BFO
BELOW	BLW
BETWEEN	BTWN
BEVERAGE	BEV
BLUE	BLU
BOTTLE	BTL
BOTTOM OF CLIMB	BOC
BOTTOM OF DESCENT	BOD
BOUNDARY	BNDRY
BOUNDARY LAYER CONTROL	BLC
BRAKE	BRK
BRAKING ACTION	BA
BREAKOFF HEIGHT	BOH
BRIGHT, NESS	BRT
BROADCAST	BCST
BUILT IN TEST EQUIPMENT	BITE
BUS TIE BREAKER	BTB
BUS TIE RELAY	BTR
BUS TRANSFER CONTACTOR	BTC

C

CABIN	CAB
CALIBRATION	CAL
CALIBRATED AIR SPEED	CAS
CALL SIGN	CS
CANCEL	CNCL

Panel Abbreviations

CAPACITY	CPTY	CLIMB SPEED MIN INITIAL	V3	CONTROL WHEEL STEERING	CWS
CAPTAIN	CAPT	CLIMB SPEED INITIAL	V4	CONTROL ZONE	CZ
CAPTURE	CAP	CLOCK	CLK	COOLER, ING	COOL
CARBURETOR	CARB	CLOCKWISE	CW	COPILOT	CP
CARD READER	CR	CLOSE, CLOSED	CLS, CLSD	CORRECT	CORR
CARGO	CRG			COST INDEX	CI
CONTINUOUS WAVE	CW	CLOUD	CLD	COUNTER CLOCKWISE	CCW
CATEGORY	CAT	COCKPIT	CKPT	COUPLE	CPL
CATHODE RAY TUBE	CRT	COCKPIT VOICE RECORDER	CVR	COUPLER	CPLR
CAUTION, ARY	CAUT	COLD	C	COURSE	CRS
CAUTION AND WARNING	CAW	COLD AIR UNIT	CAU		
CEILING	CLNG	COLLISION	COLL	COURSE DEVIATION INDICATOR	CDI
CEILING AND VIS		COLLISION AVOIDANCE	CA	COWLING	COWL
UNRESTRICTED/OK	CAVU/CAVOK	COMBINED SPEED INDICATOR	CSI	CREW MEMBER	CM
CENTIGRADE	C	COMMAND	CMD	CRITICAL ENGINE FAIL SPEED	V1
CENTIMETER	CM	COMMANDER	CMDR	CRITICAL POINT	CP
CENTER	CTR	COMMUNICATION	COM	CROSSCREW QUALIFICATION	CCQ
CENTER OF GRAVITY	CG	C0MPANY	CO	CROSS BAR	XBAR
CENTER LINE	C/L	COMPARATOR	CMPRTR	CROSS FEED	XFD
CENTRAL AIR DATA	CAD	COMPARTMENT	COMPT		
CENTRAL PROCESSOR	CP	COMPASS	COMP	CROSS LINE	XLN
CHANGE	CHG	COMPLETE	CMPL		
CHANNEL	CHAN	COMPRESSOR	CPRSR	CROSS TRACK DEVIATION	XTK
CHARGE, CHARGER	CHRG	COMPUTE	CMPT	CROSS TRACK ERROR	XTKE
CHECK	CK	COMPUTER	CMPTR	CROSS VALVE	XVLV
CHECKED	CKD	CONDITION	COND	CROSS WIND	XWND
CHECK LIST	CL	CONFIGURATION	CONFIG	CRUISE	CRZ
CHECK VALVE	CHKV	CONSTRAINT	CSTR	CRUISING SPEED DESIGN	VC
CHRONOMETER	CHR			CYLINDER	CYL
CIRCLE	CRCL	CONSTANT SPEED DRIVE	CSD		
CIRCUIT	CCT	CONTACTOR	CNTOR	**D**	
CIRCUIT BREAKER	CB	CONTINUE	CONT	DAMPER	DMPR
CIRCUIT BREAKER MONITORING	CBM			DANGER, -OUS	DNGR
CIRCULATE, CIRCULATION	CIRC	CONTINUOUS REPETITIVE CHIME	CRC	DANGER AREA	DA
CLEAR	CLR	CONTINUOUS WAVE	CW	DATA ENTR, DISPLAY PANEL	DEDP
CLEAR AIR TURBULENCE	CAT	CONTROL	CTL	DATA LINK	DL
CLEARANCE	CLNC	CONTROL AREA	CTA	DATA MANAGEMENT	DM
CLEARED	CLRD	CONTROL DISPLAY	CD	DATE	DAT
CLEARWAY	CWY	CONTROL SYSTEM ELECTRONIC	CSE	DEAD RECKONING	DR
CLIMB	CLB	CONTROL UNIT	CU	DECELERATE, ION	DECEL
				DECIBEL	DB
				DECISION ALTITUDE	DA
				DECISION HEIGHT	DH
				DECISION SPEED	V1
				DECK	DK
				DECLINATION	DEC
				DECREASE	DECR
				DEFINITION	DEF
				DEFLECTION	DEFL
				DEGREE	DEG
				DEICE	DICE
				DELAY	DLA
				DELAY MESSAGE	DEL
				DELAYED FLAP APPROACH	DFA
				DELETE	DEL
				DEPARTURE	DEP
				DEPARTURE MESSAGE	DEP
				DEPRESSURIZE, ATION	DPR
				DESCEND, DESCENT	DES
				DESIGN EYE POSITION	DEP
				DESIGNATOR	DSIG
				DESIRED TRACK	DTK
				DESTINATION	DEST
				DETECTED, TOR, TION	DET
				DEVIATION	DEV
				DIFFERENCE, TIAL	DIFF
				DIFFERENTIAL PRESSURE	dP
				DIGITAL AIDS RECORDER	DAR
				DIGITAL AIR DATA (Equipment)	DAD
				DIGITAL DISTANCE AND	
				RADIO MAGNETIC INDICATOR	DDRMI
				DIGITAL FLIGHT DATA	
				ACQUISITION UNIT	DFDAU
				DIGITAL FLIGHT DATA RECORDER	DFDR
				DIGITAL TO ANALOG	D/A
				DILUTER, ION	DIL
				DIRECT, ION, DIRECT TO	DIR
				DIRECT CURRENT	DC
				DIRECTIONAL GYRO	DG
				DIRECTION FINDER	DF
				DIRECT LIFT CONTROL	DLC
				DIRECT OPERATING COST	DOC
				DISARMED	DISRMD
				DISCHARGE	DISCH
				DISCONNECT	DISC

There are machines for printing labels with adhesive backing.

Kroy

Panel Abbreviations

DISENGAGED	DISENG	FLIGHT ENGINEER	FE	HIGH	HI
DISPLAY	DSPL	FLIGHT GUIDANCE	FG	HIGH FREQUENCY	HF
DISPLAY UNIT	DU	FLIGHT INFORMATION		HIGH LEVEL	HL
DISTANCE	DIST	CENTER	FIC	HIGH PRESSURE	HP
DISTANCE MEASURING EQUIPMENT	DME	FLIGHT INFORMATION REGION	FIR	HIGH SPEED	HSPD
DISTANCE TO GO	DTG	FLIGHT INFORMATION			
DIVERSION	DVRSN	REGION BOUNDARY	FIRB	HOLD, ING, HOLDING PATTERN	HLD
DIVERT	DVRT	FLIGHT INFORMATION SERVICE	FIS	HORIZONTAL NAVIGATION	HNAV
DOPPLER	DOPP	FLIGHT LEVEL	FL	HORIZONTAL SITUATION INDICATOR	HSI
DOUBLE SIDE BAND	DSB	FLIGHT LEVEL CHANGE	FL CH	HOT	H
DOWN	DN	FLIGHT MANAGEMENT SYSTEM	FM8	HOUR	HR
DRIFT	DFT	FLIGHT MODE ANNUNCIATOR	FMA	HYDRAULIC	HYD
DRIFT ANGLE	DA	FLIGHT NAVIGATION	FN		
DRIFT DOWN	DD	FLIGHT PATH ANGLE	FPA	**I**	
DRIVE	DRV	FLIGHT PATH TARGET	FPT		
DRY OPERATING WEIGHT	DOW	FLIGHT PATH VECTOR	FPV	ICING	ICE
		FLIGHT PLAN	FPLN	IDENTIFICATION BEACON	IBN
E		FLIGHT WARNING	FW	IDENTIFY, IER, ICATION	ID
EARTH ACCELERATION	g	FLIGHT (WEATHER) FORECAST	FIFOR		
EAST	E	FLOW	FLO	IGNITION	IGN
ECONOMIC, MY	ECON	FLUORESCENT	FLUOR	IMMEDIATE	IMM
EFFECTIVE	EFF	FOOT, FEET	FT		
ELAPSED TIME	ET	FORECAST	FCST	INBOARD	INBD
ELECTRIC, AL, ITY	ELEC	FOREIGN OBJECT DAMAGE	FOD	INBOUND	INB
ELECTROMAGNETIC INTERFERENCE	EMI	FORWARD	FWD		
ELECTRONIC	ELEX	FREEZE	FRZ	INCHES	IN(S)
ELECTRONIC ADI	EADI	FREQUENCY	FREQ	INCLUDE, SIVE	INC
ELECTRONIC CENTRALIZED		FREQUENCY MODULATION	FM		
AIRCRAFT MONITOR	ECAM	FROM	FR	INCREASE, ING	INCR
ELECTRONIC COOLING SYSTEM	ECS	FRONT	FRT	INDEX	INDX
ELECTRONIC ENGINE CONTROL	EEC	FUEL CONTROL UNIT	FCU	INDICATED AIRSPEED	IAS
ELECTRONIC FUEL CONTROL	EFC	FUEL FLOW	FF	INDICATOR, ION, ED	IND
ELECTRONIC FLIGHT INSTRUMENT	EFI	FUEL ON BOARD	FOB	INERTIAL	I
ELECTRONIC HSI	EHSI	FUEL OVER DESTINATION	FOD	INERTIAL NAVIGATION	IN
EXPAND, SION	EXP	FUEL QUANTITY	FQ	INERTIAL NAVIGATION SYSTEM	INS
EXPECTED APPROACH TIME	EAT	FUEL USED	FU	INERTIAL REFERENCE	IR
EXPEDITE	XPED	FUNCTION	FCTN	INERTIAL SENSOR	IS
EXTEND, -ED	EXTD	FUSELAGE	FUS	IN-FLIGHT ENTERTAINMENT	IFE
EXTENDED FUNCTION	XFCN			INFORMATION	INFO
EXTERNAL, EXTERIOR	EXT	**G**		INHIBITION	INHIB
EXTERNAL POWER CONTACTOR	EPC	GALLEY	GLY		
EXTINGUISH, -ER	EXTING	GALLON	GAL	INITIAL, ATION, IZATION	INIT
EXTRA FUNCTION	XFCN	GENERAL	ENL	INITIAL APPROACH	INA
EXTRACT	EXTR	GENERATOR	GEN	INITIAL APPROACH FIX	IAF
EXTRACT REFERENCE POSITION	ERP	GENERATOR CONTROL UNIT	GCU	INITIAL POINT	IP
		GENERATOR LINE CONTACTOR	GLC	INJECTION	INJ
F		GENERATOR LINE RELAY	GLR	INLET (INTAKE) GUIDE VANES	IGV
FAHRENHEIT	F	GEOGRAPHICAL REFERENCE POINT	GRP	INNER	INR
FAILURE, FAILED	FAIL	GO AROUND	GA	INNER MARKER	IM
FAN SPEED	N1	GLARESHIELD	GSHLD	INOPERATIVE	INOP
FAST/SLOW	F/S	GLIDE SLOPE	G/S		
FAULT ISOLATION MONITOR	FIM	GLIDE PATH	GP	INPUT/OUTPUT	I/0
FAULT REPORTING MANUAL	FRM	GLOBAL POSITIONING SYSTEM	GPS	INSTRUMENT	INST
FEEL SIMULATION UNIT	FSU	GRAM	G	INSTRUMENT FLIGHT RULES	IFR
FEET, FOOT	FT	GRAVITY	g	INSTRUMENT LANDING SYSTEM	ILS
FEET PER MINUTE	FPM	GREEN	GRN	INSTRUMENT METEOROLOGICAL	
FIELD ELEVATION PRESSURE	QFE	GREENWICH MEAN TIME	GMT	CONDITIONS	IMC
FIGURE	FIG	GROSS WEIGHT	GW	INTAKE	INTK
FILAMENT	FIL	GROUND	GND	INTEGRATED DRIVE GENERATOR	IDG
FILTER	FLTR	GROUND CONTROLLED APPROACH	GCA	INTERCEPT	INTCP
FINAL T 0 SAFETY SPEED	VFTO	GROUND MOVEMENT CONTROL	GMC	INTERCEPT POINT	IPT
FINAL APPROACH FIX	FAF	GROUND MOVEMENT RADAR	GMR		
FIRST OFFICER	FO	GROUND POWER UNIT	GPU	INTERCOM, INTERCOMMUNICATION	I/C
FINAL APPROACH	FNA	GROUND PROXIMITY WARNING	GPW	INTERLOCK	INTLK
FLAP	FLP	GROUND REFERENCE POINT	GRP	INTERMEDIATE FIX	IF
FLAP EXTENDED SPEED MAX	VFE	GROUND ROLL	GR	INTERMEDIATE PRESSURE	IPR
FLAP OPERATING SPEED MAX	VFO	GROUND SPEED	GS	INTERNATIONAL	INTL
FLARE	FLR	GROUND TO AIR	GA	INTERNATIONAL STANDARD	
FLEXIBLE	FLEX	GUST SPEED DESIGN MAX	VB	ATMOSPHERE	ISA
FLIGHT AUGMENTATION COMPUTER	FAC	GYRO	G	INTERPHONE	INPH
FLIGHT	FLT			INTERRUPT, ED, ION	INTRP
FLIGHT CONTROL	FC	**H**		INTERSECTION	ISCTN
FLIGHT CREW OPERATING MANUAL	FCOM	HANDLE	HNDL	INTER SYSTEM BUS	ISB
FLIGHT DATA ACQUISITION UNIT	FDAU	HEAD	HD	INVERTER	INV
FLIGHT DATA ENTRY PANEL	FDEP	HEADING	HDG	ISOLATE, ED, ION	ISOL
FLIGHT DATA RECORDER	FDR	HEADING AND ATTITUDE	HA		
FLIGHT DATA STORAGE UNIT	FDSU	HEADING SELECT	HDGS	**J**	
FLIGHT DECK	FDK	HEAD UP DISPLAY	HUD	JAMMED, -ING	JAM
FLIGHT DATA ENTRY PANEL	FDEP	HEAD WIND	HWND	JET PIPE TEMPERATURE	PT
FLIGHT DATA RECORDER	FDR	HEAT	HT	JUNCTION	JCT
FLIGHT DATA STORAGE UNIT	FDSU	HEATER	HTR		
FLIGHT DECK	F-DK	HEIGHT	H	**K**	
FLIGHT DIRECTOR	FD	HELICOPTER	HEL	KEYBOARD	KYBD
		HERTZ	HZ	KILO	K

257

Panel Abbreviations

KRUGER	KRUG	MASTER	MSTR	**O**	
KNOTS)	KT(S)	MASTER WARNING	MW	OBSERVED, -ATION	OBSV
		MAXIMUM	MAX	OBSERVER	OBSR
L		MAXIMUM CONTINUOUS THRUST	MCT	OBSTACLE, OBSTRUCTION	OBST
LANDING	LDG	MAXIMUM CRUISE	MCR	OBSTACLE CLEARANCE LIMIT	OCL
LANDING DIRECTION INDICATION	LDI	MAXIMUM LANDING WEIGHT	MLW	OFF/RESET	OFF/R
LANDING GEAR	LG	MAXIMUM RAMP WEIGHT	MRW	OFFSET	OFST
LANDING GEAR EXTENDED		MAXIMUM OPERATING SPEED MACH	MMO	OMEGA	O
SPEED MAX	VLE	MAXIMUM OPERATING SPEED AIRSPEED	VMO	OMEGA NAVIGATION	ON
LANDING GEAR OPERATING		MAXIMUM TAXI WEIGHT	MTXW	OMNI BEARING SELECTOR	OBS
SPEED MAX	VLO	MAXIMUM TAKEOFF GROSS WEIGHT	MTOGW	ON BOARD BALANCE SYSTEM	OBBS
LANDING REFERENCE SPEED	VREF	MAXIMUM TAKEOFF WEIGHT	MTOW	ON BOARD WEIGHING SYSTEM	OBWS
LANDING WEIGHT	LW	MAXIMUM ZERO FUEL WEIGHT	MZFW	ON BOARD WEIGHT AND	
LASER	LSR	MEAN AERODYNAMIC CHORD	MAC	BALANCE SYSTEM	OBWBS
LATERAL, LATITUDE	LAT	MEAN SEA LEVEL	MSL	OPEN	O
LATERAL NAVIGATION	LNAV	MEAN TIME BETWEEN FAILURE	MTBF	OPERATIONAL	OP
LATERAL REVISE	LREV	MECHANIC	MECH	OPERATIVE, ING	OPER
LATCH	LTCH	MEDIUM	MED	OPERATION	OPS
LATITUDE	LAT	MEDIUM FREQUENCY	MF	OPERATION, ING MANUAL	OM
LATITUDE/LONGITUDE	L/L	MEGA	M	OPPOSITE	OPP
LAVATORY	LAV	MEMORY/MEMORANDUM	MEM	OPTIMUM	OPT
LEADING EDGE	LE	MERCURY	Hg	OUTBOARD	OUTBD
LEAST SIGNIFICANT BIT	LSB	MESSAGE	MSG	OUTBOUND	OUTB
LEFT	L	METER	M	OUTER	OUTR
LEFT HAND (SIDE)	LH	METERING FIX	MF	OUTER MARKER	OM
LEFT LINE SELECT (KEY)	LLS	METEOROLOGICAL	MET	OUTLET	OUTL
LEFT WING DOWN	LWD	MICROPHONE	MIC	OUTPUT/INPUT	O/I
LENGTH	L	MICROWAVE LANDING SYSTEM	MLS	OUTSIDE AIR TEMPERATURE	OAT
LEVEL	LVL	MIDDLE, MIDSHIP	MID	OVERBOARD	OVBD
LEVEL CHANGE	CHG	MIDDLE MARKER	MM	OVERFLOW	OVFL
LEVER	LVR	MILE	MI	OVERHEAD	OVHD
LIGHT, LIGHTS	LT	MILES PER HOUR	MPH	OVERHEAT	OVHT
LIGHT EMITTING DIODE	LED	MILLIBAR(S)	Mb	OVERLOAD	OVLD
LIMIT	LIM	MINIMUM	MIN	OVER PRESSURE	OVPR
LIMITER	LMTR	MINIMUM APPROACH BREAKOFF HEIGHT	MABH	OVERRIDE	OVRD
LINEAR	LIN	MINIMUM CONTROL SPEED	VMC	OVERRUN	OVRN
LINE REPLACEABLE UNIT	LRU	MINIMUM DECISION/DESCENT ALTITUDE	MDA	OVERSPEED	OVSPD
LINE SELECT KEY	LSK	MINIMUM EN ROUTE ALTITUDE	MEA	OXYGEN	OXY
LIQUID OXYGEN	LOX	MINIMUM EQUIPMENT LIST	MEL		
LITER	L	MINIMUM OPERATING SPEED	VMIN	**P**	
LOAD	LD	MINIMUM REQUIREMENT LIST	MRL	PACK	PK
LOAD CLASSIFICATION NUMBER	LCN	MINIMUM SAFE ALTITUDE	MSA	PAGE	PG
LOAD RELIEF SYSTEM	LRS	MINIMUM SECTOR FUEL	MSF	PANEL	PNL
LOAD SHEDDING RELAY	LSR	MINIMUM ZERO FUEL WEIGHT	MZFW	PARALLEL TRACK	PTK
LOCAL	LCL	MINUTE	MN	PARAVISUAL DIRECTOR	PVD
LOCAL MEAN TIME	LMT	MISCELLANEOUS	MISC	PARK	PK
LOCALIZER	LOC	MISSED	MSD	PASSENGERS	PAX
LOCALIZER INERTIAL SMOOTHING	LIS	MISSED APPROACH POINT	MAP	PASSENGER ADDRESS	PA
LOCALIZER TYPE DIRECTIONAL AID	LDA	MISSED APPROACH PROCEDURE	MA PROC	PASSENGER ENTERTAINMENT	
LOCATION	LCTN	MODERATE	MOD	SYSTEM	PES
LOCATOR	LCTR	MODE CONTROL	MC	PATTERN	PAT
LOCKED	LKD	MODE SELECT PANEL	MSP	PEDAL	PEDL
LONG RANGE CRUISE	LRC	MODIFY, MODIFICATION	MOD	PEDESTAL	PED
LONG RANGE NAVIGATION	LORAN	MODULAR CONCEPT UNIT	MCU	PER CENT	%
LONGITUDE, -INAL	LONG	MONITOR, ED	MON	PERFORMANCE	PERF
LOW	LO	MOTOR	MOT	PERFORMANCE MANAGEMENT	PM
LOWER	LWR	MOUNTING	MTG	PILOT	PLT
LOWER SIDE BAND	LSB	MULTIFUNCTION DISPLAY	MFD	PILOT FLYING	PF
LOW FREQUENCY	LF	MULTIPLEX	MUX	PILOT IN COMMAND	PIC
LOW LEVEL	LL			PILOT NOT FLYING	PNF
LOW PRESSURE	LP	**N**		PITCH	PTCH
LOW SPEED	LSPD	NACELLE	NAC	PITCH AUGMENTATION COMPUTER	PAC
LOW SPEED AILERON	LSA	NAUTICAL MILES	NM	PLAN, (FLIGHT PLAN)	PLN
		NAUTICAL MILES PER HOUR (KNOTS)	KTS	PLAN POSITION INDICATOR	PPI
M		NAVIGATION/COMMUNICATION	NAVCOM	PLANNED LANDING WEIGHT	PLW
MACH	M	NAVIGATION, NAVIGATOR	NAV	PLATFORM	PLATF
MACH/AIRSPEED INDICATOR	M/ASI	NAVIGATION AID	NAVAID	PNEUMATIC	PNEU
MAGNETIC	MAG	NAVIGATION DISPLAY	ND	POINT	PT
MAGNETIC INDICATOR	MI	NAVIGATION REFERENCE POINT	NRP	POINT OF NO RETURN	PNR
MAIN LANDING GEAR	MLG	NEGATIVE	NEG	POSITION	POS
MAINTAIN, MAINTENANCE	MAINT	NEXT	NXT	POSITIVE	POS
MAINTENANCE CONTROL		NEXT PAGE	NP	POSSIBLE	PSBL
AND DISPLAY PANEL	MCDP	NEUTRAL	NEUT	POTABLE WATER	POTW
MAINTENANCE TEST PANEL	MTP	NOISE ABATE, ABATEMENT	NA	POUND	LB
MALFUNCTION	MALF	NON DIRECTIONAL BEACON	NDB	POUNDS PER SQUARE INCH	PSI
MANDATORY OCCURRENCE REPORT	MOR	NON RETURN VALVE	NRV	POWER	PWR
MANIFOLD	MANF	NORMAL	NORM	POWER CONTROL UNIT	PCU
MANEUVER SPEED DESIGN	VA	NORTH	N	POWER TRANSFER UNIT	PTU
MANEUVER SPEED RECOMMENDED	VMAN	NOSE LAUDING GEAR	NLG	PRECISION APPROACH	
MANUAL	MAN	NOSE WHEEL	NW	PATH INDICATOR	PAPI
MARKER	MKR	NOTICE TO AIRMEN	NOTAM	PRECISION APPROACH RADAR	PAR
MARKER BEACON	MB	NUMBER	#	PREDICTED, TION	PRED
				PREDICTED WIND DATA	PWD
				PREFERRED	PREFD

Panel Abbreviations

PREFLIGHT	PREFLT	REPORTING POINT	RPT	SPEED REFERENCE SYSTEM	SRS
PRESENT	PRES	REQUEST	REQ	SPOILER	SPLR
PRESENT POSITION	POS	REQUIRED	REQD	SQUELCH	SQ
PRESSURE	PRESS	REQUIRED TIME OF ARRIVAL	RTA	SQUIB	SQB
PRESSURIZATION	PRESN	RESERVE, ATION	RSV	STABILITY, IZER, IZATION	STAB
PREVIOUS	PREV	RESERVOIR	RSVR	STALL SPEED	VS
PRIMARY	PRI	RESET	RST	STANDARD	STD
PRIMARY FLIGHT DISPLAY	PFD	RESPONDER BEACON	RSP	STANDARD INSTRUMENT DEPARTURE	SID
PRINTER	PRTR	RESTRICTED, TION	REST	STANDARD OPERATING PROCEDURE	SOP
PROCEDURE	PROC	RETARD	RETD	STEADY	STDY
PROCEDURE TURN	PROCT	RETRACT, ED, ABLE	RETR	STEERING	STRG
PROFILE	PROF	RETURN	RET	STEP CLIMB	STEP CLB
PROGRAM	PRGM	REVERSE, REVISION, REVISE	REV	STEP DESCENT	STEP DES
PROGRESS	PROG	REVERSE CURRENT PROTECTION	RCP	STOPWAY	SWY
PROPELLOR	PROP	REVERSER	RVSR	STRAIGHT IN APPROACH	STA
PROTECTION	PROT	REVOLUTIONs PER MINUTE	RPM	SUBSTITUTE	SUB
PROXIMITY	PROX	RIGHT	R	SUPPLY	SPLY
PURSER	PURS	RIGHT HAND (SIDE)	RH	SURFACE	SURF
PUSH BUTTON	PB	RIGHT LINE SELECT (KEY)	RLS	SURVEILLANCE RADAR	SR
PUSH (PRESS) TO ARM	PTA	RIGHT WING DOWN	RWD	SWITCH	SW
PUSH (PRESS) TO CANCEL	PTC	ROTARY	RTRY	SYMBOL GENERATOR UNIT	SGU
PUSH (PRESS) TO RESET	PTR	ROTATION SPEED	VR	SYMMETRICAL	SYM
PUSH (PRESS) TO SET	PTS	ROTOR	RTR	SYNCHRONIZE, ATION	SYNC
PUSH (PRESS) TO TALK/TRANSMIT	PTT	ROUGH AIR SPEED	VRA	SYSTEM	SYS
PUSH (PRESS) TO TEST	PTT	ROUTE	RTE	SYSTEM DATA ANALOG CONVERTER	SDAC
PUSH (PRESS) TO INHIBIT	PTI	ROUTE SURVEILLANCE RADAR	RSR	SYSTEM DISPLAY	SD
		RUDDER	RUD		
Q		RUNWAY	RWY		
QUADRANT	QUAD	RUNWAY HEADING	QFU	**T**	
QUANTITY	QTY	RUNWAY VISUAL RANGE	RVR	TACHOMETER	TACH
QUART	QT			TACTICAL	TAC
QUICK ACCESS RECORDER	QAR	**S**		TACTICAL AIR NAVIGATION	ACAN
		SATELLITE	SAT	TAILPLANE INCIDENCE	TPI
R		SCREEN	SCRN	TAILWIND	TWND
RADIO	RAD	SEA LEVEL	SL	TAKEOFF	TO
RADAR	RDR	SEA LEVEL PRESSURE	QNH	TAKEOFF DISTANCE	TOD
RADAR HEIGHT/RADIO HEIGHT	RHT	SECONDS	SEC	TAKEOFF FINAL SAFETY SPEED	VFTO
RADIAL	R	SECOND, SECONDARY	2ND	TAKEOFF FUEL	TOF
RADIAL/DISTANCE	R/D	SECONDARY SURVEILLANCE RADAR	SSR	TAKEOFF/GO AROUND	TO/GA
RADIO ALTIMETER	RA	SECTION	SECT	TAKEOFF GROSS WEIGHT	TOGW
RADIO/INERTIAL	R/I	SECTOR	SCTR	TAKE OFF SAFETY SPEED	V2
RADIO DISTANCE MAGNETIC INDICATOR	RDMI	SEGMENT	SEG	TAKEOFF WIGHT	TOW
		SELECT, ED, ION, OR	SEL	TANK	TNK
RADIO MAGNETIC INDICATOR	RMI	SELECTIVE CALL	SELCAL	TARGET	TRGT
RADIO MANAGEMENT	RM	SENSITIVITY	SENS	TAXI WAY	TWY
RADIO NAVIGATION INDICATOR	RNI	SEPARATE, TION	SEP	TELEPHONE	TEL
RADIO RANGE	RNG	SEQUENCE, TIAL	SEQ	TELETYPEWRITER	TT
RADIO TELETYPEWRITER	RTT	SERVICE	SERV	TEMPERATURE	TEMP
RADIO TELEPHONE	RT	SERVICING	SVNG	TERMINAL	TML
RAM AIR TEMPERATURE/TURBINE	RAT	SERVICEABLE	SVBL	TERMINAL AREA SURVEILLANCE RADAR	TAR
RANDOM (AREA) NAVIGATION	RNAV	SERVO ALTIMETER	SVALT	TERMINAL (MOVEMENT) AREA	TMA
RANGE	RNG	SERVO MOTOR	SVMOT	TERMINAL FORECAST	TAF
RATE OF CLIMB	ROC	SHAKER	SHKR	TERRAIN	TERR
RATE OF DESCENT	ROD	SHED, -DING	SHED	TEST	TST
READY	RDY	SHIELD, ING	SHLD	THRESHOLD SPEED TARGET	VAT
REAR (PART)	AFT	SHOP REPLACEABLE UNIT	SRU	THROTTLE	THROT
RECALL	RCL	SHORT	SHT	THRUST	THR
RECEIVE	RCV	SHORT TAKE OFF AND LANDING	STOL	THRUST CONTROL	TC
RECEIVER	RCVR	SHORT TAKE OFF AND VERTICAL LANDING	STOVL	THRUST MANAGEMENT	TM
RECIRCULATION	RECIRC			THRUST RATING	TR
RECLEAR, ED	RCLR	SHUTTER	SHTR	THRUST REVERSER	HR REV
RECORDER	RCDR	SIGNAL	SIG	TIME	THE
RECTIFIER	RECT	SIMULATE, -TOR	SIM	TIMER	TMR
REDUCTION	REDN	SIMULTANEOUS	SIMUL	TIME TO GO	TTG
REDUNDANT, -ANCY	REDUND	SINGLE SIDE BAND	SSB	TO BE DETERMINED	TBD
REFERENCE	REF	SINGLE CHIME	SC	TOP OF CLIMB	TOC
REFERENCE EYE POSITION	REP	SINGLE STROKE CHIME	SSC	TOP OF DESCENT	TOD
REFERENCE SPEED; THRESHOLD)	VREF	SITUATION	SIT	TOTAL AIR TEMPERATURE	TAT
REFRIGERATION	FRIG	SKIP	SKP	TOUCH AND GO LANDING	TGL
REFUEL	RFL	SLAT	SLT	TOUCHDOWN	TDN
REGULATOR	REG	SLOW	S	TOUCHDOWN POINT	TDP
REGULATED TAKEOFF WEIGHT	RTOW	SMOKE	SMK	TOUCHDOWN ZONE	TDZ
REJECTED TAKEOFF	RTO	SMOKING	SMKG	TOWER	TWR
RELAY	RLY	SOUTH	S	TRACK	TRK
RELEASE	REL	SPACE	SP	TRACK ANGLE ERROR	TKE
REMAIN	RMN	SPEAKER	SPKR	TRACK MADE GOOD	TMG
REMOTE	REM	SPECIAL	SPCL	TRAFFIC	TRFC
REMOTE CONTROLLED CIRCUIT BREAKER	RCCB	SPECIFIC, ATION	SPEC	TRAILING EDGE	TE
		SPECIFIC FUEL CONSUMPTION	SFC	TRANSCEIVER	XCVR
REPEATING	RPTG	SPEED	SPD	TRANSFER	XFER
REPELLENT	RPLNT	SPEED BRAKE	SPBK	TRANSFEROFCONTROL MESSAGE	TFR
REPETITIVE CHIME	RC	SPEED CLIMB	SPD CLB	TRANSFORMER	XFMR
REPETITIVE STROKE	RS	SPEED COMMAND	SC	TRANSFORMER RECTIFIER	TR
REPORT	REP	SPEED DESCENT	SPD DES	TRANSITION	TRANS

Panel Abbreviations

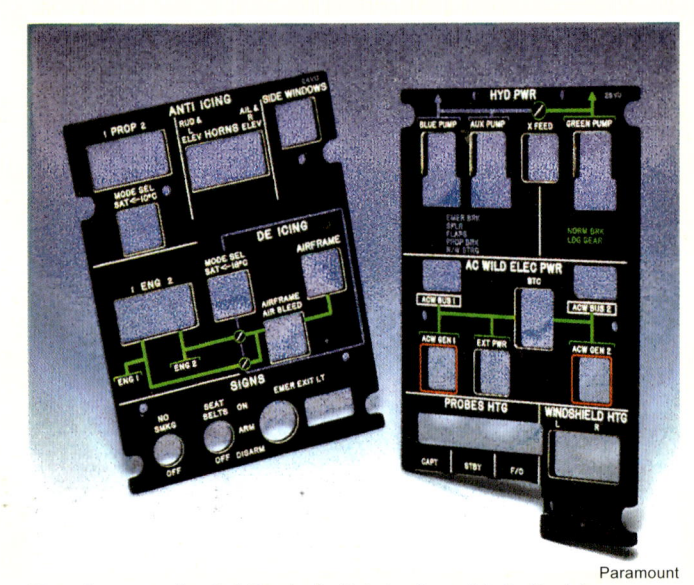

TRANSITION LEVEL	QNE
TRANSMITTER	XMTR
TRANSMITTER/RECEIVER	T/R
TRANSPONDER	XPDR
TRAVEL	TRVL
TROPOPAUSE	TROPO
TRUE	TRU
TRUE AIRSPEED	TAS
TRUE TRACK	TTK
TURBINE	TUR
TURBINE GAS TEMPERATURE	TGT
TURBINE INLET TEMPERATURE	TIT
TURBINE VIBRATION INDICATOR	TVI
TURBULENCE	TURB
TURN AND SLIP	T/S
TURN POINT	TPT
TURN RADIUS	TRAD

U

ULTRAHIGH FREQUENCY	UHF
ULTRAVIOLET	UV
UNABLE	UNABL
UNAVAILABLE	UNAVAIL
UNCOUPLED, ING	UNCPL
UNDERFLOOR	UFLOOR
UNLOCK, ED	UNLK
UNSERVICEABLE	U/S
UPPER	UPR
UPPER AIRWAY	UWY
UPPER AREA CONTROL CENTER	UAC
UPPER CONTROL AREA	UTA
UPPER SIDEBAND	USB
UTILITY	UTIL
UPPER FLIGHT INFORMATION REGION	UIR

V

VACUUM	VAC
VALVE	VLV
VARIATION, VARIABLE	VAR
VECTOR	VECT
VELOCITY	VEL
VENTILATION	VENT
VERIFY	VRFY
VERTICAL	VERT
VERTICAL BEARING	VBRG
VERTICAL DEVIATION	VDEV
VERTICAL GYRO	VG
VERTICAL NAVIGATION	VNAV
VERTICAL PROFILE	VPROF
VERTICAL REFERENCE UNIT	VRU
VERTICAL REVISE	VREV
VERTICAL SITUATION DISPLAY	VSD
VERTICAL/SHORT TAKEOFF AND LANDING	V/STOL
VERTICAL SPEED	VS
VERTICAL SPEED INDICATOR	VSI
VERTICAL TAKEOFF AND LANDING	VTOL
VERTICAL TRACK	VTK
VERTICAL TRACK ERROR	VTE
VERY HIGH FREQUENCY	VHF
VERY HIGH FREQUENCY DF STATION	VDF
VERY LOW FREQUENCY	VLF
VHF NAVIGATION	VHF/NAV
VHF OMNIDIRECTIONAL RANGE	VOR
VHF OMNITEST	VOT
VIBRATION	VIB
VISIBILITY	VIS
VISUAL APPROACH SLOPE INDICATOR	VASI
VISUAL FLIGHT RULES	VFR
VISUAL METEOROLOGICAL CONDITIONS	VMC
VLF/OMEGA	VLF/O
VOLT	V
VOLUME	VOL
VOR/DME	V/D
VOR/ILS	V/I
VOR/LOC	V/L
VOR TACTICAL AIR NAVIGATION	VORTAC

Paramount

Panels may be fabricated, finished and labelled by an outside shop specializing in this work.

W

WARNING	WARN
WARNING AND SYSTEM CONTROL PANEL	WSCP
WARNING DISPLAY	WD
WARNING LIGHTS	WL
WARNING SYSTEM	WS
WATER	WTR
WATT	W
WAYPOINT	WPT
WEATHER	WX
WEATHER RADAR	WXR
WEIGHT	WT
WEIGHT, ALTITUDE, TEMPERATURE	WAT
WEST	W
WHEEL	WHL
WHEEL WELL	WW
WIND DIRECTION	WDIR
WIND SPEED	WSPD
WIND VELOCITY	W/V
WINDSHIELD	WSHLD
WINDSHIELD GUIDANCE	WG
WING AND TAIL	WT
WINGTIP BRAKE	WTB

Y

YAW DAMPER	YD
YELLOW	YEL

Z

ZERO FUEL WEIGHT	ZFW
ZONE	ZN

Chapter 29

Test and Troubleshooting

The first step in troubleshooting an avionics problem is when the pilot describes the complaint to the technician, or fills out a squawk sheet. The quality of these reports often means the difference between a quick fix or wasted hours looking for trouble.

The typical pilot squawk is often brief, such as "My number 1 com doesn't work." This gives few clues and is only the starting point to ask questions to narrow the problem. The pilot knows much more than he thinks---if the technician asks well-placed questions.

Here are suggestions to make the diagnosis go faster. Frame your questions to learn the conditions surrounding the failure (we'll use the com as an example):

Does the problem affect *both* transmit and receive?
Does the problem happen in all directions?
Does it occur over both high and low terrain?
If there are two navcoms, are both transmitters affected?
Is communicating distance affected by the weather?
Is the problem worse during taxi and takeoff, then improves while cruising?
Have you tried a different microphone?
When was the most recent repair to any avionics equipment?

Answers to specific questions like these can narrow down the faulty area.

Technical Terms

Be certain you and the pilot are talking about the same thing. For example, one pilot said "My speaker doesn't work,"---but pointed to his *microphone*. To him, it was a "speaker" because he spoke into it.

Avoid talking "avionics tech" because pilots

TCAS Ramp Testing

Tests to verify and certify a TCAS (Traffic Alert and Collision Avoidance System) can be done outside the airplane. The ramp tester, an Aeroflex TCAS-201, communicates with the airplane via radio signals and simulates different collision conditions. Without connecting directly into airplane systems it measures signal power, frequency, interrogations and replies. The tester is programmable to perform ten different collision scenarios.

Selective Call Test

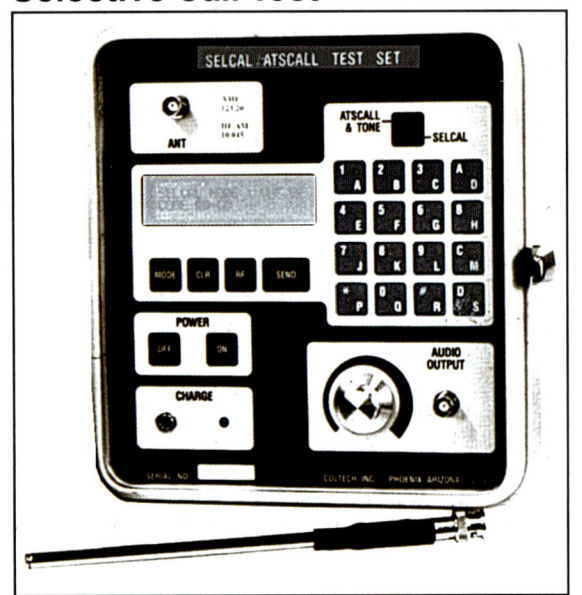

Selective calling systems aboard aircraft are checked remotely by this ramp tester, the Avtech CTS-700. It sends and receives the 16-tone code of the ARINC Selcal system. It also communicates with Atscall, a system used in some fleets, which is based on the 16 "touchtone" codes used in telephone systems. As in many ramp testers, battery power is automatically turned off if there is no activity for 15 minutes.

almost never understand it (and they don't need to).

Many pilots are ham radio operators, electrical or computer engineers and can sling technical terms. Nevertheless, don't assume they know the special lingo of avionics.

Switchology First

As you head out to the airplane, be aware that many troubles are caused by the pilot setting switches to the wrong position. This happens even though the pilot flies the airplane regularly.

I once took an airline captain for a ride in a small airplane. He pointed to a blinking light on the panel and asked, "What's that?" I told him that's the transponder reply light. (Hmmm?)

Another time, an experienced corporate pilot said he was getting no power to half his instruments. When a technician checked, he found a bad inverter (which converts battery DC to AC). The technician explained the airplane had *two* inverters and the pilot could have switched in a good one. Despite long experience the pilot was not aware that he had a "reversionary mode."

So the first item to look for when investigating most complaints is the correct setting of all switches and knobs, especially in the audio panel and for powering radios on and off.

Ramp Testers

A lot of diagnosing is done by eye, ear and knob adjustment. Radios produce many symptoms that warn of trouble if you read them. Once these clues are followed, ramp testers narrow down the problem. Let's consider informal checks on the ramp.

ADF

Compass Locator

A quick check of an ADF (Automatic Direction Finder) can be done at an airport with an ILS (instrument landing system). Many ILS's have a compass locator to mark the final approach fix about five miles from the touchdown point. Because the compass locator operates the ADF receiver, it may also serve as an approximate test signal.

You will need to know the frequency, location and ID of the compass locator. This information is on an approach chart which should be available from any IFR pilot who uses the airport. The chart also shows where the ADF needle should point from your location on the airport.

ADF Antenna Tester

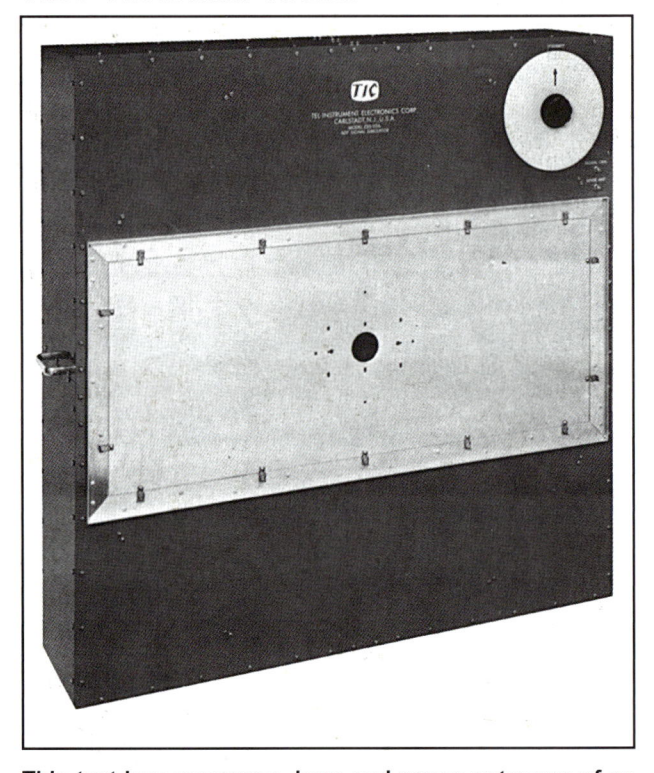

This test box measures loop and sense antennas of an ADF receiver. The antenna is mounted inside the box and stimulated by test signals. A 10-inch dial on the simulator is rotated to measure bearing within a 1/2-degree over 360 degrees. These tests once required a shielded room. The tester shown here is the TIC CES-117A.

Broadcast Station

If there are no nearby NDB stations or compass locators, make a quick check of the ADF receiver by tuning to a local AM radio station. You need to know its location to check the bearing. Most AM transmitters are out in the country, and local people often know where the tower is located.

Audio Quality

Tune to a station and turn up the volume. Is the audio distorted? Clear audio must be present for the ADF to perform its job.

An ADF operates from two antennas; a loop for finding a bearing to or from the station and a sense antenna which enables the needle to point toward (not away from) the station. Select the sense antenna, which is usually marked "Receive." The signal should be clear and strong. Next, choose the loop antenna, usually marked "ADF". The needle should swing and point to the station.

In-flight Interference

If the pilot reports rough buzzing or static in the ADF audio while in flight, it could be P-static (from precipitation). It's the build-up of static electricity on the airframe that makes raspy sounds in low-frequency radios. Check if each grounding or bonding point between the ADF system and the airframe is clean and secure. Check the condition of static wicks on the wings and tail (described in the section on Precipitation Static). The wicks bleed off charges quietly into the atmosphere.

Antennas

Doubler

This back-up plate under many aircraft antennas needs to be checked for tightness. Looseness allows moisture to enter and corrosion to form, which degrades antenna performance.

Is there still a watertight seal around the base of the antenna? Reseal this if necessary.

Coaxial Cable

Check the underside of the antenna for a clean connection where the cable connector attaches. Corrosion in the connector is a cause of poor radio performance.

If there is excess coaxial cable below the antenna, don't coil to too tightly. Inspect the cable for cracks and abrasions.

Antenna VSWR

A valuable test for determining the condition of a transmitting antenna is VSWR, for voltage standing wave ratio. It not only checks efficiency of the antenna, but cable and connectors, as well.

VSWR is a measure of how well the transmitter is applying power into the transmission line, then how efficiently the line transfers power into the antenna. For the most efficient transfer, all components must be "matched." In aviation, the standard for matching is "50 ohms," which refers to the electrical load, or "impedance" of each device. Thus, the transmitter is designed for a 50-ohm output, the cable is manufactured for 50 ohms (impedance) and the antenna is fed at a point that is electrically 50 ohms. In such a system, if the transmitter generates 10 watts of radio-frequency power, then nearly 10 watts should flow in the antenna (less a loss in the cable, which we can ignore).

This is close to a perfect system. As the antenna ages, cables crack, connectors corrode, connections no longer make good contact. Some cables are crushed or coiled too tightly and their wires lose correct spacing. At these points, the cable is no longer 50 ohms but some other value. Now when radio energy hits these areas, part of it reflects back to the transmitter and subtracts from the power going to the antenna. The relation of forward and reflected power is known as VSWR, or "voltage standing wave ratio." VSWR can be easily measured and used to determine the condition of the system; the higher the VSWR, the poorer the performance of the antenna system.

Servo Tester

Electronic Aviation Systems
The Model 101A checks autopilot servos and flight control sensors without removing them from the airplane. It measures control voltages to within 5 thousandths of a volt.

Practically speaking, if VSWR is less than about 2.5 to 1 (written as 2.5:1), this is considered acceptable in antennas operating below 200 MHz. A VSWR of 2.0:1 means about 90 percent of the power is getting through.

RF Wattmeter

A portable RF wattmeter is often used for troubleshooting antennas. It is inserted into the transmission line, where it reads forward and reflected power, and indicates VSWR. It can also isolate problems. If the antenna is disconnected and a dummy load installed, the wattmeter reads the condition of the cable. (The dummy load turns RF energy into heat.)

Forward power indicated by the RF wattmeter is from the transmitter, and should be close to the radio manufacturer's specification for this transmitter.

Autopilots

Cable Tension

This critical value is measured with a cable tension meter. Incorrect tension is the source of many autopilot complaints, and it applies to both the main and bridle cables.

Porpoising

In this autopilot problem, the airplane flies a path like a dolphin swimming in a series of arcs. Check cable tension and also the electrical adjustments on the autopilot computer as recommended by the manufacturer.

Examine the cables for broken strands. This is done by *slowly* running your fingers over the cable and feeling for sharp points.

Capstan

Look at this component for signs of wear.

Com Transceivers

During the tests described below, be sure all switches are set to the correct positions, especially those on an audio panel that controls two navcom transceivers. Turn up volume controls and select whether you will listen through the cabin speaker or headphones.

No Receive

When a pilot reports "no communication" on a VHF radio he often means he can't hear anybody on the receiver. The other half of the radio---the transmitter---may be OK. Since this is an important clue to a com problem, turn on the radio and transmit. Signs of a working transmitter include: transmit indicator lights, you hear the signal on a portable aircraft radio, a test monitor, radio-frequency wattmeter or other ramp tester.

Mobile Calibration Lab

Regulations require that ramp and bench test equipment be calibrated and certified, usually once a year. Many shops send their equipment to calibration laboratories, but another choice is a mobile cal lab. It drives to the shop and performs the work just outside the premises. It also may repair to restore equipment to required accuracy.

If the transmitter is operating it's a good sign the antenna and coaxial cable between radio and antenna are OK. Primary power is also reaching the radio, and there's no problem in the circuit breaker. So far, these signs point to a problem inside the receiver.

Squelch

Operating in the receiver, the squelch silences annoying noise when no signal is being received. Early com radios had a manual squelch that required the pilot to adjust the knob until the noise just disappeared when no signal is received. This caused a lot confusion because too-high a squelch setting caused the receiver to miss incoming signals. Much of the problem was overcome when receivers came fitted with automatic squelch circuits that are self-adjusting. However, receivers now usually include a "test" button that disables the automatic squelch, enabling you to hear the receiver at full sensitivity. If this causes a rushing noise ("static") or voices of other aircraft, it's a sign the receiver is functioning. If no noise is heard on the squelch test (and the radio transmits OK), the problem may be within the receiver. It may simply require a new setting on an internal squelch adjustment by a bench technician.

Let's say the pilot complains he can receive and transmit OK, but something keeps "breaking" the squelch---opening and closing it rapidly, with pulse-like noise. Determine whether it is happening to both navcom radios simultaneously. If it is, the problem is most likely interference from the outside. Next, find out if the problem happens wherever the pilot flies. If the noise follows the airplane, it's probably being generated aboard the airplane, with the most likely sources the magnetos, spark plugs, strobe lights or dirty alternator slip rings.

In some cases, noise breaking through the squelch is eliminated by having a bench technician raise the squelch threshold. The problem is that the radio can also miss very weak signals. The best approach is for the internal adjustment be made to the manufacturer's instructions, using a calibrated bench tester.

There are occasional reports of the squelch breaking open while taxiing on an airport surface. This is often caused when the airplane rolls over electrical cables buried underground. Unfortunately, the cables carry powerful currents to flashing lights for approach and landing guidance. Nothing can be done about this, but it lasts for only a few minutes and disappears after take-off. This underground wiring may also put "dots" on the screen of a Stormscope because they resemble lightning signals. The pilot should expect this and clear the screen. If there is actual lightning activity, it should be visible after lift off.

No Transmit

A leading cause of an inoperative transmitter is not the transmitter, It's the microphone, which contains the push-to-talk switch. The worst offender is the mike that fits in a hangar on the panel. Each time the pilot pulls it from the hangar, then hangs it up, the cable flexes two times. Although the wires are flexible, they break, so pressing the mike button no longer keys the transmitters.

Automatic Test Equipment

Aerospatiale

Automatic Test Equipment (ATE), like this ATEC 5000, checks digital circuits in airliners and military aircraft. It contains programs that automatically test hundreds of parameters. To locate intermittent problems, the test station runs continuously until a fault appears. The time and location of the fault are stored for later viewing by the technician. ATE systems like these are usually found at large airline and repair depots.

Air Data Test

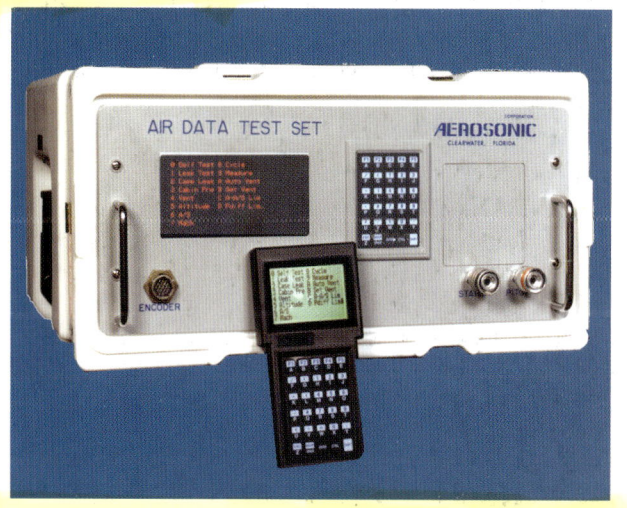

Air data test set by Aerosonic. It checks altimeter, encoding altimeter, airspeed, Mach airspeed, rate-of-climb and cabin pressure. The hand-held remote (at the bottom) enables the technician to sit in the cockpit and observe the instruments while operating the tester located elsewhere. Operating limits can be set to avoid delivering excess pressure to the instruments.

The trend among pilots is to headsets with a boom mike that is not handled. This reduces flexing, but wires are still vulnerable as the pilot turns his head. Because a few strands of wire can disable all transmitters aboard the aircraft, no pilot should fly without a spare mike.

The mike wire almost always breaks where it enters a plug or the microphone case--- because that's where it receives the greatest strain. Quite often, you can press the transmit button, wiggle the wire at these two weak points and, suddenly, the radio begins working as the broken ends touch.

The cable is repairable if the plug can be disassembled. The damaged section is cut short and the wires reconnected. Most modern plugs, however, permanently mold the wire into the cable and cannot be opened. In this case, a replacement cable is ordered from the manufacturer.

DME

Distance Measuring Equipment develops faults such as an inability to lock onto a station. Turn up the DME audio and listen for the Morse code ID (it may take up to 45 seconds to hear). This is a check on the receiver portion and proves it is properly channeled by the VOR receiver.

If you don't hear DME audio be sure your switch positions are correct; the DME may be channeled by several receivers so choose the correct one. If you still can't get DME audio, the transmitter could be defective. Because the DME receiver and transmitter use one antenna, good reception is an indication the antenna and cable are OK.

Complaints about errors in ground speed or time to station are sometimes traced to poor antenna connections. Check the coaxial cable, connectors, antenna hardware and for good bonding and grounding.

ELT-Emergency Locator Transmitter

Test in Aircraft

For the purpose of checking, you may activate an ELT in an airplane under the following conditions. First, tune an aircraft com radio to the emergency frequency of 121.5 MHz, and turn the volume to normal listening level.

1. Perform this test only on the ground, not in flight.
2. Turn on the ELT only during the first five minutes after the hour, for example; between 1:00 PM and 1:05 PM.
3. Listen for the sweeping audio tone (sounds like *"peee-owww---peee-owww"*). Allow only three sweeps to sound, then switch off the ELT.

Because the ELT antenna is within a few feet of the com receiver, the audio should be loud and strong. No audio means the ELT is dead. Weak audio suggests a spent battery, a defect in the antenna or problems in the coaxial cable. ELTs are required to carry a label for the date of battery replacement, so check if it's current.

If you hear a weak signal, turn off the ELT and listen. Because there have been so many false alarms with early ELT's you might be hearing signals from another airplane at your field.

Glideslope Receiver

There are no pilot controls for selecting a glideslope frequency. When an ILS is tuned on a VHF nav receiver, the glideslope receiver is automatically channeled to the correct frequency. Thus, a glideslope problem could be failure of the nav receiver to control the glideslope receiver.

The glideslope has no audio identifier. Because glideslope signals occupy such a narrow path to the runway, the only practical way to troubleshoot is with a ramp tester.

Glideslope signals are high in frequency and depend on good antenna and cable connections. In a light aircraft, the glideslope signal is often tapped from the VOR antenna on the tail through a splitter. Loose or dirty connections cause signal losses. The glideslope antenna in large, transport aircraft is in the nose.

The technician should warn the pilot about using a glideslope in actual instrument conditions. Many pilots don't fully understand the function of the "flag," which could be dangerous. The flag is in a small window next to the glideslope needle, usually with a red-striped symbol (a "barber pole"). If the signal is adequate, the flag is pulled, that is, it moves out of view and assures the pilot the glideslope is working. But there have been accidents where the glideslope receiver did not work---causing the needle to lie in a horizontal position. Unfortunately, this is the same indication as when flying a perfect glideslope. It is important for the pilot to not only check the flag for a good signal, but watch for small needle movements while flying the glideslope.

Lightning Strikes

The aviation industry is more concerned about lightning as airplanes convert from analog to digital avionics. Digital signals are lower in strength and more susceptible to interference. As a result, new requirements were developed to harden the new electronics against outside disturbance.

These requirements emphasize shielded enclosures, shielded wiring, bonding and grounding to protect wiring and circuit components.

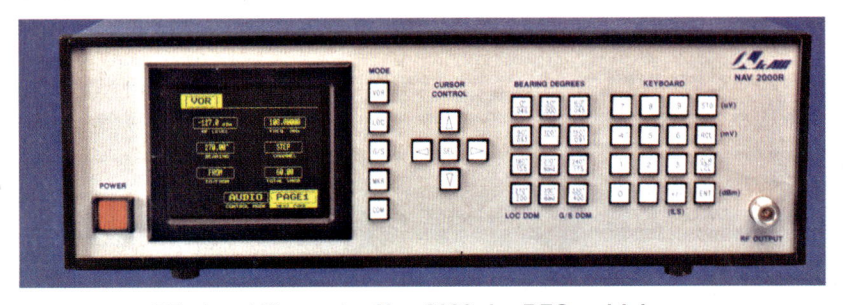

RF signal Generator Nav 2000 by BFGoodrich.

Cat III Tester

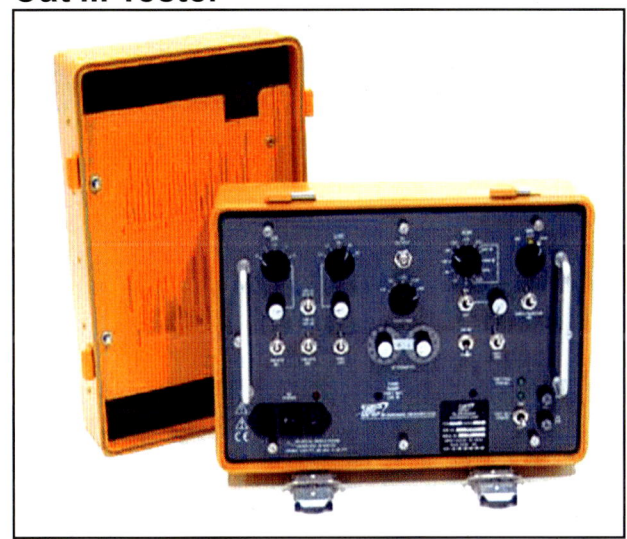

Ramp test set for checking a Category III ILS aircraft receiver. It generates a variety of signals such as VOR, glideslope, localizer, and marker beacon, as well as autopilot and flight director. The tester is operated by one person in the cockpit while the aircraft is on the ground. Shown here is the Model T-30D by TIC.

Lightning Damage

When lightning strikes an airplane, it causes a current flow in the skin. The energy generates electromagnetic fields that move inside the airplane, where they induce additional current flows in wiring and electrical equipment. Known as "lightning indirect effect," it can trip circuit breakers, disrupt digital circuits and damage other components.

Lightning may damage the skin. After landing, the pilot notices small pits or burns where lightning entered or departed. In some instances, instruments in the panel are damaged. Compasses have been known to demagnetize and swing aimlessly.

One phenomenon is when the pilot sees windshield frames start to glow. During one incident, the glow entered the cockpit, formed a ball and moved up the passenger aisle. The effect is "St. Elmo's Fire," built up by static electricity as the airplane moves through a charged atmosphere near thunderstorms. St. Elmo's is not believed to be dangerous in an airplane (but certainly an unwelcome experience for the passengers).

Loading. After an installation, the airplane is loaded with the latest software before delivery to the customer. In General Aviation aircraft, it is often done by inserting a memory card into a slot on the radio (usually for a navigation database). Future upgrades are done by the pilot.

In air transport, software is more extensive and done differently. Airliners such as the Boeing 737-600 through -900, the 747-400, the 767 and 777 use onboard loadable software. Not only can software update operational data, but upgrade avionics to meet new requirements, make improvements in design and fix errors. Because there are no hardware changes, the job might even be done during the short turn-around time at the gate.

Typical loading time for the six main (EFIS) displays on a 747-400 is 90 minutes.

Software transfer. There are several systems for loading software on an airliner. The aircraft may have a permanently-installed loader on board, or a portable loader carried by the technician and plugged into the aircraft. The Boeing 777 has a maintenance access terminal (MAT) with a mass storage device which also stores spare copies of loadable software.

If the LRU (line replaceable unit) is in a maintenance shop, loading is accomplished with automatic test equipment (ATE). Another approach is connecting to a port on the LRU that accepts a memory disk. It is common in the airlines for the avionics manufacturer to provide a spare copy of software, which is stored aboard the airplane in a binder.

Transponder

The first symptom of a malfunctioning transponder is often a complaint from an air traffic controller. If you are flying IFR or receiving flight following, your blip is on someone's radarscope. When you're told the code is wrong, immediately check the transponder knobs for the correct code selection, especially if any are caught between numbers.

Mode C, or altitude reporting, can also go wrong, showing you at the incorrect altitude. Some pilots believe the transponder altitude they report is taken from the altimeter, and that changing the baro setting will change the reported altitude. The transponder only utilizes its own altitude reference, which is preset to 29.92 inches of mercury, or sea level pressure on a standard day. When this is received by an air traffic control facility, it is automatically corrected for local sea level pressure.

There is too much room for error, it was decided, to let each pilot set his own local pressure in the transponder. Before a transponder is installed (or returned to service) the technician sets the internal pressure sensor to 29.92 inches of mercury.

In reporting a bad transponder it is helpful if the pilot can tell the technician which altitudes were said to be in error by the controller. It may simplify the repair for the technician on the workbench

Because a malfunctioning transponder can interfere with the flow of air traffic, it must be checked by a technician every two years.

The ident button of a transponder is on the front panel, but sometimes a second button is mounted in the yoke (to make it easier for the pilot). If you receive a complaint that the transponder is "identing" at the wrong time (the controller didn't ask for it) there may be a short in the yoke switch. Because the yoke is in frequent motion during flight, wires to the yoke button may fray and rub against a metal ground---which triggers the unwanted ident.

The transponder antenna is critical for good operating range. Because it mounts on the underside of the airplane, it is subject to dirt, grease and contamination. It needs to be clean and well-bonded to the airframe for high efficiency.

Transponders often have a "test" button, which mainly checks the reply lamp.

VOR Receiver

30-day Test.

To operate an aircraft under instrument flight rules (IFR) its VOR receiver must be checked within the last 30 days. The tests described below are mostly for the pilot to perform and log. The technician usually has a portable ramp tester for generating test signals, but may also make informal checks with the following resources:

VOT (VOR Test Facility)

The VOT is a special station found at large airports that transmits a test signal. Tune the VOR receiver to the VOT frequency (usually on 108 MHz) and adjust the Omnibearing Selector (OBS) until the needle (CDI, course deviation indicator) centers. The course indicator will read either 0 degrees, with the flag indicating FROM---or 180 degrees, and the flag indicating TO.

Either reading should not be greater than plus or minus 4 degrees from the desired setting (0 or 180 degrees). If the pilot makes this check in the air, the allowable error is plus or minus 6 degrees.

Ramp Data Loader

Flightline portable test set loads software on airplane through port in wheel well. It also contains fault diagnostics and flight data analysis. The tester's operating system is Windows-based.

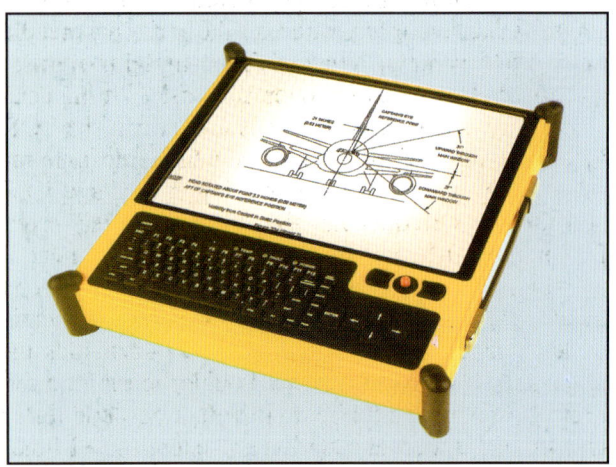

Portable Data Loader contains a mass storage cartridge to upload software directly into the aircraft. It also downloads maintenance data recorded during flight. This information can be analyzed in the shop. The loader meets the ARINC 615 spec.

VOR Receiver Checkpoint

Some large airports use a nearby VOR station (within several miles) as a reference for checking a receiver. A location on the airport is marked with a sign giving VOR identification, frequency and radial. The airplane is taxied to that point, the radio tuned and the error observed. It must not be more than plus or minus four degrees.

Two VORs

If there are no test stations like those described, it is acceptable, where possible, to tune two VOR receivers to one VOR station. The error between them should not be more than plus or minus four degrees.

If this test is done in the air, however, the acceptable error increases to 6 degrees.

Transponder Tester

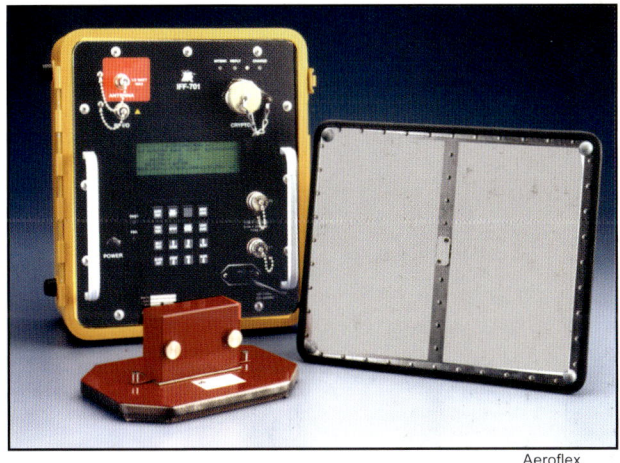

The IFF-701 is a flightline transponder tester. It performs single checks or an autotest that runs 30 tests automatically. The tester communicates with the aircraft through its antenna. The red object at the lower left is an antenna shield, which covers one aircraft antenna in installations where there are two transponders (in an anti-collision, or TCAS, system).

VOR Indicator

If the pilot reports a dead VOR needle---no movement at all---turn up the nav volume and listen for the Morse Code (or voice) identification. If none is heard and the signal is strong (the flag is "pulled," meaning not visible)---there is probably a fault inside the receiver. If ident audio is heard, but the needle won't move, that is also an internal receiver problem.

Sometimes the problem is a needle that moves erratically, darting left and right. This is often due to an internal component (a "resolver") that turns with the omnibearing selector (OBS). If the resolver gets dirty it "skips" and causes the needle to jump unpredictably. This requires a bench repair.

If everything looks good---To-From and warning (no-signal) flags behaving correctly, with good Ident audio--- the needle may have a burned-out movement.

On rare occasions, a VOR receiver may behave normally, but you hear no Ident audio. Be aware that any time a VOR ground station is being worked on, the technician turns off the audio ident. It warns pilots not to use the station for navigation.

Course Bends and Scallops

A pilot is flying and notices his VOR needle swinging from side to side. If it happens slowly, it's called "course bends"; if occurring at a rapid rate, it's "course scalloping." In many instances, the problem is at the VOR ground station. When the signal is broadcast through mountainous areas, it is bent by reflections--- and they show up on the VOR needle. Because these signal deviations cannot be completely eliminated, the government agency controlling the VOR issues a notice to airmen (notam) stating that certain radials of a VOR are unusable. However, it may be considered usable when the error is not more than 2.5 degrees for enroute navigation and 1.5 degrees for a VOR approach.

Windshield Wipers

A different problem occurs when the pilot is flying via VOR and notices the needle swinging back and forth. If he puts two VOR radios on the same station, both needles move together like windshield wipers. This is caused by propellor or rotor modulation. The signal arriving from the ground station moves through the propellor and is "chopped" in the spinning blades. If the chopping rate is close to that of the signal modulation (usually 30 times per second) the two will mix and produce a difference signal---which causes slow movement of the needles. If it's a prop plane, and the effect disappears when a VOR from a different direction is selected, that's a good sign of propellor modulation.

Rotor modulation in helicopters, on the other hand, is more troublesome because the rotor blades cover a large area. They act as electrical mirrors, causing the arriving signal to split into two parts; one is direct from the station, the other is the reflection off the rotor blades. The two signals mix and produce a third, which causes VOR needles to drift back and forth.

Avionics manufacturers recognize the problem and provide filters in radios intended for helicopter operation.

In a prop plane, be sure the VOR antennas are mounted in a location suggested by the airframe manufacturer. In some instances the windshield wiper effect may be reduced by making small changes in engine RPM.

Glideslope

Although the glideslope operates on its own receiver, it displays information on the horizontal needle of the VOR indicator. When a pilot complains of missing or erratic glideslope operation, ask if it happens only at certain airports. The reason is, when the VOR receiver is set to an ILS frequency, the glideslope receiver is "channeled" to its operating frequency. (ILS and glideslope frequencies always acts in pairs.) A defect in the VOR channeling circuit may cause this.

Unlike most ground navaids, no audio ident is transmitted for the glideslope. The GS flag---which is pulled when signal strength is adequate---provides another clue to problems. Most troubleshooting on the ramp, however, should be done with a portable tester that simulates a ground station and exercises all functions of the receiver.

Wiring and Connectors

Inspecting for Wear

After a time in service, wire bundles sag, rub and chafe, causing electrical problems--or even obstruct flight controls. A careful inspection reveals these trouble areas. Give special attention to wire supporting points, such as tie-wraps, clamps and grommets. Unsecured wires that hang in loops are dangerous.

Cut Wires

Old airplanes modified with new avionics often have wires that are simply cut off and go nowhere. The technician rarely removes old cables that are no longer used, but snips them. Such unused wires usually pose no hazard, but their ends should be insulated with crimp-on splices or tubing.

Fasteners

Nuts and bolts that secure wiring to the airplane have lock washers or other means to resist vibration. But they loosen and allow the wire bundle to move. Check and re-tighten any suspicious hardware. Reposition wiring before tightening to keep it clear of any hazards.

Oxidation

The large temperature changes in an airplane as it flies through different altitudes causes condensation and corrosion on metal fasteners. Look for powdery deposits on hardware and replace any that have deteriorated.

This is also important for a bonding or ground connection, where a cable is bolted to the metal airframe. Corrosion in that joint creates several problems that are difficult to analyze, such noisy radio reception. If hardware is discolored, it should be removed and shined with fine sandpaper or replaced if necessary.

Discoloration

Wiring insulation or any device that has darkened or changed color could be overheating. Determine the source---insufficient cooling, a radio drawing excessive current, or proximity to a hot area, for example.

Avionics Enclosures: Intermittent

A pilot reported that a panel-mounted radio "cut out" during certain times but played perfectly during others. The technician asked, "Does it happen in a climb?" When the pilot replied, "Yes," the problem was identified. During a climb, the radio case slipped out of the rear connector, then back in during level flight. (It took only a few thousandths of an inch to break the circuit.)

It is not unusual for the locking mechanism of a radio housing to loosen after many hours of flight. Thus, one of the first steps a technician might take when checking a dead radio is to simply press a hand against the face and push in. If the radio begins to play---problem is apparent. But not always. Tightening the locking screw should draw the radio into its tray and push the connector pins into their sockets. That works if everything is in perfect alignment; tray, radio and connectors. But old airplanes are not perfectly square. When re-seating the radio in a tray push the radio into the rack with your hand, and press it home into the connector. With your hand still applying pressure, tighten the locking screw. Now the screw does not have to overcome a lot of resistance and the rear connections can make good contact.

Remote Radio Racks

In large aircraft, where the radio has a remote-mounted unit, there have been many problems with trays and connectors. Because airlines found this very costly, they pressured the connector industry come up with improved designs. Not only are new locking mechanisms simple to operate, but may have clutches to prevent overtightening. Nevertheless, there are still instances of technicians failing to lock the tray in place. It's a good idea during an inspection to check remote boxes to see if they are secure in their mounting.

Some remote units will move because they shock-mounted. There are rubber mounts that absorb vibration and are mainly used for electromechanical instruments (with gyroscopes). Today, there are fewer shock-mounts in avionics because microelectronics are not as susceptible to shock. This includes later generations of gyroscopes which replace "spinning iron" wheels with laser beams or miniature accelerometers. Even small aircraft are replacing old gyro systems with solid-state sensors to operate flight instruments.

Fault Detection Device

Problems are pinpointed along a wiring run with an instrument like this Fault Detection Device. The connection is made only to one end of the wire under test. The instrument sends an electrical pulse into the wire, which reflects back from where the wire is broken, shorted or defective. This is pictured on the screen (below) which gives distance and location of the problem. Accuracy can be within 1 inch of the fault.

These testers are also known as "TDRs," for time domain reflectometry.

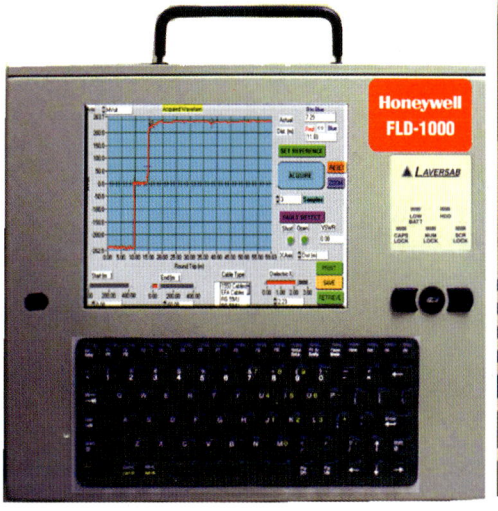

(Honeywell)

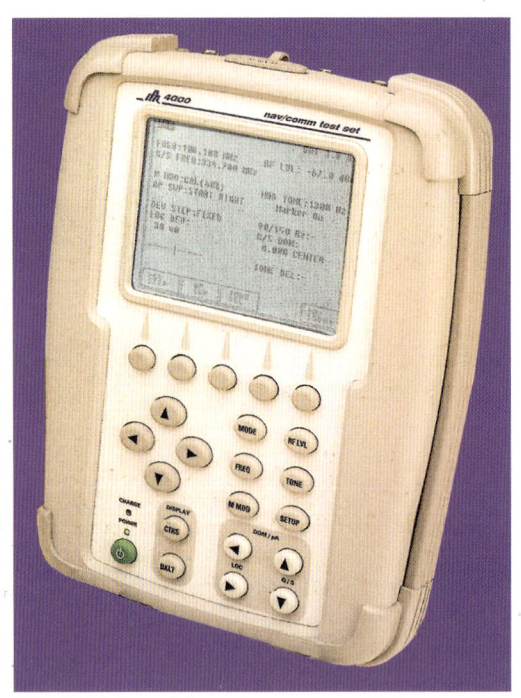

IFR 4000 portable navcom ramp tester.

Boeing

The laptop PC is part of the Boeing "Portable Maintenance Aid." It reduces time to troubleshoot by providing necessary documents, along with search capability. The system places the following at the technician's fingertips; fault isolation manual, aircraft maintenance manual, illustrated parts catalog, wiring diagrams, equipment list, maintenance tips and service letters. Nearly all Boeing aircraft are covered by the Portable Maintenance Aid and it's in use by many airlines.

Precipitation (P) Static

P-static is a form of interference caused by the buildup of electrical charges on the airplane as it flies through rain, snow, dust, ice crystals or clouds. The friction of the airframe against these particles generates static electricity. When voltage increases sufficiently, there is a "corona" discharge; a spark jumps to the surrounding atmosphere. This generates radio waves which reach aircraft antennas and are heard as noise.

Even non-metallic parts of the airplane may become charged; windshields, radomes and plastic panels, for example. This causes a discharge known as "streamering," which couples radio energy to the metal airframe and then to the antennas.

Interference is most prominent at lower frequencies, mostly affecting the Automatic Direction Finder, Loran and High Frequency (HF), but can reach up to the VHF com and nav bands.

The problem is treated with "static dischargers," popularly known as "static wicks." They're based on the principle of high voltage seeking to discharge from a sharp point. Many static wicks consist of thousands of fine carbon points which bleed off the charge at a lower voltage, thus reducing interference. Some models have only one point. The wicks also contain high resistance to keep sparks from coupling back to the airframe.

Maintenance Because static wicks are out in the airstream, they erode and need regular inspection. The manufacturer may recommend resistance checks between the tip of wick to the base of the discharger. In some models, the wick turns gray and needs to be trimmed back to expose a fresh surface.

Any static discharger depends on a good bond (ground) between its base and the airframe; the resistance must be extremely low. If corrosion appears, the mounting needs to be cleaned.

When an airplane goes to the paint shop, static wicks are often removed. After the job, the wicks may be re-installed *over* the new coat of paint! This makes the wick useless. A close inspection shows this condition, which must be corrected by removing paint under the wick mounting.

Static wicks reduce P-static by drawing the electrical charge from the airframe and discharging it quietly into the atmosphere. The construction of the wick produces a lower "corona," or discharge, voltage than the airframe.

Diagnostic Instrument

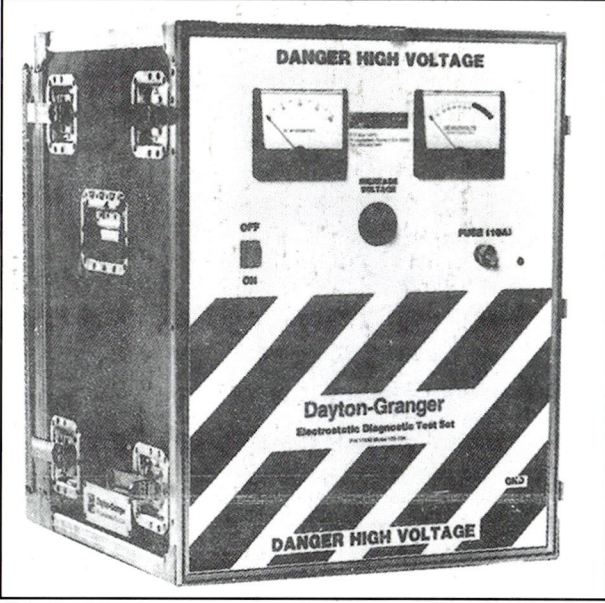

This test instrument generates high voltage to simulate static discharges built up in flight. The technician applies voltages at or near antennas, trailing edges, de-icing boots and other points that might cause sparks and other interference. He can sweep the whole airframe to determine areas where bonding is poor, a condition that encourages discharges. Windshields, radomes and other non-metallic surfaces can be charged to locate these noise sources. This electrostatic diagnostic test instrument, produced by Dayton-Granger, is used in the hangar without placing insulating blocks under the aircraft.

Static Wicks on Airplane

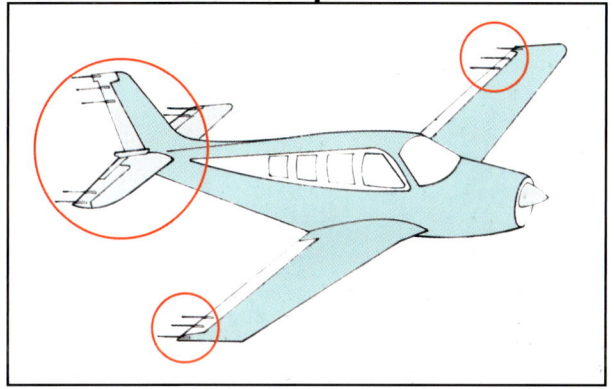

Typical locations are shown for static wicks on trailing edges of wings and tail. Manufacturers of wicks provide drawings for most aircraft that show exact locations.

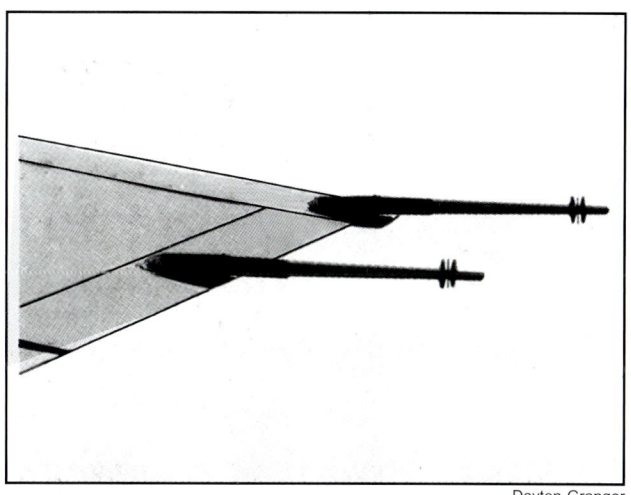

Avionics Checklist

The more symptoms of trouble provided to the technician, the quicker he can analyze problems. Here is a summary of items for both pilot and technician to help gather that information

Electrical Source
- Are there signs of a failing charging system, such as hard starting, dim lighting or changes in lamp brightness as engine rpm varies?
- Are there blown fuses or "popped" circuit breakers?
- Do you hear a whining or musical tone that varies with engine rpm? This could be alternator interference.
- Is there repeating noise in the audio, or extra dots on a Stormscope? This might caused by a strobe light or poorly shielded ignition wires.
- Is the negative terminal of the battery making a good connection with the airframe?

Com Radio
- Is the correct radio selected on the audio panel?
- Check if the volume control is at normal listening level.
- Are plugs for mikes and headsets fully seated in their jacks?
- If there is no receive audio, check if the push-to-talk switch on the mike is stuck on.
- Are mike and headphone plugs clean and shiny?
- Turn the receiver to "Test" to hear background noise.
- *If there's no audio in the headphones, switch to the cabin speaker.
- When trying to contact another station, does the operator say, "I can hear your carrier, but no modulation?"
- Check the frequency the radio is tuned to. Is it in the "Active" position?
- Many radios provide "sidetone;" as you speak. You should hear yourself in the headset.
- Does the indicator lamp light up when the mike is pressed?
- Does the same problem affect two com radios?
- Does the problem occur on more than one channel?
- Is the audio distorted, noisy?
- * Are there audio tones behind your voice when transmitting? (This could be inverter or alternator interference.)

Transponder
- When flying in a radar environment, does the reply light blink? Check if the light is not set to the dim position.
- Press the test button to check reply light.
- Turn off the DME. If the transponder problem disappears, the DME is causing interference.
- Does air traffic control say it's not receiving Mode A (identification) and/or Mode C (altitude)?
- If air traffic control reports problems, try to communicate with another radar facility to confirm it. Sometimes the problem is the radar facility.
- Is the mode selector in the correct position, and not in standby?
- Does the transponder "recycle" correctly. Turn it to standby, then turn it on. There may be about a 30 second delay before it turns on again.

Flight Control/Autopilot
- Does it follow the selected mode, such as wings level, heading hold, and track the VOR or Localizer?
- Does it track the vertical functions, such as altitude hold?
- Is the autotrim responding correctly?
- Is there "porpoising"---where the nose rises and falls.
- Can trim be adjusted?
- Check the vacuum for the gyro instruments.
- With the autopilot engaged, and the airplane on the ground, can you overpower the system with the flight controls?
- Are control cables loose?

Weather Radar
- Check the condition of the radome for cracks, water entry, splitting of layers.
- *Verify the setting of panel controls, such as brightness and sensitivity.
- Run the self-test routine
- Are display graphics sharp and clear?
- Does the radar stabilize as the aircraft maneuvers?

GPS
- Is the database current?
- If the GPS is relocated more than several hundred miles with the power off, expect it to do a sky search of five or more minutes before navigating.
- Are there enough satellites for navigation? Three is the minimum, while four are required for instrument flight.
- When on the ground, is the GPS antenna shaded by hangars or other structures?
- Check the mapping of the satellites. Are there sufficient satellites widely dispersed for good geometry?
- Check for sufficient signal strength.
- Have any warnings been issued (notices to airmen) that affect satellite coverage. Are there forecasts of severe solar storms which might affect reception?
- Does the position shown on the GPS agree with the airplane's present location?
- Is the airplane in a hangar where GPS signals may not penetrate?
- Are error messages present?

Distance Measuring Equipment (DME)
- Listen to DME audio for a Morse code identifier.
- Is the brightness turned up for the display?
- Be aware that ground speed and time-to-station function only when the airplane is flying directly to or from the ground station.
- Is the DME station off for maintenance? Check another station.

Automatic Direction Finder (ADF)
- Does interference disappear when the engine is off?
- Does the ADF needle point when "ADF" is selected and a station is tuned?
- Is the bandswitch set to the correct band?
- Do you hear audio when "Ant" is selected, a station is tuned, and there is no ADF needle movement?
- Is the bearing of the pointer accurate?
- Is the ADF needle being deflected by thunderstorms (which may be more than 100 miles away)?
- Does the test function deflect the ADF needle an appropriate number of degrees?
- Did you check for ADF action on more than one station?

Nav (VOR/ILS)
- Can you hear a Morse code and/or voice identifier?
- Does the problem appear on more than one channel?
- If reception is poor, is it true for all directions?
- Are both VOR and Localizer functions affected the same way? Can you receive one but not the other?
- Is the problem the same in a second nav radio?
- Does the VOR bearing appear to be in error?
- Center the VOR needle, then rotate the bearing selector 180 degrees. Is there an error between the two bearings (to and from)?
- Are flags indicating good signal strength?
- Compare two nav receivers to determine if both have the same problem.

Review Questions
Chapter 29 Test and Troubleshooting

29.1 A frequent problem in avionics occurs when the pilot sets switches _____.

29.2 If there are no nearby NDB (non-directional beacon) stations or compass locators, an ADF may be given a quick check by tuning to _____.

29.3 When checking an antenna, the VSWR (voltage standing wave ratio) should be less than _____.

29.4 The power output of a transmitter is measured by _____.

29.5 A quick test of whether a com receiver is working is to disable the squelch (often with the "Test" button) and listening for _____.

29.6 What is a major cause of an inoperative transmitter?

29.7 How can you check if a DME is being correctly channeled by the VOR receiver?

29.8 When can an ELT be tested?

29.9 What should you check first when the glideslope indication is missing or weak?

29.10 Damage from lightning strikes may be reduced by _____.

29.11 If the transponder code is incorrect, check for _____

29.12 What is a common problem if the transponder is reported to be "identing" at the wrong time?

29.13 Name three ways a VOR receiver can be checked for accuracy without a ramp tester.

29.14 What are (A) the cause and (B) the cure for "windshield wiper" movement of a VOR needle?

29.15 What should you suspect when a radio cuts out when airplane climbs?

29.16 How do you reduce noise from Precipitation (P) static?

About the Author

Len Buckwalter started Avionics Magazine and served as Publisher and Editor for 17 years. Earlier, he specialized in aviation and electronics, having written over 2,000 magazine articles and 22 books. His articles have been published in Air Progress, Kitplanes, AOPA Pilot, New York Times and Rotor & Wing. As one of the first to use the Internet, he started www.avionics.com in 1993 to launch *Avionics Library*, a publisher of books, CD's and software for the engineer and technician in the avionics industry.

A graduate of New York University, he served in the US Signal Corps as Communications Chief of an air-ground signal battalion, constructing and operating communications systems. Mr. Buckwalter was President of Avionics Center, an FAA-certified repair station in Leesburg, VA.

He is an instrument-rated pilot with 3000 hours' flight time and holds an FCC license for the repair of avionics equipment. His interest in the field began as a teen-ager operating an amateur ("ham") radio station, designing and building most of his equipment.

Flying his own aircraft, he attended the Experimental Aircraft Assoc. convention at Oshkosh for 15 years. He attended numerous conferences of the Aircraft Electronics Association, Airlines Electronics Engineering Committee (ARINC), and Avionics Maintenance Conference (ARINC). He can be reached at len@avionics.com

Index

A

ACARS 41
  bands and frequencies 45
  cockpit display 41
  message format 43
  messages 43
  OOOI 42
  system 42
ADF (Automatic Direction Finder) 81
  broadcast stations 85
  coastal effect 86
  compass card, fixed 81
  control-display, airline 84
  digital ADF receiver 85
  display, EFIS 86
  loop 82
  NDB station 83
  night effect 86
  quadrantal error 86
  receiver, airline LRU 84
  receiver, analog 82
  RMI (Radio Magnetic Indicator) 82
  sense 82
  system diagram 83
Aircraft earth station 35
Airways, lighted 11
altitude reporting: mode C 100
Antenna Installation 239
  ADF 246
  airline antenna locations 243
  altitude 244
  antenna types 240, 244
  antennas for light aircraft 242
  base station, mobile 252
  bonding 248
    to airframe 249
  coaxial cable 250
  combination antenna 253
  connectors 251
  couplers 251
  doubler plate 247
  duplexers 251
  gaskets 250
  GPS antennas 251
  HF (high frequency) 245
  hidden structure 246
  high-performance antennas 239
  hostile environment 241
  location 245
  mark and drill 248
  mounting 249
  selecting an antenna 243
  spec sheet, how to read 241
  typical antennas 239
  VOR blade 245
ARINC 183
  structures 183

Aviation Bands and Frequencies 230
  control and display 236
  frequency assignments 237
  future bands 234
  ground wave transmission 238
  hertz to gigahertz 234
  high frequencies 233
  Higher bands 232
  L-band 234
  line-of-sight 235
  microwave 235
  Radio frequencies (RF) 230
  radio signal 230
  skipping through ionosphere 233
  vhf band 234
Avionics
  definition of 1
Avionics master switch 173

B

Barany chair 7
Beam steering unit (BSU) 35
Bell, Alexander Graham 9
Blind flying
  first flight 2

C

Calibration 264
Cell Phones 33
Channels
  splitting 16
Clear air turbulence (CAT) 136
CNS: Communication, Navigation, Surveillance 4
Cockpit Voice and Flight Data Recorders 129
  cable assembly 133
  cockpit area mike 130, 131
  CVR basics 129
  CVR channels 130
  Flight data recorder 133
    parameters 135
    solid state 134
  flight data recorders 131
  inertial switch 130
  interconnect, CVR 132
  LRU: line replaceable unit 130
  tape drive 132
  temperature test 130
  underwater locating device 131
    tester 132
Connectors 199
  Amphenol 206
  application 201
  crimp tool 206
  crimping 202
  d-subminiature 206
  heat gun 207
  how to identify contacts 201
  Molex 206
  radio frequency 200
  reading pin connections 199
  releasing pins 207

  safety wiring 208
  solder cups 206
  soldering 202
  trends 202
  typical 200
Cooling 186

D

DME (Distance Measuring Equipment) 88
  airborne system diagram 91
  chanelling 88, 90
  EFIS display 90
  ground speed 88
  ground station 91
  jitter 89, 90
  obtaining distance 89
  pulse spacing 92
  reply frequencies 92
  scanning and agile 89
  slant range 89
  X and Y channels 92

E

EFIS: electronic flight instrument system 120
  Airbus A-320 127
  architecture 124
  electromechanical to EFIS 122
  flight deck 120
  glass cockpit 120
  MFD: multifunction display 125
  on B-747-400 126
  pictorial display 121
  replacing old instruments 121
  three-screen 123
Electrical Systems 168
  115 volt system 170
  28 volt DC 170
  AC and DC power 168
  APU (Auxiliary Power Unit) 170
  battery charge, percentage 168
  circuit breakers 173
    recessed button 176
  DC system 169
  fuses 176
  ram air turbine (RATS) 172
  switches 172
    avionics master 173
    llighted pushbutton 175
    pushbutton 174
    switch guards 175
    types 173
Electrostatic discharge 185
ELT 50
ELT (Emergency Locator Transmitter
  406 MHz ELT 51
  406 system 53
  components 52
  controls and connections 54
  Cospas-Sarsat satellites 51, 55
  direction of flight 50
  dongle 54

fleet operation 54
ground stations 51
Leosars, Geosars 52
registration 53, 55
ELT (Emergency Locator Transmitter) 266
   test in aircraft 266

G
Galileo 118
Glideslope 73
GPS/Satnav (Satellite Navigation) 108
   clock 110
   constellation 109
   EGNOS 116
   frequencies 110
   Galileo 118
   GNSS 111
   LAAS: local area augmentation system 116
     ground station 117
   launch vehicle 109
   LNAV-RNAV 116
   LPV 116
   multimode receiver 117
   position finding 112
   PPS: precise positioning service 112
   PRN code 112
   propagation corrections 112
   RAIM: receiver autonomous integrity monitoring 117
   SA: selective availability 111
   satellite signal 113
   satellite, typical 108
   SBAS 116
   second civil frequency 116
   segments 114
   space segment 109
   SPS: standard positioning service 112
   WAAS: wide area augmentation
     system 114
     system 115
Ground earth station 34

H
Hertz, Heinrich 9
HF (High Frequency)
   antenna coupler 27
   antenna mounting 27
   control panel 26
   control-display 23
   datalink 25, 26
   LRU (line replaceable unit) 25
   SSB (single sideband) 24
   system diagram 24
   transceiver 26
HSI (Horiz. Situation Indicator) 64

I
ILS (Instrument Landing System) 67
   90 Hz, 150 Hz audio tones 71
   approach lighting system 69
   categories 68

Category I 69
Category II 69
Category III 69
compass locator 72
components 68
decision height 68
glideslope 70
glideslope indications 72
glideslope, pictorial 72
glideslope receiver 73
glideslope station 73
localizer 70
localizer array 70
localizer indications 71
marker beacon 72
marker beacon ground station 74
marker beacon receiver 74
RVR (Runway Visual Range) 68
system; pictorial view 68
Inmarsat Aero System 31
Instrument panel
   first 2
interference 85

L
Lightning detection 140
Lindbergh, Charles 2
Low frequencies 232

M
Marconi, G 6
Marker beacon receiver 74
MLS: Microwave Landing System 76
   azimuth beam, diagram 77
   elevation antenna 78
   elevation beam 78
   multimode receiver 79
   space shuttle 76
   time reference scanning beam 79
   transmitting azimuth signal 77
Morse, Samuel 9
Mounting avionics 178
   airline mounting 191
   ARINC structures 183
   ATR case sizes (ARINC 404) 184
   avionics bay, corporate jet 182
   Case sizes, MCU (ARINC) 183
   cooling 186
     airline 187
     fans 187
   cutting holes 180
   electrostatic discharge 185
   indexing pins 193
   instruments 195
   instruments, airline 197
   integrated modular avionics (IMA) 194
   locking radios in racks 188
   locking systems, airline 192
   new or old installation 179
   panel-mounted radios 188
   rack, equipment cabinet 193
   radio stack 181

releasing radio 188
remote-mountiing, corporate 190
round iinstruments 196
structures 181
tray 190
tray preparation 189
Multimode receiver 117

N
NDB (Non-Directional Beacon) 83

P
Panel labels and abbreviations 254
   abbreviations 256
   engraving 254
   preprinted 254
   silk screen 254
   tape 254
   terms on labels 254
Planning the Installation 154
   :"steam gauges" 155
   4-inch EFIS 155
   basic T instrument layout 157
   connectors and pin numbers 162
   EFIS, turbine aircraft 158
   flat panel 159
   grounds 164
   installation drawings 161
   instruments and radios 156
   manuals and diagrams 161
   navcom connections, typical 166
   non-certified airplanes 154
   pin assignments 163
   schematic symbols 163
   schematics 162
   STC: Supplemental Type Certificate 154
   TC: Type Certificate 154
   type of flying 155
   typical avionics equippage 160
   viewing angle 165
   wiring diagram, reading 163

R
Radar Altimeter 104
   altitude trips 107
   antennas 105
   carrier wave 106
   components 105
   decision height 105
   display 104
   display, analog 107
   gear warning 107
   operation 106
Radio management system 21
RAIM: receiver autonomous integrity monitoring 117
RMI (Radio Magnetic Indicator) 64, 82

S
Satcom (Satellite Communications)
   aircraft earth station 30, 35
   antennas 30

conformal 31, 37
high gain 30, 31, 36
intermediate gain 33
low gain 30
beam steering unit (BSU) 35
data system 38
ground earth station 29, 34
Inmarsat 29
Inmarsat Aero System 39
radio frequency unit 35
satellite data unit (SDU) 35
space segment 32
Selcal
airborne system 48
codes 47, 48
controller 46
decoder 46
ground network 47
silk screen 254
Sperry, Elmer 7
SSB (single sideband) 24
Swift64 33

T
TCAS (Traffic Alert Collision Avoidance System 147
basic operation 148
components 151
coordinating climb and descent 150
directional interrogation 151
non-TCAS airplanes 152
RA: Resolution Advisory 150, 152
Symbols on radar display 148
TA: Threat Advisory 150, 152
tau 149
TCAS I, TCAS II 150
TCAS III 152
voice warnings 152
whisper-shout 151
Terminals 219
block 219
ring 219
Test and Troubleshooting 261
ADF antenna 262
air data test set 265
antennas 263
doubler 263
rf wattmeter 264
audio quality 263
automatic test station 265
autopilot 264
cable tension 264
porpoising 264
checklist 274
coaxial cable 263
com transceiver 264
no receive 264
no transmit 265
squelch 264
enclosures 270
fault detection device 271
glideslope 269
glideslope receiver 266

ILS (Cat III) tester 267
in-flight interference 263
lightning strike 266
loading software 267
dataloader 268
navcom ramp tester 272
P (precipitation) static 273
static wicks 273
Portable maintenance aid 272
reporting trouble 261
RF signal generator 266
selective call 262
servo 263
switchology 262
TCAS ramp test 261
technical terms 261
transponder 267
VOR checkpoint 268
VOR receiver 268
course bends 269
VOR test facility (VOT) 268
VSWR 263
wiring and connectors 270
Transponder 94
aircraft address 98
ATCRBS and Mode S 95
code selection 101
control-display 94
control-display (airline) 98
ID code 101
interrogator, ground 95
LRU (line replaceable unit) 98
mode A interrogation 99
mode C interrogation 99
mode S 96
mode S: all call 102
mode S: interrogations and replies 102
mode S: selective address 102
panel-mounted 96
reserved codes 101
squawk 94
System diagram 97
Turn and Bank 8

V
VDR: VHF Data Radio 16
VHF Com 16
acceptable radios 17
basic connections 18
control panel 20
line replaceable unit 20
radio management system 21
splitting channels 21
system diagram 19
VOR 57
course indicator 63
coverage 58
HSI (Horizontal Situation Indicator) 64
Nav control-display 65
navigation 63
principles 59

receiver diagram 62
reference and variable phase 61
RMI (Radio Magnetic Indicator) 64
service volume 58
signal components 59
signal structure 60
VOR-DME station 57

W
Weather detection 136
clear air turbulence 136
datalink 141, 145
lightning detection 140
sensors 138
Stormscope 141, 143
weather radar
color coding display 137
control panel 140
radome 142
receiver-transmitter 139
system components 139
thunderstorms 138, 139
windshear 143
computer 144
Weather radar 137
antenna 141
Wind shear 143
Wiring the Airplane 210
Adel clamp 222
bending coaxial cable 228
chafing and abrasion 224
clamping 223, 226
conduit 225
corrosive chemicals 225
ducts 225
electromagnetic interference 222
grounding 227
Harnessing 222
high grade wire 213
high risk area 211
high temperature 225
intervals 220
lacing 223
length 216
marking 220
methods 221
moisture 225
nicked and broken wires 217
precut 217
PVC 212
selecting 213
service loops 229
splicing 217
knife 218
location of splices 218
stranded vs. solid 216
stripping 216
SWAMP area 210
tie wraps (cable ties) 223
wire and cable types 215
Wright Brothers 1, 6